Lecture Notes in Computer Science   10856

Commenced Publication in 1973
Founding and Former Series Editors:
Gerhard Goos, Juris Hartmanis, and Jan van Leeuwen

More information about this series at http://www.springer.com/series/7407

Jon Lee · Giovanni Rinaldi
A. Ridha Mahjoub (Eds.)

Combinatorial Optimization

5th International Symposium, ISCO 2018
Marrakesh, Morocco, April 11–13, 2018
Revised Selected Papers

 Springer

Editors
Jon Lee
Department of Industrial and Operations
  Engineering
University of Michigan
Ann Arbor, MI
USA

A. Ridha Mahjoub
LAMSADE, Université Paris-Dauphine
Paris
France

Giovanni Rinaldi
Informatica (IASI)
Istituto di Analisi dei Sistemi ed
Rome
Italy

ISSN 0302-9743   ISSN 1611-3349   (electronic)
Lecture Notes in Computer Science
ISBN 978-3-319-96150-7   ISBN 978-3-319-96151-4   (eBook)
https://doi.org/10.1007/978-3-319-96151-4

Library of Congress Control Number: 2018947875

LNCS Sublibrary: SL1 – Theoretical Computer Science and General Issues

This Springer imprint is published by the registered company Springer Nature Switzerland AG
The registered company address is: Gewerbestrasse 11, 6330 Cham, Switzerland

Preface

This volume contains the regular papers presented at ISCO 2018, the 5th International Symposium on Combinatorial Optimization, held in Marrakesh, Morocco during April 11–13, 2018. ISCO 2018 was preceded during April 9–10 by the Spring School on "Advanced Mixed Integer Programming Formulation Techniques" given by Juan Pablo Vielma and Joye Huchette (MIT, USA). ISCO is a new biennial symposium. The first edition was held in Hammamet, Tunisia, in March 2010, the second in Athens, Greece, in April 2012, the third in Lisbon, Portugal, in March 2014, and the fourth in Vietri Sul Mare, Italy, in May 2016. The symposium aims to bring together researchers from all the communities related to combinatorial optimization, including algorithms and complexity, mathematical programming, operations research, stochastic optimization, multi-objective optimization, graphs, and combinatorics. It is intended to be a forum for presenting original research on all aspects of combinatorial optimization, ranging from mathematical foundations and theory of algorithms to computational studies and practical applications, and especially their intersections.

In response to the call for papers, ISCO 2018 received 75 regular submissions. Each submission was reviewed by at least three Program Committee (PC) members with the assistance of external reviewers. The submissions were judged on their originality and technical quality and the PC had to discuss in length the reviews and make tough decisions. As a result, the PC selected 35 regular papers to be presented in the symposium giving an acceptance rate of 46% (69 short papers were also selected from both regular and short submissions). Four eminent invited speakers, Friedrich Eisenbrand (EPFL, Lausanne, Switzerland), Marica Fampa (Federal University of Rio de Janeiro, Brazil), Bernard Gendron (University of Montreal, Canada), and Franz Rendl (University of Klagenfurt, Graz, Austria), also gave talks at the symposium. The revised versions of the accepted regular papers and extended abstracts of the invited talks are included in this volume.

We would like to thank all the authors who submitted their work to ISCO 2018, and the PC members and external reviewers for their excellent work. We would also like to thank our invited speakers as well as the speakers of the Spring School for their exciting lectures. They all much contributed to the quality of the symposium.

Finally, we would like to thank the Organizing Committee members for their dedicated work in preparing this conference, and we gratefully acknowledge our sponsoring institutions for their assistance and support.

May 2018

Jon Lee
A. Ridha Mahjoub
Giovanni Rinaldi

Organization

Program Committee Co-chairs

Abdellatif El Afia	Mohammed V University, Rabat, Morocco
Jon Lee	University of Michigan, USA
A. Ridha Mahjoub	Paris Dauphine University, France
Giovanni Rinaldi	IASI, Rome, Italy

Steering Committee

M. Baïou	CNRS, Clermont-Auvergne University, France
P. Fouilhoux	Pierre and Marie Curie University, Paris, France
L. Gouveia	University of Lisbon, Portugal
N. Maculan	Federal University of Rio de Janeiro, Brazil
A. R. Mahjoub	Paris Dauphine University, France
V. Paschos	Paris Dauphine University, France
G. Rinaldi	IASI, Rome, Italy

Program Committee

M. Anjos	Polytechnique Montreal, Canada
F. Barahona	IBM, T. J. Watson, New York, USA
A. Basu	Johns Hopkins University, USA
Y. Benadada	ENSIAS, Mohammed V University, Rabat, Morocco
V. Bonifaci	IASI, Rome, Italy
F. Bonomo	University of Buenos Aires, Argentina
F. Carrabs	University of Salerno, Italy
R. Cerulli	University of Salerno, Italy
K. Chandrasekaran	University of Illinois, Urbana-Champaign, USA
J. Correa	University of Chile, Santiago, Chile
F. Eisenbrand	EPFL, Lausanne, Switzerland
I. Elhallaoui	Polytechnique Montreal, Canada
A. Elhilali Alaoui	Sidi Mohamed Ben Abdellah University, Fez, Morocco
M. Ettaouil	Sidi Mohamed Ben Abdellah University, Fez, Morocco
M. Fampa	Federal University of Rio de Janeiro, Brazil
S. Fujishige	RIMS, Kyoto University, Japan
R. Fukasawa	University of Waterloo, Canada
T. Fukunaga	JST, PRESTO, Japan
B. Gendron	University of Montreal, Canada
C. Gentile	IASI, Rome, Italy
M. Gentili	University of Salerno, Italy
E. Gourdin	Orange Labs, Paris, France

V. Goyal	Columbia University, USA
A. Gupte	Clemson University, USA
M. Haouari	University of Qatar, Qatar
I. Kacem	LCOMS, University of Lorraine, Metz, France
V. Kaibel	Otto-von-Guericke Universität, Magdeburg, Germany
N. Kamiyama	Kyushu University, Japan
Y. Kobayashi	University of Tsukuba, Japan
M. Labbé	Free University, Brussels, Belgium
M. Lampis	Paris-Dauphine University, France
A. Letchford	Lancaster University, UK
L. Liberti	LIX, Ecole Polytechnique, Paris, France
I. Ljubic	ESSEC, Paris, France
M. Lübbecke	RWTH Aachen University, Germany
S. Mattia	IASI, Rome, Italy
S. T. McCormick	UBC, Vancouver, Canada
V. Nagarajan	University of Michigan, USA
P. Pesneau	University of Bordeaux, France
F. Rendl	University of Klagenfurt, Graz, Austria
J. J. Salazar Gonzalez	Universidad de La Laguna, Tenerife, Spain
A. Toriello	Georgia Institute of Technology, USA
P. Toth	University of Bologna, Italy
E. Uchoa	UFF Rio de Janeiro, Brazil
P. Ventura	IASI, Rome, Italy
S. Weltge	ETHZ Zürich, Switzerland
H. Yaman	Bilkent University, Ankara, Turkey

Organizing Committee

Rdouan Faizi (Chair)	Mohammed V University, Rabat, Morocco
Abdellatif El Afia	Mohammed V University, Rabat, Morocco
Raddouane Chiheb	Mohammed V University, Rabat, Morocco
Sanaa El Fkihi	Mohammed V University, Rabat, Morocco
Pierre Fouilhoux	Pierre and Marie Curie University, Paris, France
Rohaifa Khaldi	Mohammed V University, Rabat, Morocco
A. Ridha Mahjoub	Paris Dauphine University, France
Sébastien Martin	Lorraine University, Metz, France
Fatima Ouzayd	Mohammed V University, Rabat, Morocco

Additional Reviewers

Arbib, Claudio	Bansal, Manish
Arslan, Okan	Bayram, Vedat
Au, Gary	Bazgan, Cristina
Bacci, Tiziano	Bentz, Cédric

Boecker, Sebastian
Bornstein, Claudson
Chakrabarty, Deeparnab
Da Fonseca, Guilherme D.
De Souza, Cid C.
Deza, Antoine
El Housni, Omar
Galtier, Jerome
Goerigk, Marc
Gouleakis, Themis
Grappe, Roland
Hassani, Rachid
Ignatius, Joshua
Kerivin, Herve
Khosravian Ghadikolaei, Mehdi
Kocuk, Burak
Lange, Julia
Leite Bulhões Júnior, Teobaldo
Macambira, Ana Flavia
Marinelli, Fabrizio

Messaoudi, Mayssoun
Mommessin, Clément
Mömke, Tobias
Nederlof, Jesper
Neto, Jose
Oliveira Colares, Rafael
Oriolo, Gianpaolo
Ouzineb, Mohamed
Paat, Joseph
Poss, Michael
Pralat, Pawel
Raiconi, Andrea
Raupp, Fernanda
Ruthmair, Mario
Segev, Danny
Simonetti, Luidi
Souza, Uéverton
Stamoulis, Georgios
Stecco, Gabriella
Yang, Boshi

Sponsoring Institutions

ENSIAS, Morocco
LAMSADE, Université Paris Dauphine, CNRS, France
Université Mohammed V de Rabat, Morocco

Plenary Lectures

Proximity Results and Faster Algorithms for Integer Programming Using the Steinitz Lemma

Friedrich Eisenbrand

EPFL, Lausanne, Switzerland
friedrich.eisenbrand@epfl.ch

We consider integer programming problems in standard form $max\{cTx : Ax = b, x \geq 0, x \in Z^n\}$ where $A \in Z^{m \times n}$, $b \in Z^m$ and $c \in Z^n$. We show that such an integer program can be solved in time $(m\delta)O(m) \cdot ||b||_{\infty}^2$, where δ is an upper bound on each absolute value of an entry in A. This improves upon the longstanding best bound of Papadimitriou (1981) of $(m \cdot \delta)^{O(m^2)}$, where in addition, the absolute values of the entries of b also need to be bounded by δ. Our result relies on a lemma of Steinitz that states that a set of vectors in R^m that is contained in the unit ball of a norm and that sum up to zero can be ordered such that all partial sums are of norm bounded by m. We also use the Steinitz lemma to show that the ℓ_1-distance of an optimal integer and fractional solution, also under the presence of upper bounds on the variables, is bounded by $m \cdot (2m \cdot \delta + 1)^m$. Here δ is again an upper bound on the absolute values of the entries of A. The novel strength of our bound is that it is independent of n. We provide evidence for the significance of our bound by applying it to general knapsack problems where we obtain structural and algorithmic results that improve upon the recent literature.

Challenges in MINLP: The Euclidean Steiner Tree Problem in $\mathbb{R}^n$

Marcia Fampa

Federal University of Rio de Janeiro, Brazil
fampa@cos.ufrj.br

The Euclidean Steiner Tree Problem (ESTP) is to find a network of minimum length interconnecting p points in $\mathbb{R}^n$, known as terminals. Such network may be represented by a tree, where the nodes are given by the terminals, and possibly by additional points, known as Steiner points. The length of the network is defined as the sum of the Euclidean lengths of the edges in the tree. Without allowing Steiner points, the problem is the easily solved Minimum Spanning Tree Problem, nevertheless, the possibility of using Steiner points, makes its solution very difficult, particularly when n is greater than 2.

An interesting feature about the ESTP is its history. It traces back to the 17th century when the French mathematician Pierre de Fermat proposed a challenge where three points were given in the plane, and the goal was to find a fourth point such that the sum of its distance to the three given points was at minimum. The challenge was solved with the works of Torricelli (1640), Cavalieri (1647) and Heinen (1834). Two generalizations of Fermat's challenge were later presented in the famous book "What is mathematics?", by Courant and Robbins (1941). The first is the problem of finding a point such that the sum of its distance to p given points is at minimum, which was introduced as the Fermat Problem. The second generalization is the ESTP.

The solution of Fermat's challenge made it possible to identify important properties satisfied by an optimal solution of the ESTP, a Steiner Minimal Tree (SMT). These properties have been used in the construction of mathematical models and algorithms for the problem. Maculan, Michelon and Xavier (2000) presented a mixed integer nonlinear programming (MINLP) formulation for the ESTP. The formulation (MMX) describes topologies of SMTs with linear constraints and binary variables, which indicate the edges in the tree. The objective function of the model is a non-convex function on the binary variables and also on continuous variables that represent the position of the Steiner points. Even though a great improvement was observed on the performance of global-optimization solvers in the last decade, the results that we obtain when applying well known solvers, such as Scip and Couenne, to MMX are frustrating.

In this work, we are interested in identifying characteristics of the ESTP that make it such a great challenge for MINLP solvers. By identifiyng these characteristics, we propose alternative formulations with the objective of modeling it as a convex problem, strengthening the formulation with the development of valid inequalities, eliminating isomorphic topologies present in the feasible set of MMX, and handling the non-differentiability of the square root used in the computation of the Euclidean distance. We finally note that these characteristics are also found in other optimization problems, especially with geometric properties, and we use them not only to advance in the numerical solution of the ESTP, but also to point directions for improvements on MINLP solvers. Motivated by our smoothing strategy for the square root, for example, a new feature was incorporated into Scip, for handling piecewise-smooth univariate functions that are globally concave.

Lagrangian Relaxations and Reformulations for Network Design

Bernard Gendron

University of Montreal, Canada
bernard.gendron@cirrelt.ca

We consider a general network design model for which we compare theoretically different Lagrangian relaxations. Fairly general assumptions on the model are proposed, allowing us to generalize results obtained for special cases. The concepts are illustrated on the fixed-charge multicommodity capacitated network design problem, for which we present different Lagrangian relaxations: the well-known shortest path relaxation (that decomposes by commodity) and knapsack relaxation (that decomposes by arc), and new relaxations that decompose by node. Three such node-based relaxations are presented: the first one is based on a partial relaxation of the flow conservation equations, the second one is based on the same relaxation, but also uses Lagrangian decomposition (also called variable splitting), while the third one exploits solely Lagrangian decomposition. We show that these three new relaxations define Lagrangian duals that can improve upon the linear programming relaxation bounds, contrary to the shortest path and knapsack relaxations, which provide the same lower bound as the linear programming relaxation. Dantzig-Wolfe reformulations are derived for each of these Lagrangian relaxations and bundle methods are proposed for solving these reformulations. The different Lagrangian relaxations are also used as a basis for developing Lagrangian matheuristics that solve restricted mixed-integer linear programming models, including intensification and diversification mechanisms. Computational results on a large set of benchmark instances are presented, demonstrating that the node-based Lagrangian relaxation bounds are significantly better than linear programming relaxation bounds and that the Lagrangian matheuristics are competitive with state-of-the-art heuristic methods for the fixed-charge multicommodity capacitated network design problem.

Order Through Partition: A Semidefinite Programming Approach

Franz Rendl

University of Klagenfurt, Graz, Austria
franz.rendl@aau.at

Ordering Problems on n objects involve pairwise comparison among all objects. This typically requires $\binom{n}{2}$ decision variables.

In this talk we investigate the idea of partitioning the objects into k groups (k-partition) and impose order only among the partition blocks.

We demonstrate the efficiency of this approach in connection with the bandwidth minimization on graphs. We consider relaxations of the partition model with the following characteristics:

(1) The weakest model is formulated in the space of symmetric $n \times n$ matrices and has the Hoffman-Wielandt theorem in combination with eigenvalue optimization as a theoretical basis.

(2) We also consider semidefinite relaxations in the space of $n \times n$ matrices, involving k semidefinite matrix variables. The idea here is to linearize the quadratic terms using eigenvalue decompositions.

(3) Finally, the strongest model is formulated in the space of symmetric $nk \times nk$ matrices. It is based on the standard reformulation-linearization idea.

We present theoretical results for these relaxations, and also some preliminary computational experience in the context of bandwidth minimization.

Co-authors: Renata Sotirov (Tilburg, Netherlands) and Christian Truden (Klagenfurt, Austria)

Contents

Cluster Editing with Vertex Splitting

Faisal N. Abu-Khzam[1,2]([⊠]), Judith Egan[1], Serge Gaspers[3,4], Alexis Shaw[5]([⊠]),
and Peter Shaw[6]([⊠])

[1] Charles Darwin University, Darwin, Australia
[2] Lebanese American University, Beirut, Lebanon
faisal.abukhzam@lau.edu.lb
[3] The University of New South Wales, Sydney, Australia
[4] Data61, CSIRO, Sydney, Australia
[5] Centre for Quantum Computation and Communication Technology,
Centre for Quantum Software and Information,
Faculty of Engineering and Information Technology,
University of Technology Sydney, Sydney, NSW, Australia
alexis.shaw@gmail.com
[6] Massey University, Manawatu, New Zealand
p.shaw@massey.ac.nz

Abstract. In the CLUSTER EDITING problem, a given graph is to be
transformed into a disjoint union of cliques via a minimum number of
edge editing operations. In this paper we introduce a new variant of
Cluster Editing whereby a vertex can be divided, or split, into two or
more vertices thus allowing a single vertex to belong to multiple clusters.
This new problem, CLUSTER EDITING WITH VERTEX SPLITTING, has
applications in finding correlation clusters in discrete data, including
graphs obtained from Biological Network analysis. We initiate the study
of this new problem and show that it is fixed-parameter tractable when
parameterized by the total number of vertex splitting and edge editing
operations. In particular we obtain a $4k(k + 1)$ vertex kernel for the
problem.

1 Introduction

Given a graph G and a non-negative integer k, the CLUSTER EDITING problem
asks whether G can be turned into a disjoint union of cliques by a sequence of
at most k edge-editing operations. The problem is known to be $\mathcal{NP}$-Complete
since the work of Křivánek and Morávek in [20], and does not seem to have any
reasonable polynomial-time approximation unless the number of clusters is at
most two [24].

We assume that the reader is familar with fixed-parameter tractability
and kernelization [9,12,16,22]. The Cluster Editing problem is fixed-parameter
tractable when parameterized by k, the total number of edge editing operations
[6,17]. Over the last decade, Cluster Editing has been well studied from both theoretical and practical perspectives (see, for example, [4,7,8,10,11,13,14,18,19]).

© Springer International Publishing AG, part of Springer Nature 2018
J. Lee et al. (Eds.): ISCO 2018, LNCS 10856, pp. 1–13, 2018.
https://doi.org/10.1007/978-3-319-96151-4_1

In general, clustering results in a partition of the input graph, thus it forces each and every data element to be in one and only one cluster. This can be a limitation when a data element plays roles in multiple clusters. This situation, (i.e. the existence of *hubs*), is recorded in work on gene regulatory networks [1], where enumeration of maximal cliques was considered a viable alternative to clustering. Moreover, the existence of hubs can effectively hide clique-like structures and also greatly increase the computational time required to obtain optimum correlation clustering solutions [23,25]. Improved solutions (for correlation clustering) can be obtained using the MULTI-PARAMETERIZED CLUSTER-EDITING problem [2] which further restricts the number of false positive and/or false negative correlations (add and delete edge-edits incident to a vertex) that can be ascribed to a given variable. However, the need to identify variables that lie in the intersection of multiple clusters could further complicate this multi-parameterized model.

The CLUSTER EDITING WITH VERTEX SPLITTING problem (CEVS) is introduced in this paper in an attempt to allow for overlapping clusters in graphs that are assumed to be noisy in the sense that edges are assumed to have been perturbed after the clusters overlap. CEVS can be viewed as an extended version of the CLUSTER EDITING problem.

In addition to introducing CEVS, we investigate its parameterized complexity and obtain a polynomial kernel for the problem. In doing so we employ the notion of a critical clique, as introduced in [21], and applied to the CLUSTER EDITING problem in [18]. Our proof technique is based on a novel clean edit sequence approach which could be of interest by itself.

This paper is structured as follows. Section 2 overviews some background material. Section 3 introduces the edit sequence approach while Sect. 4 is devoted to critical cliques. In Sect. 5 we obtain a quadratic kernel. We conclude in Sect. 6 with a summary and future directions.

2   Preliminaries

We assume familiarity with basic graph theoretic terminology. All graphs in this work are simple, unweighted and undirected. The vertex and edge sets of a graph G are denoted by $V(G)$ and $E(G)$ respectively. For a subset V' of $V(G)$, we denote by $G[V']$ the subgraph of G that is induced by V'.

A *kernelization* or *kernel* for a parameterized problem P is a polynomial time function that maps an instance (I, k) to an instance (I', k') of P such that:

- (I, k) is a YES instance for P if and only if (I', k') is a YES instance;
- $|I'| < f(k)$ for some computable function f;
- $k' < g(k)$ for some computable function g.

A *proper* kernelization is a kernelization such that $g(k) < k$ [3]. The function $f(k)$ is also called the size of the kernel. A problem has a kernel if and only if it is FPT [12], however not every FPT problem has a kernel of polynomial size [5].

A *k-partition* of a set S is a collection of pairwise disjoint sets $S_1, S_2, \ldots S_k$ such that $S = \bigcup_{i=1}^{k} S_i$. A *k-covering* of a set S is a collection of sets $S_1, S_2, \ldots S_k$ such that $S = \bigcup_{i=1}^{k} S_i$. A *cluster graph* is a graph in which the vertex set of each connected component induces a clique.

Problem Definition. The CLUSTER EDITING WITH VERTEX SPLITTING Problem (henceforth CEVS) is defined as follows. Given a graph $G = (V, E)$ and an integer k, can a cluster graph G' be obtained from G by a k-edit-sequence $e_1 \ldots e_k$ of the following operations:

1. do nothing,
2. add an edge to E,
3. delete an edge from E, and
4. an *inclusive vertex split*, that is for some $v \in V$ partition the vertices in $N(v)$ into two sets U_1, U_2 such that $U_1 \cup U_2 = N(v)$, then remove v from the graph and add two new vertices v_1 and v_2 with $N(v_1) = U_1$ and $N(v_2) = U_2$.

A vertex $v \in V(G)$ is said to *correspond* to a vertex $v' \in V(G')$, constructed from G by an edit-sequence S if v' is a leaf on the division-tree T for v defined as follows:

(i) v is the root of the tree, and
(ii) if an edit sequence operation splits a vertex u which lies on the tree then the two vertices that result from the split are children of u.

As noted earlier, Cluster Editing corresponds to the special case where no vertex splitting is permitted. So it would appear that CLUSTER EDITING WITH VERTEX SPLITTING is $\mathcal{NP}$-Hard due to the $\mathcal{NP}$-hardness of the CLUSTER EDITING problem. Moreover, suppressing vertex splitting is not an option in the definition of CEVS. The $\mathcal{NP}$-hardness of the problem is not obvious, so we pose it as an open problem at this stage.

Our main focus in this paper is on the parameterized complexity of CEVS. We shall present a quadratic-size kernel for the problem, which proves it to be fixed-parameter tractable.

A similar problem has been defined and studied in [15] where a vertex is allowed to be part of at most s clusters. In this case s is either treated as constant or as a different parameter, which makes the $\mathcal{NP}$-hardness proof easy since the case $s = 1$ corresponds to the Cluster Editing problem. In our work we model the overlap via another editing operation so we do not set a separate parameter for the number of splittings per vertex. We are able to design a kernelization algorithm that achieves a quadratic-size kernel (while the best kernel bound achieved in [15] is cubic for the special case where $s = 2$).

3   The Edit-Sequence Approach

Defining CEVS in terms of edit-sequences is based on looking for the closest graph which is a cluster graph, where distance is defined by the shortest edit-sequence. An edit-sequence may however include a lot of redundancy (for example, swap two edge additions). In this section we show how to eliminate redundancy, first by showing that we can consider a specific form of edit sequence, and then showing that we can efficiently compute the shortest edit-sequence between two graphs. This will provide some much needed structure to the problem and provide a base for subsequent proofs.

3.1   Restricted Re-ordering of the Edit-Sequence

Two edit sequences $S = e_1 \ldots e_k$ and $S' = e'_1 \ldots e'_k$ are said to be *equivalent* if:

- G_S and $G_{S'}$, the graphs obtained from G by S and S' respectively are isomorphic to each other with isomorphism $f : V(G_S) \to V(G_{S'})$, and
- if $u_S \in V(G_S)$ and $u_{S'} = f(u_S)$ then the division tree which u_S is contained in and the division tree which $u_{S'}$ is contained in share a common root. In other words, u_S and $u_{S'}$ correspond to the same vertex of the original graph.

Lemma 1. *For any edit-sequence $S = e_1 \ldots e_i e_{i+1} \ldots e_k$ where e_i is an edge deletion and e_{i+1} is an edge addition, there is an equivalent edit-sequence $S' = e_1 \ldots e'_i e'_{i+1} \ldots e_k$ of the same length where either e'_i is an edge addition and e'_{i+1} is an edge deletion, or both e'_i and e'_{i+1} are do-nothing operations.*

Proof. We begin by noting that we only have to consider the edits e_i and e_{i+1} as we can think of the edit-sequence being a sequence of functions composed with each other, thus if e_i deletes edge uv and e_{i+1} adds edge wx then the graph immediately after applying the two operations in the opposite order will be the same in all cases except that where $uv = wx$ whereby the net effect is that nothing happens, as required.

Lemma 2. *For any edit-sequence $S = e_1 \ldots e_i e_{i+1} \ldots e_k$ where e_i is a vertex splitting and e_{i+1} is an edge deletion there is an equivalent edit-sequence $S' = e_1 \ldots e'_i e'_{i+1} \ldots e_k$ where either e'_i is an edge deletion and e_{i+1} is a vertex splitting or e'_i is a do-nothing operation and e'_{i+1} is a vertex splitting.*

Proof. If the edge deleted by e_{i+1} is not incident to one of the resulting vertices of the splitting e_i then swapping the two operations produces the required edit-sequence E'. Otherwise let e_i split vertex v and e_{i+1} delete edge uv_i. Then if e_i has associated covering U_1, U_2 of $N(v)$ and without loss of generality $u \in U_1$ then if $u \notin U_2$ then the edit-sequence with e'_i being a deletion operation deleting uv and e'_i being the vertex splitting and $U'_i = U_i \setminus \{u\}$ and $U'_1 = U_2$ is equivalent to E. Otherwise, $u \in U_1 \cap U_2$. Without loss of generality, suppose uv_2 is deleted by e_{i+1}. Then the sequence where e'_i is a do-nothing operation and where e'_{i+1} is a vertex splitting on v with covering U'_1, U'_2 with $U'_1 = U_1$ and $U'_2 = U_2 \setminus \{u\}$ is equivalent.

Lemma 3. *For any edit-sequence $S = e_1 \ldots e_i e_{i+1} \ldots e_k$ where e_i is a vertex splitting and e_{i+1} is an edge addition there exists an equivalent sequence $S' = e_1 \ldots e_i' e_{i+1}' \ldots e_k$ where either e_i' is an edge addition and e_{i+1}' is a vertex splitting, or e_i' is a do-nothing operation and e_{i+1}' is a vertex splitting.*

Proof. If the edge added by e_{i+1} is not incident to one of the resulting vertices of the splitting e_i then simply swap the two operations to produce the required edit-sequence E'. Otherwise, without loss of generality, let e_i divide vertex v on covering U_1, U_2 and e_{i+1} add vertex $w v_1$. Then let e_i' be the operation that adds the edge wv, if wv does not exist at that point, otherwise e_i' is a do-nothing operation and let e_{i+1}' split vertex v on covering $U_1' = U_1 \cup \{w\}, U_2' = U_2$. The resulting edit-sequence is equivalent to E.

Lemma 4. *For any edit-sequence $S = e_1 \ldots e_k$ where e_i is a do-nothing operation, the edit-sequence $S' = e_1 \ldots e_{i-1} e_{i+1} \ldots e_k$ is equivalent to it and has strictly smaller length.*

3.2   Edit Sequences in Add-Delete-Split Form

From the above lemmas we can deduce the following theorem.

Theorem 1. *For every edit-sequence $S = e_1 \ldots e_k$ there is an edit-sequence $S' = e_1' \ldots e_{k'}'$ with equal or lesser length such that*

1. *if e_i' is an edge addition and e_j' is an edge deletion or a vertex splitting, then $i < j$,*
2. *if e_i' is an edge deletion and e_j' is a vertex splitting, then $i < j$, and*
3. *S' contains no do-nothing operations.*

We refer to an edit-sequence satisfying the statement of Theorem 1 as an edit-sequence in the add-delete-split form. We will now consider only these edit-sequences, as for any equivalence class of edit-sequences, there is a minimal member of that equivalence class which is in add-delete-split form. In fact, the equivalence class of an add-delete-split edit-sequence is the intersection of an equivalence class of edit-sequences and the set of edit-sequences in add-delete-split form. A minimal member of any such equivalence class is an edit-sequence in add-delete-split form.

Uniqueness of the Pre-splitting Edge Modification Graph Corresponding to Any Add-Delete-Split Edit-Sequence Equivalence Class. It is now necessary to prove that in any equivalence class the graph obtained after the addition and deletion of edges and before splitting vertices is fixed. By doing so we provide a significant amount of structure to the problem, and do away with the direct use of edit-sequences altogether when searching for a solution.

The approach we adopt is to work on time-reversed edit-sequences, taking the final graph of the edit-sequence and the relation between the vertices in the initial graph and the final graph, and proving that we always arrive at the same graph. In preparation for this we define the *split relation*, $f : V \to 2^{V'}$ for a given solution S to CEVS for a graph G and edit-distance k as a function.

The split relation for such a solution S, graph G and distance k is a function $f : V \to 2^{V'}$ defined such that when $G' = (V', E')$ is derived from G by S the following properties hold on f

1. For a vertex $v \in V$: $v' \in f(v)$ if and only if v corresponds to v' under S,
2. For any $u, v \in V$ that $f(u) \cap f(v) = \emptyset$, and
3. For any $u \in V$ that $f(u) \neq \emptyset$.

A simple consequence of this definition is that two edit-sequences are equivalent if and only if the resulting graphs are isomorphic and the split relation is equivalent under that isomorphism.

In order to talk about time-reversed vertex-splitting sequences we define a *merge graph* as being a graph $G' = (V', E')$ derived from another graph $G = (V, E)$ by a sequence of vertex merge operations, that is there is a relation $f : V' \to 2^V$ called the *merge relation* which partitions the vertex set V on members of V' such that $u'v' \in E'$ if and only if $uv \in E$ for some $u \in f(u')$ and some $v \in f(v')$. A *vertex merge* operation constructs a merge graph with a merge relation such that there is only one v' such that $|f(v')| \neq 1$ and for that value $f(v') = u, v$; we call this the *merger* of u and v. A *k-merge-graph* G' of G is a merge graph for which there is a sequence of exactly k vertex merges $G^1 \ldots G^k = G'$ such that for all $i = 1 \ldots k$ and defining $G^0 = G$, we have G^i is derived from G^{i-1} by a vertex merge operation.

Lemma 5. *For any graph $G = (V, E)$ and merge-graph $G' = (V', E')$ of G with merge relation $R : V' \to 2^V$ we have that*

$$E' = \{u, v \in V' : \exists u' \in R(u) \ \exists v' \in R(v) \text{ such that } u'v' \in E\} .$$

Proof. If $V = V'$ then no merge has occurred and so this is trivially so, otherwise we proceed by induction on $k = |V| - |V'|$.

Suppose that $G' = (V', E')$ is a k-merge-graph of G with merge relation $R^k : V' \to 2^V$ and suppose that G'' is a 1-merge-graph of G' and a $(k+1)$-merge-graph of G with relations $R' : V'' \to 2^{V'}$ and $R^{k+1} : V'' \to 2^V$. Without loss of generality suppose that G'' was produced by merging vertices u and v of G' into w. Then, by definition,

$$E'' = E' \setminus (\{vx \in E' | x \in V'\} \cup \{ux \in E' | x \in V'\}) \cup \{wx | x \in (N_{G'}(u) \cup N_{G'}(v))\}) .$$

Therefore we deduce that the edge set E'' is the same as E' except on those edges incident to u and v, which were merged into W. However, $R'(x) = \{x\}$ for all vertices $x \neq w$ and $R'(w) = \{u, v\}$ and so,

$$E'' = \{u'', v'' \in V'' : \exists u' \in R'(u'') \ \exists v' \in R'(v'') \ u'v' \in E'\}.$$

By our induction hypothesis we also know that

$$E' = \{u', v' \in V' : \exists u \in R^k(u') \ \exists v \in R^k(v') \ uv \in E\}.$$

By merging these two equations together (noting that we can disclose the initial definition of u and v in the second equation) to give

$$E'' = \{u'', v'' \in V'' : \exists u' \in R'(u'') \ \exists v' \in R'(v'') \ \exists u \in R^k(u') \ \exists v \in R^k(v') \ uv \in E\}.$$

By re-ordering the order of the existential operators we can obtain

$$E'' = \{u'', v'' \in V'' : \exists u' \in R'(u'') \ \exists u \in R^k(u') \ \exists v' \in R'(v'') \ \exists v \in R^k(v') \ uv \in E\}.$$

This is the same as saying,

$$E'' = \{u'', v'' \in V'' : \exists u \in \bigcup_{u' \in R'(u'')} R^k(u') \ \exists v \in \bigcup_{v' \in R'(v'')} R^k(v') \ uv \in E\}.$$

By the definition or R^{k+1} this means

$$E'' = \{u'', v'' \in V'' : \exists u \in R^{k+1}(u'') \ \exists v \in R^{k+1}(v'') \ uv \in E\}$$

as required.

3.3 Representation of Edit-Sequences as Resultant Graphs and Merge Relations

We can see that a vertex merge is the time-reversed image of a vertex splitting, in as much as that any sequence of splits will correspond to a time-reversed sequence of merges and the converse. Thus we can use vertex mergers to prove the following.

Lemma 6. *For any collection of edit-sequences in add-delete-split form which are equivalent to some edit-sequence S, the graph G_{R_S} immediately preceding the vertex splitting is the same for all members of the class in that form. Further if the split relation for this equivalence class is R and the graph $G' = (V'E')$ resulting from S are known, then*

$$E' = \{u, v \in V' : \exists u' \in R(u) \ \exists v' \in R(v) \ such \ that \ u'v' \in E\} \ .$$

Proof. This follows directly from Lemma 5, and Theorem 1.

We are now ready to prove the following lemma:

Lemma 7. *For any graph $G = (V, E)$ there is a computable bijection between pairs $(G' = (V', E'), f : V \to 2^{V'})$ of resultant graphs and split-relations and equivalence classes of edit-sequences. Further there is an algorithm to compute a min-edit-sequence from the resultant graph/split-relation pair for this class in $O((|V'| - |V|)\Delta(G) + |V| + |E| + |V'| + |E'|)$ time.*

Proof. The edit-sequence to graph/relation direction has been proved above, further we have proved that if two edit-sequences have the same graph/split-relation pair then they are equivalent. Thus all that remains to be proved is that we can always construct a min-edit-sequence from an input graph to a valid resultant graph/split-relation pair. We note:

- As the split-relation f can be represented as a merge-relation it is possible by Lemma 5 to construct a graph G_R such that G_R has the same vertex set as G and there is an edit-sequence consisting of only vertex divisions from G_R to G' and with relation-relation f. Further this can be done in $O((|V'| - |V|)\Delta(G) + |V| + |E| + |V'| + |E'|)$ time, and
- As G_R shares the same vertex set as G it is possible to construct an optimal edit-sequence from G to G_R with only edge additions and deletions (by looking at the edge sets of G and G_R). Further we can do this in $O(|E|)$ time.

So we can construct an edit-sequence from G to G' with split-relation f in $O((|V'| - |V|)\Delta(G) + |V| + |E| + |V'| + |E'|)$ time. Further by Lemma 6 this graph G_R is fixed for all edit-sequences in add-delete-divide form, of which one is minimal. Thus as the initial add-divide sequence is minimal by construction, and all division sequences are minimal, this sequence is a min-edit-sequence from G to G' with split relation f as required.

3.4   Representation of Optimal Cluster Graph Edit-Sequences by Coverings

Consider the CEVS problem for a graph G and an edit distance k. If there is a solution edit-sequence S for this problem, we may also represent the resulting graph G' by a covering of the vertices in the original graph. As G' is a cluster graph every pair of vertices from a clique are joined by an edge, and we can reconstruct G' and $f : V(G) \to 2^{V(G')}$, the corresponding vertex relation. And so by Lemma 7 we can represent the search space for optimal CEVS edit-sequences by coverings of the vertex set of G, and evaluate the min-edit distance for each of them in $O((|V'| - |V|)\Delta(G) + |V| + |E| + |V'| + |E'|)$ time.

4 Critical Cliques

Originally introduced by Lin et al. [21], critical cliques provide a useful tool in understanding the clusters in graphs. A *critical clique* of a graph $G = (V, E)$ is a maximal induced subgraph C of G such that:

- C is a complete graph.
- There is some subset $U \subseteq V$ such that for every $v \in V(C)$, $N[v] = U$.

It was shown in [21] that each vertex is in exactly one critical clique. Let the critical clique containing a vertex v be denoted by $CC(v)$. The critical clique graph $CC(G)$ can then also be defined as a graph with vertices being the critical cliques of G, having edges wherever there is an edge between the members of the critical cliques in the original graph [21]. That is to say, that the critical clique graph $G' = (V', E')$ related to the graph $G = (V, E)$ is the graph with $V' = CC(G)$ and edges $E' = \{uv | \forall x \in V_C(u).\forall y \in V_C(v).xy \in E\}$ Furthermore, the vertices in G' are given as a weight the number of vertices they represent in the original graph, similarly for the edges.

The following lemma, dubbed "the critical clique lemma" is adapted from Lemma 1 in [18], with a careful restatement in the context of this new problem.

Lemma 8. *Any covering $C = (S_1 \ldots S_l)$ corresponding to a solution to* CEVS *for a graph $G = (V, E)$ that minimizes k will always satisfy the following property: for any $v \in G$, and for any $S_i \in C$ either $CC(v) \subseteq S_i$ or $CC(v) \cap S_i = \emptyset$.*

Proof omitted for length reasons.

By the critical clique lemma, the CEVS problem is equivalent to a weighted version of the problem on the critical clique graph.

Lemma 9. *If there is a solution to* CEVS *on (G, k) then there are at most $4k$ non-isolated vertices in $CC(G)$. Moreover, there are at most $3k+1$ vertices in any connected component of $CC(G)$ and there are at most k connected components in $CC(G)$ which are non-isolated vertices.*

Proof. We follow an approach similar to that taken by Fellows et al. and Gou. Let S_{opt} be an optimal solution of CEVS and partition the vertex set of $CC(G)$ into 4 sets W, X, Y and Z. Let W be the set of vertices which are the endpoint of some edge added by S_{opt}, Let X be the subset of vertices which are the endpoint of some edge deleted by S_{opt} and not in W, Let Y be the subset of vertices which are split by S_{opt} and not in $W \cup X$, finally let Z be all other non-isolated vertices in G. As each vertex in W, X, and Y is affected by some operation in S_{opt} and any operation in S_{opt} can affect at most 2 vertices, if $|S_{opt}| < k$ then $|W \cup X \cup Y| < 2k$. Let us now consider Z. Suppose that $u, v \in V_{S_{[opt]}} \cap Z$ are in the same clique in $G_{S_{opt}}$, then as they are in Z they are adjacent to exactly every vertex in Y as they are not involved in any edge addition, deletion or vertex splitting. However as they are adjacent to each other and have the same neighborhood, apart from each other, they are in a critical clique together.

i.e. $u = v$. Thus there can be at most one vertex in Z for any connected component of $G_{S_{opt}}$.

Now every vertex in Z is adjacent to a vertex in $W \cup X \cup Y$. To see this suppose that $z \in Z$ is not, then by the above lemma it is the only vertex in Z in its clique in $G_{S_{opt}}$, however as it is in Z then it has neither been split nor been severed from any vertex, thus it is an isolated vertex of $CC(G)$. This is a contradiction.

As each connected component of $G_{S_{opt}}$, which is not an isolated vertex in $CC(G)$, contains at least one vertex of $W \cup X \cup Y$ there are at most $2k$ vertices in Z and there are at most $4k$ non-isolated vertices in $CC(G)$.

As each connected component $CC(G)$ has to be separated into at most $k+1$ cliques in $G_{S_{opt}}$ there can be at most $k + 1$ elements of Z in any connected component of $CC(G)$, thus there can be at most $3k+1$ vertices in any connected component of $CC(G)$.

As no clique in $G_{S_{opt}}$ has members from two connected components of G, and as if any connected component of G is not an isolated vertex in $CC(G)$ there is at least one edit performed to some vertex in that connected component there can be at most k such connected components which are non isolated vertices in $CC(G)$.

5   A $4k(k + 1)$ Vertex Kernel

From the result in Sect. 4 we can devise a polynomial size kernel for CEVS. To achieve this we propose three reduction rules, prove that they are valid, and that their application gives a kernel as required.

Reduction Rule 1. *Remove all isolated Cliques.*

Lemma 10. *Reduction Rule 1 is sound.*

Proof. As no optimal solution has a final clique which bridges two connected components of a graph G and an isolated clique needs no edits to make it complete, the clique will remain in all optimal solutions and as such can be removed without affecting the result.

Reduction Rule 2. *Reduce all critical cliques with more than $k+1$ vertices to $k + 1$ vertices.*

Lemma 11. *Reduction Rule 2 is sound.*

Proof. As no solution clique in an optimal solution partially contains a critical clique and the cost to delete the edges incident to, or add edges to all of, or to divide the vertices of such a critical clique is greater than k is too high to be allowed we only need to maintain this invariant. Thus we can remove vertices from any critical clique with more than $k + 1$ vertices until there are at most $k + 1$ vertices in a critical clique and this will not affect the result.

Reduction Rule 3. *If there are more than $4k$ non-isolated critical cliques reduce the graph to a P_3 and set $k = 0$ in this case.*

Lemma 12. *Reduction Rule 3 is sound.*

Proof. As proved in Lemma 9 if there are more than $4k$ non-isolated critical cliques then there is no solution, thus we can emit a trivial No-instance.

Theorem 2. *There exists a polynomial-time reduction procedure that takes an arbitrary instance of the* CLUSTER EDITING WITH VERTEX SPLITTING *problem and produces an equivalent instance whose order (number of vertices) is bounded above by $4k(k+1)$. In other words, CEVS admits a quadratic-order kernel.*

Proof. As shown in the previous lemmas, Reduction Rules 1, 2 and 3 are well founded, and as after having applied them exhaustively, there are at most $4k$ critical cliques in the input graph. Due to Reduction Rule 2, there is no critical clique with more than $k + 1$ vertices, and therefore, the application of these reduction rules results in a $4k(k+1)$ vertex kernel.

6   Conclusion

By allowing a vertex to split into two vertices we extend the notion of Cluster Editing in an attempt to better-model clustering problems where a data element may have roles in more than one cluster. The corresponding new version of Cluster Editing is shown to be fixed-parameter tractable via a new approach that is based on edit-sequence analysis.

The vertex splitting operation may also be applicable to other classes of target graphs, including bipartite graphs, disjoint complete bipartite graphs (bi-cluster graphs), chordal-graphs, comparability graphs, and perfect graphs. The results in Sect. 3 are directly applicable to these other classes, and it may be possible to find an analog to the critical clique lemma for bi-cluster graphs.

The work reported in this paper did not consider the exclusive version of vertex splitting, where the two vertices which result from a split must additionally have disjoint neighborhoods. This is ongoing research at this stage.

Acknowledgments. Serge Gaspers is the recipient of an Australian Research Council (ARC) Future Fellowship (FT140100048) and acknowledges support under the ARC's Discovery Projects funding scheme (DP150101134). Alexis Shaw is the recipient of an Australian Government Research Training Program Scholarship.

References

1. Abu-Khzam, F.N., Baldwin, N.E., Langston, M.A., Samatova, N.F.: On the relative efficiency of maximal clique enumeration algorithms, with applications to high-throughput computational biology. In: International Conference on Research Trends in Science and Technology (2005)
2. Abu-Khzam, F.N.: On the complexity of multi-parameterized cluster editing. J. Discrete Algorithms **45**, 26–34 (2017)

3. Abu-Khzam, F.N., Fernau, H.: Kernels: annotated, proper and induced. In: Bodlaender, H.L., Langston, M.A. (eds.) IWPEC 2006. LNCS, vol. 4169, pp. 264–275. Springer, Heidelberg (2006). https://doi.org/10.1007/11847250_24
4. Böcker, S.: A golden ratio parameterized algorithm for cluster editing. J. Discrete Algorithms 16, 79–89 (2012)
5. Bodlaender, H.L., Downey, R.G., Fellows, M.R., Hermelin, D.: On problems without polynomial kernels. J. Comput. Syst. Sci. 75(8), 423–434 (2009)
6. Cai, L.: Fixed-parameter tractability of graph modification problems for hereditary properties. Inf. Process. Lett. 58(4), 171–176 (1996)
7. Chen, J., Meng, J.: A 2k kernel for the cluster editing problem. J. Comput. Syst. Sci. 78(1), 211–220 (2012)
8. Chen, J., Huang, X., Kanj, I.A., Xia, G.: Strong computational lower bounds via parameterized complexity. J. Comput. Syst. Sci. 72(8), 1346–1367 (2006)
9. Cygan, M., et al.: Parameterized Algorithms. Springer, Cham (2015). https://doi.org/10.1007/978-3-319-21275-3
10. D'Addario, M., Kopczynski, D., Baumbach, J., Rahmann, S.: A modular computational framework for automated peak extraction from ion mobility spectra. BMC Bioinform. 15(1), 25 (2014)
11. Dehne, F., et al.: The cluster editing problem: implementations and experiments. In: Bodlaender, H.L., Langston, M.A. (eds.) IWPEC 2006. LNCS, vol. 4169, pp. 13–24. Springer, Heidelberg (2006). https://doi.org/10.1007/11847250_2
12. Downey, R.G., Fellows, M.R.: Fundamentals of Parameterized Complexity. Texts in Computer Science. Springer, London (2013). https://doi.org/10.1007/978-1-4471-5559-1
13. Fadiel, A., Langston, M.A., Peng, X., Perkins, A.D., Taylor, H.S., Tuncalp, O., Vitello, D., Pevsner, P.H., Naftolin, F.: Computational analysis of mass spectrometry data using novel combinatorial methods. AICCSA 6, 8–11 (2006)
14. Fellows, M., Langston, M., Rosamond, F., Shaw, P.: Efficient parameterized preprocessing for cluster editing. In: Csuhaj-Varjú, E., Ésik, Z. (eds.) FCT 2007. LNCS, vol. 4639, pp. 312–321. Springer, Heidelberg (2007). https://doi.org/10.1007/978-3-540-74240-1_27
15. Fellows, M.R., Guo, J., Komusiewicz, C., Niedermeier, R., Uhlmann, J.: Graph-based data clustering with overlaps. Discrete Optim. 8(1), 2–17 (2011). Parameterized Complexity of Discrete Optimization
16. Flum, J., Grohe, M.: Parameterized Complexity Theory. Springer, Heidelberg (2006). https://doi.org/10.1007/3-540-29953-X
17. Gramm, J., Guo, J., Hüffner, F., Niedermeier, R.: Graph-modeled data clustering: exact algorithms for clique generation. Theory Comput. Syst. 38(4), 373–392 (2005)
18. Guo, J.: A more effective linear kernelization for cluster editing. Theoret. Comput. Sci. 410(8–10), 718–726 (2009)
19. Hüffner, F., Komusiewicz, C., Moser, H., Niedermeier, R.: Fixed-parameter algorithms for cluster vertex deletion. Theory Comput. Syst. 47(1), 196–217 (2010)
20. Křivánek, M., Morávek, J.: NP-hard problems in hierarchical-tree clustering. Acta Informatica 23(3), 311–323 (1986)
21. Lin, G.-H., Kearney, P.E., Jiang, T.: Phylogenetic k-Root and Steiner k-Root. In: Goos, G., Hartmanis, J., van Leeuwen, J., Lee, D.T., Teng, S.-H. (eds.) ISAAC 2000. LNCS, vol. 1969, pp. 539–551. Springer, Heidelberg (2000). https://doi.org/10.1007/3-540-40996-3_46
22. Niedermeier, R.: An Invitation to Fixed-Parameter Algorithms. Oxford University Press, Oxford (2006)

23. Radovanović, M., Nanopoulos, A., Ivanović, M.: Hubs in space: popular nearest neighbors in high-dimensional data. J. Mach. Learn. Res. **11**(Sep), 2487–2531 (2010)
24. Shamir, R., Sharan, R., Tsur, D.: Cluster graph modification problems. Discrete Appl. Math. **144**(1–2), 173–182 (2004)
25. Tomašev, N., Radovanović, M., Mladenić, D., Ivanović, M.: The role of hubness in clustering high-dimensional data. In: Huang, J.Z., Cao, L., Srivastava, J. (eds.) PAKDD 2011. LNCS, vol. 6634, pp. 183–195. Springer, Heidelberg (2011). https://doi.org/10.1007/978-3-642-20841-6_16

Compact MILP Formulations
for the p-Center Problem

Zacharie Ales[1,2(✉)] and Sourour Elloumi[1,2]

[1] ENSTA-ParisTech/UMA, 91762 Palaiseau, France
[2] Laboratoire CEDRIC, Paris, France
{zacharie.ales,sourour.elloumi}@ensta-paristech.fr

Abstract. The p-center problem consists in selecting p centers among M to cover N clients, such that the maximal distance between a client and its closest selected center is minimized. For this problem we propose two new and compact integer formulations.

Our first formulation is an improvement of a previous formulation. It significantly decreases the number of constraints while preserving the optimal value of the linear relaxation. Our second formulation contains less variables and constraints but it has a weaker linear relaxation bound.

We besides introduce an algorithm which enables us to compute strong bounds and significantly reduce the size of our formulations.

Finally, the efficiency of the algorithm and the proposed formulations are compared in terms of quality of the linear relaxation and computation time over instances from OR-Library.

Keywords: p-center · Discrete location · Equivalent formulations Integer programming

1 Introduction

We consider N clients $\{C_1, ..., C_N\}$ and M potential facility sites $\{F_1, ..., F_M\}$. Let d_{ij} be the distance between C_i and F_j. The objective of the p-center problem is to open up to p facilities such that the maximal distance (called *radius*) between a client and its closest selected site is minimized.

This problem is very popular in combinatorial optimization and has many applications. We refer the reader to the recent survey [2]. Very recent publications include [6,7] which provide heuristic solutions and [3] on an exact solution method.

In this paper, we will focus on mixed-integer linear programming formulations of the p-center problem.

© Springer International Publishing AG, part of Springer Nature 2018
J. Lee et al. (Eds.): ISCO 2018, LNCS 10856, pp. 14–25, 2018.
https://doi.org/10.1007/978-3-319-96151-4_2

Let $\mathcal{M}$ and $\mathcal{N}$ respectively be the sets $\{1, ..., M\}$ and $\{1, ..., N\}$. The most classical formulation, denoted by (P_1), for the p-center problem (see for example [4]) considers the following variables:

- y_j is a binary variable equal to 1 if and only if F_j is open;
- x_{ij} is a binary variable equal to 1 if and only if C_i is assigned to F_j;
- R is the radius.

$$(P_1) \begin{cases} \min\ R & \text{(a)} \\ \text{s.t.} \sum_{j=1}^{M} y_j \leq p & \text{(b)} \\ \sum_{j=1}^{M} x_{ij} = 1 & i \in \mathcal{N} & \text{(c)} \\ x_{ij} \leq y_j & i \in \mathcal{N}, j \in \mathcal{M} & \text{(d)} \\ \sum_{j=1}^{M} d_{ij}\, x_{ij} \leq R & i \in \mathcal{N} & \text{(e)} \\ x_{ij}, y_j \in \{0,1\} & i \in \mathcal{N}, j \in \mathcal{M} \\ r \in \mathbb{R} \end{cases} \qquad (1)$$

Constraint (1b) ensures that no more than p facilities are opened. Each client is assigned to exactly one facility through Constraints (1c). Constraints (1d) link variables x_{ij} and y_j while (1e) ensure the coherence of the objective.

A more recent formulation, denoted by (P_2), was proposed in [5]. Let $D^0 < D^1 < ... < D^K$ be the different d_{ij} values $\forall i \in \mathcal{N}\ \forall j \in \mathcal{M}$. Note that, if many distances d_{ij} have the same value, K may be significantly lower than $M \times N$. Let $\mathcal{K}$ be the set $\{1, ..., K\}$. Formulation (P_2) is based on the variables y_j, previously introduced, and one binary variable z^k, for each $k \in \mathcal{K}$, equals to 1 if and only if the optimal radius is greater than or equal to D^k:

$$(P_2) \begin{cases} \min\ D^0 + \sum_{k=1}^{K} (D^k - D^{k-1})\, z^k & \text{(a)} \\ \text{s.t.}\ 1 \leq \sum_{j=1}^{M} y_j \leq p & \text{(b)} \\ z^k + \sum_{j\,:\,d_{ij}<D^k} y_j \geq 1 & i \in \mathcal{N}, k \in \mathcal{K} & \text{(c)} \\ y_j, z^k \in \{0,1\} & j \in \mathcal{M}, k \in \mathcal{K} \end{cases} \qquad (2)$$

Constraints (2c) ensure that if no facility located at less than D^k of client C_i is selected, then the radius must be greater than or equal to D^k.

This formulation has been proved to be tighter than (P_1) [5]. However, its size strongly depends on the value K (i.e., the number of distinct distances d_{ij}).

It also has recently been adapted to the p-dispersion problem which consists in selecting p facilities among N such that the minimal distance between two selected facilities is maximized [8].

A last formulation, that can be deduced from (P_2) by a change of variables, has been recently introduced [3] and named (P_4). It contains, for all $k \in \mathcal{K}$, a binary variable u_k equal to 1 if and only if the optimal radius is D^k (i.e., $u_k = z^k - z^{k+1}$ and $z^k = \sum_{q=k}^{K} u_q$):

$$(P_4) \begin{cases} \min \sum_{k=1}^{K} D^k u_k & \text{(a)} \\ \text{s.t.} \\ \displaystyle\sum_{j\,:\,d_{ij} \leq D^k} y_j \geq \sum_{q=1}^{k} u_q & i \in \mathcal{N}, k \in \mathcal{K} & \text{(b)} \\ \displaystyle\sum_{k=1}^{K} u_k = 1 & \text{(c)} \\ y_j, u_k \in \{0,1\} & j \in \mathcal{M}, k \in \mathcal{K} \end{cases} \quad (3)$$

They also proposed a weaker version of this formulation, called (P_3), obtained by replacing the left-hand side of constraints (3b) by u_k. They proved that (P_4) leads to the same linear relaxation bound and has the same size as (P_2).

The rest of the paper is organized as follows. Section 2 presents our two new formulations. In Sect. 3 we introduce an algorithm. Finally, Sect. 4 describes numerical results on instances from the OR-Library.

2 Our New Formulations

2.1 Formulation (CP_1)

In (P_2), for all $k \in \mathcal{K}$, variable z^k is equal to 1 if and only if the optimal radius is greater than or equal to D^k. As a consequence, the following constraints are valid

$$z^k \geq z^{k+1} \quad k \in \{1, ..., K-1\}. \tag{4}$$

We first show that these inequalities are redundant for (P_2). Let (P_2') be the formulation obtained when constraints (4) are added to (P_2) and let $v(\overline{F})$ be the optimal value of the linear relaxation of a given formulation F. We now prove that adding constraints (4) does not improve the quality of the linear relaxation.

Proposition 1. $v(\overline{P_2'}) = v(\overline{P_2})$

Proof. We show that an optimal solution $(\tilde{y}, \tilde{z})$ of the relaxation of (P_2) satisfies (4). For each distance D^k there exists a client $i(k)$ such that

$$\tilde{z}^k + \sum_{j\,:\,d_{i(k)j} < D^k} \tilde{y}_j = 1 \tag{5}$$

otherwise $\tilde{z}^k$ can be decreased and $(\tilde{y}, \tilde{z})$ is not optimal.

We now assume that $\tilde{z}^{k-1} < \tilde{z}^k$ for some index $k \in \{2, ..., K\}$. It follows that

$$\tilde{z}^{k-1} + \sum_{j \,:\, d_{i(k)j} < D^{k-1}} \tilde{y}_j < \tilde{z}^k + \sum_{j \,:\, d_{i(k)j} < D^k} \tilde{y}_j = 1$$

The last equality follows from (5). Therefore, constraints (2c) for $i(k)$ and $k-1$ is violated. $\qquad\square$

We now prove that a large part of constraints (2c) are redundant in (P_2').

Let N_i^k be the set of facilities located at less than D^k from client C_i. We can observe that N_i^k is included in N_i^{k+1}, for all $k \in \mathcal{K}$. Moreover, N_i^k is equal to N_i^{k+1} if and only if there is no facility at distance D^k from client C_i. Let S_i be the set of indices $k \in \{1, ..., K-1\}$ such that N_i^k is different from N_i^{k+1}. Observe that $|S_i| \leq \min(M, K)$.

We define Formulation (CP_1) as Formulation (P_2') where only the constraints (2c) such that $k \in S_i$ or $k = K$ are kept.

$$(CP_1) \begin{cases} \min \; D^0 + \sum_{k=1}^{K} (D^k - D^{k-1}) \, z^k & \text{(a)} \\[2mm] \text{s.t.,} \\[1mm] \quad z^k + \sum_{j \,:\, d_{ij} < D^k} y_j \geq 1 & i \in \mathcal{N}, \; k \in S_i \cup \{K\} \quad \text{(b)} \\[2mm] \quad y_j, z^k \in \{0,1\} & j \in \mathcal{M}, k \in \mathcal{K} \end{cases} \quad (6)$$

The number of constraints is dominated by the number of constraints (6b). This number is bounded by both NM and NK.

The following proposition proves that (CP_1) is a valid formulation.

Proposition 2. (CP_1) *is a valid formulation of the p-center problem.*

Proof. We show that the constraints removed from (P_2') are dominated. If $N_i^k = N_i^{k+1}$, then $\sum_{j \,:\, d_{ij} < D^k} y_j = \sum_{j \,:\, d_{ij} < D^{k+1}} y_j$. Since $z^k \geq z^{k+1}$, we have:

$$z^k + \sum_{j \,:\, d_{ij} < D^k} y_j \geq z^{k+1} + \sum_{j \,:\, d_{ij} < D^{k+1}} y_j \geq 1.$$

As a consequence, the constraint (2c) associated with i and k is dominated by the one associated with i and $k+1$. $\qquad\square$

We now prove that Formulations (P_2) and (CP_1) lead to the same bound by linear relaxation.

Proposition 3. $v(\overline{CP_1}) = v(\overline{P_2})$.

Proof. The arguments used in the proof of Proposition 2 can be used again to show that the constraints removed from (P_2') do not impact the value of the linear relaxation. $\qquad\square$

To sum up, (CP_1) is a valid formulation that has the same LP bound as (P_2). However, as detailed in Table 1, Formulation (CP_1) is much smaller since it reduces the number of constraints by a factor of up to N.

2.2   Formulation (CP_2)

We now introduce a second formulation, denoted by (CP_2), which contains less variables and constraints than (CP_1).

We replace the K binary variable z^k with a unique general integer variable r which represents the index of a radius:

$$(CP_2) \begin{cases} \min\ r \\ \text{s.t.} \\ \quad r + k \sum_{j\,:\,d_{ij}<D^k} y_j \geq k \qquad i \in \mathcal{N}, k \in S_i \cup \{K\} \qquad \text{(a)} \\ \quad y_j \in \{0,1\} \qquad\qquad\qquad\qquad j \in \mathcal{M} \\ \quad r \in \{0,...,K\} \end{cases} \qquad (7)$$

Constraints (7a) play a similar role to Constraints (6b).

Formulation (CP_2) does not directly provide the value of the optimal radius R but its index r such that $D^r = R$. We now prove that Formulation (CP_2) is valid.

Proposition 4. (CP_2) *is a valid formulation of the p-center problem.*

Proof. Let $(\tilde{y}, \tilde{z})$ be an integer solution of (CP_1). We first show that there exists an integer solution $(\overline{y}, \overline{r})$ of (CP_2) which provides the same radius by setting $\overline{y} = \tilde{y}$ and $\overline{r} = \sum_{k=1}^{K} \tilde{z}^k$. We need to prove that constraints (7a) are satisfied. We know that

$$\tilde{z}^k + \sum_{j\,:\,d_{ij}<D^k} \tilde{y}_j \geq 1$$

is satisfied for any client C_i and any distance D^k.

If $\tilde{z}^k$ is equal to 0, the corresponding Constraint (7a) is satisfied, as $\sum_{j\,:\,d_{ij}<D^k} \tilde{y}_j \geq 1$. Otherwise, the same result is obtained since the $\tilde{z}^k$ variables are ordered in decreasing order which leads to $\overline{r} \geq k$. These two solutions provide the same radius as $D^0 + \sum_{k=1}^{K}(D^k - D^{k-1})\,\tilde{z}^k = D^{\sum_{k=1}^{K} \tilde{z}^k}$.

We now prove that for any solution $(\tilde{y}, \tilde{r})$ of (CP_2) there exists an equivalent solution $(\overline{y}, \overline{z})$ of (CP_1). We set $\overline{y} = \tilde{y}$ and $\overline{z}^k = 1$ if and only if $\tilde{r} \geq k$. Constraint

$$\tilde{r} + k \sum_{j\,:\,d_{ij}<D^k} \tilde{y}_j \geq k \qquad (8)$$

is satisfied for any $k \in \mathcal{K}$. If $\tilde{r}$ is lower than k, then at least one variable $\tilde{y}_j$ from Eq. (8) is equal to 1 and the corresponding constraint (6b) is satisfied. Otherwise, $\overline{z}^k$ is equal to 1 and the same conclusion is reached. $\square$

We now prove that the linear relaxation of (CP_1) is stronger than the one of (CP_2).

Assumption 1. *We shall suppose $D^0 = 0$ and $\forall k \in \mathcal{K}$, $D^k - D^{k-1} = 1$.*

This assumption is not restrictive, one can transform any instance by replacing any distance D^k by its rank k. The transformed problem is equivalent as if the optimal radius is D^{k^*}, then the optimal solution of the transformed problem is k^*.

Under this assumption, problems (CP_1) and (CP_2) have the same optimal values, both of them compute the rank of the optimal radius.

Proposition 5. *Let $\overline{CP_1}$ and $\overline{CP_2}$ respectively be the LP relaxation of (CP_1) and (CP_2), $v(\overline{CP_1}) \geq v(\overline{CP_2})$ under Assumption 1.*

Proof. Let $(\tilde{y}, \tilde{z})$ be a solution of $\overline{CP_1}$. We build a solution $(\overline{y}, \overline{r})$ of $\overline{CP_2}$ with the same value. We take $\overline{y} = \tilde{y}$ and $\overline{r} = \sum_{k=1}^{K} \tilde{z}^k$.

We need to prove that constraints (7a) are satisfied.

Since the z^k variables are ordered in decreasing order by Constraints 4, it follows that $\overline{r} \geq k\tilde{z}^k \ \forall k \in \mathcal{K}$. This and Constraints (2c) imply that Constraints (7a) are satisfied. □

Table 1 summarizes the size of the previously mentioned formulations.

Table 1. Size of the four formulations ($K \leq NM$).

Formulation	# of variables	# of constraints
(P_1)	$\mathcal{O}(NM)$	$\mathcal{O}(NM)$
(P_2), (P_3), (P_4)	$\mathcal{O}(M + K)$	$\mathcal{O}(NK)$
(CP_1)	$\mathcal{O}(M + K)$	$\mathcal{O}(\min(NM, NK))$
(CP_2)	$\mathcal{O}(M)$	$\mathcal{O}(\min(NM, NK))$

3 A Two-Step Resolution Algorithm

We present, in this section, a two-step algorithm to solve more efficiently the p-center problem.

Let lb be a lower bound of the optimal radius. We suppose that lb is one of the distances D^k since, otherwise, lb can be set to the next distance. All the distances d_{ij} lower than lb can be replaced by lb.

Similarly, all the distances d_{ij} greater than an upper bound ub can be replaced by $ub + 1$ in order to discard solutions of value greater than ub.

The size of Formulations (P_2) and (CP_1) strongly depends on K. This value can be reduced by identifying lower and upper bounds. Such bounds can easily be obtained, as mentioned in [5].

Our resolution algorithm, depicted in Fig. 1, can be applied to any formulation F of the p-center problem including (P_1), (P_2), (P_3), (P_4), (CP_1) and (CP_2). It is mainly based on the idea that whenever the optimal value $\overline{v}$ of the linear relaxation of F is not equal to an existing distance, then there exists $k \in K$ such that $D^{k-1} < \overline{v} < D^k$. In that case, D^k constitutes a stronger lower bound than

$\overline{v}$ and the linear relaxation can be solved again. This process is repeated until an existing distance is obtained as the optimal value of the linear relaxation. This constitutes Step 1 of the algorithm.

The bound obtained when applying this algorithm over (P_2) or (CP_1) corresponds to the one called LB^*, computed by a binary search algorithm in [5].

Step 1 can be further improved by introducing the notion of *dominated clients* and *dominated facilities* within some reduction rules. A facility F_a is dominated if there exists another facility F_b such that $d_{ia} \geq d_{ib}$ for all clients i. Such a facility can be removed as it will always be at least as interesting to assign a client to F_b than to F_a. Similarly, a client C_a is said to be dominated if there exists another client C_b such that $d_{aj} \leq d_{bj}$ for all facilities j. Dominated clients can also be ignored.

Instructions 3 and 4 are repeated since new dominated clients and facilities may be found when a bound is improved, and vice versa.

Step 2 of Algorithm 1 consists in solving Formulation F to optimality with the improved bounds lb and ub computed in Step 1.

Algorithm 1:
F: formulation of the p-center problem
p: maximal number of centers
d: distances
lb, ub: initial bounds
Result: The optimal radius
// Step 1
1 **repeat**
2   **repeat**
3   Remove dominated clients and facilities // Reduction rules
4         $(lb, ub) \leftarrow$ Compute bounds
5   **until** *lb and ub are not improved and no more dominated clients or facilities have been found*
6     $\overline{v} \leftarrow$ SolveLinearRelaxation(F, lb, ub)
7     $lb \leftarrow \min_k\{D^k \ : \ \overline{v} \leq D^k\}$
8 **until** $\overline{v} = lb$ // *until $\overline{v}$ is one of the existing distances*
// Step 2
9 $r^* \leftarrow$ SolveOptimally(F, lb, ub)
10 **return** r^*

Fig. 1. Algorithm used to solve the p-center problem through F, a p-center formulation.

4   Numerical Results

We implement Formulations (P_1), (P_2), (CP_1) and (CP_2) as well as Algorithm 1 on an Intel XEON E3-1280 with 3,5 GHz and 32 GB of RAM with the Java API of CPLEX 12.7. Following several authors, we consider instances from the OR-Library [1].

4.1 Comparing Sizes and Computation Times on 5 Instances

Table 2 presents a comparison of the sizes of the four formulations on the five first instances of the OR-Library with $N = M = 100$. We use the initial lower bound $LB_0 = \max_{i \in \mathcal{N}} \min_{j \in \mathcal{M}} d_{ij}$ and initial upper bound $UB_0 = \min_{j \in \mathcal{M}} \max_{i \in \mathcal{N}} d_{ij}$ introduced in [5].

As expected, the number of variables in (CP_1) and (P_2) are equal and are significantly lower than in (P_1). Formulation (P_2) has more constraints than Formulation (P_1). Formulation (CP_1) has by far less constraints than (P_2). All this explains why (CP_1) has the best performances in every aspect.

Formulation (CP_2) is the most compact but this does not fully compensate the poor quality of its LP bound.

Table 2. Size and resolution times (1 thread) of the formulations for the five first OR-Library instances with $lb = LB_0$ and $ub = UB_0$.

		(P_1)	(P_2)	(CP_1)	(CP_2)
Instance 1	Number of variables	10101	286	286	101
	Number of constraints	12209	18602	6089	5903
$(LB_0 = 0)$	LP bound	97,57	106,54	106,54	83,62
$(UB_0 = 186)$	Resolution time (s)	9,14	251,28	**3,16**	14,94
Instance 2	Number of variables	10101	277	277	101
	Number of constraints	12473	17702	6094	5917
$(LB_0 = 0)$	LP bound	76,72	85,68	85,68	70,19
$(UB_0 = 178)$	Resolution time (s)	15,69	47,31	**2,99**	19,80
Instance 3	Number of variables	10101	305	305	101
	Number of constraints	11293	20502	6852	6647
$(LB_0 = 0)$	LP bound	73,24	83,28	83,28	68,92
$(UB_0 = 205)$	Resolution time (s)	11,68	21,02	**2,85**	10,99
Instance 4	Number of variables	10101	299	299	101
	Number of constraints	12009	19902	6403	6204
$(LB_0 = 0)$	LP bound	54,55	64,16	64,16	52,42
$(UB_0 = 204)$	Resolution time (s)	3,19	43,02	**1,64**	12,90
Instance 5	Number of variables	10101	270	270	101
	Number of constraints	11777	17002	6263	6093
$(LB_0 = 0)$	LP bound	30,37	37,82	37,82	29,29
$(UB_0 = 169)$	Resolution time (s)	1,93	25,10	**1,66**	11,65

Table 3. Comparison of the different formulations with $lb = LB_1$ and $ub = UB_1$. For each instance, the smallest time appears in bold. Symbol "-" means that the instance was not solved within 1 h.

	N	p	opt	lb	ub	(P_1)		(P_2)		(CP_1)		(CP_2)	
						b	t	b	t	b	t	b	t
1	100	5	127	59	133	98	2,4	107	75,3	107	**1,0**	85	4,0
2	100	10	98	56	117	77	2,9	86	7,3	86	**0,5**	71	5,2
3	100	10	93	55	116	74	2,9	84	2,5	84	**0,2**	69	3,1
4	100	20	74	41	127	55	0,7	65	7,9	65	**0,6**	53	3,4
5	100	33	48	23	87	31	0,8	38	1,0	38	**0,1**	30	1,5
6	200	5	84	38	94	68	35,9	75	106,7	75	**2,7**	59	47,1
7	200	10	64	34	79	51	20,5	58	100,2	58	**1,8**	46	26,1
8	200	20	55	30	72	41	20,7	48	87,2	48	**1,6**	38	19,6
9	200	40	37	22	73	28	8,9	33	14,9	33	**1,4**	27	29,8
10	200	67	20	11	44	15	1,6	18	0,8	18	**0,3**	14	5,5
11	300	5	59	34	67	50	99,0	54	30,4	54	**6,2**	44	68,1
12	300	10	51	30	72	43	229,7	48	71,0	48	**7,2**	39	98,7
13	300	30	36	20	56	28	114,0	33	44,6	33	**4,7**	26	106,9
14	300	60	26	14	60	19	157,1	23	33,4	23	**12,9**	18	151,7
15	300	100	18	10	42	13	8,6	16	9,4	16	**0,9**	13	30,2
16	400	5	47	26	51	41	403,2	45	25,3	45	**3,3**	36	54,5
17	400	10	39	21	47	33	737,8	36	35,0	36	**24,9**	29	149,2
18	400	40	28	16	50	22	664,7	25	96,4	25	**22,1**	20	431,4
19	400	80	18	10	40	14	226,2	16	81,4	16	**18,5**	13	116,9
20	400	133	13	7	32	10	9,0	12	3,0	12	**0,9**	10	22,5
21	500	5	40	23	48	35	2581,0	37	118,3	37	**13,6**	31	194,6
22	500	10	38	21	49	31	-	35	924,4	35	**24,6**	28	507,8
23	500	50	22	13	38	17	1375,8	20	212,2	20	**38,4**	16	481,8
24	500	100	15	9	35	12	573,7	14	51,0	14	**29,6**	11	209,2
25	500	167	11	6	27	8	57,2	10	5,1	10	**2,0**	8	23,1
26	600	5	38	21	43	32	3093,6	35	106,0	35	**13,6**	28	152,4
27	600	10	32	18	39	28	3118,9	30	104,3	30	**48,3**	25	341,5
28	600	60	18	10	33	14	-	16	176,2	16	**103,3**	13	-
29	600	120	13	7	36	10	-	12	130,7	12	**77,8**	9	893,6
30	600	200	9	5	29	7	106,5	8	**12,4**	8	15,7	7	89,8
31	700	5	30	16	34	27	1793,8	28	68,8	28	**12,5**	24	139,9
32	700	10	29	16	35	25	-	27	718,7	27	**127,3**	22	944,5
33	700	70	15	9	26	13	-	14	155,1	14	**76,0**	12	890,1
34	700	140	11	6	30	9	2617,9	10	168,7	10	**32,8**	8	464,9
35	800	5	30	16	32	27	-	29	23,0	29	**13,0**	23	170,6
36	800	10	27	16	34	24	-	26	**130,3**	26	821,7	21	1056,6
37	800	80	15	8	26	12	-	14	222,5	14	**90,9**	11	1706,9
38	900	5	29	15	35	25	-	27	68,8	27	**19,0**	21	300,1
39	900	10	23	13	28	20	-	22	**348,4**	22	1190,0	18	1786,4
40	900	90	13	7	22	10	-	12	551,0	12	**129,5**	10	1059,9
				Total			57699		5129		**2991**		16390

Table 4. Results obtained with Algorithm 1 of Fig. 1 with $lb = LB_1$ and $ub = UB_1$.

	N	p	opt	(CP_1)		(CP_2)	
				t1	t2	t1	t2
1	100	5	127	0,2	**0,3**	0,3	0,7
2	100	10	98	0,2	**0,2**	0,3	0,4
3	100	10	93	0,2	**0,3**	0,3	0,4
4	100	20	74	0,3	**0,4**	0,4	0,5
5	100	33	48	0,1	**0,2**	0,3	0,4
6	200	5	84	1,9	**2,7**	5,2	6,3
7	200	10	64	1,1	**1,4**	3,0	3,4
8	200	20	55	0,8	**1,0**	2,8	3,0
9	200	40	37	2,0	**2,7**	4,5	5,4
10	200	67	20	0,4	**0,6**	0,9	1,1
11	300	5	59	0,8	**0,9**	2,2	2,2
12	300	10	51	3,4	**4,6**	10,2	12,5
13	300	30	36	3,6	**4,6**	8,8	9,8
14	300	60	26	3,5	**4,5**	14,8	17,5
15	300	100	18	1,5	**2,1**	3,3	3,7
16	400	5	47	1,4	**1,4**	6,4	6,4
17	400	10	39	3,3	**4,3**	9,5	10,6
18	400	40	28	5,8	**8,3**	29,1	33,3
19	400	80	18	4,1	**6,2**	9,8	12,1
20	400	133	13	2,5	**3,0**	4,0	5,0
21	500	5	40	3,1	**4,0**	9,7	10,3
22	500	10	38	16,6	**26,5**	38,6	48,3
23	500	50	22	7,0	**9,9**	31,5	37,1
24	500	100	15	7,6	**11,4**	18,5	23,7
25	500	167	11	3,7	**4,6**	7,5	9,0
26	600	5	38	4,6	**5,3**	19,3	20,7
27	600	10	32	9,5	**12,5**	23,0	26,2
28	600	60	18	14,4	**17,5**	42,0	48,7
29	600	120	13	23,4	**32,7**	91,0	111,4
30	600	200	9	10,5	**15,1**	17,4	21,9
31	700	5	30	8,2	**9,3**	15,8	17,5
32	700	10	29	18,8	**71,8**	33,8	109,8
33	700	70	15	10,2	**14,3**	25,4	34,4
34	700	140	11	34,2	**46,4**	90,1	107,6
35	800	5	30	2,2	**2,2**	11,8	12,0
36	800	10	27	20,0	**30,3**	40,5	53,1
37	800	80	15	21,8	**27,8**	50,2	60,9
38	900	5	29	12,2	**12,7**	29,7	30,3
39	900	10	23	36,6	**49,7**	45,5	153,4
40	900	90	13	21,8	**31,2**	50,3	70,7
Total					**484**		1142

4.2   Relaxation and Computation Times on the 40 OR-Library Instances

In Table 3, we perform a larger comparison with stronger bounds *lb* and *ub* equal to the bounds LB_1 and UB_1 introduced in [5]. The resolution is then performed by CPLEX with its default parameters but with a maximal CPU time of 1 h.

The first column is the instance number. The three following columns provide N, p and the optimal value of the instances ($N = M$ in these instances). Columns 5 and 6 contain the initial bounds LB and UB. For each formulation, column "b" corresponds the optimal value of the linear relaxation and column "t" to the resolution time in seconds.

We can first observe that Formulations (CP_1) and (P_2) solve all the 40 instances within 1 h while ten instances are not solved with (P_1) and one instance is not solved with (CP_2). We can even observe that (CP_1) solves the whole set of instances in less than 50 min and (P_2) in less than 85 min.

Formulation (P_2) outperforms (CP_1) mainly on instances 36 and 39. This is possibly due to some difficulty of the solver to find good feasible solutions.

4.3   Results of Algorithm 1

Table 4 presents the results of Algorithm 1 with formulations (CP_1) and (CP_2). Columns "t1" and "t2" respectively correspond to the time of the first phase and the total time.

Formulation (CP_2) is now able to solve all the instances within 1 h. We observe that the total time to solve the 40 instances is reduced by approximately 6 times for (CP_1) and 14 times for (CP_2) if compared to Table 3.

5   Conclusion

We introduced two new compact formulations of the p-center problem. We theoretically compared the quality of their LP bounds and their sizes to existing formulations. Numerical experiments confirmed these results and highlighted the fact that our new formulation (CP_1) outperforms the previously known formulations (P_1) and (P_2) at all levels. Our more compact formulation (CP_2) suffers from the poor quality of its linear relaxation. Another aspect of our work was to embed the formulations within a two-step algorithm in order to obtain better computation times.

Our future work will focus on improving our compact formulation through polyhedral studies.

References

1. Beasley, J.E.: OR-library: distributing test problems by electronic mail. J. Oper. Res. Soc. **41**, 1069–1072 (1990)
2. Calik, H., Labbé, M., Yaman, H.: p-center problems. In: Laporte, G., Nickel, S., da Gama, F.S. (eds.) Location Science, pp. 79–92. Springer, Cham (2015). https://doi.org/10.1007/978-3-319-13111-5_4
3. Calik, H., Tansel, B.C.: Double bound method for solving the p-center location problem. Comput. Oper. Res. **40**(12), 2991–2999 (2013)
4. Daskin, M.S.: Network and Discrete Location Analysis. Wiley, New York (1995)
5. Elloumi, S., Labbé, M., Pochet, Y.: A new formulation and resolution method for the p-center problem. INFORMS J. Comput. **16**(1), 84–94 (2004)

6. Ferone, D., Festa, P., Napoletano, A., Resende, M.G.C.: A new local search for the p-center problem based on the critical vertex concept. In: Battiti, R., Kvasov, D.E., Sergeyev, Y.D. (eds.) LION 2017. LNCS, vol. 10556, pp. 79–92. Springer, Cham (2017). https://doi.org/10.1007/978-3-319-69404-7_6

7. Ferone, D., Festa, P., Napoletano, A., Resende, M.G.C.: On the fast solution of the p-center problem. In: 2017 19th International Conference on Transparent Optical Networks (ICTON), pp. 1–4, July 2017

8. Sayah, D., Irnich, S.: A new compact formulation for the discrete p-dispersion problem. Eur. J. Oper. Res. **256**(1), 62–67 (2017)

The Next Release Problem: Complexity, Exact Algorithms and Computations

José Carlos Almeida Jr., Felipe de C. Pereira, Marina V. A. Reis, and Breno Piva[✉]

Departamento de Computação, Universidade Federal de Sergipe, Av. Marechal Rondon, s/n, Jd. Rosa Elze, São Cristóvão, Sergipe 49100-000, Brasil
{fcpereira,marinavar,brenopiva}@dcomp.ufs.br

Abstract. The Next Release Problem (NRP) is an important problem in Software Engineering. Several papers investigate the NRP, most of them considering heuristics to solve the problem. However, the literature lacks a more theoretical approach to this problem, specially regarding its complexity, approximability and more powerful exact algorithms. In this paper we aim to help filling this gap.

Keywords: Strong NP-hardness · FPTAS · Integer programming Branch-and-cut · Search Based Software Engineering

1 Introduction

Search-Based Software Engineering (SBSE) is concerned with solving optimization problems in Software Engineering (SE), or treating SE problems as search problems [6]. Through the software development cycle, several problems can be formulated as search problems. From requirements to testing and all the way through releasing, there are problems that can be thought of as search/optimization problems.

The Next Release Problem (NRP) was first formalized by Bagnall et al. in [1] as a description of a company's next release plan considering its involvement in the development and maintenance of large, complex systems to a set of clients that have different needs and different values for the company.

The input of NRP can be formalized as been composed by a set R of requirements, a set of clients C, a budget $B \in \mathbb{Z}^+$ and a directed graph $D = (R \cup C, A)$ indicating the association between requirements and between requirements and clients. The set of arcs in D is $A \subseteq R \times (R \cup C)$. There are also two functions $\omega : R \to \mathbb{Z}^+$ and $\delta : C \to \mathbb{Z}^+$ indicating, respectively, the cost of each requirement and the value of each client.

In a valid input of NRP, D is an acyclic and transitive graph, meaning that if there is a path from vertex a to vertex b, there is no path from b to a. Also, if there is a path from a to b, there is also an arc $(a, b) \in A$. Figure 1 illustrates an NRP instance.

B. Piva—The authors would like to thank PIBIC/UFS for the support.

© Springer International Publishing AG, part of Springer Nature 2018
J. Lee et al. (Eds.): ISCO 2018, LNCS 10856, pp. 26–38, 2018.
https://doi.org/10.1007/978-3-319-96151-4_3

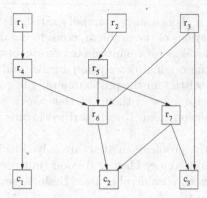

Fig. 1. An NRP instance with 3 clients and 7 requirements. The transitive arcs are omitted for clarity reasons.

A solution of NRP is given by a set $R' \subseteq R$ to be implemented and a set $C' \subseteq C$ of clients that are satisfied by R'. The cost of the solution is given by the sum of the costs of the requirements in R', i.e., $\sum_{r \in R'} \omega(r)$. Meanwhile, the value of the solution is the sum of the values of the clients in C', i.e., $\sum_{c \in C'} \delta(c)$. For a solution to be valid it must have cost at most B (the budget), if a requirement r is implemented, all of its prerequisites, i.e., all the requirements s for which $(s, r) \in A$, must also be implemented. At last, a client c can only be considered satisfied if all of its requirements, i.e., all the requirements r for which $(r, c) \in A$, are implemented.

The objective of NRP is to find a solution that satisfies the properties above and has a maximum value. This description can be formalized using the following integer linear program (IP) formulation due to Bagnall et al. [1]:

$$\text{(NRP)} \qquad z = \max \sum_{c \in C} \delta(c) y_c \tag{1}$$

$$\text{subject to} \qquad \sum_{r \in R} \omega(r) x_r \leq B \tag{2}$$

$$x_{r'} \leq x_r \qquad \forall\, (r, r') \in A \tag{3}$$

$$y_c \leq x_r \qquad \forall\, (r, c) \in A \tag{4}$$

$$x_r, y_c \in \{0, 1\}; r \in R, c \in C \tag{5}$$

in this model variables x_r indicate whether a requirement is implemented or not, meanwhile, y_c indicate if a client is satisfied. Inequality (2) guarantees that the cost of the solution is not greater than the budget. The constraints described by (3) and (4) enforce that the proper requirements are implemented.

There are different versions of the problem in the literature, for instance, [2] considers a version of the NRP that is equivalent to a knapsack problem. In [14] the authors propose a bi-objective version of the problem where the goal is to maximize the solution value and minimize its cost, however, in their version of the problem, a client can be partially satisfied. van den Akker et al. [8] explore different variations of the problem, considering different scenarios like having a single pool of developers or multiple teams etc. Several other papers are developed considering the original description by Bagnall et al., among them, [3, 7] and [12] present heuristics for the problem and [4, 9] compare heuristics and exact methods, just to cite a few of them. In this paper, whenever we mention the NRP we mean the version from [1] as described in this section.

Motivation. Typically, the problems in SBSE are solved using heuristics and the reason is that in SE usually we need fast and good (but not necessarily optimal) solutions for instances with considerable size. Besides, several of the problems are proven to be $\mathcal{NP}$-hard.

Nonetheless, developing exact algorithms is important to produce optimal solutions for instances of reasonable size, thus providing a basis for comparison and possibly new insights for the improvement of heuristics. However a search in [13] reveals that very few works in the field even consider a comparison with exact methods, a fact pointed out in [5] and much less the use of approximative algorithms. In fact, some theoretical aspects of the problems seem to be completely neglected, like the possible existence of pseudo-polynomial algorithms or polynomial algorithms with approximation guarantees.

In this work we consider a classical problem in SBSE, i.e., NRP, but instead of simply trying to solve it or produce a faster algorithm, we analyze the problem from a more theoretical point of view, although producing a faster algorithm is certainly a good thing. In this analysis we consider the strong $\mathcal{NP}$-hardness of the problem, the existence of a fully polynomial time approximation scheme (FPTAS) [10] and we use some of the theory of integer linear programming to try to improve a previously known algorithm.

Our Contribution. In this paper we prove the strong $\mathcal{NP}$-hardness of NRP. We show that the NRP does not admit an FPTAS. We present a new family of valid inequalities for the NRP IP formulation from [1]. We present a separation heuristic routine for this family of valid inequalities and use this routine to construct a Branch-and-Cut (B&C) algorithm. We create and make available a new set of bigger instances for the NRP based on instances from the literature. Finally, the performance of the B&C algorithm is compared against that of a Branch-and-Bound (B&B) algorithm that is currently the faster algorithm in the experiments found in the literature.

Organization of the Text. The remaining of the text is organized as follows: Sect. 2 presents complexity results for the NRP. In Sect. 3 the new family of valid inequalities and separation routine are described, while in Sect. 4 the results of the computational experiments are presented and analyzed. Finally, in Sect. 5 the conclusions and future directions of research are discussed.

2   The Complexity of NRP

In [1], Bagnall et al. showed that the NRP is $\mathcal{NP}$-hard. The proof is based on the fact that NRP generalizes the well-known 0-1 knapsack problem and, therefore, there is an obvious reduction from this problem to the NRP. That is, for each item in the knapsack instance we create a pair of client and requirement for that item where the requirement has the same cost as the item and the client has the same value as the item. The budged of the NRP instance is the capacity of the knapsack instance.

The 0-1 knapsack problem is a weakly $\mathcal{NP}$-hard problem [10], thus having pseudo-polynomial algorithms and also fully polynomial time approximation schemes (FPTAS) [10]. However, no similar results are known for NRP. In this section we show that NRP is strongly $\mathcal{NP}$-hard and for this reason it cannot have a FPTAS.

According to the definition of strong $\mathcal{NP}$-hardness, a problem is strongly $\mathcal{NP}$-hard if there is no pseudo-polynomial time algorithm to solve it unless $\mathcal{P} = \mathcal{NP}$, i.e., an algorithm that has a polynomial time complexity if the problem has an unary encoding. Therefore, in order to show that NRP is strongly $\mathcal{NP}$-hard we show that a pseudo-polynomial algorithm for this problem cannot exist unless $\mathcal{P} = \mathcal{NP}$. This is done through a reduction from the k-clique problem.

In the k-clique problem we are given a simple graph G and a number k and we must decide whether there is a clique of size k in G. Since a clique is a complete subgraph and G is simple, a clique with k vertices must have $k(k-1)/2$ edges. The idea of the reduction is to construct an instance I of NRP such that if there is a solution of value $k(k-1)/2$ for I, then there is a clique with k vertices in G.

Proposition 1. *The* NRP *is strongly* $\mathcal{NP}$-*hard.*

Proof. Let $G = (V, E)$ be a simple graph and k an integer. Thus, G and k form an instance of the k-Clique problem. Now, let us use G and k to construct an instance I for NRP. For each vertex v in V we create a requirement r_v in I with cost 1 and for each edge uv in E we create a client c_{uv} in I with value 1. For each client c_{uv} in I we associate the requirements r_u and r_v. Finally, we define the budget of I as k.

We must observe that an optimal solution for I provides a solution to the existence of a clique with size k in G. There are two cases to consider, the first is when the answer for the k-clique instance is positive, i.e., there is a clique of size k in G and the other is when the maximum clique in G is smaller than k.

If there is a clique of size k in G then, there are $k(k-1)/2$ edges connecting these k vertices in G. There are also k requirements in I corresponding to the vertices and $k(k-1)/2$ clients corresponding to the edges in the clique. Since each client is associated with the two requirements related to the endpoints of its corresponding edge, it is clear that a solution that implements the k requirements related to the k vertices of the clique, which is possible since each requirement has cost 1 and the budget is k, will satisfy the $k(k-1)/2$ clients corresponding to the edges of the clique and since every client has value 1, $k(k-1)/2$ will also be the value of the solution. Notice that since the budget is k and each client has a unique pair of requirements, it is impossible to have a better solution.

If, on the other hand, there is no clique of size k in G, that means that every subgraph containing k vertices has less than $k(k-1)/2$ edges. Suppose there is a solution for I with value $k(k-1)/2$. Since every client has value 1, there are $k(k-1)/2$ clients satisfied and since each client has a unique pair of requirements, there are at least k requirements implemented. Considering the budget is k and each requirement costs 1, there must be exactly k requirements implemented. Now, we know by the construction of I that each of the k requirements corresponds to a vertex in G and each of the clients corresponds to an edge connecting a pair of these k vertices, therefore, there must be a subgraph in G with k vertices and $k(k-1)/2$ edges, a contradiction.

From the two cases described it is possible to conclude that G has a clique of size k if and only if I has a solution of value $k(k-1)/2$. That is, the construction of I describes a reduction from k-clique to NRP. It is trivial to notice that the size of I is polynomial in the size of G and k and the construction of I can be done in a time that is also polynomial in the size of G and k. Hence, we have a polynomial time reduction.

Now, suppose there is a pseudo-polynomial algorithm for NRP. In the construction described above, the only numerical data in I are the values of the clients and costs of requirements, all of them equal to 1, therefore, an encoding of I in unary has a size that is polynomial in the size of a binary encoding so, a pseudo-polynomial algorithm for NRP solves I in polynomial time and by consequence, also the k-clique problem. Hence, such algorithm cannot exist unless $\mathcal{P} = \mathcal{NP}.\square$

Corollary 1. *The* NRP *does not admit an* FPTAS, *unless* $\mathcal{P} = \mathcal{NP}$.

3   Integer Programming Approach

Considering that the NRP is $\mathcal{NP}$-hard, the most common approach to find solutions for this problem has been through the use of heuristics. This is reasonable since every known exact algorithm for the problem has exponential time complexity. Moreover, the problem is of practical importance and, thus, solutions for real instances must be found despite its hardness.

Nonetheless, as previously stated, having exact algorithms is still important to produce optimal solutions to evaluate the quality of heuristics. Besides, there are cases in the literature where the time necessary to run an exact algorithm is experimentally competitive with the one of heuristics, thus making them a good option in practice. In [9], Veerapen et al. showed that this might be the case for NRP.

Veerapen et al. [9] re-implemented an IP based algorithm using the formulation described in [1] and tested it using the set of instances generated by Xuan et al. [12]. Xuan et al. generated two sets of instances, the first set named *Classic Instances* was generated using the description given in [1]. The second set called *Realistic Instances* was generated by mining a database of bug reports for two well known softwares. Xuan et al. proposed a Backbone-based Multilevel

Algorithm (BMA) for NRP and compared the results of this algorithm with other heuristics from literature for NRP, concluding that for most of the instances, the BMA presented a better performance, both in quality of the solution and speed. However, in [9], the authors showed that the IP based algorithm is able to solve all the instances tested by Xuan et al. to optimality in a few seconds while the best heuristic takes several minutes to find a solution without any guarantees of quality.

In the remaining of this section we describe a new valid inequality for the formulation proposed in [1] and a separation procedure to be used in a B&C algorithm.

Cover Inequalities. In the knapsack problem we are given a set $N = \{a_1, a_2, ..., a_n\}$ of items, a weight limit W and functions $w : N \to \mathbb{R}^+$ and $v : N \to \mathbb{R}^+$ representing, respectively the weight and the value of each item. The set of valid solutions for an instance of the knapsack problem can be represented as $S = \{X \in \mathbb{B}^n | \sum_{i \in \{1...n\}} x_i w(a_i) \leq W\}$, therefore, an optimal solution is given by $\max\{\sum_{i \in \{1...n\}} x_i v(a_i) | (x_1, x_2, ..., x_n) \in S\}$.

A cover in the context of the knapsack problem is a set $H \subseteq \{1, ..., n\}$ such that $\sum_{i \in H} w(a_i) > W$. Therefore, the inequality $\sum_{i \in H} x_i \leq |H| - 1$ is valid for the formulation of the problem under consideration. A minimal cover is a cover that for any $j \in H$, $H \setminus \{j\}$ is not a cover. An extended cover of a cover H is given by $E(H) = H \cup \{j \in \{1, ..., n\} \setminus H | w(a_j) \geq w(a_i) \forall i \in H\}$ and the inequality $\sum_{i \in E(H)} x_i \leq |H| - 1$ is also valid for this formulation [11].

Given that the weight limit in the knapsack problem and the budget in the NRP have similar roles, it is possible to consider cover inequalities for the NRP. Therefore, it is easy to see that the inequalities $\sum_{i \in H} x_{r_i} \leq |H| - 1$ and $\sum_{i \in E(H)} x_{r_i} \leq |H| - 1$ are valid for (NRP) where $H \subseteq \{1, ..., |R|\}$ such that $\sum_{i \in H} \omega(r_i) > B$ and $E(H) = H \cup \{j \in \{1, ..., |R|\} \setminus H | \omega(r_j) \geq \omega(r_i) \forall i \in H\}$.

The cover inequalities for the NRP shown above take into consideration only variables related to requirements (x variables), hence we can try to strengthen them by adding variables related to clients (y variables). Since the extended cover inequalities are lifts of cover inequalities, they are stronger than the latter, so we are now going to consider a lift over these inequalities. In order to facilitate the understanding of these liftings we are going to consider the example in Fig. 1 with budget 19 and the following costs and values, respectively for the requirements and clients: $\omega(r_1) = 6$, $\omega(r_2) = 11$, $\omega(r_3) = 6$, $\omega(r_4) = 7$, $\omega(r_5) = 6$, $\omega(r_6) = 8$, $\omega(r_7) = 1$ and $\delta(c_1) = 4$, $\delta(c_2) = 6$, $\delta(c_3) = 3$.

Let $E(H)$ be an extended cover for NRP, in the example we could have $E(H) = \{1, 2, 3, 4\}$. Let $\chi(C')$ denote an ordering of a set of clients C' according to some criterion, for instance, $\chi(C) = (c_2, c_1, c_3)$. And denote by $\chi(C')[0]$ the index of the first client in this ordering, in our example, $\chi(C)[0] = 2$. Now, let $F(c, H') = \{i \in \{1, ..., |R|\} \setminus H' | (r_i, c) \in A\}$ denote the set of requirements whose indexes are not in H' that are prerequisites of client c, in the current example, for instance, $F(c_2, E(H)) = \{5, 6, 7\}$. Finally, a client cover over sets H' of requirements indexes and C' of clients can be defined as follows: $CC(H', C') =$ $\chi(C')[0] \cup CC(H' \cup F(\chi(C')[0], H'), C' \setminus \{\chi(C')[0]\})$ if $\sum_{j \in F(c, H')} \omega(r_j) \geq$ $\omega(r_i)$ for all $i \in E(H)$ or $CC(H', C') = CC(H', C' \setminus \{\chi(C')[0]\})$ otherwise, moreover, $CC(H', \emptyset) = \emptyset$. This definition results recursively in the following values: $CC(\{1, ..., 4\}, \{1, 2, 3\}) = \{2\} \cup CC(\{1, ..., 7\}, \{1, 3\})$ and $CC(\{1, ..., 7\}, \{1, 3\}) = CC(\{1, ..., 7\}, \{3\}) = CC(\{1, ..., 7\}, \emptyset) = \emptyset$, therefore, $CC(\{1, ..., 4\}, \{1, 2, 3\}) = \{2\}$. With these definitions we can now define the extended client cover inequalities as $\sum_{i \in E(H)} x_{r_i} + \sum_{j \in CC(E(H), C)} y_{c_j} \leq |H| - 1$.

The validity of these inequalities follows from the same arguments as for the extended cover inequality problem for the knapsack problem. Moreover, the cost of each client is calculated based on sets of requirements that are disjoint from any other set of requirement considered for the costs of other clients.

Separation Heuristic Procedure. In order to use the extended client cover inequalities in a B&C algorithm we must first define a way of finding an inequality of this family to incorporate to the model. This can be achieved through a separation routine. The goal of a separation routine is given a point p and a set of valid solutions P (a polyhedron), find an inequality that is valid for P but not satisfied by p. Therefore, if we have a solution of the linear relaxation s of NRP, we could use a separation routine to find an inequality that is valid for the set of valid solutions of NRP but not for s, thus obligating the solution of the linear relaxation of the model with the new inequality to be closer to a valid solution for NRP than the previous solution.

Next we describe a routine that heuristically tries to find a separating extended client cover inequality for the current linear relaxation solution. In that description we consider the relative weight u_r of requirement r in a solution S for the linear relaxation as been the cost of the requirement $(\omega(r))$ multiplied by the value of the corresponding variable x_r, i.e., $u_r = \omega(r)x_r$.

Separation Routine
1. Order R decreasingly by the relative weight of its elements in the solution.
2. Insert requirements in order in H until a cover is formed.
3. Remove from H the requirement with smallest cost until a minimal cover is obtained.
4. Define RHS = |H|-1.
5. Extend cover H with requirements having a cost greater than any requirement in H, obtaining E(H).
6. Define the ordering of clients decreasingly by the value of its corresponding variable in the solution multiplied by the sum of costs of its exclusive prerequisites not in E(H).
7. Get CC(E(H), C).
8. Check if the obtained inequality is violated by the current solution.

With this separation heuristic routine in hand we can now construct a B&C algorithm by executing this routine at each node of the B&B tree. At each execution of the routine, a corresponding inequality is added to the IP model if it is violated by the current linear relaxation solution.

4   Computational Results

In this section we describe the experiments performed to determine the effectiveness of the proposed B&C algorithm. To do so we compare a B&B algorithm constructed using the formulation from [1] against the B&C algorithm described in Sect. 3. In order to make a fair comparison both algorithms were implemented and executed using the same computational environment and restricted to 1 h (3,600 s) of computation.

Computational Environment. All the experiments were performed using an Intel(R) Core(TM) i3-4005U CPU @ 1.70 GHz with 4 GB of RAM memory. The programming language used was C/C++ and we used GUROBI-C++ API v6.5.1 IP solver. All the cuts, heuristics and preprocessing were turned off and only one thread was used to avoid confusing variables in the analysis.

Instances. Three sets of instances were used in the experiments: the two sets from [12], *Classic Instances* and *Realistic Instances*, and a new set constructed based on the set of *Classic Instances* but with all the parameters multiplied by some factor. This set is called *Big Instances*.

There are five *Classic Instance* groups called nrp-1 to nrp-5. Each of these instances was generate following the rules in Table 1. In that table the symbol "/" indicates a separation between levels, therefore, according to the first line, nrp-1, for example, has three levels with 20, 40 and 80 requirements, respectively at the first, second and third levels. The second line of the table indicate the range of the costs of each requirement at each level. The meaning of the remaining lines can be inferred from its description column. All the details missing are fulfilled randomly. To obtain an instance of NRP we still need a budget so, for each instance group three budget values (and therefore, three instances) are generated corresponding to 30% of the sum of all the requirements costs, 50% and 70%. These same budget ratios were used in all instance sets.

The *Big Instances* set can be divided into two sets, the x1.6 set and the x2.2 set. For each of them, the name of the set indicate the multiplication factor used to generate the instance groups. Tables 2 and 3 summarize the generation rules for these instance groups. These instances can be downloaded from the address http://gpto.dcomp.ufs.br.

Analysis of the Results. In Table 5 it is possible to see the results obtained from the B&B and B&C algorithms. In that table, the column with header "Instance" indicates the name of the instance been solved, the columns with header "Time", "Bound" and "Value" indicate respectively the total time (in seconds) to obtain

Table 1. Generation rules for *Classic Instances*.

Instance group	nrp-1	nrp-2	nrp-3	nrp-4	nrp-5
# Reqs./ level	20/40/80	20/40/80/160/320	250/500/750	250/500/750/1000/750	500/500/500
Cost of req.	1–5/2–8/5–10	1–5/2–7/3–9/4–10/5–15	1–5/2–8/5–10	1–5/2–7/3–9/4–10/5–15	1–3/2/3–5
# Max child reqs.	8/2/0	8/6/4/2/0	8/2/0	8/6/4/2/0	4/4/0
# Rqts. of client	1–5	1–5	1–5	1–5	1
# Clients	100	500	500	750	1000
Value of client	10–50	10–50	10–50	10–50	10–50

Table 2. Generation rules for instances in set x1.6.

Instance group	nrpx1.6-1	nrpx1.6-2	nrpx1.6-3	nrpx1.6-4	nrpx1.6-5
# Reqs./ level	32/64/128	32/64/128/254/512	400/800/1200	400/800/1200/1600/1200	800/800/800
Cost of req.	1–5/2–8/5–10	1–5/2–7/3–9/4–10/5–15	1–5/2–8/5–10	1–5/2–7/3–9/4–10/5–15	1–3/2/3–5
# Max child reqs.	12/3/0	12/9/6/3/0	12/3/0	12/9/6/3/0	6/6/0
# Rqts. of client	1–6	1–6	1–6	1–6	1
# Clients	160	800	800	1200	1600
Value of client	10–50	10–50	10–50	10–50	10–50

the solution, the best bound found and the value of the best valid solution. There are two sets of these three columns, one for the B&B and one for the B&C algorithm. Finally, the column with header "Root_obj" gives the value of the objective function at the root for both the B&B and B&C tree. This value can be useful to analyze how much we had to improve in order to get an optimal valid solution. Due to space constraints we omit the results for the *Realistic Instances* in this table.

Since both algorithms are exact, there is no sense in comparing the values of the solutions obtained. We can, however analyze how many of the problems were solved to optimality by each algorithm within the given time limit, and we can compare the times necessary to solve the instances. For the *Realistic Instances* both algorithms solved all the instances to optimality within the time limit. The same is true for the *Classic Instances*. For the *Big Instances* of set x1.6, the B&B algorithm was able to solve all the instances while B&C was unable to solve one of the instances (nrpx1.6-2-0.5). Regarding set x2.2, both algorithms were unable to solve 5 out of 15 instances. It is noteworthy that the unsolved instances are the same for both algorithms.

Table 3. Generation rules for instances in set x2.2.

Instance group	nrpx2.2-1	nrpx2.2-2	nrpx2.2-3	nrpx2.2-4	nrpx2.2-5
# Reqs./ level	44/88/176	44/88/176/352/704	550/1100/1650	550/1100/1650/2200/1650	1100/1100/ 1100
Cost of req.	1–5/2–8/5–10	1–5/2–7/3–9/4–10/5–15	1–5/2–8/5–10	1–5/2–7/3–9/4–10/5–15	1–3/2/3–5
# Max child reqs.	17/4/0	17/13/8/4/0	17/4/0	17/13/8/4/0	8/8/0
# Rqts. of client	1–8	1–8	1–8	1–8	1
# Clients	220	1100	1100	1650	2200
Value of client	10–50	10–50	10–50	10–50	10–50

In order to compare the times we calculate the ratio between the time spent by B&C over the time spent by B&B. Therefore, if this ratio is bigger than 1, it means that the B&C was slower while a ratio between 0 and 1 means it was faster. After calculating the ratio for each instance we calculated the geometric mean. Since we are dealing with ratios, this kind of mean has the nice property that the inverse of the mean is equal to the mean of the inverses, therefore, our results are not affected by the choice of numerator and denominator. We calculate the mean for each set of instances and also, the mean considering only the instances solved to optimality by both algorithms. Table 4 summarize these results.

Notice from Table 4 that on average, the B&C algorithm is slower than B&B for the *Classic Instances* and the *Realistic Instances* but is faster for the *Big Instances*. One possible explanation for this is that the instances in the first two sets are too easy, therefore, the time spent by the separation routine is not compensated. Meanwhile, the sets of *Big Instances* are hard enough for making the use of cuts worth.

Table 4. Geometric means of the ratio time (B&C)/time(B&B) considering all the instances of each group and only the instances solved to optimality by both algorithms.

Instance set	Classic	Realistic	Big x1.6	Big x2.2
GMean	1.248	1.741	0.943	0.801
GMeanOpt	1.248	1.741	0.885	0.716

Table 5. Results of B&B and B&C for *Classic Intances* and *Big Instances*.

Instance	B&B			B&C			Root_obj
	Time	Bound	Value	Time	Bound	Value	
nrp-1-0.3	0.11	1100	1100	0.4	1100	1100	1116
nrp-1-0.5	0.13	1742	1742	0.21	1742	1742	1789
nrp-1-0.7	0.07	2454	2454	0.1	2454	2454	2458
nrp-2-0.3	1.06	5453	5453	1.76	5453	5453	5456
nrp-2-0.5	3.44	8679	8679	4.13	8679	8679	8716
nrp-2-0.7	0.57	11938	11938	0.66	11938	11938	11946
nrp-3-0.3	1.57	8379	8379	0.49	8379	8379	8380
nrp-3-0.5	0.96	12399	12399	0.42	12399	12399	12401
nrp-3-0.7	0.44	15007	15007	1.04	15007	15007	15008
nrp-4-0.3	13.06	11585	11585	21.11	11585	11585	11587
nrp-4-0.5	3.45	17311	17311	5.56	17311	17311	17314
nrp-4-0.7	1.58	21851	21851	1.91	21851	21851	21852
nrp-5-0.3	0.25	20626	20626	0.29	20626	20626	20626
nrp-5-0.5	0.52	26248	26248	0.46	26248	26248	26248
nrp-5-0.7	0.13	29318	29318	0.18	29318	29318	29318
nrpx1.6-1-0.3	0.21	1796	1796	0.11	1796	1796	1849
nrpx1.6-1-0.5	0.27	2906	2906	0.37	2906	2906	2947
nrpx1.6-1-0.7	0.07	4027	4027	0.06	4027	4027	4030
nrpx1.6-2-0.3	295.81	7591	7591	307.85	7591	7591	8290
nrpx1.6-2-0.5	1573.72	12608	12608	3600.34	12695	12603	13395
nrpx1.6-2-0.7	2.98	18302	18302	3.7	18302	18302	18497
nrpx1.6-3-0.3	13.37	11203	11203	12.5	11203	11203	11222
nrpx1.6-3-0.5	2.52	17479	17479	1.19	17479	17479	17479
nrpx1.6-3-0.7	0.47	22718	22718	0.61	22718	22718	22718
nrpx1.6-4-0.3	3544.15	14410	14410	2611.9	14410	14410	14985
nrpx1.6-4-0.5	180.04	23898	23898	278.49	23898	23898	24060
nrpx1.6-4-0.7	16.85	32529	32529	19.45	32529	32529	32529
nrpx1.6-5-0.3	3.76	32585	32585	1.19	32585	32585	32585
nrpx1.6-5-0.5	0.7	41944	41944	0.69	41944	41944	41944
nrpx1.6-5-0.7	0.42	48093	48093	0.41	48093	48093	48093
nrpx2.2-1-0.3	0.91	2108	2108	0.54	2108	2108	2214
nrpx2.2-1-0.5	3.11	3450	3450	4.23	3450	3450	3638
nrpx2.2-1-0.7	1.32	4930	4930	1.18	4930	4930	5059
nrpx2.2-2-0.3	3600.64	9872	8835	3600.64	9934	8850	10871
nrpx2.2-2-0.5	3600.56	16906	15208	3600.83	16937	15250	17846
nrpx2.2-2-0.7	3600.62	24025	23440	3600.66	24071	23393	24821
nrpx2.2-3-0.3	28.75	14040	14040	45.8	14040	14040	14042
nrpx2.2-3-0.5	24.17	21967	21967	8.92	21967	21967	21972
nrpx2.2-3-0.7	3.5	29364	29364	4.06	29364	29364	29365
nrpx2.2-4-0.3	3620.89	18172	16354	3621.07	18519	16174	18671
nrpx2.2-4-0.5	3620.83	29745	28845	3621.47	29753	28860	30327
nrpx2.2-4-0.7	198.2	41972	41972	222.3	41972	41972	41982
nrpx2.2-5-0.3	15.06	42891	42891	1.67	42891	42891	42891
nrpx2.2-5-0.5	7.74	56521	56521	9.16	56521	56521	56522
nrpx2.2-5-0.7	3.09	65220	65220	1.53	65220	65220	65221

5   Conclusions and Future Works

NRP is strongly $\mathcal{NP}$-hard and does not admit an FPTAS, however, it is still not clear whether there is a PTAS or if there is a limit on its approximability.

Despite having exponential time complexities, IP based algorithms have shown their usefulness. Furthermore, the B&C algorithm suggests that there is still room for improvement, since this algorithm was able to solve bigger instances quicker than a simple B&B algorithm. A first step to better understanding these algorithms would be to perform a polyhedral study and determine the strength of the known valid inequalities.

References

1. Bagnall, A.J., Rayward-Smith, V.J., Whittley, I.M.: The next release problem. Inf. Softw. Technol. **43**(14), 883–890 (2001)
2. Baker, P., Harman, M., Steinhofel, K., Skaliotis, A.: Search based approaches to component selection and prioritization for the next release problem. In: 2006 22nd IEEE International Conference on Software Maintenance, pp. 176–185, September 2006
3. Botelho, G., Rocha, A., Britto, A., Silva, L.: Investigating bioinspired strategies to solve large scale next release problem. In: SET 15 in CIbSE 2015 - Ibero American Conference on Software Engineering (2015)
4. Freitas, F.G., Coutinho, D.P., Souza, J.T.: Software next release planning approach through exact optimization. Int. J. Comput. Appl. **22**(8), 1–8 (2011)
5. Freitas, F., Silva, T., Carmo, R., Souza, J.: On the applicability of exact optimization in search based software engineering. In: Cohen, M.B., Ó Cinnéide, M. (eds.) SSBSE 2011. LNCS, vol. 6956, p. 276. Springer, Heidelberg (2011). https://doi.org/10.1007/978-3-642-23716-4_29
6. Harman, M.: Search based software engineering. In: Alexandrov, V.N., van Albada, G.D., Sloot, P.M.A., Dongarra, J. (eds.) ICCS 2006. LNCS, vol. 3994, pp. 740–747. Springer, Heidelberg (2006). https://doi.org/10.1007/11758549_100
7. Jiang, H., Zhang, J., Xuan, J., Ren, Z., Hu, Y.: A hybrid ACO algorithm for the next release problem. CoRR, abs/1704.04777 (2017)
8. van den Akker, M., Brinkkemper, S., Diepen, G., Versendaal, J.: Flexible release composition using integer linear programming. Technical report UU-CS-2004-063, Institute of Information and Computing Sciences, Utrecht University, December 2004. In English, 16 pages
9. Veerapen, N., Ochoa, G., Harman, M., Burke, E.K.: An integer linear programming approach to the single and bi-objective next release problem. Inf. Softw. Technol. **65**(Supplement C), 1–13 (2015)
10. Williamson, D.P., Shmoys, D.B.: The Design of Approximation Algorithms, 1st edn. Cambridge University Press, New York (2011)
11. Wolsey, L.A.: Integer Programming. Wiley, New York (1998)

12. Xuan, J., Jiang, H., Ren, Z., Luo, Z.: Solving the large scale next release problem with a backbone-based multilevel algorithm. IEEE Trans. Softw. Eng. **38**(5), 1195–1212 (2012)
13. Zhang, Y., Harman, M., Mansouri, A.: The SBSE repository: a repository and analysis of authors and research articles on search based software engineering. http://crestweb.cs.ucl.ac.uk/resources/sbse_repository/
14. Zhang, Y., Harman, M., Mansouri, A.S.: The multi-objective next release problem. In: Proceedings of the 9th Annual Conference on Genetic and Evolutionary Computation, GECCO 2007, New York, NY, USA, pp. 1129–1137. ACM (2007)

Polytope Membership in High Dimension

Evangelos Anagnostopoulos[1](✉), Ioannis Z. Emiris[1,3],
and Vissarion Fisikopoulos[2]

[1] National and Kapodistrian University of Athens, Athens, Greece
{aneva,emiris}@di.uoa.gr
[2] Oracle Corp., Athens, Greece
vissarion.fisikopoulos@oracle.com
[3] ATHENA Research Center, Maroussi, Greece

Abstract. We study the fundamental problem of polytope membership aiming at convex polytopes in high dimension and with many facets, given as an intersection of halfspaces. Standard data-structures and brute force methods cannot scale, due to the curse of dimensionality. We design an efficient algorithm, by reduction to the approximate Nearest Neighbor (ANN) problem based on the construction of a Voronoi diagram with the polytope being one bounded cell. We thus trade exactness for efficiency so as to obtain complexity bounds polynomial in the dimension, by exploiting recent progress in the complexity of ANN search. We present a novel data structure for boundary queries based on a Newton-like iterative intersection procedure. We implement our algorithms and compare with brute-force approaches to show that they scale very well as the dimension and number of facets grow larger.

Keywords: Geometric optimization · Convex polytope
Membership oracle · Approximation algorithms · General dimension
Nearest-neighbor search

1 Introduction

In geometric optimization, convex polytopes are very important objects appearing also as feasible regions in linear programming. Let us consider a convex polytope P in H-representation, that is as the intersection of a finite set of linear inequalities: $P = \{x \in \mathbb{R}^d \mid Ax \le b, A \in \mathbb{R}^{n \times d}, b \in \mathbb{R}^n\}$. An important question on such a polytope is that of point membership. We wish to preprocess P in order to obtain a membership data structure which, given a query point q, efficiently decides whether q lies inside or outside P. A decision can be reached by testing all n inequalities for a complexity of $O(nd)$. This trivial approach is often a plausible exact solution, especially in the high-dimensional case. In order to design a more efficient algorithm in high dimension, we will focus on the approximate polytope membership problem where the membership data structure is allowed to answer incorrectly for points lying very close to the boundary of the polytope. A formal definition will be provided later in Sect. 2.2.

© Springer International Publishing AG, part of Springer Nature 2018
J. Lee et al. (Eds.): ISCO 2018, LNCS 10856, pp. 39–51, 2018.
https://doi.org/10.1007/978-3-319-96151-4_4

Algorithms used to solve combinatorial optimization problems, such as the ellipsoid, interior point or randomized methods (for the latter see [1]), usually rely on randomly sampling convex polytopes. The inner loop of such algorithms needs access to a membership or a boundary oracle, where the latter is the procedure that computes the intersection of a ray with the boundary of the polytope and is equivalent to membership via binary search. The oracle specification means that we are not interested in how the solution is computed or of its computational complexity. Grötschel et al. [2] proposed the oracle model of computation and among other results they prove the polynomial time equivalence of basic oracles such as optimization, separation, and membership. This has become a commonly employed tool in combinatorial optimization mainly for studying the computational complexity of problems. Another important example of application is volume approximation [3,4] which has also an established connection to combinatorial optimization. For example, the volume of order polytopes gives the number of linear extentions of the associated partial order set.

From a practical point of view opening the oracle black box, in particular membership, and improving their complexity, implies improvements to the applicability of the aforementioned algorithms. For example, the first implementation of randomized algorithms that scale in high dimension appeared in [5]. Their approach relies on the standard random walks known as hit-and-run, which require a boundary oracle. Notice that, although this software can handle polytopes in spaces whose dimension goes up to 200, it cannot scale as efficiently for specific classes of polytopes with a large number of facets. In particular, it cannot approximate the volume of cross-polytopes of dimension 20 or more.

Here, we radically shift the aforementioned paradigm and, moreover, improve upon the complexity of membership and boundary data structures, when dimension d is an input parameter. We exploit the approximate setting and allow ourselves to answer correctly within some approximation error ϵ and with some success probability. Our new paradigm uses a reduction to the Approximate Nearest Neighbor (ANN) problem, which is the most fundamental problem among those today with a practical, poly-time solution in high-dimensions.

Previous Work. There are two classical results for the approximate membership problem, both based on creating ϵ-approximating polytopes and answering membership on them. Any convex body is ϵ-approximated by a polytope with $O(1/\epsilon^{(d-1)/2})$ facets, which is asymptotically tight in the worst case [6]. This leads to a membership data structure with space and query complexity in $O(1/\epsilon^{(d-1)/2})$. Using a d-dimensional grid, membership takes constant time (assuming a model of computation that supports the floor function) and space grows to $O(1/\epsilon^{d-1})$ [7].

A relevant line of work on approximate membership in fixed d uses space-time trade-offs [8] to achieve a space of $O(1/\epsilon^{(d-1)(1-(2\lfloor \log t \rfloor -2)/t)})$ with query time $O(\log(1/\epsilon)/\epsilon^{(d-1)/t})$, for trade-off parameter $t \geq 4$. In [9], again for fixed d, they opt for a hierarchy of ellipsoids selected by a sampling process on classical structures from the theory of convexity defined on the polytope. They achieve space $O(1/\epsilon^{(d-1)/2})$ with an optimal query time of $\log(1/\epsilon)$.

We present state-of-the-art approaches to ANN as we build atop of those for our oracles. There are many solutions to this problem, but in principle, methods that scale polynomially with d belong to two categories. First, the well studied Locality Sensitive Hashing (LSH) [10]. The other category focuses on random projections [11], then uses fast algorithms in fixed dimension. Both achieve sublinear query time with (near-)linear storage, while scaling polynomially in d, and both have a probability of success p.

Our Contribution. We describe a simple constructive reduction from the polytope membership problem to ANN, then show under which conditions this reduction holds for the respective approximate versions of the problems. This gives us the flexibility to exploit advances in the research of ANN in order to offer, the first (as far as the authors are aware) practical approximate polytope membership data structure in high dimension with complexity bounds polynomial in the dimension d and sublinear in the number of inequalities n. This is our main result, in Theorem 5. We also present an application of this membership data structure for creating boundary data structures for H-polytopes. We implement and experimentally examine our algorithms; we illustrate that they scale well as dimension and number of facets grow larger. Our implementation is linked to the software of [5] for polytope volume, so as to provide faster oracles.

The rest of the paper is organized as follows. The next section discusses (approximate) membership and the reduction to ANN. Section 3 considers the boundary data structures. The implementation and experiments are in Sect. 4. We conclude with open questions. Certain proofs are omitted due to lack of space; they can be found in our arXiv technical report.

2 Approximate Polytope Membership

We assume that the given H-polytope P is full dimensional and that its representation is minimal, i.e. that it does not contain redundant inequalities.

We denote the i-th (in)equality of P as $a_i x \leq b_i, 1 \leq i \leq n$. We associate each facet of the polytope with a corresponding (in)equality and denote it as F_i. Formally: $F_i = \{x \in P \mid a_i x = b_i\}$, $1 \leq i \leq n$. The hyperplanes that define non-empty F_i's, i.e. for which $F_i \neq \emptyset$ are called non-redundant or supporting and we extend that label to their inequalities. We denote as ∂P the boundary of P: $\partial P = \{x \in P \mid \exists i, \ 1 \leq i \leq n \ \text{s.t.} \ x \in F_i\}$.

2.1 Exact Polytope Membership Oracle

A reduction from the exact polytope membership problem to the exact nearest neighbor problem was established in [12], where it was shown that there is a connection between the boundaries of polytopes in $\mathbb{R}^d$ and power diagrams in $\mathbb{R}^{d-1}$. Power diagrams define a partition of the Euclidean space into a cell complex based on a set of spheres. Each sphere identifies a specific cell and that cell consists of all the points whose power distance is minimized for that sphere.

The power diagram is a generalized Voronoi diagram, and coincides with the Voronoi diagram of the sphere centers if all spheres have equal radii.

Theorem 1 *[12, Theorem 4]. For any polyhedron $P \in \mathbb{R}^d$, which is expressible as the intersection of upper halfspaces, there exists an affinely equivalent power diagram in hyperplane $h_0 : x_d = 0$.*

A cell complex C and a polyhedron $P \subset \mathbb{R}^{d+1}$ are said to be affinely equivalent if there exists a central or parallel projection ϕ such that, for each face f of C, $f = \phi(g)$ holds for some face g of P. This provides a reduction from ray shooting in a polyhedron to point location in a polyhedral complex. In the case of polytope membership, the polyhedral complex becomes a single cell (the polytope) and the power diagram becomes a Voronoi diagram. This provides a reduction from polytope membership to Nearest neighbor.

Corollary 2. *Let $P \subset \mathbb{R}^d$ be a convex polytope described as the intersection of n non-redundant halfspaces. For every point $p^* \in P \setminus \partial P$ it is possible to compute a set S of $n+1$ points such that, $p^* \in S$ and, given a query point q, the exact Polytope Membership test for a query point q reduces to finding the Nearest Neighbor of q among these $n+1$ points.*

Proof. We initialize $S = \{p^*\}$. We will describe for completeness the procedure to compute the remaining n points of S such that the corresponding Voronoi diagram of these n points and p^* will have the polytope P as the voronoi cell of p^*. These $n+1$ points will be the points of the corollary (Fig. 1).

For each facet F_i and its corresponding hyperplane $H_i := a_i x = b_i$, $1 \leq i \leq n$, we compute the projection of p^* on H_i and denote it as f_i. Then, we compute the point p_i, $1 \leq i \leq n$, such that the line segment (p^*, p) is perpendicular to H_i and $d(p^*, H_i) = ||p^* - f_i||_2 = d(p_i, H_i)$, where $d(p, S) = \min_{x \in S} ||p - x||_2$. Equivalently, $p_i = f_i + (f_i - p^*)$.

We now have a set of points $S = \{p^*, p_1, \ldots, p_n\}$ of $n+1$ points that have the following property. In the Voronoi diagram of S, by construction, the cell that corresponds to p^* is precisely the input polytope P. By the Voronoi property, the following holds: $q \in P \Leftrightarrow ||p^* - q||_2 \leq ||q - s||_2$, $\forall s \in S$. Polytope membership returns "YES" iff the nearest neighbor of q is p^*. □

Remark. A nearest neighbor computation or data structure on these $n+1$ points of Corollary 2 provides us with an exact Membership Oracle for the polytope P. We also emphasize that the choice of $p^* \in P$ is arbitrary. This means that a set S satisfying the Corollary can be computed for each point $p^* \in P \setminus \partial P$.

2.2   Approximate Polytope Membership Oracle

Let us consider the following relaxation.

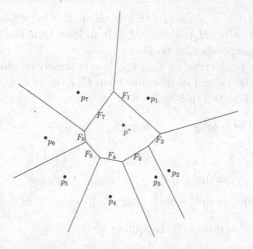

Fig. 1. A conceptual presentation of the constructive proof in the case of d=2. Each p_i corresponds to the symmetric point of p^* about the facet F_i.

Definition 3 (Approximate Polytope Membership Problem). *Given a convex polytope $P \subset \mathbb{R}^d$ and an approximation parameter $\epsilon \in (0,1)$, an ϵ-approximate polytope membership query decides whether a query point $q \in \mathbb{R}^d$ lies inside or outside of P, but may return either answer if q's distance from the boundary of P is at most $\epsilon \cdot diam(P)$.*

We define $P^{-\epsilon} = \{x \in P \mid d(x, \partial P) > \epsilon \cdot diam(P)\}$. Obviously the aforementioned problem makes sense only when $P^{-\epsilon} \neq \emptyset$. Otherwise, we can always return "NO" for a query point q and be correct.

Theorem 4 (Approximate Membership). *Approximate Polytope Membership for an H-polytope P and an approximation parameter ϵ, such that $P^{-\epsilon} \neq \emptyset$, reduces to the ANN problem on the pointset $S = \{p^*, p_i : 1 \leq i \leq n\}$, where $p^* \in P^{-\epsilon}$ and the remaining p_i are computed as in the proof of Corollary 2.*

Proof. Let $p^* \in P^{-\epsilon}$ and S be the corresponding pointset of Corollary 2 for P. Let $\Delta(P) = \max\limits_{p_i \in S \setminus \{p^*\}} \|p_i - p^*\|_2$. By construction, the following holds for $\Delta(P)$: $2\epsilon \cdot diam(P) < \Delta(P) < 2diam(P)$. Let $q \in \mathbb{R}^d$ be a query point such that $\|q - p^*\| < \frac{\Delta(P)}{2\epsilon}$. For any other $q' \in \mathbb{R}^d$, we return "NO", because $\|q' - p^*\|_2 \geq \frac{\Delta(P)}{2\epsilon} \Rightarrow \|q' - p^*\| > diam(P) \Rightarrow q' \notin P$. We distinguish two cases when $q \in P^{-\epsilon}$ and $q \in \{\mathbb{R}^d \mid q \notin P \ \wedge \ d(q, \partial P) > \epsilon \cdot diam(P)\}$.

– Let $q \in P^{-\epsilon}$, we wish to select an ϵ' for the ANN problem such that:

$$(1 + \epsilon') < \|p_i - q\|_2 / \|p^* - q\|_2 \tag{1}$$

Essentially, this would imply that p^* is the nearest neighbor of q, while every $p_i \in S \setminus \{p^*\}$ is not an ϵ'-NN of q.

Let $r_i = d(p^*, H_i) \geq \epsilon \cdot diam(P)$, where H_i is the hyperplane defining facet F_i. By construction, $d(p^*, H_i) = d(p_i, H_i)$. It follows that the segment $p^* p_i$ has length $2r_i$, as it is perpendicular to H_i.

Next, we define the projection of q on the line spanned by the segment $p^* p_i$ as $q_i = (p_i - p^*) \cdot q / ||p_i - p^*||_2$ and its distance from H_i as $a_i = d(q_i, H_i) \geq \epsilon \cdot diam(P)$ Obviously now, as depicted in Fig. 2:

$$||p_i - q_i||_2 = r_i + a_i, \ ||p^* - q_i||_2 = r_i - a_i$$

Therefore,

$$||p_i - q||_2^2 = ||p_i - q_i||_2^2 + ||q - q_i||_2^2 = (r_i + a_i)^2 + k_i^2$$
$$||p^* - q||_2^2 = ||p^* - q_i||_2^2 + ||q - q_i||_2^2 = (r_i - a_i)^2 + k_i^2,$$

where $k_i = ||q - q_i||_2^2 < diam(P)$. It follows that,

$$\frac{||p_i - q||_2^2}{||p^* - q||_2^2} = \frac{(r_i + a_i)^2 + k_i^2}{(r_i - a_i)^2 + k_i^2} = 1 + \frac{4 r_i a_i}{(r_i - a_i)^2 + k_i^2}$$

$$\geq 1 + \frac{4\epsilon^2 (diam(P))^2}{(r_i - a_i)^2 + k_i^2} \geq 1 + \frac{4\epsilon^2 (diam(P))^2}{2(diam(P))^2} \geq 1 + 2\epsilon^2$$

Substituting in (1), yields: $(1 + \epsilon') < \sqrt{1 + 2\epsilon^2} \Rightarrow \epsilon' < \sqrt{1 + 2\epsilon^2} - 1$.

– Let $q \in \{\mathbb{R}^d \mid q \notin P \ \wedge \ d(q, \partial P) > \epsilon \cdot diam(P)\}$. Assume the nearest neighbor of q is $p_i \in S \setminus \{p^*\}$. Similarly, we are looking for an ϵ' such that:

$$(1 + \epsilon') < ||p^* - q||_2 / ||p_i - q||_2$$

This means p^* cannot be an ANN of q. Now, like before:

$$\frac{||p^* - q||_2^2}{||p_i - q||_2^2} = \frac{(r_i + a_i)^2 + k_i^2}{(r_i - a_i)^2 + k_i^2} = 1 + \frac{4 r_i a_i}{(r_i - a_i)^2 + k_i^2}$$

$$\geq 1 + \frac{4(\epsilon \cdot diam(P))^2}{(r_i - a_i)^2 + k_i^2} \geq 1 + \frac{4(\epsilon \cdot diam(P))^2}{2 \left(\frac{2\Delta(P)}{2\epsilon}\right)^2}$$

$$\geq 1 + \frac{4\epsilon^4 \cdot diam^2(P)}{2\Delta^2(P)} > 1 + \frac{4\epsilon^4 \cdot diam^2(P)}{4 \cdot diam(P)}$$

$$> 1 + e^4 \cdot diam(P)$$

It follows that, $\epsilon' < \sqrt{e^4 \cdot diam(P)} - 1$.

Choosing $\epsilon' = \min\{\sqrt{e^4 \cdot diam(P)} - 1, \sqrt{1 + 2\epsilon^2} - 1\}$ and answering ϵ'-ANN queries on this set solves the original problem, because if a query point $q \in P^{-\epsilon}$, then we have ensured that the ϵ'-ANN data structure will correctly identify p^* as the only approximate nearest neighbor of q. Similarly in a symmetric argument, for every $q \notin P$, such that $d(q, \partial P) > \epsilon \cdot diam(P)$, p^* will not be an approximate nearest neighbor of q. Lastly, if $d(q, \partial P) \leq \epsilon \cdot diam(P)$ the response from the ANN data structure does not matter. Therefore, the reduction is complete. □

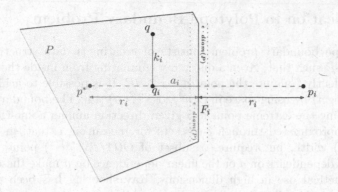

Fig. 2. p_i corresponds to the symmetric point of p^* about the facet F_i. We decompose the distances $||p^* - q||_2$ and $||p_i - q||_2$ and express them in terms of a_i and k_i. Notice how $q \in P^{-\epsilon} \Rightarrow a_i \geq \epsilon \cdot diam(P)$ and how $k_i < diam(P)$, as q cannot be a vertex.

We now employ approaches for ANN to obtain a bound polynomial in the dimension by introducing a probability of success. Below, $\tilde{O}$ omits logarithmic factors.

Theorem 5 *[Approximate Membership in High Dimension]. For an H-polytope $P \subset \mathbb{R}^d$ and an approximation parameter ϵ, such that $P^{-\epsilon} \neq \emptyset$, we can solve the Approximate Polytope membership problem on P by building a data structure on P answering queries in $\tilde{O}(dn^{\rho+o(1)})$ time and using $\tilde{O}(n^{1+\rho+o(1)} + dn)$ space, with a high probability of success, where $\rho = 1/(2(1 + \epsilon')^2 - 1)$ and $\epsilon' = \min\{\sqrt{e^4 \cdot diam(P)} - 1, \sqrt{1 + 2\epsilon^2} - 1\}$.*

Proof. The Chebyshev center of a polytope P is the center of the largest inscribed ball. Formally: $\arg \min_{x \in P} \max_{y \in P} ||x - y||_2^2$. Let c be the Chebyshev center of P with radius r and assume $c \notin P^{-\epsilon}$, in order to deduce an absurdity.

$$c \notin P^{-\epsilon} \Rightarrow r < \epsilon \cdot diam(P) \tag{2}$$

Take a point $c' \in P^{-\epsilon}$, as $P^{-\epsilon} \neq \emptyset$.

$$d(c', F_i) \geq \epsilon \cdot diam(P), \quad 1 \leq i \leq n \Rightarrow B(c', \epsilon \cdot diam(P)) \subset P \tag{3}$$

Combining (2) and (3) produces an absurdity as we have found a larger inscribed ball in P, contradicting the property of c. Therefore, $c \in P^{-\epsilon}$. We use $p^* = c$ as the starting point of the construction of the pointset S in the proof of Theorem 4. Answering ANN queries on S using the LSH data structure of [13], completes this proof. $\square$

Remark. Any high-dimensional ANN solution can be utilized in the last step of Theorem 5 and we can inherit its complexity and its properties.

3   Application to Polytope Boundary Problem

The polytope boundary problem consists of creating a data structure for an H-polytope P such that, given a query ray emanating from inside the polytope, we can efficiently compute the point $p = r \cap \partial P$. It is possible to achieve query time in $O(\log n)$ by using space in $O(n^d / \log^{\lfloor d/2 \rfloor} n)$ [14]. The boundary oracle is dual to finding the extreme point in a given direction among a known pointset. This is ϵ-approximated through ϵ-coresets for measuring extent, in particular (directional) width, but requires a subset of $O((1/\epsilon)^{(d-1)/2})$ points [15]. The exponential dependence on d or the linear dependence on n make these methods of little practical use in high dimensions. Ray shooting has been studied in practice only in low dimensions, as well.

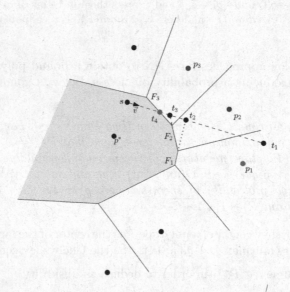

Fig. 3. An example of the boundary oracle converging to a solution. The query ray is $r = (s, v)$ and $t_4 = r \cap \partial P$ is the solution. t_1, t_2, t_3, t_4 were computed in sequence.

Exact Polytope Boundary Oracle. We now describe an iterative procedure for P based on an exact nearest neighbor data structure E_MEM defined on the pointset S of Corollary 2 that we described in Sect. 2.2. This exact nearest neighbor data structure will act as the exact membership oracle for the polytope P. We call this algorithm BoundaryOracle.

Finding the Starting Point. The first step is to find a starting point t_1 such that $t_1 \in r$ and $t_1 \notin P$. We may use the intersection of r with a bounding box around P. A bounding box of P can be readily computed by solving $2d$ linear programs to compute the farthest points on P along the coordinate directions.

Finding the Intersection Point. We obtain an efficient method following a derivative-like approach. Given starting point $t_1 \notin P$: let p_i be the nearest neighbor of t_1 using the data structure defined for membership: $p_i = \text{E\_MEM}(t_1)$. Let H_i be the hyperplane supporting the facet F_i used to define p_i; F_i separates the cell of p_i from P in the Voronoi diagram. Let $t_2 = (H_i \cap r)$. Iterate by computing $t_3, t_4, \ldots$, until membership decides $t_n \in P$. This procedure is illustrated in Fig. 3.

Lemma 6 (Correctness of algorithm BoundaryOracle**).** *BoundaryOracle always converges to a solution for the boundary problem for a given polytope P.*

Approximate Polytope Boundary Oracle. Now, we define an approximate version of the polytope boundary problem.

Definition 7 (Approximate Polytope Boundary Problem). *Given a convex H-polytope $P \subset \mathbb{R}^d$ and an approximation parameter $\epsilon \in (0, 1)$, preprocess P into a data structure such that, given a query ray $r \subset \mathbb{R}^d$ emanating from inside P, it is possible to efficiently compute a point $r^* \in r$ such that $d(r^*, \partial P) \leq \epsilon \cdot diam(P)$.*

We make two additional changes to the algorithm presented in the previous section. First, we compare t_i's and t_{i+1}'s distance from the ray's source point s. If the distance is not improved, then we discard the current t_{i+1} and set it as $t_{i+1} = (t_i - s) - \frac{v}{\|v\|_2}\epsilon$. In other words, in this case we take an ϵ-step from t_i towards the ray's apex. The second change concerns termination. Now we stop when the approximate membership oracle identifies a point t_i as being inside the polytope, or when the point t_i lies in the opposite direction of the ray.

Algorithm 1. Approximate Boundary Oracle

```
Input: H-polytope P⊂ℝᵈ, ray r (pair (s,v)), ε
Output: t∈ℝᵈ s.t. t∈r and d(t,∂P)≤εdiam(P)

A_MEM = approximate membership oracle for P
Q = bounding_box(P)
t = Q∩r;
do
   pᵢ = A_MEM(t);
   if pᵢ==p then return t+ v/||v||₂ ε; end
   tₚᵣₑᵥ = t
   H = Hᵢ //facet corresponding to pi
   t = H∩r
   if ||t-s||₂ ≥ ||tₚᵣₑᵥ-s||₂ then t = (tₚᵣₑᵥ-s) - v/||v||₂ ε; end
   if (t-s)·v < 0 then return s+ v/||v||₂ ε; end
while True;
```

Lemma 8 (Correctness of Algorithm 1). *Algorithm 1 always converges to a solution for the approximate boundary problem.*

4   Implementation and Experiments

Implementation. All of our code[1] is linked to the software of [5]. It is written in C++11 based on using the CGAL[2] library for the readily available data structures of d-dimensional objects, Eigen3 for some linear algebra computations and FALCONN [16] for the approximate nearest neighbor data structure. We remind the reader at this point that for a polytope $P(d, n, i)$ we compute $n + 1$ points, out of which one point $p^* \in P$ while all remaining n points $p_i \notin P, 1 \leq i \leq n$. FALCONN offers LSH only for angular distances so in order to take advantage of that we use it in the following manner. We consider our pointset already centered around the internal point, in our case the origin. We build a FALCONN data structure using the Hyperplane LSH family and setting $k = 11, l = 1$, number of probes=40, when the number of facets $n \geq 10000$. Otherwise, we set them to $l = 1$, $k = 8$ and number of probes=150. l corresponds to the number of hash tables built, k corresponds to the number of hash functions used per hash table and number of probes is a parameter for the multi-probe LSH scheme [17]. The data structure is built for every computed point besides the internal one. Then, assuming that for a query q FALCONN returns an approximate nearest neighbor guess x_i, we compare $d(x_i, q)$ to $d(p^*, q)$ and return the point closest to q out of x_i, p^*. The parameters for FALCONN were selected manually, while trying to maintain a 90% success rate for membership.

Datasets. We experiment on a synthetic dataset consisting of high-dimensional polytopes with a large number of facets. In particular, for the following set of possible dimensions $\boldsymbol{d} = \{40, 100, 500, 1000\}$ and the following set of possible number of facets $\boldsymbol{n} = \{5000, 10000, 20000, 50000, 100000, 500000, 1000000\}$, we generate 5 polytopes for every combination of $\boldsymbol{d} \times \boldsymbol{n}$. Each polytope $P(d, n, i), d \in \boldsymbol{d}, n \in \boldsymbol{n}, i \in \{1, 2, 3, 4, 5\}$ lives in a d-dimensional Euclidean space and is described by n inequalities of the form: $a_j x \leq 1000, 1 \leq j \leq n$, where $a_j \sim mod(U(0, 32767), 1000)$. The notation $U(i, j)$ denotes the uniform real distribution over $[i, j]$. By construction, each polytope contains the origin 0, which we use as the internal point needed by the approximate membership oracle. If that assumption was not satisfied, we could have computed an internal point either by solving a linear program or by computing an important point of the polytope, like the Chebyshev center.

Evaluation Protocol. For both oracles we report pre-processing time, total query time, and success rate vs n and d as n and d vary in their respective sets $\boldsymbol{n}, \boldsymbol{d}$. Specifically for the boundary oracle we also report the average number of steps that it required in order to reach a solution and we also compute the min,max and average distances of the point returned from our approximate boundary bracle to the actual point that the exact ray shooting problem should have computed. We compare the query time to the naive approach of checking all n facets of P. For the membership oracle we sample 1000 query points inside the polytope

---

[1] https://github.com/van51/volume_approximation.
[2] http://www.cgal.org/.

via the popular hit-and-run paradigm and then move these points sufficiently far from the origin so that they lie outside the polytope. This generates another 1000 points to form a total of 2000 points. Similarly for the boundary oracle we use 1000 query points in total.

Results. Table 1 depicts the total time in seconds for creating the approximate membership oracle on random polytopes for different values of d, n. Figures 4 and 5 depict total time in seconds for all queries to be completed. Parameters were tuned such that the membership oracle achieved an accuracy of $>90\%$, i.e. at least 9 out of 10 queries succeed on average. The results matched our expectations with regards to the behaviour of the oracles in high dimension, where we can see a huge difference in the query time, especially as the number of facets grows larger as well.

Table 1. Preprocessing time in seconds for membership oracle. This includes computing the $n + 1$ pointset and creating the ANN data structure on top of it.

		Number of facets						
		5000	10000	20000	50000	100000	500000	1000000
Dimension	40	0.006 s	0.013 s	0.027 s	0.057 s	0.125 s	0.518 s	0.795 s
	100	0.015 s	0.035 s	0.057 s	0.121 s	0.230 s	1.005 s	1.885 s
	500	0.055 s	0.108 s	0.193 s	0.419 s	0.717 s	3.396 s	6.744 s
	1000	0.101 s	0.192 s	0.342 s	0.783 s	1.470 s	5.500 s	10.770 s

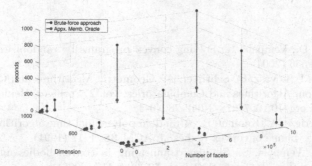

Fig. 4. Average timing results for 2000 queries for varying n and d. Half of the queries were inside the random polytopes and half were outside.

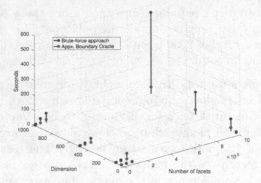

Fig. 5. Average timing results for 1000 ray queries for varying n and d. The approximate boundary oracle took on average at most 4 steps.

5   Future Work

For the membership oracle it would be nice to see how the choice of the internal point affects ϵ' of the ANN. The choice of the Chebyshev center as internal point should be optimal. For the boundary oracle we would like to bound its convergence rate. The experiments demonstrate that it adapts well and converges very fast. The holy grail of our efforts is to incorporate the high dimensional version of the boundary oracle in sampling approaches.

Acknowledgements. The first two authors are partially supported by the European Union's H2020 research and innovation programme under grant agreement No 734242.

References

1. Bertsimas, D., Vempala, S.: Solving convex programs by random walks. J. ACM **51**(4), 540–556 (2004)
2. Grötschel, M., Lovász, L., Schrijver, A.: Geometric Algorithms and Combinatorial Optimization. Algorithms and Combinatorics, vol. 2. Springer, Heidelberg (1988). https://doi.org/10.1007/978-3-642-78240-4
3. Dyer, M., Frieze, A., Kannan, R.: A random polynomial-time algorithm for approximating the volume of convex bodies. J. ACM **38**, 1–17 (1991)
4. Lovász, L., Vempala, S.: Simulated annealing in convex bodies and an $O^*(n^4)$ volume algorithm. J. Comp. Syst. Sci. **72**, 392–417 (2006)
5. Emiris, I., Fisikopoulos, V.: Efficient random-walk methods for approximating polytope volume. In: Proceedings of Symposium on Computational Geometry, Kyoto, pp. 318–325 (2014). Final version to appear in ACM Trans. Math. Soft
6. Dudley, R.: Metric entropy of some classes of sets with differentiable boundaries. J. Approximation Theory **10**, 227–236 (1974)
7. Bentley, J., Preparata, F., Faust, M.: Approximation algorithms for convex hulls. Commun. ACM **25**, 64–68 (1982)

8. Arya, S., da Fonseca, G.D., Mount, D.: Polytope approximation and the Mahler volume. In: Proceedings of ACM-SIAM Symposium on Discrete Algorithms (SODA) (2012)
9. Arya, S., da Fonseca, G.D., Mount, D.: Optimal approximate polytope membership. In: Proceedings of ACM-SIAM Symposium on Discrete Algorithms (2017)
10. Indyk, P., Motwani, R.: Approximate nearest neighbors: towards removing the curse of dimensionality. In: Proceedings of STOC (1998)
11. Anagnostopoulos, E., Emiris, I., Psarros, I.: Low-quality dimension reduction and high-dimensional approximate nearest neighbor. In: Proceedings of Symposium on Computational Geometry, pp. 436–450 (2015). Final version to appear in ACM Trans. Alg
12. Aurenhammer, F.: Power diagrams: properties, algorithms and applications. SIAM J. Comput. **16**, 78–96 (1987)
13. Andoni, A., Razenshteyn, I.: Optimal data-dependent hashing for approximate near neighbors. In: Proceedings of ACM STOC (2015)
14. Ramos, E.: On range reporting, ray shooting and k-level construction. In: Proceedings of Symposium on Computational Geometry (1999)
15. Agarwal, P., Har-Peled, S., Varadarajan, K.: Geometric approximation via coresets. In: Combinatorial and Computational Geometry (MSRI) (2005)
16. Andoni, A., Indyk, P., Laarhoven, T., Razenshteyn, I., Schmidt, L.: Practical and optimal LSH for angular distance. In: Proceedings of Conference on NIPS (2015)
17. Lv, Q., Josephson, W., Wang, Z., Charikar, M., Li, K.: Multi-probe LSH: efficient indexing for high-dimensional similarity search. In: Proceedings of Conference on VLDB (2007)

Graph Orientation with Splits

Yuichi Asahiro[1], Jesper Jansson[2(✉)], Eiji Miyano[3], Hesam Nikpey[4],
and Hirotaka Ono[5]

[1] Department of Information Science, Kyushu Sangyo University, Fukuoka, Japan
asahiro@is.kyusan-u.ac.jp
[2] Department of Computing, The Hong Kong Polytechnic University, Hung Hom,
Kowloon, Hong Kong
jesper.jansson@polyu.edu.hk
[3] Department of Systems Design and Informatics, Kyushu Institute of Technology,
Iizuka, Japan
miyano@ces.kyutech.ac.jp
[4] Department of Computer Engineering, Sharif University of Technology,
Tehran, Iran
hnikpey@ce.sharif.edu
[5] Graduate School of Informatics, Nagoya University, Nagoya, Japan
ono@i.nagoya-u.ac.jp

Abstract. The *Minimum Maximum Outdegree Problem* (MMO) is to
assign a direction to every edge in an input undirected, edge-weighted
graph so that the maximum weighted outdegree taken over all vertices
becomes as small as possible. In this paper, we introduce a new variant
of MMO called the *p-Split Minimum Maximum Outdegree Problem* (*p*-
Split-MMO) in which one is allowed to perform a sequence of p split
operations on the vertices before orienting the edges, for some specified
non-negative integer p, and study its computational complexity.

Keywords: Graph orientation · Maximum flow · Vertex cover
Partition · Algorithm · Computational complexity

1 Introduction

An *orientation* of an undirected graph is an assignment of a direction to each of
its edges. The computational complexity of constructing graph orientations that
optimize various criteria has been studied, e.g., in [1–5,7,9,12,14], and positive
as well as negative results are known for many variants of these problems.

For example, the *Minimum Maximum Outdegree Problem* (MMO) [4–7,14]
takes as input an undirected, edge-weighted graph $G = (V, E, w)$, where V,
E, and w denote the set of vertices of G, the set of edges of G, and an edge-
weight function $w : E \to \mathbb{Z}^+$, respectively, and asks for an orientation of G
that minimizes the resulting maximum weighted outdegree taken over all vertices in the oriented graph. In general, MMO is strongly NP-hard and cannot be
approximated within a ratio of $3/2$ unless P = NP [4]. However, in the special

© Springer International Publishing AG, part of Springer Nature 2018
J. Lee et al. (Eds.): ISCO 2018, LNCS 10856, pp. 52–63, 2018.
https://doi.org/10.1007/978-3-319-96151-4_5

case where all edges have weight 1, MMO can be solved exactly in polynomial time [14]. MMO has applications to load balancing, resource allocation, and data structures for fast vertex adjacency queries in sparse graphs [6, 7] based on the technique of placing each edge in the adjacency list of exactly one of its two incident vertices. E.g., if G is a planar graph then G admits an orientation in which every vertex has outdegree at most 3 and such an orientation can be found in linear time [7], which means that for a planar graph, any adjacency query can be answered in $O(1)$ time after linear-time preprocessing. As an additional example of a graph orientation problem, finding an orientation that maximizes the number of vertices with outdegree 0 is the Maximum Independent Set Problem [2], which cannot be approximated within a ratio of n^ϵ for any constant $0 \leq \epsilon < 1$ in polynomial time unless P = NP [15]. Similarly, finding an orientation that minimizes the number of vertices with outdegree at least 1 is the Minimum Vertex Cover Problem and minimizing the number of vertices with outdegree at least 2 is the problem of finding a smallest subset of the vertices in G whose removal leaves a pseudoforest [2], both of which admit polynomial-time 2-approximation algorithms [10].

In this paper, we introduce a new variant of MMO called the *p-Split Minimum Maximum Outdegree Problem* (*p*-Split-MMO), where p is a specified nonnegative integer, and study its computational complexity. Here, one is allowed to perform a sequence of p *split operations* on the vertices before orienting the edges. When thinking of MMO as a load balancing problem, the split operation can be interpreted as a way to alleviate the burden on the existing machines by adding an extra machine.

The paper is organized as follows. Section 2 gives the formal definition of *p*-Split-MMO. Section 3 presents an $O((n + p)^p \cdot poly(n))$-time algorithm for the unweighted case of the problem, where n is the number of vertices in the input graph, while Sect. 4 proves that if p is unbounded then the problem becomes NP-hard even in the unweighted case. On the other hand, for the edge-weighted case, Sect. 5 shows that *p*-Split MMO with weighted edges is weakly NP-hard even if restricted to $p = 1$. Finally, Sect. 6 proves that the most general case of the problem, i.e., with weighted edges as well as unbounded p, is strongly NP-hard. See Table 1 for a summary of the new results.

Table 1. Overview of the computational complexity of *p*-Split MMO. Note that in the edge-weighted case, the edge weights are included in the input so it is possible to further classify the NP-hardness results as either weakly NP-hard or strongly NP-hard.

	Unweighted graphs	Edge-weighted graphs
Constant p	$O((n + p)^p \cdot poly(n))$ time (Sect. 3, Theorem 1)	Weakly NP-hard (Sect. 5, Theorem 3)
Unbounded p	NP-hard (Sect. 4, Theorem 2)	Strongly NP-hard (Sect. 6, Theorem 4)

2 Definitions

Let $G = (V, E, w)$ be an undirected, edge-weighted graph with vertex set V, edge set E, and edge weights defined by the function $w : E \to \mathbb{Z}^+$. An *orientation* Λ of G is an assignment of a direction to every edge $\{u, v\} \in E$, i.e., $\Lambda(\{u, v\})$ is either (u, v) or (v, u). For any orientation Λ of G, the *weighted outdegree* of a vertex u is

$$d_\Lambda^+(u) = \sum_{\substack{\{u,v\} \in E: \\ \Lambda(\{u,v\}) = (u,v)}} w(\{u, v\})$$

and the *cost* of Λ is

$$c(\Lambda) = \max_{u \in V}\{d_\Lambda^+(u)\}.$$

Let MMO be the following optimization problem, previously studied in [4–7, 14].

> **The Minimum Maximum Outdegree Problem (MMO):**
>
> Given an undirected, edge-weighted graph $G = (V, E, w)$, where V, E, and w denote the set of vertices of G, the set of edges of G, and an edge-weight function $w : E \to \mathbb{Z}^+$, output an orientation Λ of G with minimum cost.

Next, for any $v \in V$, the set of vertices in V that are neighbors of v is denoted by $\Gamma[v]$ and the set of edges incident to v is denoted by $E[v]$. A *split operation* on a vertex v_i in G is an operation that transforms: (i) the vertex set of G to $(V \setminus v_i) \cup \{v_{i,1}, v_{i,2}\}$, where $v_{i,1}$ and $v_{i,2}$ are two new vertices; and (ii) the edge set of G to $(E \setminus E[v_i]) \cup \{\{v_{i,1}, s\} : s \in S\} \cup \{\{v_{i,2}, s'\} : s' \in \Gamma[v_i] \setminus S\}$ for some subset $S \subseteq \Gamma[v_i]$. For any non-negative integer p, a *p-split* on G is a sequence of p split operations successively applied to G. Note that in a p-split, a new vertex resulting from a split operation may in turn be the target of a later split operation.

The problem that we study in this paper generalizes MMO above and is defined as follows for any non-negative integer p.

> **The p-Split Minimum Maximum Outdegree Problem (p-Split-MMO):**
>
> Given an undirected, edge-weighted graph $G = (V, E, w)$, where V, E, and w denote the set of vertices of G, the set of edges of G, and an edge-weight function $w : E \to \mathbb{Z}^+$, output a graph G' and an orientation Λ' of G' such that: (i) G' is obtained by a p-split on G; (ii) Λ' has minimum cost among all orientations of all graphs obtainable by a p-split on G.

See Fig. 1 for an example. Throughout the paper, we denote the number of vertices and edges in the input graph G by n and m, respectively. Any orientation of a graph G', where G' can be obtained by applying a p-split to G, will be referred to as a *p-split orientation of G*. The decision version of p-Split-MMO, denoted by p-Split-MMO(W), asks whether or not the input graph G has a p-split orientation Λ' with $c(\Lambda') \leq W$ for a specified integer W.

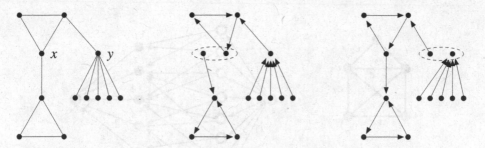

Fig. 1. Consider the instance of 1-Split-MMO on the left (here, all edge weights are 1). If the split operation is applied to the vertex x as shown in the middle figure, the resulting instance of MMO can be oriented with maximum outdegree equal to 1, so this is an optimal solution. Observe that if the vertex y had been split instead, the minimum maximum outdegree would have been 2. This shows that greedily applying the split operations to the highest degree nodes will not necessarily yield an optimal solution.

3   An Algorithm for Unweighted Graphs

This section presents an algorithm for p-Split-MMO on graphs with unweighted edges (equivalently, where all edge weights are equal to 1). Its time complexity is $O((n + p)^p \cdot poly(n))$, which is polynomial when $p = O(1)$.

Our basic strategy is to transform p-Split-MMO to the maximum flow problem on directed networks with edge capacities: (i) We first select an integer W as an upper bound on the cost of a p-split orientation. (ii) Next, we construct a flow network $\mathcal{N}$ based on the input graph G and the integer W. (iii) By computing a maximum network flow in $\mathcal{N}$, we solve p-Split-MMO(W), i.e., determine whether p-Split-MMO(W) admits a feasible solution or not. (iv) By refining W according to a binary search while repeating steps (ii) and (iii), we find the minimum possible value of W and retrieve an optimal p-split orientation of G from the corresponding flow network.

We now describe the details. (Refer to Fig. 2 for an example of the construction.) Let $G = (V, E)$ be the input graph and p any non-negative integer. For any positive integer W and multisubset S of V (i.e., a subset of V in which repetitions are allowed) of cardinality p, define the following flow network $\mathcal{N}_{W,S} = (V_{\mathcal{N}}, E_{\mathcal{N}})$:

$$V_{\mathcal{N}} = V \cup E \cup \{s, t\}$$
$$E_{\mathcal{N}} = \bigcup_{e=\{u,v\}\in E} \{(s, e), (e, u), (e, v)\} \cup \bigcup_{v \in V} \{(v, t)\}$$

where s and t are newly created vertices. Note that $|V_{\mathcal{N}}| = n + m + 2$ and $|E_{\mathcal{N}}| = n + 3m$. The capacity $cap(u, v)$ of each edge $(u, v) \in E_{\mathcal{N}}$ is set to:

- $cap(s, e) = 1$ for every $e \in E$;
- $cap(e, u) = cap(e, v) = 1$ for every $e = \{u, v\} \in E$; and

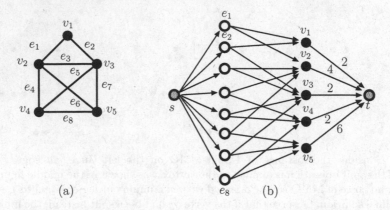

Fig. 2. (a) An input graph G and (b) the flow network $\mathcal{N}_{W,S}$ constructed from G when $p = 3$, $W = 2$, and $S = \{v_2, v_5, v_5\}$. For clarity, only edge capacities in $\mathcal{N}_{W,S}$ greater than 1 are displayed.

- $cap(v, t) = W + W \cdot occ(v)$ for every $v \in V$, where $occ(v)$ is defined as the number of occurrences of v in S.

Consider any maximum flow in $\mathcal{N}_{W,S}$. Since the edge capacities are integers, we can assume that the maximum flow is integral by the integrality theorem (see, e.g., [8]). Then we have:

Lemma 1. *The maximum directed flow from vertex s to vertex t in $\mathcal{N}_{W,S}$ equals $|E|$ if and only if G has a p-split orientation with cost at most W obtained after doing $occ(v)$ split operations on each $v \in V$.*

Proof. ($\Rightarrow$) Let F be a maximum directed flow from s to t with integer values and assume it is equal to $|E|$. Since there are $|E|$ units of flows leaving s in F, exactly one edge among (e, u) and (e, v) for every $e = \{u, v\} \in E$ has one unit of flow in $\mathcal{N}_{W,S}$. We construct a p-split orientation Λ of G by first orienting each edge $e = \{u, v\} \in E$ as (u, v) if (e, u) is using one unit of flow in F and (e, v) is using zero units of flow in F, or as (v, u) otherwise. At this point, each vertex $v \in V$ has outdegree at most $W + W \cdot occ(v)$ because there are at most this many units of flow entering v in $\mathcal{N}_{W,S}$. Next, for each $v \in V$, do $occ(v)$ split operations on v and distribute its outgoing edges evenly among each v and its resulting new vertices so that every vertex has outdegree at most W. Since $\sum_{v \in V} occ(v) = p$, the resulting Λ is a p-split orientation of G.

($\Leftarrow$) Suppose there is a p-split orientation of G with cost at most W obtained by doing $occ(v)$ split operations on each $v \in V$. Then we can construct a flow in $\mathcal{N}_{W,S}$ that has $|E|$ units of flow by using: (i) all $|E|$ edges of the form (s, e); (ii) $|E|$ edges of the form (e, u) where $e = \{u, v\} \in E$ (either (e, u) or (e, v) depending on if $\{u, v\}$ was oriented as (u, v) or (v, u)); and (iii) at most $|V|$ edges of the form (v, t). Observe that for (iii), each $v \in V$ has at most $W + W \cdot occ(v)$

units of flow entering it in $\mathcal{N}_{W,S}$, which is within the capacity limit of its outgoing edge (v,t), so in total, we have $|E|$ units of flow from s to t. $\square$

Lemma 2. *p-Split-MMO can be solved in $O((n+p)^p \cdot n^2 \cdot T(|V_{\mathcal{N}}|, |E_{\mathcal{N}}|) \cdot \log n)$ time, where $T(|V_{\mathcal{N}}|, |E_{\mathcal{N}}|)$ is the running time for solving the maximum network flow problem on a directed graph with vertex set $V_{\mathcal{N}}$ and edge set $E_{\mathcal{N}}$.*

Proof. For any candidate value of W, we can identify a p-split orientation of G with cost at most W or determine that none exists, by evaluating every multisubset S of V of cardinality p, constructing $\mathcal{N}_{W,S}$, computing a maximum directed flow in $\mathcal{N}_{W,S}$, and applying Lemma 1. The number of multisubsets is at most $\binom{n-1+p}{p} = O((n+p)^p)$, constructing each $\mathcal{N}_{W,S}$ takes $O(n+m) = O(n^2)$ time, and each maximum network flow instance is solved in $T(|V_{\mathcal{N}}|, |E_{\mathcal{N}}|)$ time.

Since the graph G is unweighted, W is upper-bounded by the maximum degree of a vertex. Therefore, applying binary search to obtain the minimum possible value of W (i.e., the smallest W for which the maximum flow is still $|E|$ for some multisubset S of V) increases the running time by a factor of $O(\log n)$. The total time complexity is $O((n+p)^p \cdot n^2 \cdot T(|V_{\mathcal{N}}|, |E_{\mathcal{N}}|) \cdot \log n)$. $\square$

Since $|V_{\mathcal{N}}| = O(m)$ and $|E_{\mathcal{N}}| = O(m)$, plugging in $T(|V_{\mathcal{N}}|, |E_{\mathcal{N}}|) = O(m^2)$ (see [13]) yields:

Theorem 1. *p-Split-MMO for unweighted graphs can be solved in $O((n+p)^p \cdot n^2 m^2 \log n)$ time.*

4   Unweighted Graphs, Unbounded p

We now prove the NP-hardness of p-Split-MMO for unbounded p, even when restricted to unweighted graphs. Recall that p-Split-MMO(W) is the decision version of p-Split-MMO which asks if G has a p-split orientation of cost at most W. The main result of this section is:

Theorem 2. *p-Split-MMO(3) for unweighted graphs and unbounded p is NP-complete.*

Proof. p-Split-MMO(3) is in NP because a nondeterministic algorithm can guess a p-split of G and an orientation of the resulting graph in polynomial time and check if this orientation has cost at most 3.

To prove the NP-hardness, we give a polynomial-time reduction from the decision version of the Minimum Vertex Cover Problem, VC(k), defined as: Given an undirected graph $G = (V, E)$ and a positive integer k, determine if there is a subset $V' \subseteq V$ with $|V'| \le k$ such that for each $\{u, v\} \in E$, at least one of u and v belongs to V'. It is known that VC(k) remains NP-complete even if restricted to graphs of degree at most three [11].

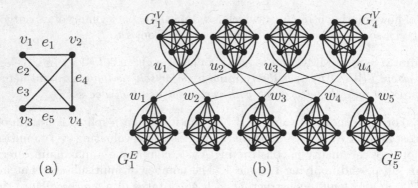

Fig. 3. Illustrating the reduction from VC(k) to p-Split-MMO(3). (a) An instance of VC(k) with four vertices and five edges. (b) The instance of p-Split-MMO(3) constructed from (a).

The reduction is as follows. (See Fig. 3 for an example.) Suppose we are given an instance $G = (V, E)$ of VC(k), where G has degree at most three. Write $V = \{v_1, v_2, \ldots, v_n\}$ and $E = \{e_1, e_2, \ldots, e_m\}$. We construct an instance G' of p-Split-MMO(3) by defining: (i) a set $U = \{u_1, u_2, \ldots, u_n\}$ of n vertices, where each u_i corresponds to $v_i \in V$; and (ii) a set $W = \{w_1, w_2, \ldots, w_m\}$ of m vertices, where each w_j corresponds to $e_j \in E$. In addition, we prepare: (iii) $n + m$ complete graphs with six vertices each, denoted by G_1^V through G_n^V and G_1^E through G_m^E. Let $V(G_i^V) = \{u_{i,1}, u_{i,2}, \ldots, u_{i,6}\}$ for each $i \in \{1, 2, \ldots, n\}$ and $V(G_j^E) = \{w_{j,1}, w_{j,2}, \ldots, w_{j,6}\}$ for each $j \in \{1, 2, \ldots, m\}$. The vertex set of G' is thus $U \cup W \cup V(G_1^V) \cup V(G_2^V) \cup \cdots \cup V(G_n^V) \cup V(G_1^E) \cup V(G_2^E) \cup \cdots \cup V(G_m^E)$. Next, insert the following edges into the edge set of G' (which already includes the edges of G_1^V through G_n^V and G_1^E through G_m^E): (iv) edges $\{u_h, w_j\}$ and $\{u_i, w_j\}$ if $e_j = \{u_h, u_i\} \in E$ for each $j \in \{1, 2, \ldots, m\}$; (v) an edge $\{u_i, u_{i,h}\}$ for each $i \in \{1, 2, \ldots, n\}$ and each $h \in \{1, 2, \ldots, 6\}$; and (vi) an edge $\{w_j, w_{j,h}\}$ for each $j \in \{1, 2, \ldots, m\}$ and each $h \in \{1, 2, \ldots, 5\}$. Note that each u_i in G' has degree equal to (6 + the degree of v_i in G) and every w_j in G' has degree 7. Finally, we set $p = k$. This completes the reduction.

Next, we show that G has a vertex cover with size at most p if and only if G' has a p-split orientation whose cost is at most three.

($\Rightarrow$) Suppose that G has a vertex cover C of size p. Let $C' \subseteq U$ be the p vertices in G' that correspond to vertices in C. Apply a split operation on each $u_i \in C'$ to transform it into a pair of vertices u_i and u_i^*, the first one (u_i) being adjacent to all six vertices from G_i^V and the second one (u_i^*) being adjacent to the at most three neighbors from W. Let G'' be the resulting graph. By definition, G'' is obtained by applying a p-split to G' and we will now show that G'' admits an orientation of cost three.

First, every G_i^V forms a K_7 (a complete graph with seven vertices) together with u_i in G''. Orient each such K_7 so that all of its vertices have outdegree three, e.g., by applying Proposition 2 in [3]. Secondly, orient the (at most three) edges incident to each u_i^*-vertex away from u_i^*. Since C is a vertex cover, every w_j-vertex in G' will be incident to at most one unoriented edge of the form $\{u_i, w_j\}$ after this step is done. Next, for each w_j, if there is one unoriented edge of the form $\{u_i, w_j\}$ then orient it away from w_j. Finally, every w_j and G_j^E form a K_7 with one edge incident to w_j missing; orient this subgraph as above, but let w_j have one less outgoing edge than the other vertices so that the outdegree of each such vertex is at most three. This yields an orientation of G'' of cost three.

($\Leftarrow$) Suppose G' has a p-split orientation of cost at most three. If some vertex $u_{i,h}$ in G_i^V was split then we obtain another p-split orientation of cost at most three by not splitting $u_{i,h}$ but splitting u_i instead and orienting the edges of the resulting K_7 as described above, and similarly for vertices in G_j^E. We may therefore assume that every vertex that is split comes from $U \cup W$. Next, if some vertex w_j in W is split and it has an incident u_i-vertex that is not split then we replace the split operation on w_j by a split operation on u_i; by doing so and orienting the edge between u_i and w_j towards w_j, the cost of the orientation will not increase. This produces a p-split orientation of G' in which every vertex from W is incident to at least one vertex from the set of (at most p) vertices from U that were split, which then gives a vertex cover of G of size at most p. $\square$

Corollary 1. *For any constant $\varepsilon > 0$, it is NP-hard to approximate p-Split-MMO to within a factor of $\frac{4}{3} - \varepsilon$, even for unweighted graphs.*

Proof. In the reduction in the proof of Theorem 2, there always exists a p-split orientation Λ' of G' satisfying $c(\Lambda') \leq 4$, as can be seen by ignoring all available split operations and just orienting the at most two edges of the form $\{u_i, w_j\}$ for each w_j away from w_j and all other edges as in the first part of the proof of Theorem 2. Since there exists a p-split orientation Λ' with $c(\Lambda') \leq 3$ if and only if the given instance of $VC(k)$ has a vertex cover with size at most k, the above reduction is a gap-introducing one, i.e., if there existed a polynomial-time $(\frac{4}{3} - \varepsilon)$-approximation algorithm for p-split-MMO(3), then $VC(k)$ could be solved in polynomial time. $\square$

5   Edge-Weighted Graphs, Bounded p

In this section, we prove that p-Split-MMO on edge-weighted graphs is weakly NP-hard even if restricted to $p = 1$. To do so, we give a polynomial-time reduction from the Partition Problem, defined as follows: Given a set $S = \{s_1, s_2, \ldots, s_n\}$ of n positive integers, determine if there exists a subset $S' \subseteq S$ such that $\sum_{s_i \in S'} s_i = \sum_{s_j \in S \setminus S'} s_j$. The Partition Problem is weakly NP-hard and admits a pseudopolynomial-time solution [11].

Theorem 3. *1-Split-MMO is weakly NP-hard even if the input is restricted to edge-weighted wheel graphs.*

Proof. We construct an edge-weighted, undirected graph $G = (V, E, w)$ from any given instance $S = \{s_1, s_2, \ldots, s_n\}$ of the Partition Problem. Define $K = \frac{\sum_{i=1}^n s_i}{2}$

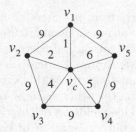

Fig. 4. Let $S = \{1,2,4,5,6\}$ be an instance of the Partition Problem. The reduction in the proof of Theorem 3 sets $K = 9$ and constructs the edge-weighted graph G above.

and assume without loss of generality that $s_i \leq K$ for all $s_i \in S$. The vertex set V consists of: (i) n vertices representing the integers in S and denoted by $v_1, v_2, \ldots, v_n$; and (ii) one special vertex, denoted by v_c. The edge set E consists of: (iii) the n edges $\{v_1, v_2\}, \{v_2, v_3\}, \ldots, \{v_n, v_1\}$ forming a cycle; and (iv) the n edges $\{v_c, v_1\}, \{v_c, v_2\}, \ldots, \{v_c, v_n\}$ forming a star. (Hence, G is a wheel graph.) For every edge e of type (iii), assign $w(e) = K$. For every edge $\{v_c, v_i\}$ of type (iv), assign $w(\{v_c, v_i\}) = s_i$. An example is shown in Fig. 4.

Below, we show that the answer to the given instance S of the Partition Problem is yes if and only if G has a 1-split orientation whose cost is at most K.

($\Rightarrow$) If there exists an $S' \subseteq S$ such that $\sum_{s_i \in S'} s_i = \sum_{s_j \in S \setminus S'} s_j$ then apply a split operation on the vertex v_c and let the two resulting vertices $v_{c,1}$ and $v_{c,2}$ be adjacent to the set of vertices of type (i) representing S' and $S \setminus S'$, respectively. For $i \in \{1, 2\}$, orient every edge that involves $v_{c,i}$ away from $v_{c,i}$. Orient the remaining n edges so that they form a directed cycle $v_1 \rightarrow v_2 \rightarrow \cdots \rightarrow v_n \rightarrow v_1$. This way, the weighted outdegree of every vertex is at most K.

($\Leftarrow$) Let Λ' be a 1-split orientation of G of cost at most K. If S contains a single element equal to K then the answer to the given instance of the Partition Problem is trivially yes. On the other hand, if $s_i \neq K$ for all $s_i \in S$ then we claim that the vertex in G to which the split operation was applied is v_c. To prove the claim, suppose the split operation was applied to some other vertex v_j, where $j \in \{1, 2, \ldots, n\}$, thereby replacing v_j by two vertices $v_{j,1}$ and $v_{j,2}$. Each of the n edges not involving v_c has weight K, so at most one of the $n + 1$ vertices in $\{v_1, v_2, \ldots, v_n, v_{j,1}, v_{j,2}\} \setminus \{v_j\}$ can orient its edge involving v_c towards v_c. Let the weight of this edge be s_k. Then the weighted outdegree of v_c is $2K - s_k > K$ because $s_i < K$ for all $s_i \in S$, contradicting that the cost of Λ' is at most K. This proves the claim. Now, since the split operation was applied to v_c (thus replacing v_c by two vertices $v_{c,1}$ and $v_{c,2}$) and the cost of Λ' is at most K, each of the n vertices in $\{v_1, v_2, \ldots, v_n\}$ has one of the n edges of weight K oriented away from it. This means that every edge of the form $\{v_{c,i}, v_j\}$ is oriented away from $v_{c,i}$, and since the sum of these edges' weights is $2K$, each of $v_{c,1}$ and $v_{c,2}$ must have weighted outdegree exactly equal to K. Let S' be the set of weights of the edges incident to $v_{c,1}$. Then $\sum_{s_i \in S'} s_i = \sum_{s_j \in S \setminus S'} s_j = K$ and the answer to the given instance of the Partition Problem is yes.   □

Corollary 2. *For every fixed integer $p \geq 1$, p-Split-MMO on edge-weighted graphs is weakly NP-hard.*

6 Edge-Weighted Graphs, Unbounded p

Here, we prove that p-Split-MMO with weighted edges is strongly NP-hard if p is sufficiently large, i.e., $p = \Omega(n)$. This result is obtained via a polynomial-time reduction from the 3-Partition Problem: Given a multiset $S = \{s_1, s_2, \ldots, s_{3n}\}$ of positive integers and an integer B such that $B/4 < s_i < B/2$ for every $i \in \{1, 2, \ldots, 3n\}$ and $\sum_{s_i \in S} s_i = n \cdot B$ hold, determine if S can be partitioned into n multisets $S_1, S_2, \ldots, S_n$ so that $|S_j| = 3$ and $\sum_{s_i \in S_j} s_i = B$ for every $j \in \{1, 2, \ldots, n\}$. The 3-Partition Problem is known to be strongly NP-hard [11].

Theorem 4. *p-Split-MMO is strongly NP-hard even if the input is restricted to edge-weighted cactus graphs.*

Proof. We construct an edge-weighted, undirected graph $G = (V, E, w)$ from any given instance (S, B) of the 3-Partition Problem, where $S = \{s_1, s_2, \ldots, s_{3n}\}$. Let $p = n - 1$ and recall that $B = \frac{\sum_{i=1}^{3n} s_i}{n}$ by definition. G consists of:

- $3n$ subgraphs, G_1 through G_{3n}, each of which is associated with an element in S. For each $i \in \{1, 2, \ldots, 3n\}$, G_i contains three vertices u_i, v_i, and w_i and three edges $\{u_i, v_i\}$, $\{u_i, w_i\}$, and $\{v_i, w_i\}$ (i.e., G_i is a triangle graph). The weight of every edge in G_i is set to B.
- One special vertex v_c.
- For $i \in \{1, 2, \ldots, 3n\}$, an edge $\{v_c, v_i\}$ of weight s_i that connects G_i to v_c.

The constructed graph is a cactus graph. This completes the description of the reduction. See Fig. 5 for an illustration.

Now we show that the answer to the 3-Partition Problem on input S is yes if and only if the constructed graph G has a p-split orientation of cost B.

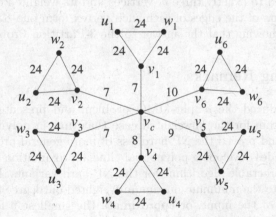

Fig. 5. An instance of the 3-Partition Problem with $S = \{7, 7, 7, 8, 9, 10\}$ and $B = 24$ yields the graph G shown above. In the construction, $n = 2$ and $p = 2 - 1 = 1$.

($\Rightarrow$) If the answer to the 3-Partition Problem is yes, divide the elements of S into n multisets $S_1, S_2, \ldots, S_n$ where every S_j has the sum B and $|S_j| = 3$. Then, do p split operations on v_c so that each of the resulting $p + 1 = n$ vertices, called *center vertices*, becomes adjacent to exactly three vertices v_x, v_y, and v_z, where $\{s_x, s_y, s_z\}$ is one of the S_j-sets. By orienting all $3n$ edges involving center vertices away from the center vertices, and for each $i \in \{1, 2, \ldots, 3n\}$, orienting the three edges $\{u_i, v_i\}$, $\{u_i, w_i\}$, and $\{v_i, w_i\}$ as (u_i, v_i), (w_i, u_i), and (v_i, w_i), we obtain a p-split orientation of G of cost B.

($\Leftarrow$) Consider any p-split orientation Λ' of G with cost B. Let σ be the total number of split operations in this p-split that were done on vertices in the G_i-subgraphs. First, we show by contradiction that $\sigma = 0$. Suppose $\sigma \geq 1$. If we start from G and apply a sequence of $p - \sigma$ split operations to v_c and the new vertices created by these operations, v_c will be replaced by a set of $p - \sigma + 1 = n - \sigma$ vertices, henceforth denoted by C. Call the $3n$ edges that contain a vertex from C *center edges*. Due to the weights of the edges in each G_i-subgraph, if no split operations are done on u_i, v_i, or w_i then the center edge between v_i and C must be oriented away from C, but each split operation applied to a vertex of the form u_i, v_i, or w_i will allow at most one center edge to become oriented towards C. Let W' be the sum of the weights of the center edges that were oriented away from C in Λ'. By definition, the weight of every center edge is less than $\frac{B}{2}$, so $W' > n \cdot B - \sigma \cdot \frac{B}{2}$. According to the pigeonhole principle, at least one vertex in C must have weighted outdegree at least $W'/(n - \sigma)$. However, $W'/(n - \sigma) > (n \cdot B - \sigma \cdot \frac{B}{2})/(n - \sigma) > (n \cdot B - \sigma \cdot B)/(n - \sigma) = B$, which is a contradiction because the cost of the p-split orientation was B. Thus, $\sigma = 0$ and $|C| = p + 1 = n$. Next, note that if a vertex x in C was connected to four or more v_i-vertices then since these edges must be oriented away from C and each of them has weight strictly larger than $\frac{B}{4}$, the weighted outdegree of x would be strictly larger than B, which is impossible. Finally, since each of the n vertices in C can be connected to at most three v_i-vertices and there are $3n$ v_i-vertices in total, it must be connected to exactly three v_i-vertices and its weighted outdegree is B. Letting the weights of the edges of each such vertex form one S_j-set then gives a partition of S showing that the answer to the 3-Partition Problem is yes. $\square$

7   Concluding Remarks

This paper introduced the p-Split-MMO problem and presented a maximum flow-based algorithm for the unweighted case that runs in polynomial time for any constant p, and proved the NP-hardness of more general problem variants. Future work includes developing polynomial-time approximation algorithms and fixed-parameter tractable algorithms for the NP-hard variants. E.g., one could try to approximate the minimum maximum weighted outdegree for a value of p specified as part of the input, or approximate the smallest p for which some specified upper bound on the maximum weighted outdegree is attainable.

Also, it would be interesting to study how the computational complexity of p-Split-MMO changes if the output orientation is required to be *acyclic* or *strongly connected.* Borradaile *et al.* [5] recently showed that unweighted MMO with either the acyclicity constraint or the strongly connectedness constraint added remains solvable in polynomial time. In contrast, the closely related problem of outputting a *minimum lexicographic orientation* of an input graph, which is solvable in polynomial time for unconstrained orientations, becomes NP-hard for acyclic orientations [5] while its computational complexity for strongly connected orientations is still unknown.

Acknowledgments. The authors would like to thank Richard Lemence and Edelia Braga for some discussions related to the topic of this paper. This work was partially supported by JSPS KAKENHI Grant Numbers JP17K00016 and JP17K00024, and JST CREST JPMJR1402.

References

1. Asahiro, Y., Jansson, J., Miyano, E., Ono, H.: Graph orientation to maximize the minimum weighted outdegree. Int. J. Found. Comput. Sci. **22**(3), 583–601 (2011)
2. Asahiro, Y., Jansson, J., Miyano, E., Ono, H.: Graph orientations optimizing the number of light or heavy vertices. J. Graph Algorithms Appl. **19**(1), 441–465 (2015)
3. Asahiro, Y., Jansson, J., Miyano, E., Ono, H.: Degree-constrained graph orientation: maximum satisfaction and minimum violation. Theory Comput. Syst. **58**(1), 60–93 (2016)
4. Asahiro, Y., Jansson, J., Miyano, E., Ono, H., Zenmyo, K.: Approximation algorithms for the graph orientation minimizing the maximum weighted outdegree. J. Comb. Optim. **22**(1), 78–96 (2011)
5. Borradaile, G., Iglesias, J., Migler, T., Ochoa, A., Wilfong, G., Zhang, L.: Egalitarian graph orientations. J. Graph Algorithms Appl. **21**(4), 687–708 (2017)
6. Brodal, G.S., Fagerberg, R.: Dynamic representations of sparse graphs. In: Dehne, F., Sack, J.-R., Gupta, A., Tamassia, R. (eds.) WADS 1999. LNCS, vol. 1663, pp. 342–351. Springer, Heidelberg (1999). https://doi.org/10.1007/3-540-48447-7_34
7. Chrobak, M., Eppstein, D.: Planar orientations with low out-degree and compaction of adjacency matrices. Theor. Comput. Sci. **86**(2), 243–266 (1991)
8. Cormen, T., Leiserson, C., Rivest, R.: Introduction to Algorithms. The MIT Press, Cambridge (1990)
9. Deming, R.W.: Acyclic orientations of a graph and chromatic and independence numbers. J. Comb. Theory Ser. B **26**, 101–110 (1979)
10. Fujito, T.: A unified approximation algorithm for node-deletion problems. Discrete Appl. Math. **86**(2–3), 213–231 (1998)
11. Garey, M., Johnson, D.: Computers and Intractability - A Guide to the Theory of NP-Completeness. W. H. Freeman and Company, New York (1979)
12. Khoshkhah, K.: On finding orientations with the fewest number of vertices with small out-degree. Discrete Appl. Math. **194**, 163–166 (2015)
13. Orlin, J.B.: Max flows in $O(nm)$ time, or better. In: Proceedings of STOC 2013, pp. 765–774. ACM (2013)
14. Venkateswaran, V.: Minimizing maximum indegree. Discrete Appl. Math. **143**(1–3), 374–378 (2004)
15. Zuckerman, D.: Linear degree extractors and the inapproximability of Max Clique and Chromatic Number. Theory Comput. **3**(1), 103–128 (2007)

The Stop Number Minimization Problem: Complexity and Polyhedral Analysis

Mourad Baïou, Rafael Colares$^{(\boxtimes)}$, and Hervé Kerivin

CNRS and Université Clermont Auvergne, Clermont-Ferrand, France
{baiou,colares,kerivin}@isima.fr

Abstract. The Stop Number Minimization Problem arises in the management of a dial-a-ride system with small autonomous electric vehicles. In such a system, clients request for a ride from an origin point to a destination point, and a fleet of capacitated vehicles must satisfy all requests. The goal is to minimize the number of pick-up/drop-off operations. In [18], a special case was conjectured to be NP-Hard. In this paper we give a positive answer to this conjecture for any fixed capacity greater than or equal to 2. Moreover, we introduce a set of non-trivial instances that can be solved in polynomial time for capacity equal to 2, but is NP-Hard for higher capacities. We also present a new family of valid inequalities that are facet-defining for a large set of instances. Based on these inequalities, we derive a new efficient branch-and-cut algorithm.

Keywords: Autonomous vehicles · Branch-and-cut · Complexity

1 Introduction

Significant changes are arising in the way people travel from point to point. Tightening CO_2 regulations together with a consistent growth in the usage of smartphones and web services are inducing new technology-driven trends in transportation. All in all, four major trends deserve to be highlighted: diverse mobility, autonomous driving, electrification, and connectivity [11].

In this scenario arises the VIPAFLEET project which attempts to contribute to the development of innovative and sustainable urban mobility solutions. The VIPA (a French acronym that stands for Autonomous Individual Passenger Vehicle) is an electrical vehicle designed by Ligier [15] and EasyMile [9] to operate without any driver, notably in closed and semi-closed sites like industrial and commercial areas, medical complexes and campuses.

VIPAFLEET is only one among many projects derived from the changes in transportation sector. The new trends are giving birth to new challenges, not only in terms of modern technologies but also in terms of unexplored problems in the management of such mobility systems (see [17] for a survey on the subject). Among such systems we can mention ride-sharing [1], car-sharing [3,21] and dial-a-ride (DAR) [6] systems.

© Springer International Publishing AG, part of Springer Nature 2018
J. Lee et al. (Eds.): ISCO 2018, LNCS 10856, pp. 64–76, 2018.
https://doi.org/10.1007/978-3-319-96151-4_6

The VIPA may work in many modes. In this paper, we focus on the 'tram mode' which asks for the management of a specific dial-a-ride system. In such a mode, the circuit and its stations are predefined. A special station is denoted the depot. Customers use their smartphones or a 'call terminal' to request for a ride from an origin station to some destination station of their choice. For its part, the fleet of capacitated vehicles travels around the circuit (always in the same direction, say counterclockwise) and stops at a station if requested.

It is essential to mention that unlike usual applications, stations are not part of the circuit but attached to it (see Fig. 1). Therefore, in order to pick-up or drop-off some customer, a deviation on its regular route becomes necessary. If we suppose that each deviation takes about the same amount of time to be travelled through, then we may state that the global travel time is essentially related to how many deviations the fleet is forced to execute. For this reason, improving the service quality fairly corresponds to minimizing the total number of stops performed by the fleet of vehicles.

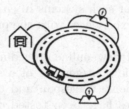

Fig. 1. Circuit scheme

The *Stop Number Minimization Problem* (SNMP) consists of assigning each client demand to a vehicle such that no vehicle gets overloaded, and the total number of vehicles' stops is minimized. For this, one may use as many vehicles as desired. Notice that a vehicle is allowed to make several tours before serving a demand and that a demand may request several seats on a single vehicle.

Typically, in practice, the system should be able to respond on-demand through *online* procedures. However, in order to evaluate such responsive algorithms, the static case (*offline*) must be benchmarked. Furthermore, the knowledge and insights obtained from the resolution and difficulties of the static case are fundamental to the design of better suited *online* procedures. A natural choice is thus to investigate where lies the real difficulty of SNMP on the static case.

From this point of view, we focus on the constrained version of SNMP where each demand can request only one seat at a vehicle and the fleet must respond to all requests in a single tour. This restricted problem is denoted *Unit Stop Number Minimization Problem*, hereafter denoted USNMP. In this paper we give new results on the complexity of USNMP answering the conjecture given in [18]. On the other hand, we also discuss the combinatorics behind USNMP by analyzing the integer linear program formulation proposed in [18]. In this sense, we give a new family of important facet-inducing inequalities and introduce symmetry-breaking constraints capable of improving computational performances.

The paper is organized as follows: Previous and related work are presented in Sect. 2. In Sect. 3, we define and formalize the USNMP. In Sect. 4, the complexity of USNMP is discussed, and in Sect. 5, polyhedral approaches are studied. Finally, our computational results are presented in Sect. 6.

2   Previous and Related Work

Since autonomous and electric vehicles represent a rupture from the old traditional way of travelling, resulting in less pollution, safer traffic and a more rational usage of public space, many studies from a wide range of perspectives can be found in the literature [17]. Here we focus on presenting the work related to VIPA vehicles.

Even if there remains a great deal of challenges before the deployment of fully automated transport systems, the first steps have already been made towards it. A certification procedure for such systems in cities and urban environments is described in [22]. The first experiences with the procedure are also mentioned with VIPA vehicles.

From the practical point of view, online procedures are developed and implemented in order to handle the management of a similar request system, but with different objective functions (makespan and total length for example) in [4]. In addition, VIPA vehicles have been tested on real situations and their performances have been reported in [20].

The number of stops was considered as a key performance indicator for the first time in [18], where the SNMP is first introduced. Both exact and heuristic approaches are proposed through branch-and-price and GRASP procedures. In the same paper, the complexity of SNMP is defined to be NP-Hard with a simple reduction from *Partition Problem* [12]. The USNMP variant is also mentioned but it is not further investigated. Its complexity is conjectured to be NP-Hard but remains an open question. In this paper we answer this question.

Arising from telecommunication industry, the *C-Edge Partition* (*C*-EP) problem has many similarities to the USNMP. In [13], its complexity is investigated and approximation algorithms are conceived. In Sect. 4, we define this related problem and detail its correlation to USNMP.

3   The Unit Stop Number Minimization Problem

Throughout this paper the edge set of an undirected graph G is given by $E(G)$ or simply E if the graph in question is implicit. Similarly, the vertex set of G is denoted $V(G)$ or simply V. Finally, let $V(E)$ denote the set of vertices spanned by the edge set E.

Since only one tour is allowed, the notion of circuit can now be replaced by an ordered line from the depot to the last station on the circuit. Thereby, demands can be represented by intervals (from its origin to its destination) on this line.

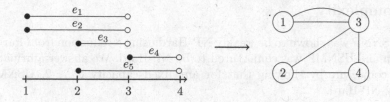

Fig. 2. Construction of the associated graph $G_\mathcal{I}$ from instance $\mathcal{I}$

Let $V = \{1, \ldots, n\}$ denote the set of stations sequentially ordered as they appear in the circuit network and $E = \{e_1, \ldots, e_m\}$ denote the set of unit load dial-a-ride demands such that each demand $e \in E$ is specified by an origin station $o_e \in V$ and a destination station $d_e \in V$, that is, $e = (o_e, d_e)$. Also, let K denote the set of available identical vehicles each with capacity $C \in \mathbb{Z}_+$. It may be easily seen that $|E|$ is an upper bound on the number of available vehicles. Therefore, we always set $|K| = |E|$.

Given an instance $\mathcal{I} = (V, E, C)$ of USNMP, the graph associated to $\mathcal{I}$ is denoted $G_\mathcal{I} = (V, E)$ (or simply G if the instance in question is implicit). Notice that parallel edges are allowed. Figure 2 depicts the construction of $G_\mathcal{I} = (V, E)$ from instance $\mathcal{I}$.

Let $\Delta_E(v) = \{e \in E : o_e \leq v \text{ and } d_e \geq v + 1\}$, that is, the set of demands that cross or starts at station v. We say that demand $e \in E$ intersects station $v \in V$ if $e \in \Delta_E(v)$. Then a feasible solution to USNMP is a partition of E into $|K|$ subsets $\{E_1, \ldots, E_{|K|}\}$ (*i.e.*, $\bigcup_{i=1}^{|K|} E_i = E$ and $E_i \cap E_j = \emptyset$ for $i \neq j$), such that $|\Delta_{E_i}(v)| \leq C$ for any $i \in K$ and $v \in V$. Remark that there always exists a trivial feasible solution where each demand is assigned to a different vehicle (*i.e.*, $E_i = \{e_i\}$ for each $i \in \{1, \ldots, |K|\}$).

Given a feasible solution $\{E_1, \ldots, E_{|K|}\}$, the vehicle i stops in every station of $V(E_i)$. Therefore, the cost of this solution is $\sum_{i=1}^{|K|} |V(E_i)|$, and the problem USNMP is to find a feasible solution of minimum cost. To fix ideas, consider the instance $\mathcal{I}$ described in Fig. 2 with $C = 2$. Then a feasible solution of cost 5 is $E_1 = \{e_1, e_2\}$, $E_2 = \{e_3, e_4, e_5\}$, and $E_i = \emptyset$ for $i \in \{3, 4, 5\}$.

An interesting particular case arises when there exists some station v' wherein all demands intersect, that is, $\Delta_E(v') = E$. Thus each vehicle may take at most C demands, that is, each subset E_i must have at most C edges. This case is denoted *Intersection*-USNMP. To illustrate it, consider again instance $\mathcal{I}$ described in Fig. 2. $\mathcal{I}$ is not an instance of *Intersection*-USNMP. However, if we define $\mathcal{I}'$ by removing e_4 from $\mathcal{I}$, then every demand intersects station 2, that is, $E = \Delta_E(2)$.

In this case, each station is either an origin station or a destination station, and v' is chosen to be the last origin station. Notice that, consequently, the associated graph $G_{\mathcal{I}'}$ is bipartite: the set V can be divided into two independent sets S and T, where $S = \{v \in V : v \leq v'\}$ is the set of origin stations and $T = \{v \in V : v \geq v' + 1\}$ is the set of destination stations. In the next section we show that USNMP is NP-Hard even in this restricted case.

4   Complexity

In [18], SNMP was shown to be weakly NP-Hard using a reduction from Partition Problem and USNMP was conjectured to be NP-Hard. We answer affirmatively to this conjecture by showing that for any fixed capacity $C \geq 2$, USNMP is strongly NP-Hard.

Theorem 1. *USNMP is NP-Hard for any fixed capacity $C \geq 2$.*

Proof. *Path Traffic Grooming* (PTG) problem arises in optical networks and is defined in [2] as follows. An instance $\mathcal{J} = (P, R, g)$ of PTG is given by a simple path P on n vertices, an integer g called grooming factor and a simple graph R on the same vertices as P, $R = (V(P), E)$. We say that an edge $r = (i, j)$ of R overlaps edge $e \in E(P)$ if e lies in the subpath between i and j in P. The PTG problem is to find a partition of edges E into subsets $\{R_1, \ldots, R_{|E|}\}$ that minimizes $\sum_{i=1}^{|E|} |V(R_i)|$, where the number of edges in each subset R_i overlapping edge $e \in E(P)$ is at most g. Recall that $V(R_i)$ is the set of end-nodes of R_i.

The problem PTG is NP-Hard. In fact, PTG is showed in [2] to be APX-Complete for any fixed grooming factor $C \geq 2$, that is, an NP-Hard problem that can be approximated within a constant factor but do not admits an approximation factor of $1 + \epsilon$ unless $P = NP$.

Given an instance $\mathcal{J} = (P, R, g)$ of PTG and an instance $\mathcal{I} = (V, E, C)$ of USNMP, such that there are no parallel demands in E, it is easy to see that both problems are equivalent. Indeed, path P may be seen as the sequence of stations on the circuit of USNMP, the edges of R are the demands E and the grooming factor g can be translated into the capacity of the vehicles C. USNMP is thus a generalization of PTG. $\square$

Corollary 1. *USNMP is APX-Complete.*

We now study the particular case where there exists a station wherein all demands intersect. For this, we define the *C-Edge Partition* problem (C-EP), first introduced in [13]. Given an instance $\mathcal{J} = (H, C)$, where $H = (V, E)$ is a simple graph and C is an integer, find a partition of E into subsets $\{S_1, \ldots, S_{|E|}\}$ that minimizes $\sum_{i=1}^{|E|} |V(S_i)|$, such that $|S_i| \leq C$. Notice that given an instance $\mathcal{J} = (H, C)$ of C-EP and an instance $\mathcal{I} = (V, E, C)$ of Intersection-USNMP, both problems are equivalent when H is bipartite and there are no parallel demands in E.

For $C = 2$, it has been shown in [13] that C-EP problem reduces to the *Maximum 2-chain Packing* problem, which can be solved in $O(m)$ [16]. Below we show that Intersection-USNMP can also be solved in polynomial time when $C = 2$, even if there are parallel demands.

Theorem 2. *Intersection-USNMP can be solved in polynomial time when $C = 2$.*

Proof. Let $G_{\mathcal{I}} = (V, E)$ be the graph associated with an instance $\mathcal{I} = (V, E, 2)$ of *Intersection*-USNMP. We define a partition $\{E_1, \ldots, E_{|K|}\}$ of E as follows.

Let each subset E_i be composed of two parallel demands for $0 \leq i \leq k'$, such that the set of remaining demands $E' = E \setminus \{\bigcup_{i=0}^{k'} E_i\}$ does not contain any pair of parallel demands. Moreover, let $\{E_{k'+1}, \ldots, E_{|K|}\}$ be the solution given by solving the maximum 2-chain packing problem on the simple graph $G' = (V, E')$. We show that $\{E_1, \ldots, E_{|K|}\}$ is an optimal solution.

For this, let $\mathrm{OPT}_2(H')$ be the optimal value of *Intersection*-USNMP for some graph $H' = (V', E')$ and capacity $C = 2$. Moreover, let $H'' = (V', E'')$ be the graph obtained from H' by adding two parallel demands a and b to E', that is, $E'' = E' \cup \{a, b\}$. We claim that $\mathrm{OPT}_2(H'') = \mathrm{OPT}_2(H') + 2$ and thus, in order to obtain the optimal solution for H'', one can just put a and b together in the same subset (vehicle) E_i'', and solve the problem on H'.

Suppose that a and b are in different subsets E_a'' and E_b'' in the optimal solution (otherwise the claim is true) and that $\mathrm{OPT}_2(H'') \leq \mathrm{OPT}_2(H') + 1$. By definition, $V(E_a'')$ and $V(E_b'')$ have at least 2 vertices (the end-nodes of a and b). If both $V(E_a'')$ and $V(E_b'')$ have strictly more than 2 vertices, then $\mathrm{OPT}_2(H'') \geq \mathrm{OPT}_2(H') + 2$. Hence, at least one of them must have exactly 2 vertices, say $V(E_a'')$. Therefore, either E_a'' has a single edge a or it has another edge c parallel to a. In both cases, we can put a and b on the same subset to obtain another solution with the same cost, which proves $\mathrm{OPT}_2(H'') = \mathrm{OPT}_2(H') + 2$.   □

Next we show that, surprisingly, *Intersection*-USNMP is NP-Hard for $C = 3$ even when the associated bipartite graph G is planar.

Theorem 3. *The Intersection-USNMP is NP-Hard for $C = 3$ even when G is restricted to the class of planar graphs.*

Proof. Given a graph $G = (V, E)$, the *Connected-k-Edge Partition* problem is to find a connected k-partition of E, that is, a partition of E, where each subset induces a connected subgraph having exactly k edges. In [8], this problem is proved to be NP-Hard for any fixed $k \geq 3$, even when graph G is a planar bipartite graph.

We reduce the *Connected-3-Edge Partition* to *Intersection*-USNMP for $C = 3$. Given a simple planar bipartite graph $B = (S, T, E)$, where $|S \cup T| = n$ and $|E| = 3m$, we construct the following *Intersection*-USNMP instance: Let $V = |S \cup T|$ be the set of stations. We number the stations in S from 1 to $|S|$ and stations in T from $|S| + 1$ to n. The set of demands is defined by edges $E(B)$. Notice that each demand has its origin in S and its destination in T. Clearly, all demands intersect each other at station $|S|$. We show that B has a connected-3-edge partition if and only if the constructed instance has a solution with $4m$ stops.

Given a *Intersection*-USNMP solution $\{E_1, \ldots, E_{k'}, \ldots, E_{|K|}\}$, where all subsets $E_i \neq \emptyset$ for $i \leq k'$ and $E_i = \emptyset$ for $k' + 1 \leq i \leq |K|$, we define $\rho_i = \frac{|E_i|}{|V(E_i)|}$ the density of subset (vehicle) $i \in 1, \ldots, k'$. From our construction, $G = (V, E)$ is a simple planar bipartite graph. Thus, the highest density ρ^* a vehicle with capacity 3 can get is $\frac{3}{4}$, which can be achieved if and only if its 3 assigned demands are connected.

A solution with $4m$ stops is optimal and all subgraphs are connected since

$$\sum_{i=1}^{k} |V_i| = \sum_{i=1}^{k} \frac{|E_i|}{\rho_i} \geq \sum_{i=1}^{k} \frac{|E_i|}{\rho^*} = \frac{3m}{\rho^*} = 4m.$$

Thus, if the optimal solution stops $4m$ times, we can derive a connected-3-edge partition of B. Reversely, a connected-3-edge partition of B clearly gives a solution with $4m$ stops.   □

For fixed $C \geq 3$, C-EP is NP-Hard for general graphs [13]. From Theorem 3, this result may be extended to the more restricted class of planar bipartite graphs. This is summarized in the theorem below.

Theorem 4. *C-EP is NP-Hard for $C = 3$ even when G is restricted to the class of planar bipartite graphs.*

5   Polyhedral Approach

The USNMP was formulated in [18] as the following integer linear problem:

$$\min \sum_{v \in V} \sum_{i \in K} y_v^i \tag{1}$$

subject to

$$\sum_{i \in K} x_e^i = 1 \qquad\qquad\qquad \forall e \in E, \tag{2}$$

$$\sum_{e \in \Delta_E(v)} x_e^i \leq C \qquad\qquad\qquad \forall v \in V, i \in K, \tag{3}$$

$$x_e^i \leq y_v^i \qquad\qquad \forall i \in K, e \in E, v \in \{o_e, d_e\}, \tag{4}$$

$$x_e^i \geq 0 \qquad\qquad\qquad \forall e \in E, i \in K, \tag{5}$$

$$x_e^i \in \{0, 1\} \qquad\qquad\qquad \forall e \in E, i \in K, \tag{6}$$

$$y_v^i \in \{0, 1\} \qquad\qquad\qquad \forall v \in V, i \in K. \tag{7}$$

The variable x_e^i expresses the fact that demand e is assigned or not to vehicle i, that is, $x_e^i = 1$ if $e \in E_i$, otherwise $x_e^i = 0$. The variable y_v^i expresses whether or not vehicle i stops at station v, that is, $y_v^i = 1$ if $v \in V(E_i)$, otherwise $y_v^i = 0$. The objective function (1) aims at minimizing the total number of stops. The assignment constraints (2) guarantee that each demand is assigned to exactly one single vehicle. The capacity constraints (3) ensures the vehicle's capacity is not violated at any station of the circuit. Stop constraints (4) forces a vehicle to stop at the departure and arrival stations of a demand assigned to it. Finally, constraints (5), (6) and (7) apply bounds and domains to variables.

Notice that, in practice, (7) can be dropped from the integer program since the integrality of variables the y is assured by (1), (4) and (6). Furthermore, we suppose that for each station $v \in V$, there is at least one demand that stops at it, otherwise one may just remove it from the set of stations.

From primal-dual relations, we show that the linear relaxation defined by (1)–(5) always provides the trivial lower bound of value $|V|$, that is, each station $v \in V$ must be visited at least once.

Theorem 5. *The linear relaxation, defined by (1)–(5), provides a lower bound of $|V|$ stops.*

This result shows how the given formulation is weak. To illustrate it, consider an instance where all demands go from station 1 to station 2: the linear relaxation provides a lower bound of 2 stops, while the integer optimal solution value is $2 \left\lceil \frac{|E|}{C} \right\rceil$. Therefore, it is not a surprise that traditional branch-and-bound approaches have failed in solving this integer program efficiently. In order to overcome this, we investigate new valid inequalities that improve this formulation.

Firstly, we introduce two natural valid inequalities (8) and (9) that reinforce the capacity constraints by ensuring that a vehicle can take at most C demands starting or finishing at a station if and only if it stops at this station:

$$\sum_{e \in \delta_E^-(v)} x_e^i \leq C y_v^i \qquad \forall v \in V, i \in K, \qquad (8)$$

$$\sum_{e \in \delta_E^+(v)} x_e^i \leq C y_v^i \qquad \forall v \in V, i \in K, \qquad (9)$$

where $\delta_E^+(v) = \{e \in E : o_e = v\}$ and $\delta_E^-(v) = \{e \in E : d_e = v\}$ is the set of demands that have v as their origin and destination, respectively.

Next we present a new important family of valid inequalities, called *k-tree inequalities* that considerably improve the value of the linear relaxation even when inequalities (8) and (9) are added. To define these inequalities we need some additional notations.

For a given graph G, let $d_G(v)$ denote the degree of vertex $v \in V$. Given a subset of edges $T \subseteq E$, the undirected graph induced by edges in T is denoted by $G[T]$. Finally, T is called a *k-tree* if $|T| = k$ and $G[T]$ is a tree.

Consider the family of *k-tree inequalities*,

$$\sum_{e \in T} x_e^i \leq \sum_{v \in V(G[T])} (d_{G[T]}(v) - 1) y_v^i \qquad \forall i \in K, j \in V, T \subseteq \Delta_E(j), \qquad (10)$$

such that T is a (C+1)-tree.

Theorem 6. *The k-tree inequalities are valid.*

Proof Sketch. We prove that *k-tree inequalities* are valid by showing that they are Chvátal-Gomory inequalities of rank 1 [5,14]. More specifically, we show that with a conical combination of inequalities (3) and (4), we obtain the following inequality:

$$(C+1)\sum_{e\in T} x_e^i \le (C+1) \sum_{v\in V(G[T])} (d_{G[T]}(v) - 1)y_v^i + C,$$

which can be divided by $C+1$ and the independent term $\frac{C}{C+1}$ rounded down:

$$\sum_{e\in T} x_e^i \le \sum_{v\in V(G[T])} (d_{G[T]}(v) - 1)y_v^i + \left\lfloor \frac{C}{C+1} \right\rfloor.$$

□

The inclusion of constraints (8) and (9) are quite effective when the graph is dense with respect to C, that is, nodes often have degree greater than C. However, when the graph is sparse they fail to reinforce the formulation. The worst case arises notably when the degree of each node is bounded by C. In this case, such constraints become redundant and do not strengthen the formulation at all. On the other hand, *k-tree inequalities* (10) are very important on sparse graphs. Consider the example given by Fig. 3 where $C = 2$. In such example, the lower bound provided by the linear program (1)–(9) is $|V|$, while the integer optimal solution value is $\frac{3|V|}{2}$. The addition of *k-tree inequalities* enables the linear relaxation to find the integer optimal solution. Indeed, we show that *k-tree inequalities* are actually facet-defining.

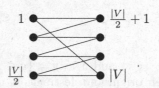

Fig. 3. Example of instance where *k-tree inequalities* are important

Theorem 7. *The k-tree inequalities are facet-defining under the following necessary and sufficient conditions:*

1. *If $G[T]$ is not a star.*
2. *$G[T]$ contains more than 2 non-leaf nodes or $E \setminus T$ does not contain an edge (u, v) such that u and v are non-leaf nodes of $G[T]$.*

Unfortunately, *k-tree inequalities* appear in exponential number and thus, must be treated through a separation procedure. However, its separation problem is NP-Hard and cannot be solved exactly without spending substantial computational time. As a consequence, we solve the separation problem heuristically.

Theorem 8. *The separation problem of k-tree inequalities is NP-Hard even when all variables y_v^i have the same value for all $v \in V(G[T])$ and a given $i \in K$.*

Proof Sketch. We give a reduction from the *k-Minimum Spanning Tree* problem (k-MST) which is defined as follows. Given a graph $G = (V, E)$ with edge weights w_e, find a tree $T = (V', E')$ spanning exactly k edges (or $k+1$ vertices) of weight at most B. The k-MST can be solved in polynomial time when G is itself a tree or k is a fixed constant, but it is NP-Hard in the general case [10, 19].   □

6   Computational Results

The polyhedral theory behind valid inequalities and how they are capable of sculpting the convex hull of integer solutions is quite elegant. From the optimization point of view however, our goal is clear: finding an optimal solution. Consequently, it is not relevant whether or not an inequality is valid, as long as we guarantee that there is at least one optimal solution in the feasible region.

Many assignment problems hide a natural symmetry issue that slows down typical branch and bound applications. Symmetric solutions can be seen as different solutions with the same objective function value. In [7], this issue is addressed by introducing symmetry-breaking constraints. Such constraints are intentionally not valid inequalities in the sense that they attempt to remove some integer feasible solutions from the feasible region, while keeping a symmetric solution for each solution removed.

We have adapted these constraints to fit USNMP. Constraints (11) reinforces constraints (2) in a way that the *i-th* demand must be assigned to one of the first i vehicles. Constraints (12) assures that demand e can be assigned to vehicle i only if vehicle $i - 1$ serves at least one of the first $e - 1$ demands.

$$\sum_{i=1}^{e} x_e^i = 1 \qquad\qquad \forall e \in E \qquad (11)$$

$$\sum_{j=i}^{e} x_e^j \leq \sum_{u=i-1}^{e-1} x_u^{i-1} \qquad\qquad \forall e \in E, i \in K : e \geq j \qquad (12)$$

Table 1 provides a comparison of performances between the formulation (1)–(7) given in [18] solved by CPLEX 12.7 (column CPLEX) and the reinforced formulation (1)–(9) with the addition of symmetry-breaking constraints (11)–(12) and *k-tree* inequalities (10) embedded on the branch-and-cut procedure (column B&C). The instances were generated randomly with the number of demands $|E|$ ranging from 30 to 50, the number of stations $|V|$ from 10 to 30 and the capacity was fixed at 5. For each set of parameters, 3 instances were tested (column i). The time of resolution (CPU) is displayed in seconds. Time limit was set to one hour and the instances that were not solved within 1 hour are marked with an asterisks. The number of nodes on the branch-and-bound tree is displayed under column *Nodes*. The gap percentage between the best integer solution found and the lower bound provided by the linear relaxation is displayed under column *GAP*. Finally, *Time on Cuts* shows the relative amount of time spent on solving the separation problem.

As Table 1 shows, some of the instances that CPLEX was not able to solve within an hour, could be solved by the reinforced formulation. Another way to understand the strength of the added constraints is by comparing the respective optimality gaps and the sizes of the branch-and-bound trees. However, there is still place for improvements as the separation problem requires a great amount of time to be solved.

Table 1. Computational results

Instances				CPLEX			B & C			
$|E|$	$|V|$	C	i	CPU(s)	Node($\times$1000)	GAP (%)	CPU (s)	Nodes($\times$1000)	GAP (%)	Time on Cuts (%)
30	10	5	1	3.90	2.98	27.91	0.78	0.09	13.30	34.62
30	10	5	2	2.41	1.34	21.55	0.82	0.07	7.45	28.05
30	10	5	3	43.75	35.7	28.85	3.49	0.79	15.23	37.82
30	20	5	1	1.52	1.12	16.09	1.10	0.17	10.31	54.55
30	20	5	2	6.75	4.11	17.22	0.89	0.12	10.78	33.71
30	20	5	3	0.99	0.35	10.78	0.40	0.02	6.66	5.00
30	30	5	1	0.22	0.07	5.13	0.13	0.001	2.13	7.69
30	30	5	2	0.19	0.01	2.70	0.08	0	0	12.50
30	30	5	3	0.33	0.02	5.56	0.28	0.02	3.17	17.86
40	10	5	1	3600*	1073	39.59	225.1	15.57	16.70	63.91
40	10	5	2	70.84	30.2	35.07	14.18	2.70	18.30	51.48
40	10	5	3	3600*	1091	37.79	191.7	16.59	23.04	63.31
40	20	5	1	89.79	42.8	19.55	34.08	3.44	14.33	67.31
40	20	5	2	151.4	61.1	18.03	16.89	0.99	10.81	56.72
40	20	5	3	466.1	194.7	21.84	63.55	6.29	15.29	66.55
40	30	5	1	17.11	9.32	11.63	2.72	0.08	6.72	29.78
40	30	5	2	48.74	60.1	10.64	3.91	0.43	6.77	55.24
40	30	5	3	74.52	65.0	12.64	4.20	0.45	7.61	59.05
50	10	5	1	3600*	963.7	43.88	3600*	207.4	25.82	55.16
50	10	5	2	3600*	876.3	44.35	1210	69.67	25.81	47.91
50	10	5	3	3600*	773.9	43.21	3600*	133.4	18.94	68.46
50	20	5	1	3600*	848.9	33.93	3600*	60.66	21.39	90.90
50	20	5	2	3600*	868.6	31.19	3600*	128.0	22.13	73.06
50	20	5	3	3600*	930.6	31.02	3600*	112.3	23.94	85.88
50	30	5	1	3600*	1275	19.30	260.3	14.51	11.93	73.31
50	30	5	2	3600*	1051	18.94	311.2	17.60	13.25	71.11
50	30	5	3	3600*	1549	15.25	206.5	8.95	9.37	57.09

7   Conclusion

In this paper we investigated the USNMP, a constrained variant of SNMP. We proved that USNMP is NP-Hard for any fixed capacity $C \geq 2$. We also, studied a special case where there exists a station wherein all demands intersect and showed that this case is easy to solve when $C = 2$, but is surprisingly NP-Hard for $C \geq 3$. In addition, we introduced new families of valid inequalities. Finally, we showed that these inequalities have a great impact in the resolution of the problem.

Acknowledgements. We want to thank the laboratory of excellence IMobS3 for its financial support.

References

1. Agatz, N., Erera, A., Savelsbergh, M., Wang, X.: Optimization for dynamic ride-sharing: a review. Eur. J. Oper. Res. **223**(2), 295–303 (2012)
2. Amini, O., Pérennes, S., Sau, I.: Hardness and approximation of traffic grooming. Theoret. Comput. Sci. **410**(38–40), 3751–3760 (2009)
3. Boyacı, B., Zografos, K.G., Geroliminis, N.: An optimization framework for the development of efficient one-way car-sharing systems. Eur. J. Oper. Res. **240**(3), 718–733 (2015)
4. Bsaybes, S., Quilliot, A., Wagler, A.K.: Fleet management for autonomous vehicles. arXiv preprint arXiv:1609.01634 (2016)
5. Chvátal, V.: Edmonds polytopes and a hierarchy of combinatorial problems. Discrete Math. **4**(4), 305–337 (1973)
6. Cordeau, J.-F., Laporte, G.: The dial-a-ride problem: models and algorithms. Ann. Oper. Res. **153**(1), 29–46 (2007)
7. Denton, B.T., Miller, A.J., Balasubramanian, H.J., Huschka, T.R.: Optimal allocation of surgery blocks to operating rooms under uncertainty. Oper. Res. **58**(4-part-1), 802–816 (2010)
8. Dyer, M.E., Frieze, A.M.: On the complexity of partitioning graphs into connected subgraphs. Discrete Appl. Math. **10**(2), 139–153 (1985)
9. Easymile (2017). http://www.easymile.com
10. Fischetti, M., Hamacher, H.W., Jørnsten, K., Maffioli, F.: Weighted k-cardinality trees: complexity and polyhedral structure. Networks **24**(1), 11–21 (1994)
11. Gao, P., Kaas, H.-W., Mohr, D., Wee, D.: Automotive revolution-perspective towards 2030 how the convergence of disruptive technology-driven trends could transform the auto industry. Advanced Industries McKinsey & Company (2016)
12. Garey, M.R., Johnson, D.S.: Computers and Intractability, vol. 29. W.H. Freeman, New York (2002)
13. Goldschmidt, O., Hochbaum, D.S., Levin, A., Olinick, E.V.: The sonet edge-partition problem. Networks **41**(1), 13–23 (2003)
14. Gomory, R.E., et al.: Outline of an algorithm for integer solutions to linear programs. Bull. Am. Math. Soc. **64**(5), 275–278 (1958)
15. Ligier (2017). http://www.ligier.fr
16. Masuyama, S., Ibaraki, T.: Chain packing in graphs. Algorithmica **6**(1–6), 826–839 (1991)
17. Pelletier, S., Jabali, O., Laporte, G.: 50th anniversary invited articlegoods distribution with electric vehicles: review and research perspectives. Transp. Sci. **50**(1), 3–22 (2016)
18. Pimenta, V., Quilliot, A., Toussaint, H., Vigo, D.: Models and algorithms for reliability-oriented dial-a-ride with autonomous electric vehicles. Eur. J. Oper. Res. **257**(2), 601–613 (2017)

19. Ravi, R., Sundaram, R., Marathe, M.V., Rosenkrantz, D.J., Ravi, S.S.: Spanning treesshort or small. SIAM J. Discrete Math. **9**(2), 178–200 (1996)
20. Royer, E., Marmoiton, F., Alizon, S., Ramadasan, D., Slade, M., Nizard, A., Dhome, M., Thuilot, B., Bonjean, F.: Retour d'expérience après plus de 1000 km en navette sans conducteur guidée par vision. In: Proceedings of RFIA (2016)
21. Shaheen, S.A., Cohen, A.P.: Carsharing and personal vehicle services: worldwide market developments and emerging trends. Int. J. Sustain. Transp. **7**(1), 5–34 (2013)
22. Van Dijke, J., Van Schijndel, M., Nashashibi, F., De La Fortelle, A.: Certification of automated transport systems. Proc. Soc. Beh. Sci. **48**, 3461–3470 (2012)

Maximum Concurrent Flow
with Incomplete Data

Pierre-Olivier Bauguion[1]($\boxtimes$), Claudia D'Ambrosio[2], and Leo Liberti[2]

[1] IRT SystemX, 8 Avenue de la Vauve, 91120 Palaiseau, France
pierre-olivier.bauguion@irt-systemx.fr
[2] CNRS LIX, Ecole Polytechnique, 91128 Palaiseau, France
{dambrosio,liberti}@lix.polytechnique.fr

Abstract. The Maximum Concurrent Flow Problem (MCFP) is often used in the planning of transportation and communication networks. We discuss here the MCFP with incomplete data. We call this new problem the Incomplete Maximum Concurrent Flow Problem (IMCFP). The main objective of IMCFP is to complete the missing information assuming the known and unknown data form a MCFP and one of its optimal solutions. We propose a new solution technique to solve the IMCFP which is based on a linear programming formulation involving both primal and dual variables, which optimally decides values for the missing data so that they are compatible with a set of scenarios of different incomplete data sets. We prove the correctness of our formulation and benchmark it on many different instances.

Keywords: Maximum concurrent flow
Multi-commodity flow problems · Incomplete data · Unknown data
Uncertainty · Inverse optimization · Transportation systems

1 Introduction

Network flows have been introduced long ago (see, e.g., [12], enhanced later by [11]) to tackle single commodity flow problems, such as the max-flow problem. Since then, these models have been generalized for multiple commodities [20] and grouped under the label of multi-commodity flow models. Nowadays, multi-commodity flow formulations are extensively used in many contexts for their ability to capture the movements of different types of commodities in various real-world activities such as people in transportation models, data in telecommunication networks, water flows... (see, e.g., [1]). These formulations are generally used to help make the best cost-effective solution for allocating resources; this leads to optimize a cost function.

In real-world applications, the available data are often uncertain or incomplete, and their actual values may only be revealed at a time when the overall decision strategy has already been chosen. This is often the case in transportation systems where the parameters are time-dependent and event-sensitive. Statistical inference and data mining represent convenient ways to deal with this

© Springer International Publishing AG, part of Springer Nature 2018
J. Lee et al. (Eds.): ISCO 2018, LNCS 10856, pp. 77–88, 2018.
https://doi.org/10.1007/978-3-319-96151-4_7

uncertainty. One of the best known inference models in transportation systems is the Four Step Model [18], which is an algorithm that iterates over time according to an equilibrium criterion. More recently, a lot of attention has been devoted to machine learning approaches, which generally performs better on large scale datasets. In this context, [21] proposes bayesian networks and [17] uses a deep learning approach to forecast flow in transportation systems.

However, optimization methods that deal with uncertainty actually do exist. To the best of our knowledge, [10] was the first to propose a stochastic approach to tackle incompleteness of input data. It assumes that uncertain data follow some given probability distribution, and that the objective of this approach is not to optimize a certain cost, but an expected cost instead. On the other hand, [22] proposed a complementary approach by optimizing a robust criterion such as the worst case or the maximum regret. This particular method received renewed attention from [6,7], while [5] applies this approach to multi-commodity flow problems by considering a polyhedral uncertainty set of demands. Later, [3] mixed the recourse variables introduced in [10] with the robust approach for a network flow and design problem.

Optimization methods can also be used to optimally fit experimental measurements. In [15], multi-commodity flow optimization is used to model a gas transportation network while retrieving missing data. The problem discussed in [15] consists in recomposing the flow on each arc, knowing only the global amount of incoming and outgoing flows for each node. The problem of finding a minimal adjustment of the cost function to ensure the optimality of a given solution generated a particular interest with [9] under the label of inverse optimization. For example [2,23] apply this concept to multi-commodity flow problems (especially min cost flow problem). The survey [14] on this subject includes situations where the inverse problem seeks parameters other than objective function coefficients.

The Maximum Concurrent Flow Problem (MCFP) has been extensively studied over time [4,8,20], but in this paper we present a new approach for finding optimal maximum concurrent flows using incomplete data. Our method seeks optimal solutions and completes the partial input. This problem typically arises when we have insights about the global behavior of a system while data are partially unknown [15]. Symmetrically it can validate/invalidate a hypothetical behavior by comparing it with the observed data. This is particularly relevant in transportation when the routing strategy of passengers is known while data are incomplete. We call this problem Incomplete Maximum Concurrent Flow Problem (IMCFP).

The rest of this paper is organized as follows: Sect. 2 recalls the MCFP and presents the IMCFP. In Sect. 3 we propose a formulation for the IMCFP by integrating both primal and dual formulations of the MCFP, and prove its correctness based on the complementary slackness conditions of Linear Programming (LP), which we recall for convenience [16]. Then, in Sect. 4 we present preliminary experiments for these formulations and their practical interest. Finally, Sect. 5 concludes the paper.

2   The Maximum Concurrent Flow

Consider the following (well-known) problem.

> MCFP. Given a simple directed graph $G = (V, A)$, an arc capacity function $c : A \to \mathbb{R}_+$, a set K of triplets $k = (o, d, D) \in V^2 \times \mathbb{R}_+$, find: a scalar $\gamma \geq 0$ (called *threshold*) and a set of flows f^k on G for each $k \in K$ such that (i) for each arc $a \in A$, the *arc load* of f on a (i.e., the sum of the flows on a) does not exceed the arc capacity c_a; (ii) each "$o - d$" flow f^k has value γD for each $k = (o, d, D) \in K$; (iii) γ is maximum.

We recall that, for a given $k = (o, d, D) \in K$, a flow f having value D in a graph G from node o to node d is a non-negative arc function $f : A \to \mathbb{R}_+$ such that $\sum\limits_{j \in N^-(d)} f_{jd} - \sum\limits_{j \in N^+(d)} f_{dj} = D$ and the following flow balance equations hold (we omit k index for clarity):

$$\forall i \in V \smallsetminus \{o, d\} \qquad \sum_{j \in N^-(i)} f_{ji} = \sum_{j \in N^+(i)} f_{ij}. \tag{1}$$

We also recall that $N^-(i)$ is the set of nodes j such that $(j, i) \in A$ and $N^+(i)$ such that $(i, j) \in A$, for each $i \in V$.

The MCFP was introduced in [20]. It can be formulated as follows using LP:

$$\left.\begin{array}{c} \max\limits_{\gamma \geq 0, f \geq 0} \qquad \gamma \\[4pt] \forall k = (o, d, D) \in K \qquad \sum\limits_{j \in N^-(d)} f^k_{jd} = \gamma D + \sum\limits_{j \in N^+(d)} f^k_{dj} \\[6pt] \forall k = (o, d, D) \in K, i \in V \smallsetminus \{o, d\} \qquad \sum\limits_{j \in N^-(i)} f^k_{ji} = \sum\limits_{j \in N^+(i)} f^k_{ij} \\[6pt] \forall a = (i, j) \in A \qquad \sum\limits_{k \in K} f^k_a \leq c_a. \end{array}\right\} \tag{2}$$

Its dual is (one can choose appropriate inequalities instead of equalities for the first and second sets of constraints of (2)):

$$\left.\begin{array}{c} \min\limits_{p \geq 0, u \geq 0} \qquad \sum\limits_{a \in A} u_a c_a \\[4pt] \forall k = (o, d, D) \in K, \forall a = (i, j) \in A, \qquad p^k_i + u_a \geq p^k_j \\[4pt] \forall k = (o, d, D) \in K, \qquad p^k_o = 0 \\[4pt] \sum\limits_{k = (o, d, D) \in K} p^k_d D \geq 1 \end{array}\right\} \tag{3}$$

where u_a, (for $a \in A$) are the dual variables associated to the capacity constraints (last set of constraints of (2)) and p^k_i (for $k \in K, i \in V$) are the dual variables associated to the flow conservation constraints (first and second sets of constraints of (2)). This implies that the MCFP is polynomial-time solvable (for example with an interior point algorithm). The MCFP is also strongly polynomial-time solvable [19], but it appears to be common knowledge that, for instance sizes of current practical interest, it is empirically more efficient to use a good LP solver on Problem (2) rather than the algorithm in [19]. The MCFP is often used in real-life applications in order to design networks or evaluate the arcs with highest risk of becoming saturated [13]. The main applied interest in the MCFP is that, through the ratio variable γ, it ensures a fairness of arc capacity utilization over all flows.

Our motivation for studying this problem stems from transportation networks, be they road or rail-oriented. As a critical increase of the load can induce a decrease of the quality of services, an hypothesis consists of assuming that the passengers traffic tends naturally to balance itself to an equilibrium [18]. One can model this problem by minimizing the maximum capacity utilization, and the latter can be reformulated as a MCFP [20]. In our context, we have historical traffic data including a partial observation of the arc loads for a certain subset of arcs. For some networks, we are also given a subset $A' \subset A$ of arcs with known capacities.

In general, however, we do not know the arc capacities. The problem we are interested in is the MCFP with incomplete arc capacities. The MCFP in LP formulation (2) without the capacity constraints is clearly an unbounded LP. To avoid this situation, we employ a given set S of scenarios from our historical arc load database. Each scenario $s = (A^s, \ell^s, K^s) \in S$ consists of a subset $A^s \subset A$ of arcs, a partial arc load function $\ell^s : A^s \to \mathbb{R}_+$, and a set of commodities K^s. We require that: (i) missing capacities should be estimated so as to allow the maximum known arc loads over all scenarios, (ii) arc loads from computed flows should be as close as possible to the loads given in the scenarios, and (iii) each flow solution for a scenario should describe an optimal solution of the MCFP w.r.t. capacity and commodity values. We therefore define the following problem, which is new as far as we could ascertain.

IMCFP. Given a graph $G = (V, A)$, a subset $A' \subset A$, a partial arc capacity function $c : A' \to \mathbb{R}_+$, and a set S of scenarios $(A^s \subset A, \ell^s, K^s)$ where $\ell^s : A^s \to \mathbb{R}_+$ and K^s is a set of triplets $k = (o, d, D) \in V^2 \times \mathbb{R}_+$, find: a threshold function $\gamma : S \to \mathbb{R}_+$, a complementary arc capacity function $c : A \setminus A' \to \mathbb{R}_+$, and a set of flows f^{sk} (for $s \in S$ and for $k \in K^s$) such that (i) for each arc $a \in A$ and for each $s \in S$, the arc load of f on a is bounded above by c_a; (ii) for each $s \in S$ and arc $a \in A^s$, the arc load of f on a is as close as possible to the arc load $\gamma_s \ell^s_a$; (iii) for each $s \in S$, the flows f^s and γ^s should be optimal with respect to an MCFP defined over the capacities c_a over all $a \in A$ and the commodities K^s.

Although the IMCFP is natively cast in a multi-objective fashion (see condition (ii)), in practice we minimize a max norm over all arcs and all scenarios. We remark that condition (iii) is only apparently recursive: we want to decide f, c at the same time and also require that every f^s should be optimal flows w.r.t. a putative MCFP instance defined over the values of the c variables and the K^s parameters. We shall see below that the IMCFP can be formulated by means of a Mixed-Integer Linear Programming formulation that combines both primal and dual variables.

3   The IMCFP

In this section we shall first introduce a Mixed-Integer Linear formulation for the IMCFP, and then prove its correctness.

3.1   Formulation

- *Sets*:
- V: set of nodes
- A: set of arcs
- S: set of scenarios
- K^s: set of commodities for scenario s
- A': subset of arcs from which the capacity is known
- A^s: subset of arcs from which the load is known.

- *Parameters*:
- $k = (o, d, D)$ for $k \in K^s, s \in S$: commodity data (origin $o \in V$, destination $d \in V$, demand value $D \in \mathbb{R}_+$)
- $\ell^s : A^s \to \mathbb{R}_+$: load function over the arcs A^s for each scenario $s \in S$
- $c : A' \to \mathbb{R}_+$: capacity function over the arcs A'
- M^w: "Big M" parameter associated to the binary weights w
- M^f: "Big M" parameter associated to the binary flows x.

- *Decision variables*:
- $f_a^{sk} \geq 0$ for $a \in A, s \in S, k \in K^s$: flow variable of arc $a \in A$, for scenario $s \in S$ and demand $k \in K^s$
- $\gamma_s \geq 0$ for $s \in S$: threshold variable for scenario $s \in S$
- $p_i^{sk} \geq 0$ for $i \in V, s \in S, k \in K^s$: potential variable (dual variable from MCFP's conservation constraint) for node i, scenario s, and demand k
- $u_a^s \geq 0$ for $a \in A, s \in S$: weight variable (dual variable from MCFP's capacity constraint) for arc $a \in A$ and scenario $s \in S$
- $w_a^s \in \{0, 1\}$ for $a \in A, s \in S$: binary variable that allows for a corresponding weight u_a^s to be greater than 0 or not
- $x_a^{sk} \in \{0, 1\}$ for $a \in A, s \in S, k \in K^s$: binary variable that allows for a corresponding flow f_a^{sk} to be greater than 0 or not
- $\Delta \geq 0$: maximal difference between the load parameters with the computed ones
- $c_a \geq 0$ for $a \in A \setminus A'$: capacity variable of arc a.

- *Objective function*:

$$\min_{f, \gamma, p, u, w, x, c} \Delta. \tag{4}$$

- *Constraints*:
- flow conservation:

$$\forall s \in S, \forall k = (o, d, D) \in K^s, \forall i \in V \setminus \{o, d\} \quad \sum_{j \in N^-(i)} f_{ji}^{sk} = \sum_{j \in N^+(i)} f_{ij}^{sk} \tag{5}$$

- demand satisfaction:

$$\forall s \in S, \forall k = (o, d, D) \in K^s \quad \sum_{j \in N^-(d)} f_{jd}^{sk} - \sum_{j \in N^+(d)} f_{dj}^{sk} = \gamma_s D \qquad (6)$$

- min-cost node access:

$$\forall s \in S, \forall k \in K^s, \forall (i, j) \in A \quad p_i^{sk} + u_{ij}^s - p_j^{sk} \geq 0 \qquad (7)$$

- min-cost path condition:

$$\forall s \in S, \forall k \in K^s, \forall (i, j) \in A \quad p_i^{sk} + u_{ij}^s - p_j^{sk} \leq 1 - x_{ij}^{sk} \qquad (8)$$

- origin access:

$$\forall s \in S, \quad \forall k = (o, d, D) \in K^s \quad p_o^{sk} = 0 \qquad (9)$$

- complementary slackness condition on capacity:

$$\forall s \in S, \forall a \in A \quad c_a \leq \sum_{k \in K^s} f_a^{sk} + M^w (1 - w_a^s) \qquad (10)$$

- binary flow constraint:

$$\forall s \in S, \forall k \in K^s, \forall a \in A \quad f_a^{sk} \leq M^f x_a^{sk} \qquad (11)$$

- binary weight constraint:

$$\forall s \in S, \forall a \in A \quad u_a^s \leq w_a^s \qquad (12)$$

- dual weights constraint:

$$\forall s \in S \quad \sum_{a \in A} u_a^s = 1 \qquad (13)$$

- threshold bound:

$$\forall s \in S, \forall a \in A^s \quad c_a \geq \gamma_s \ell_a^s \qquad (14)$$

- feasibility bound:

$$\forall s \in S, \forall a \in A \quad \sum_{k \in K^s} f_a^{sk} \leq c_a \qquad (15)$$

- max norm (i):

$$\forall s \in S, \forall a \in A^s \quad \Delta \geq \gamma_s \ell_a^s - \sum_{k \in K^s} f_a^{sk} \qquad (16)$$

- max norm (ii):

$$\forall s \in S, \forall a \in A^s \quad \Delta \geq \sum_{k \in K^s} f_a^{sk} - \gamma_s \ell_a^s. \qquad (17)$$

The model aims at minimizing the maximal error Δ, which is the absolute value of the difference for each scenario and each arc between the load of f and the load $\gamma_s \ell_a^s$ (see constraints (16) and (17)) as stated in condition (ii) of the IMCFP. Constraints (5) and (6) are flow conservation constraints already mentioned in (1). Constraints (7), (8), (9), and (11) provide the system that ensures the optimality of the routing in the sense of the weights u, which correspond to the dual variables in (3) of capacity constraint in MCFP (2). Constraints (10) and (12) describe the slackness conditions for the dual variables of the capacities. Constraint (13) ensures a feasible dual solution of (3) for each scenario. Finally, Constraints (15) and (14) guarantee the feasibility of (2) as condition (i) of IMCFP. Note that c_a of (14), (15), and (10) can either be a parameter or a variable, depending whether a is in A' or not. The condition (iii) of IMCFP follows by Proposition 1 and Theorem 2 (below).

3.2 Correctness

Proposition 1. *Let $f = (f_a^k \mid a \in A, k \in K) \geq 0$ and $\gamma \geq 0$ be a feasible flow solution of the MCFP (2) and $p = (p_i^k \mid k \in K, i \in V) \geq 0$ and $u = (u_a \mid a \in A) \geq 0$ be a feasible solution of its dual (3). These two assertions are equivalent:*

1. *(a) $\forall a \in A$, we have $u_a(c_a - \sum_{k \in K} f_a^k) = 0$*

   (b) $\forall a = (i,j) \in A, \forall k \in K$, we have $f_a^k(p_i^k + u_a - p_j^k) = 0$

   (c) $\sum_{a \in A} u_a > 0$

2. *f and γ are optimal for the MCF (2), and p and u (both scaled by $\sum_{k=(o,d,D) \in K} p_d^k D$) are optimal for its dual (3).*

Proof. $2 \Rightarrow 1$: If f and γ (resp. p and u) are optimal for (2) (resp. for (3)), the complementary slackness conditions [16] state immediately the first two equations. The third complementary slackness condition of MCFP states $\sum_{k \in K} p_d^k D = 1$. Therefore, (3) ensures there exists a so that $u_a > 0$, meaning $\sum_{a \in A} u_a > 0$.

$1 \Rightarrow 2$: If $\sum_{a \in A} u_a > 0$ and $\forall a \in A, u_a \geq 0$, there exists $a \in A$ so that $u_a > 0$.

The first condition implies that for this a we have $c_a = \sum_{k \in K} f_a^k$. We can assume the arcs have a non-zero capacity (otherwise the arc should not have existed), meaning there exists a $k = (o, d, D)$ for which $f_a^k > 0$. The second condition implies that going through this arc satisfies the "min cost path condition", meaning $p_d^k \geq u_a > 0$. We can then construct $Q = \sum_{k=(o,d,D) \in K} p_d^k D > 0$. Choose u_a' so that $u_a' = \frac{u_a}{Q}$, and, for each $k = (o, d, D) \in K$, $p_d'^k = \frac{p_d^k}{Q}$. Then we trivially have $\sum_{k=(o,d,D) \in K} p_d'^k D = 1$. Substituting u by u' and p by p', one can observe that

$\forall a \in A, \ u_a(c_a - \sum_{k \in K} f_a^k) = 0$ and $\forall a = (i,j) \in A, \forall k \in K \ f_a^k(p_i^k + u_a - p_j^k) = 0$

still hold and (u,p) becomes a feasible solution of (3). The proposition follows from the LP complementary slackness conditions [16]. □

Theorem 2. *Given a graph $G = (V, A)$ and a feasible solution (f, c) of IMCFP, for each $s \in S$, f^s corresponds to an optimal solution of an MCFP with respect to c and K^s.*

Proof. Let us choose a $s \in S$. f^s is feasible for MCFP (2) with parameters c and K^s due to (5), (6) and (15). p^s and u^s describes a feasible solution of (3) with parameters c and K^s due to (13) and (7) (one can scale the solution to ensure $\sum_{k=(o,d,D) \in K^s} p_d^{sk} D \geq 1$). The combination of (10) and (12) ensure the first slackness condition, as $u_a^s = 0$ if the $c_a - \sum_{k \in K^s} f_a^{sk} > 0$. Constraints (7), (8) and (11) validate the second slackness condition. The third slackness condition is obtained by (13). Proposition 1 validates the optimality of f^s for MCFP (2) with parameters c and K^s. As it holds for each $s \in S$, it concludes the proof. □

4   Numerical Results

We can solve IMCFP instances by simply formulating them as the Mixed-Integer Linear Program in Sect. 3.1 and solving them using an off-the-shelf solver. By the polynomial number of constraints and variables of our formulation, we know that the decision version of IMCFP is in **NP**. Although the MCFP is polynomial-time solvable, our IMCFP formulation introduces the need of binary variables to ensure optimality among the different scenarios, meaning it is still an open question whether IMCFP is **NP**-complete or not. To solve all the following instances, we solved the IMCFP model with the MILP solver IBM CPLEX 12.6 with default settings and a time limit of 1200 CPU seconds on a personal computer (Intel Core i7-6820HQ 2.70 GHz, 16 GB DDR3 RAM). All graphs used in the following experiments are based on the topology of the Paris subway network, often restricted to the left bank. Therefore, each node represents a connection between one or more different metro lines, and each arc represents a section of a line. The network is strongly connected, meaning we can generate complete sets of demands $(|K| = (|V| - 1)|V|)$. The demand value for each commodity $k \in K$ is an integer uniformly chosen in the interval $[1, 10]$.

Firstly, to evaluate the prediction performance of the proposed formulations, we generated integer capacities for the MCFP in the interval $[1, 15]$ and then calculated an optimal solution of MCFP for each scenario of demands, keeping the same capacities C among them. It allows us to give the total configuration of loads ℓ_a^s and K^s as input where an optimal solution with zero Δ value exists. This let us compare the computed capacities in $A \setminus A'$, with those we chose to construct our instances of MCFP (the generated ones). Secondly, we generated ℓ_a^s, K^s, and c_a according to a feasible flow solution (in terms of conservation and capacity constraints) which is not MCFP compliant (i.e., whose input data

Table 1. IMCFP, Paris left bank subway network topology ($|V| = 13, |A| = 38$).

$C(\%)$	$|S| = 3$				$|S| = 4$				$|S| = 5$			
	$c(\%)$	Gap	$T(s)$	Δ	$c(\%)$	Gap	$T(s)$	Δ	$c(\%)$	Gap	$T(s)$	Δ
0	42.11	13.00	2.95	0.00	42.11	14.00	4.31	0.00	63.16	7.00	6.11	0.00
10	60.53	9.72	2.64	0.00	65.79	11.00	3.07	0.00	63.16	9.00	7.43	0.00
20	68.42	11.00	2.12	0.00	71.05	10.00	118.00	0.00	73.68	10.00	4.82	0.00
30	60.53	14	1.83	0.00	78.95	8.58	4.48	0.00	76.32	9.00	*	0.04
40	44.74	12.83	2.18	0.00	81.58	6.00	3.00	0.00	73.68	13.00	5.40	0.00
50	84.21	10.46	2.14	0.00	86.84	6.00	3.31	0.00	78.95	11.77	6.80	0.00
60	81.58	10.00	2.59	0.00	81.58	7.00	4.12	0.00	89.47	13.00	14.09	0.00
70	81.58	9.26	2.01	0.00	89.47	9.00	2.11	0.00	84.21	12.72	6.05	0.00
80	86.84	12.00	2.56	0.00	97.37	6.00	3.09	0.00	94.74	11.25	6.63	0.00
90	100.00	0.00	1.90	0.00	97.37	0.69	3.46	0.00	92.11	4.00	9.95	0.00
100	100.00	0.00	1.42	0.00	100.00	0.00	2.47	0.00	100.00	0.00	3.74	0.00

do not follow an optimal MCFP solution pattern). The second set of instances aims at observing how our model deals with data based on a wrong hypothesis (structure of an optimal solution of MCFP) and, hence, how the objective value is impacted. Moreover, we studied the impact of the quantity of known and unknown data by giving a fixed percentage of capacities for MCFP ($\frac{|A'|}{|A|} \times 100$) as an input.

The resulting tables are organized as follows. The number of nodes ($|V|$) and arcs ($|A|$) are reported in the caption of each table. The proportion of known capacities is specified in the first column "$C(\%)$". The rest of the table is divided in 3 subsets of columns, 4 for each value of cardinality of S. The four columns report: (i) the amount of capacities "$c(\%)$" that has been successfully predicted (the percentage of successfully predicted capacities may be lower than the given ones due to the truncation process of "$C(\%)|A'|$"); (ii) the maximal absolute gap ("Gap") observed between the predicted capacities and the generated ones; (iii) the CPU time in seconds "$T(s)$" (we denote termination due to time limit by *); and (iv) objective value "Δ".

The most striking thing one can note is the effectiveness of our methodology in predicting arc capacities (see Table 1). Even if without insights of capacities the prediction correctness remains low, it skyrockets to more than 50% of additionnal correct predictions with only three scenarios. This can be explained by the fact that even with few capacities the γ_s become strongly bounded. This proportion of corrected predictions tends naturally to arise when more and more capacity parameters are known, but the number of scenarios seems to remain a strong source of insights for the network, especially when the number of known capacities is low (see Table 2). However, the number of scenarios seems more

Table 2. IMCFP, Paris subway network topology ($|V| = 57, |A| = 209$).

$C(\%)$	$|S| = 3$				$|S| = 4$				$|S| = 5$			
	$c(\%)$	Gap	$T(s)$	Δ	$c(\%)$	Gap	$T(s)$	Δ	$c(\%)$	Gap	$T(s)$	Δ
0	13.40	12.50	*	0.28	14.83	9.44	*	3.66	64.60	12.11	*	1.00
10	15.31	13.78	*	0.33	17.70	11.84	*	2.21	68.42	11.04	*	3.15
20	67.46	12.53	*	10.54	69.86	10.77	*	5.20	60.29	13.21	*	1.09
30	76.55	13.44	*	2.00	74.16	12.45	*	0.80	75.12	13.29	*	0.71
40	73.68	13.41	*	1.00	79.43	11.00	*	1.52	76.56	12.04	*	1.03
50	80.38	9.82	*	1.00	54.07	12.54	390.44	0.00	84.21	11.54	*	2.36
60	78.95	12.90	*	0.33	86.12	11.43	*	2.65	83.73	12.02	*	9.00
70	73.2	13.02	*	0.04	90.90	10.45	*	3.00	90.43	9.16	*	3.85
80	90.43	9.50	*	2.00	92.34	10.61	*	1.39	92.82	11.26	*	6.03
90	90.43	12.94	*	0.82	94.26	11.71	*	5.00	97.13	12.10	*	3.13
100	100.00	0.00	*	3.00	100.00	0.00	*	1.00	100.00	0.00	*	1.83

and more important as the size of the instance grows. Indeed, the average percentage of correct predictions is quite close compared from $|S| = 4$ to $|S| = 5$ (around 81% for both) in the instances of Table 1 as if this increase did not bring further information. But this average percentage rises from 70% to 79% between $|S| = 4$ and $|S| = 5$ in the instances of Table 2. This could mean that the larger the instance, the higher the number of required scenarios to reach a similar prediction performance. It is also interesting to note that some of these results are obtained without even reaching optimality, especially on the total metro network instances, meaning that the routing problems are quite hard to solve even when all capacities are known. This is confirmed by the amount of time consumed by the instances when all the capacities are set. This suggests that MCFPs are also quite hard to solve. This is a well-known result that motivated the design of approximation schemes to solve it (see e.g. [8]). As one can expect, confronted with inconsistent data (input data that do not follow a structure of an optimal solution of the MCFP), the prediction effectiveness does not perform as well as the previous case (see Table 3). Nevertheless, our model often tends to bring a solution that fits significantly the input loads, especially when few capacities are known. Generally speaking, this means that a low objective value does not guarantee necessarily that the MCFP pattern hypothesis on our input data is right. However, when all capacities are set ($A' = A$), this seems to constraint drastically our model whose objective value skyrockets. Therefore, it suggests that our model can still be used to validate/invalidate this hypothesis on condition that all capacities are known.

Table 3. IMCFP, Paris left bank subway network topology ($|V| = 13, |A| = 38$). Input data not MCFP compliant.

	$|S| = 3$				$|S| = 4$				$|S| = 5$			
$C(\%)$	$c(\%)$	Gap	$T(s)$	Δ	$c(\%)$	Gap	$T(s)$	Δ	$c(\%)$	Gap	$T(s)$	Δ
0	0	13.40	1.87	0.00	10.53	13.54	*	0.2	5.26	11.97	8.25	0.00
10	15.79	12.62	*	0.34	10.53	13.18	121.12	0.00	10.53	10.24	*	0.15
20	23.68	14.06	*	0.30	28.95	9.67	4.48	0.00	21.05	12.61	5.07	0.00
30	28.95	12.64	*	0.00	28.95	13.66	4.31	0.00	31.58	13.53	4.1	0.00
40	39.47	11.36	*	2.00	42.11	12.26	*	0.95	50.00	12.89	*	1.00
50	55.26	8.95	*	3.17	50.00	11.93	6.50	0.00	57.89	10.09	*	1.00
60	57.90	12.63	*	0.43	57.89	9.55	*	1.12	65.79	11.53	5.16	1.00
70	68.42	11.89	859.08	1.00	68.42	12.93	2.67	0.00	73.68	9.00	7.28	1.00
80	81.58	7.49	2.21	1.00	78.95	9.47	4.00	2.00	81.58	8.69	5.16	1.00
90	89.47	12.08	1.16	0.50	92.10	12.08	1.07	1.00	89.47	9.89	6.33	1.35
100	100.00	0.00	0.52	14.00	100.00	0.00	0.66	11.67	100.00	0.00	0.80	14.13

5   Conclusion and Future Work

In this paper we studied a new problem called the IMCFP and proposed a solution technique to tackle it. The purpose of the new formulation proposed is to find a routing of commodities that fits input data at best (namely loads and demand matrix) as [15]. But, in addition, the solution has to follow an optimal structure of MCFP regarding a set of known and unknown data. The practical interest of the problem we discussed arises in transportation systems when it comes to recompose unknown data assuming hypothesis of data structure (namely a MCFP and one of its optimal solutions). Moreover, this formulation can validate/invalidate hypothesis of a MCFP's optimal routing by confronting it to observed data. More theoretically, we showed that this problem can be tackled by embedding primal and dual formulations and complementary slackness conditions all together into a single Mixed Integer Linear Program. We proved the correctness of such models, and led experiments to emphasize empirical behaviours and computational hardness. Future work will focus on computational performance improvements so that larger graphs can be treated.

References

1. Ahuja, R.K., Magnanti, T.L., Orlin, J.B.: Network Flows: Theory, Algorithms, and Applications. Prentice Hall Inc., Upper Saddle River (1993)
2. Ahuja, R.K., Orlin, J.B.: Inverse optimization, part I: Linear programming and general problem (1998)
3. Atamturk, A., Zhang, M.: Two-stage robust network flow and design under demand uncertainty. Oper. Res. **55**(4), 662–673 (2007)
4. Bauguion, P.O., Ben-Ameur, W., Gourdin, E.: Efficient algorithms for the maximum concurrent flow problem. Networks **65**(1), 56–67 (2015)

5. Ben-Ameur, W., Kerivin, H.: Routing of uncertain traffic demands. Optim. Eng. **3**, 283–313 (2005)
6. Ben-Tal, A., Goryashko, A., Guslitzer, E., Nemirovski, A.: Adjustable robust solutions of uncertain linear programs. Math. Program. **99**(2), 351–376 (2004)
7. Bertsimas, D., Sim, M.: The price of robustness. Oper. Res. **52**(1), 35–53 (2004)
8. Bienstock, D., Raskina, O.: Asymptotic analysis of the flow deviation method for the maximum concurrent flow problem. Math. Program. Ser. B **91**, 479–492 (2002)
9. Burton, D., Toint, P.L.: On an instance of the inverse shortest paths problem. Math. Program. **53**(1), 45–61 (1992)
10. Dantzig, G.B.: Linear programming under uncertainty. Manage. Sci. **1**(3–4), 197–206 (1955)
11. Edmonds, J., Karp, R.M.: Theoretical improvements in algorithmic efficiency for network flow problems. J. ACM **19**(2), 248–264 (1972)
12. Ford, L.R., Fulkerson, D.R.: Maximal flow through a network. Can. J. Math. **8**, 399–404 (1956)
13. Gerla, M.: A cut saturation algorithm for topological design of packet-switched communication networks. In: Proceedings of the National Telecommunication Conference, pp. 1074–1085 (1974)
14. Heuberger, C.: Inverse combinatorial optimization: a survey on problems, methods, and results. J. Comb. Optim. **8**(3), 329–361 (2004)
15. Hooker, J.N.: Inferring network flows from incomplete information with application to natural gas flows. Oak Ridge National Laboratory (1980)
16. Kuhn, H.W., Tucker, A.W.: Nonlinear programming. In: Proceedings of the Second Berkeley Symposium on Mathematical Statistics and Probability, pp. 481–492. University of California Press, Berkeley (1951)
17. Lv, Y., Duan, Y., Kang, W., Li, Z., Wang, F.Y.: Traffic flow prediction with big data: a deep learning approach. IEEE Trans. Intell. Transp. Syst. **16**(2), 865–873 (2015)
18. Manheim, M.L.: Fundamentals of Transportation Systems Analysis. MIT Press, Cambridge (1979)
19. Norton, C., Plotkin, S., Tardos, E.: Using separation algorithms in fixed dimension. J. Algorithms **13**, 79–98 (1992)
20. Shahrokhi, F., Matula, D.W.: The maximum concurrent flow problem. J. ACM **37**(2), 318–334 (1990)
21. Shiliang, S., Changshui, Z., Guoqiang, Y.: A bayesian network approach to traffic flow forecasting. IEEE Trans. Intell. Transp. Syst. **7**(1), 124–132 (2006)
22. Soyster, A.L.: Convex programming with set-inclusive constraints and applications to inexact linear programming. Oper. Res. **21**, 1154–1157 (1973)
23. Zhang, J., Liu, Z.: Calculating some inverse linear programming problems. J. Comput. Appl. Math. **72**(2), 261–273 (1996)

Characterising Chordal Contact B_0-VPG Graphs

Flavia Bonomo[1,2], María Pía Mazzoleni[3,4(✉)], Mariano Leonardo Rean[1,2],
and Bernard Ries[5]

[1] Facultad de Ciencias Exactas y Naturales, Departamento de Computación,
Universidad de Buenos Aires, Buenos Aires, Argentina
fbonomo@dc.uba.ar, marianorean@gmail.com
[2] Instituto de Investigación en Ciencias de la Computación (ICC),
CONICET-Universidad de Buenos Aires, Buenos Aires, Argentina
[3] Facultad de Ciencias Exactas, Departamento de Matemática,
Univesidad Nacional de La Plata, La Plata, Argentina
pia@mate.unlp.edu.ar
[4] CONICET, Buenos Aires, Argentina
[5] University of Fribourg, Fribourg, Switzerland
bernard.ries@unifr.ch

Abstract. A graph G is a B_0-*VPG graph* if it is the vertex intersection
graph of horizontal and vertical paths on a grid. A graph G is a *contact
B_0-VPG graph* if the vertices can be represented by interiorly disjoint
horizontal or vertical paths on a grid and two vertices are adjacent if
and only if the corresponding paths touch. In this paper, we present a
minimal forbidden induced subgraph characterisation of contact B_0-VPG
graphs within the class of chordal graphs and provide a polynomial-time
algorithm for recognising these graphs.

Keywords: Vertex intersection graphs · Contact B_0-VPG graphs
Forbidden induced subgraphs · Chordal graphs
Polynomial-time algorithm

1 Introduction

Golumbic et al. introduced in [2] the concept of *vertex intersection graphs of
paths on a grid* (referred to as *VPG graphs*). An undirected graph $G = (V, E)$ is
called a VPG graph if one can associate a path in a rectangular grid with each
vertex such that two vertices are adjacent if and only if the corresponding paths
intersect at at least one grid-point. It is not difficult to see that VPG graphs are
equivalent to the well known class of string graphs, i.e., intersection graphs of
curves in the plane (see [2]).

A particular attention was paid to the case where the paths have a limited
number of *bends* (a bend is a 90 degrees turn of a path at a grid-point). An
undirected graph $G = (V, E)$ is then called a B_k-*VPG graph*, for some integer

© Springer International Publishing AG, part of Springer Nature 2018
J. Lee et al. (Eds.): ISCO 2018, LNCS 10856, pp. 89–100, 2018.
https://doi.org/10.1007/978-3-319-96151-4_8

$k \geq 0$, if one can associate a path with at most k *bends* on a rectangular grid with each vertex such that two vertices are adjacent if and only if the corresponding paths intersect at at least one grid-point. Since their introduction in 2012, B_k-VPG graphs, $k \geq 0$, have been studied by many researchers and the community of people working on these graph classes is still growing (see [1, 2, 4–8, 11, 12]).

These classes are shown to have many connections to other, more traditional, graphs classes such as interval graphs (which are clearly B_0-VPG graphs), planar graphs (recently shown to be B_1-VPG graphs (see [12])), string graphs (as mentioned above equivalent to VPG graphs), circle graphs (shown to be B_1-VPG graphs (see [2])) and grid intersection graphs (GIG) (equivalent to bipartite B_0-VPG graphs (see [2])). Unfortunately, due to these connections, many natural problems are hard for B_k-VPG graphs. For instance, colouring is NP-hard even for B_0-VPG graphs and recognition is NP-hard for both VPG and B_0-VPG graphs [2]. However, there exists a polynomial-time algorithm for deciding whether a given chordal graph is B_0-VPG (see [4]).

A related notion to intersection graphs are *contact graphs*. Such graphs can be seen as a special type of intersection graphs of geometrical objects in which objects are not allowed to cross but only to touch each other. In the context of VPG graphs, we obtain the following definition. A graph $G = (V, E)$ is called a *contact VPG graph* if the vertices can be represented by interiorly disjoint paths (i.e., if an intersection occurs between two paths, then it occurs at one of their endpoints) on a grid and two vertices are adjacent if and only if the corresponding paths touch. If we limit again the number of bends per path, we obtain *contact B_k-VPG graphs*. These graphs have also been considered in the literature (see for instance [5, 9, 13]). It is shown in [9] that every planar bipartite graph is a contact B_0-VPG graph. Later, in [5], the authors show that every K_3-free planar graph is a contact B_1-VPG graph. The authors in [13] consider the special case in which whenever two paths touch on a grid point, this grid point has to be the endpoint of one of the paths and belong to the interior of the other path. It is not difficult to see that in this case, the considered graphs must all be planar.

In this paper, we will consider contact B_0-VPG graphs and we will present a minimal forbidden induced subgraph characterisation of contact B_0-VPG graphs restricted to chordal graphs. This characterisation allows us to derive a polynomial-time recognition algorithm for the class of chordal contact B_0-VPG graphs. Recall that chordal B_0-VPG graphs can also be recognised in polynomial time (see [4]), even though no structural characterisation of them is known so far. Our results can be considered as a first step to obtain a better understanding of contact B_0-VPG graphs and their structure.

Our paper is organised as follows. In Sect. 2, we give definitions and notations that we will use throughout the paper. We also present some first observations and results that will be useful in the remaining of the paper. In Sect. 3, we consider chordal graphs and characterise those that are contact B_0-VPG by minimal forbidden induced subgraphs. Section 4 presents a polynomial-time algorithm for recognising chordal contact B_0-VPG graphs based on the characterisation mentioned before. Finally, in Sect. 5, we present conclusions and future work.

2   Preliminaries

For concepts and notations not defined here we refer the reader to [3]. All graphs that we consider here are simple (i.e., without loops or multiple edges). Let $G = (V, E)$ be a graph. If $u, v \in V$ and $uv \notin E$, uv is called a *nonedge* of G. We write $G - v$ for the subgraph obtained by deleting vertex v and all the edges incident to v. Similarly, we write $G - e$ for the subgraph obtained by deleting edge e without deleting its endpoints.

Given a subset $A \subseteq V$, $G[A]$ stands for the *subgraph of G induced by A*, and $G \backslash A$ denotes the induced subgraph $G[V \backslash A]$.

For each vertex v of G, $N_G(v)$ denotes the *neighbourhood* of v in G and $N_G[v]$ denotes *closed neighbourhood* $N_G(v) \cup \{v\}$.

A *clique* is a set of pairwise adjacent vertices. A vertex v is *simplicial* if $N_G(v)$ is a clique. A *stable set* is a set of vertices no two of which are adjacent. The *complete graph* on n vertices corresponds to a clique on n vertices and is denoted by K_n. nK_1 stands for a stable set on n vertices. K_4-e stands for the graph obtained from K_4 by deleting exactly one edge.

Given a graph H, we say that G *contains no induced H* if G contains no induced subgraph isomorphic to H. If $\mathcal{H}$ is a family of graphs, a graph G is said to be *$\mathcal{H}$-free* if G contains no induced subgraph isomorphic to some graph belonging to $\mathcal{H}$.

Let $\mathcal{G}$ be a class of graphs. A graph belonging to $\mathcal{G}$ is called a *$\mathcal{G}$-graph*. If $G \in \mathcal{G}$ implies that every induced subgraph of G is a $\mathcal{G}$-graph, $\mathcal{G}$ is said to be *hereditary*. If $\mathcal{G}$ is a hereditary class, a graph H is a *minimal forbidden induced subgraph of $\mathcal{G}$*, or more briefly, *minimally non-$\mathcal{G}$*, if H does not belong to $\mathcal{G}$ but every proper induced subgraph of H is a $\mathcal{G}$-graph.

We denote as usual by C_n, $n \geq 3$, the *chordless cycle* on n vertices and by P_n the *chordless path* or *induced path* on n vertices. A graph is called *chordal* if it does not contain any chordless cycle of length at least four. A *block graph* is a chordal graph which is $\{K_4$-$e\}$-free.

An undirected graph $G = (V, E)$ is called a *B_k-VPG graph*, for some integer $k \geq 0$, if one can associate a path with at most k *bends* (a bend is a 90° turn of a path at a grid-point) on a rectangular grid with each vertex such that two vertices are adjacent if and only if the corresponding paths intersect at atleast one grid-point. Such a representation is called a *B_k-VPG representation*. The horizontal grid lines will be referred to as *rows* and denoted by $x_0, x_1, \dots$ and the vertical grid lines will be referred to as *columns* and denoted by $y_0, y_1, \dots$. We are interested in a subclass of B_0-VPG graphs called contact B_0-VPG. A *contact B_0-VPG representation* $\mathcal{R}(G)$ of G is a B_0-VPG representation in which each path in the representation is either a horizontal path or a vertical path on the grid, such that two vertices are adjacent if and only if the corresponding paths intersect at at least one grid-point without crossing each other and without sharing an edge of the grid. A graph is a *contact B_0-VPG graph* if it has a contact B_0-VPG representation. For every vertex $v \in V(G)$, we denote by P_v the corresponding path in $\mathcal{R}(G)$ (see Fig. 1). Consider a clique K in G. A path P_v representing a vertex $v \in K$ is called a *path of the clique K*.

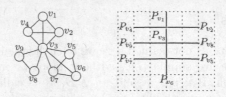

Fig. 1. A graph G and a contact B_0-VPG representation.

Let us start with some easy but very helpful observations.

Observation 1. *Let G be a contact B_0-VPG graph. Then the size of a biggest clique in G is at most 4, i.e., G is K_5-free.*

Let G be a contact B_0-VPG graph, and K be a clique in G. A vertex $v \in K$ is called an *end* in a contact B_0-VPG representation of K, if the grid-point representing the intersection of the paths of the clique K corresponds to an endpoint of P_v.

Observation 2. *Let G be a contact B_0-VPG graph, and K be a clique in G of size four. Then, every vertex in K is an end in any contact B_0-VPG representation of K.*

Next we will show certain graphs that are not contact B_0-VPG graphs and that will be part of our characterisation. Let H_0 denote the graph composed of three K_4's that share a common vertex and such that there are no other edges (see Fig. 3).

Lemma 1. *If G is a contact B_0-VPG graph, then G is $\{K_5, H_0, K_4\text{-}e\}$-free.*

Proof. Let G be a contact B_0-VPG graph. It immediately follows from Observation 1 that G is K_5-free.

Now let v, w be two adjacent vertices in G. Then, in any contact B_0-VPG representation of G, P_v and P_w intersect at a grid-point P. Clearly, every common neighbour of v and w must also contain P. Hence, v and w cannot have two common neighbours that are non-adjacent. So, G is $\{K_4\text{-}e\}$-free.

Finally, consider the graph H_0 which consists of three cliques of size four, say A, B and C, with a common vertex x. Suppose that H_0 is contact B_0-VPG. Then, it follows from Observation 2 that every vertex in H_0 is an end in any contact B_0-VPG representation of H_0. In particular, vertex x is an end in any contact B_0-VPG representation of A, B and C. In other words, the grid-point representing the intersection of the paths of each of these three cliques corresponds to an endpoint of P_x. Since these cliques have only vertex x in common, these grid-points are all distinct. But this is a contradiction, since P_x has only two endpoints. So we conclude that H_0 is not contact B_0-VPG, and hence the result follows.   □

3   Chordal Graphs

In this section, we will consider chordal graphs and characterise those that are contact B_0-VPG. First, let us point out the following important observation.

Observation 3. *A chordal contact B_0-VPG graph is a block graph.*

This follows directly from Lemma 1 and the definition of block graphs.

The following lemma states an important property of minimal chordal non contact B_0-VPG graphs that contain neither K_5 nor K_4-e.

Lemma 2. *Let G be a chordal $\{K_5, K_4\text{-}e\}$-free graph. If G is a minimal non contact B_0-VPG graph, then every simplicial vertex of G has degree exactly three.*

Proof. Since G is K_5-free, every clique in G has size at most four. Therefore, every simplicial vertex has degree at most three. Let v be a simplicial vertex of G. Assume first that v has degree one and consider a contact B_0-VPG representation of $G - v$ (which exists since G is minimal non contact B_0-VPG). Let w be the unique neighbour of v in G. Without loss of generality, we may assume that the path P_w lies on some row of the grid. Now clearly, we can add one extra column to the grid between any two consecutive vertices of the grid belonging to P_w and adapt all paths without changing the intersections (if the new column is added between column y_i and y_{i+1}, we extend all paths containing a grid-edge with endpoints in column y_i and y_{i+1} in such a way that they contain the new edges in the same row and between column y_i and y_{i+2} of the new grid, and any other path remains the same). But then we may add a path representing v on this column which only intersects P_w (adding a row to the grid and adapting the paths again, if necessary) and thus, we obtain a contact B_0-VPG representation of G, a contradiction. So suppose now that v has degree two, and again consider a contact B_0-VPG representation of $G - v$. Let w_1, w_2 be the two neighbours of v in G. Then, w_1, w_2 do not have any other common neighbour since G is $\{K_4\text{-}e\}$-free. Let P be the grid-point corresponding to the intersection of the paths P_{w_1} and P_{w_2}. Since these paths do not cross and since w_1, w_2 do not have any other common neighbour (except v), there is at least one grid-edge having P as one of its endpoints and which is not used by any path of the representation. But then we may add a path representing v by using only this particular grid-edge (or adding a row/column to the grid that subdivides this edge and adapting the paths, if the other endpoint of the grid-edge belongs to a path in the representation). Thus, we obtain a contact B_0-VPG representation of G, a contradiction. We conclude therefore that v has degree exactly three. $\square$

Let v be a vertex of a contact B_0-VPG graph G. An endpoint of its corresponding path P_v is *free* in a contact B_0-VPG representation of G, if P_v does not intersect any other path at that endpoint; v is called *internal* if there exists no representation of G in which P_v has a free endpoint. If in a representation of G a path P_v intersects a path P_w but not at an endpoint of P_w, v is called a *middle neighbour* of w.

In the following two lemmas, we associate the fact of being or not an internal vertex of G with the contact B_0-VPG representation of G.

Due to lack of space, the proof of the following lemma is ommited.

Lemma 3. *Let G be a chordal contact B_0-VPG graph and let v be a non internal vertex in G. Then, there exists a contact B_0-VPG representation of G in which all the paths representing vertices in $G - v$ lie to the left of a free endpoint of P_v (by considering P_v as a horizontal path).*

Lemma 4. *Let G be a chordal contact B_0-VPG graph. A vertex v in G is internal if and only if in every contact B_0-VPG representation of G, each endpoint of the path P_v either corresponds to the intersection of a representation of K_4 or intersects a path P_w, which represents an internal vertex w, but not at an endpoint of P_w.*

Proof. The if part is trivial. Assume now that v is an internal vertex of G and consider an arbitrary contact B_0-VPG representation of G. Let P be an endpoint of the path P_v and K the maximal clique corresponding to all the paths containing the point P. Notice that clearly v is an end in K by definition of K. First, suppose there is a vertex w in K which is not an end. Then, it follows from Observation 2 that the size of K is at most three. Without loss of generality, we may assume that P_v lies on some row and P_w on some column. If w is an internal vertex, we are done. So we may assume now that w is not an internal vertex in G. Consider $G \setminus (K \setminus \{w\})$, and let C_w be the connected component of $G \setminus (K \setminus \{w\})$ containing w. Notice that w is not an internal vertex in C_w either. By Lemma 3, there exists a contact B_0-VPG representation of C_w with all the paths lying to the left of a free endpoint of P_w. Now, replace the old representation of C_w by the new one such that P corresponds to the free endpoint of P_w in the representation of C_w (it might be necessary to refine –by adding rows and/or columns– the grid to ensure that there are no unwanted intersections) and P_w uses the same column as before. Finally, if K had size three, say it contains some vertex u in addition to v and w, then we proceed as follows. Similar to the above, there exists a contact B_0-VPG representation of C_u, the connected component of $G \setminus (K \setminus \{u\})$ containing u, with all the paths lying to the left of a free endpoint of P_u, since u is clearly not internal in C_u. We then replace the old representation of C_u by the new one such that the endpoint of P_u that intersected P_w previously corresponds to the grid-point P and P_u lies on the same column as P_w (again, we may have to refine the grid). This clearly gives us a contact B_0-VPG representation of G. But now we may extend P_v such that it strictly contains the grid-point P and thus, P_v has a free endpoint, a contradiction (see Fig. 2). So w must be an internal vertex.

Now, assume that all vertices in K are ends. If $|K| = 4$, we are done. So we may assume that $|K| \leq 3$. Hence, there is at least one grid-edge containing P, which is not used by any paths of the representation. Without loss of generality, we may assume that this grid-edge belongs to some row x_i. If P_v is horizontal, we may extend it such that it strictly contains P. But then v is not internal anymore, a contradiction. If P_v is vertical, then we may extend P_w, where $w \in K$ is such that P_w is a horizontal path. But now we are again in the first case discussed above.   □

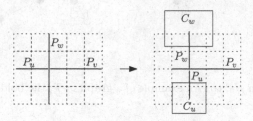

Fig. 2. Figure illustrating Lemma 4.

In other words, Lemma 4 tells us that a vertex v is an internal vertex in a chordal contact B_0-VPG graph if and only if we are in one of the following situations:

- v is the intersection of two cliques of size four (we say that v is of type 1);
- v belongs to exactly one clique of size four and in every contact B_0-VPG representation, v is a middle neighbour of some internal vertex (we say that v is of type 2);
- v does not belong to any clique of size four and in every contact B_0-VPG representation, v is a middle neighbour of two internal vertices (we say that v is of type 3).

Notice that two internal vertices of type 1 cannot be adjacent (except when they belong to a same K_4). Furthermore, an internal vertex of type 1 cannot be the middle-neighbour of some other vertex.

Let $\mathcal{T}$ be the family of graphs defined as follows. $\mathcal{T}$ contains H_0 (see Fig. 3) as well as all graphs constructed in the following way: start with a tree of maximum degree at most three and containing at least two vertices; this tree is called the *base tree*; add to every leaf v in the tree two copies of K_4 (sharing vertex v), and to every vertex w of degree two one copy of K_4 containing vertex w (see Fig. 3). Notice that all graphs in $\mathcal{T}$ are chordal.

Lemma 5. *The graphs in $\mathcal{T}$ are not contact B_0-VPG.*

Proof. By Lemma 1, the graph H_0 is not contact B_0-VPG. Consider now a graph $T \in \mathcal{T}$, $T \neq H_0$. Suppose that T is contact B_0-VPG. Denote by $B(T)$ the base tree of T and consider an arbitrary contact B_0-VPG representation of T. Consider the base tree $B(T)$ and direct an edge uv of it from u to v if the path P_v contains an endpoint of the path P_u (this way some edges might be directed both ways). If a vertex v has degree $d_B(v)$ in $B(T)$, then by definition of the family $\mathcal{T}$, v belongs to $3 - d_B(v)$ K_4's in T. Notice that P_v spends one endpoint in each of

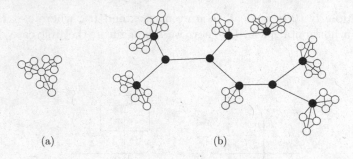

(a)   (b)

Fig. 3. (a) The graph H_0. (b) An example of a graph in $\mathcal{T}$; the bold vertices belong to the base tree.

these K_4's. Thus, any vertex v in $B(T)$ has at most $2 - (3 - d_B(v)) = d_B(v) - 1$ outgoing edges. This implies that the sum of out-degrees in $B(T)$ is at most $\sum_{v \in B(T)} (d_B(v) - 1) = n - 2$, where n is the number of vertices in $B(T)$. But this is clearly impossible since there are $n - 1$ edges in $B(T)$ and all edges are directed. □

Using Lemmas 2–5, we are able to prove the following theorem, which provides a minimal forbidden induced subgraph characterisation of chordal contact B_0-VPG graphs.

Theorem 4. *Let G be a chordal graph. Let $\mathcal{F} = \mathcal{T} \cup \{K_5, K_4\text{-}e\}$. Then, G is a contact B_0-VPG graph if and only if G is $\mathcal{F}$-free.*

Proof. Suppose that G is a chordal contact B_0-VPG graph. It follows from Lemmas 1 and 5 that G is $\mathcal{T}$-free and contains neither a K_4-e nor a K_5.

Conversely, suppose now that G is chordal and $\mathcal{F}$-free. By contradiction, suppose that G is not contact B_0-VPG and assume furthermore that G is a minimal non contact B_0-VPG graph. Let v be a simplicial vertex of G (v exists since G is chordal). By Lemma 2, it follows that v has degree three. Consider a contact B_0-VPG representation of $G - v$ and let $K = \{v_1, v_2, v_3\}$ be the set of neighbours of v in G. Since G is $\{K_4$-$e\}$-free, it follows that any two neighbours of v cannot have a common neighbour which is not in K. First suppose that all the vertices in K are ends in the representation of $G - v$. Thus, there exists a grid-edge not used by any path and which has one endpoint corresponding to the intersection of the paths $P_{v_1}, P_{v_2}, P_{v_3}$. But now we may add the path P_v using exactly this grid-edge (we may have to add a row/column to the grid that subdivides this grid-edge and adapt the paths, if the other endpoint of the grid-edge belongs to a path in the representation). Hence, we obtain a contact B_0-VPG representation of G, a contradiction.

Thus, we may assume now that there exists a vertex in K which is not an end, say v_1. Notice that v_1 must be an internal vertex. If not, there is a contact B_0-VPG representation of $G - v$ in which v_1 has a free end. Then, using similar arguments as in the proof of Lemma 4, we may obtain a representation of $G - v$ in which all vertices of K are ends. As described previously, we can add P_v to obtain a contact B_0-VPG representation of G, a contradiction. Now, by Lemma 4, v_1 must be of type 1, 2 or 3. Let us first assume that v_1 is of type 1. But then v_1 is the intersection of three cliques of size 4 and thus, G contains H_0, a contradiction. So v_1 is of type 2 or 3. But this necessarily implies that G contains a graph $T \in \mathcal{T}$. Indeed, if v_1 is of type 2, then v_1 corresponds to a leaf in $B(T)$ (remember that v_1 already belongs to a K_4 containing v in G); if v_1 is of type 3, then v_1 corresponds to a vertex of degree two in the base tree of T. Now, use similar arguments for an internal vertex w adjacent to v_1 and for which v_1 is a middle neighbour: if w is of type 2, then it corresponds to a vertex of degree two in $B(T)$; if w is of type 3, then it corresponds to a vertex of degree three in $B(T)$; if w is of type 1, it corresponds to a leaf of $B(T)$. In this last case, we stop. In the other two cases, we simple repeat the arguments for an internal vertex adjacent to w and for which w is a middle neighbour. We continue this process until we find an internal vertex of type 1 in the procedure which then gives us, when all vertices of type 1 are reached, a graph $T \in \mathcal{T}$. Since G is finite, we are sure to find such a graph T.   □

Interval graphs form a subclass of chordal graphs. They are defined as being chordal graphs not containing any asteroidal triple, i.e., not containing any three pairwise non adjacent vertices such that there exists a path between any two of them avoiding the neighbourhood of the third one. Clearly, any graph in $\mathcal{T}$ for which the base tree has maximum degree three contains an asteroidal triple. On the other hand, H_0 and every graph in $\mathcal{T}$ obtained from a base tree of maximum degree at most two are clearly interval graphs. Denote by $\mathcal{T}'$ the family consisting of H_0 and the graphs of $\mathcal{T}$ whose base tree has maximum degree at most two. We obtain the following corollary which provides a minimal forbidden induced subgraph characterisation of contact B_0-VPG graphs restricted to interval graphs.

Corollary 1. *Let G be an interval graph and $\mathcal{F}' = \mathcal{T}' \cup \{K_5, K_4\text{-}e\}$. Then, G is a contact B_0-VPG graph if and only if G is $\mathcal{F}'$-free.*

4   Recognition Algorithm

In this section, we will provide a polynomial-time recognition algorithm for chordal contact B_0-VPG graphs which is based on the characterisation given in Sect. 3. This algorithm takes a chordal graph as input and returns YES if the graph is contact B_0-VPG and, if not, it returns NO as well as a forbidden induced subgraph. We will first give the pseudo-code of our algorithm and then explain the different steps.

Input: a chordal graph $G = (V, E)$;
Output: YES, if G is contact B_0-VPG; NO and a forbidden induced subgraph, if G is not contact B_0-VPG.

1. list all maximal cliques in G;
2. if some edge belongs to two maximal cliques, return NO and $K_4 - e$;
3. if a maximal clique contains at least five vertices, return NO and K_5;
4. label the vertices such that $l(v) = $ number of K_4's that v belongs to;
5. if for some vertex v, $l(v) \geq 3$, return NO and H_0;
6. if $l(v) \leq 1 \ \forall v \in V \setminus \{w\}$ and $l(w) \leq 2$, return YES;
7. while there exists an unmarked vertex v with $2 - l(v)$ outgoing arcs incident to it, do
   7.1 mark v as internal;
   7.2 direct the edges that are currently undirected, uncoloured, not belonging to a K_4, and incident to v towards v;
   7.3 for any two incoming arcs $wv, w'v$ such that $ww' \in E$, colour ww';
8. if there exists some vertex v with more than $2 - l(v)$ outgoing arcs, return NO and find $T \in \mathcal{T}$; else return YES.

Steps 1–5 can clearly be done in polynomial time (see for example [10] for listing all maximal cliques in a chordal graph). Furthermore, it is obvious to see how to find the forbidden induced subgraph in steps 2, 3 and 5. Notice that if the algorithm has not returned NO after step 5, we know that G is $\{K_4-e, K_5, H_0\}$-free. So we are left with checking whether G contains some graph $T \in \mathcal{T}, T \neq H_0$. Since each graph $T \in \mathcal{T}$ contains at least two vertices belonging to two K_4's, it follows that if at most one vertex has label 2, G is $\mathcal{T}$-free (step 6), and thus we conclude by Theorem 4 that G is contact B_0-VPG.

During step 7, we detect those vertices in G that, in case G is contact B_0-VPG, must be internal vertices (and mark them as such) and those vertices w that are middle neighbours of internal vertices v (we direct the edges wv from w to v). Furthermore, we colour those edges whose endpoints are middle neighbours of a same internal vertex.

Consider a vertex v with $2 - l(v)$ outgoing arcs. If a vertex v has $l(v) = 2$, then, in case G is contact B_0-VPG, v must be an internal vertex (see Lemma 4). This implies that any neighbour of v, which does not belong to a same K_4 as v, must be a middle neighbour of v. If $l(v) = 1$, this means that v belongs to one K_4 and is a middle neighbour of some internal vertex. Thus, by Lemma 4 we know that v is internal. Similarly, if $l(v) = 0$, this means that v is a middle neighbour of two distinct internal vertices. Again, by Lemma 4 we conclude that v is internal. Clearly, step 7 can be run in polynomial time.

So we are left with step 8, i.e., we need to show that G is contact B_0-VPG if and only if there exists no vertex with more than $2 - l(v)$ outgoing arcs. First notice that only vertices marked as internal have incoming arcs. Furthermore, notice that every maximal clique of size three containing an internal vertex has two directed edges of the form $wv, w'v$ and the third edge is coloured, where v is the first of the three vertices that was marked as internal. This is because the graph is (K_4-e)-free and the edges of a K_4 are neither directed nor coloured.

Thanks to the marking process described in step 7 and the fact that only vertices marked as internal have incoming arcs, we can make the following observation.

Observation 5. *Every vertex marked as internal in step 7 has either label 2 or is the root of a directed induced tree (directed from the root to the leaves) where the root w has degree $2 - l(w)$ and every other vertex v has degree $3 - l(v)$ in that tree, namely one incoming arc and $2 - l(v)$ outgoing arcs.*

Let us show that the tree mentioned in the previous observation is necessarily induced. Suppose there is an edge not in the tree that joins two vertices of the tree. Since the graph is a block graph, the vertices in the resulting cycle induce a clique, so in particular there is a triangle formed by two edges of the tree and an edge not in the tree. But, as observed above, in every triangle of G having two directed edges, the edges point to the same vertex (and the third edge is coloured, not directed). Since no vertex in the tree has in-degree more than one, this is impossible.

Based on the observation, it is clear now that if a vertex has more than $2 - l(v)$ outgoing arcs, then that vertex is the root of a directed induced tree (directed from the root to the leaves), where every vertex v has degree $3 - l(v)$, i.e., a tree that is the base tree $B(T)$ of a graph $T \in \mathcal{T}$. Indeed, notice that every vertex v in a base tree has degree $3 - l(v)$. The fact that tree is induced can be proved the same way as above. This base tree can be found by a breadth-first search from a vertex having out-degree at least $3 - l(v)$, using the directed edges. Thanks to the labels, representing the number of K_4's a vertex belongs to, it is then possible to extend the $B(T)$ to an induced subgraph $T \in \mathcal{T}$. This can clearly be implemented to run in polynomial time.

To finish the proof that our algorithm is correct, it remains to show that if G contains an induced subgraph in $\mathcal{T}$, then the algorithm will find a vertex with at least $3 - l(v)$ outgoing arcs. This, along with Theorem 4, says that if the algorithm outputs YES then the graph is contact B_0-VPG (given that the detection of K_5, K_4-e and H_0 is clear). Recall that we know that G is a block graph after step 2. Notice that if a block of size 2 in a graph of $\mathcal{T}$ is replaced by a block of size 4, we obtain either H_0 or a smaller graph in $\mathcal{T}$ as an induced subgraph. Moreover, adding an edge to a graph of $\mathcal{T}$ in such a way that now contains a triangle, then we obtain a smaller induced graph in $\mathcal{T}$. Let G be a block graph with no induced K_5 or H_0. By the observation above, if G contains a graph in $\mathcal{T}$ as induced subgraph, then G contains one, say T, such that no edge of the base tree $B(T)$ is contained in a K_4 in G, and no triangle of G contains two edges of $B(T)$. So, all the edges of $B(T)$ are candidates to be directed or coloured.

In fact, by step 7 of the algorithm, every vertex of $B(T)$ is eventually marked as internal, and every edge incident with it is either directed or coloured, unless the algorithm ends with answer NO before. Notice that by the observation about the maximal cliques of size three and the fact that no triangle of G contains two edges of $B(T)$, if an edge vw of $B(T)$ is coloured, then both v and w have an outgoing arc not belonging to $B(T)$. So, in order to obtain a lower bound on the out-degrees of the vertices of $B(T)$ in G, we can consider only the arcs of $B(T)$ and we can consider the coloured edges as bidirected edges. With an argument similar to the one in the proof of Lemma 5, at least one vertex has out-degree at least $3 - l(v)$.

5   Conclusions and Future Work

In this paper, we presented a minimal forbidden induced subgraph characterisation of chordal contact B_0-VPG graphs and provide a polynomial-time recognition algorithm based on that characterisation. In order to obtain a better understanding of what contact B_0-VPG graphs look like, the study of contact B_0-VPG graphs within other graph classes is needed. It would also be interesting to investigate contact B_0-VPG graph from an algorithmic point of view and analyse for instance the complexity of the colouring problem or the stable set problem in that graph class.

References

1. Alcón, L., Bonomo, F., Mazzoleni, M.P.: Vertex intersection graphs of paths on a grid: characterization within block graphs. Graphs Comb. **33**(4), 653–664 (2017)
2. Asinowski, A., Cohen, E., Golumbic, M., Limouzy, V., Lipshteyn, M., Stern, M.: Vertex intersection graphs of paths on a grid. J. Graph Algorithms Appl. **16**, 129–150 (2012)
3. Bondy, J.A., Murty, U.S.R.: Graph Theory. Springer, New York (2007)
4. Chaplick, S., Cohen, E., Stacho, J.: Recognizing some subclasses of vertex intersection graphs of 0-Bend paths in a grid. In: Kolman, P., Kratochvíl, J. (eds.) WG 2011. LNCS, vol. 6986, pp. 319–330. Springer, Heidelberg (2011). https://doi.org/10.1007/978-3-642-25870-1_29
5. Chaplick, S., Ueckerdt, T.: Planar graphs as VPG-graphs. J. Graph Algorithms Appl. **17**(4), 475–494 (2013)
6. Cohen, E., Golumbic, M.C., Ries, B.: Characterizations of cographs as intersection graphs of paths on a grid. Discrete Appl. Math. **178**, 46–57 (2014)
7. Cohen, E., Golumbic, M.C., Trotter, W.T., Wang, R.: Posets and VPG graphs. Order **33**(1), 39–49 (2016)
8. Felsner, S., Knauer, K., Mertzios, G.B., Ueckerdt, T.: Intersection graphs of L-shapes and segments in the plane. Discrete Appl. Math. **206**, 48–55 (2016)
9. de Fraysseix, H., Ossona de Mendez, P., Pach, J.: Representation of planar graphs by segments. Intuitive Geom. **63**, 109–117 (1991)
10. Galinier, P., Habib, M., Paul, C.: Chordal graphs and their clique graphs. In: Nagl, M. (ed.) WG 1995. LNCS, vol. 1017, pp. 358–371. Springer, Heidelberg (1995). https://doi.org/10.1007/3-540-60618-1_88
11. Golumbic, M.C., Ries, B.: On the intersection graphs of orthogonal line segments in the plane: characterizations of some subclasses of chordal graphs. Graphs Comb. **29**(3), 499–517 (2013)
12. Gonçalves, D., Insenmann, L., Pennarum, C.: Planar Graphs as L-intersection or L-contact graphs, arXiv:1707.08833v2 (2017)
13. Nieke, A., Felsner, S.: Vertex contact representations of paths on a grid. J. Graph Algorithms Appl. **19**(3), 817–849 (2015)

Approximating the Caro-Wei Bound
for Independent Sets in Graph Streams

Graham Cormode, Jacques Dark, and Christian Konrad$^{(\boxtimes)}$

Department of Computer Science, Centre for Discrete Mathematics
and its Applications (DIMAP), University of Warwick, Coventry, UK
{g.cormode,j.dark,c.konrad}@warwick.ac.uk

Abstract. The Caro-Wei bound states that every graph $G = (V, E)$ contains an independent set of size at least $\beta(G) := \sum_{v \in V} \frac{1}{\deg_G(v)+1}$, where $\deg_G(v)$ denotes the degree of vertex v. Halldórsson et al. [1] gave a randomized one-pass streaming algorithm that computes an independent set of expected size $\beta(G)$ using $O(n \log n)$ space. In this paper, we give streaming algorithms and a lower bound for approximating the Caro-Wei bound itself.

In the edge arrival model, we present a one-pass c-approximation streaming algorithm that uses $O(\overline{d} \log(n)/c^2)$ space, where $\overline{d}$ is the average degree of G. We further prove that space $\Omega(\overline{d}/c^2)$ is necessary, rendering our algorithm almost optimal. This lower bound holds even in the *vertex arrival model*, where vertices arrive one by one together with their incident edges that connect to vertices that have previously arrived. In order to obtain a poly-logarithmic space algorithm even for graphs with arbitrarily large average degree, we employ an alternative notion of approximation: We give a one-pass streaming algorithm with space $O(\log^3 n)$ in the vertex arrival model that outputs a value that is at most a logarithmic factor below the true value of β and no more than the maximum independent set size.

1 Introduction

For very large graphs, the model of streaming graph analysis, where edges are observed one by one, is a useful lens. Here, we assume that the graph of interest is too large to store in full, but some representative summary is maintained incrementally. We seek to understand how well different problems can be solved in this model, in terms of the size of the summary, the time taken to process each edge and answer a query, and the accuracy of any approximation obtained. Variants arise in the model depending on whether edges can be removed as well as added, and whether edges arrive grouped in some order, and so on.

The work of GC is supported in part by European Research Council grant ERC-2014-CoG 647557, The Alan Turing Institute under EPSRC grant EP/N510129/1 the Yahoo Faculty Research and Engagement Program and a Royal Society Wolfson Research Merit Award; JD is supported by a Microsoft Research Studentship; and CK by EPSRC grant EP/N011163/1.

© Springer International Publishing AG, part of Springer Nature 2018
J. Lee et al. (Eds.): ISCO 2018, LNCS 10856, pp. 101–114, 2018.
https://doi.org/10.1007/978-3-319-96151-4_9

Independent Sets and the Caro-Wei Bound. We study questions pertaining to *independent sets* within graphs. Independent sets play a fundamental role in graph theory, and have many applications in optimization and scheduling problems. Given a graph, an independent set is a set of nodes such that there is no edge between any pair. One important objective is to find a *maximum independent set*, i.e., an independent set of maximum cardinality. This is a challenging task even in the offline setting: Computing a maximum independent set is NP-hard on general graphs [2], and remains hard to approximate within a factor of $n^{1-\epsilon}$ for any $\epsilon > 0$ [3,4].

Despite this strong intractability result, there is substantial interest in computing independent sets of non-trivial sizes. The best polynomial time algorithm for maximum independent set was given by Feige and has an approximation factor of $O(\frac{n \log^2(\log n)}{\log^3 n})$ [5]. Since no substantial improvements on this bound are possible, many works give approximation guarantees or absolute bounds on the size of an independent set in terms of the degrees of the vertices of the input graph. For example, it is known that the Greedy algorithm, which iteratively picks a node of minimum degree and then removes all neighbors from consideration, has an approximation factor of $(\Delta+2)/3$, where Δ is the maximum degree of the input graph [6]. The Greedy algorithm also achieves the *Caro-Wei* bound [7,8], which is the focus of this paper: Caro [9] and Wei [7] independently proved that every graph G contains an independent set of size

$$\beta(G) := \sum_{v \in V} \frac{1}{\deg_G(v) + 1}. \tag{1}$$

The quantity $\beta(G)$ is an attractive bound. It is known that it gives polylogarithmic approximation guarantees on graphs that are of *polynomially bounded-independence* [10], which means (informally) that the size of a maximum independent set in an r-neighborhood around a node is bounded in size by a polynomial in r. For example, on unit disc graphs, which are of polynomially bounded-independence, $\beta(G)$ is a $O\left((\frac{\log n}{\log \log n})^2\right)$ approximation to the size of a maximum independent set. In distributed computing, the Caro-Wei bound is particularly interesting, since an independent set of size $\beta(G)$ can be computed in a single communication round [10]. Very relevant to the present work is a result by Halldórsson et al. [1], who showed that an independent set of expected size $\beta(G)$ can be computed space efficiently in the data streaming model.

Independent Sets in the Streaming Model. Due to the aforementioned computational hardness of the maximum independent set problem, every streaming algorithm that approximates a maximum independent set within a polynomial factor n^δ, for any constant $\delta < 1$, requires exponential time, unless $\mathsf{P} = \mathsf{NP}$. By sampling a subset of vertices $V' \subseteq V$, storing all edges between vertices V' while processing the stream, and outputting a maximum independent set in the subgraph induced by V' (using an exponential time computation), it is possible to obtain a randomized one-pass c-approximation streaming algorithm for maximum independent set with $\tilde{O}(\frac{n^2}{c^2})$ space[1]. Halldórsson et al. [11] showed that this is best possible: They proved that even for the seemingly simpler task of approximating the size of a maximum independent set, every c-approximation streaming algorithm requires $\tilde{\Omega}(\frac{n^2}{c^2})$ space. In order to circumvent both the large space lower bound and the exponential time computations required, in a different work, Halldórsson et al. [1] relaxed the desired quality guarantee and gave one-pass streaming algorithms for computing independent sets with expected sizes that match the Caro-Wei bound. These algorithms use $O(n \log n)$ space and have constant update times.

Approximating the Solution Size. In this paper, we ask whether we can reduce the space requirements of $O(n \log n)$ even further, if, instead of computing an independent set whose size is bounded by the Caro-Wei bound, we approximate the size of such an independent set, i.e., the Caro-Wei bound itself. This objective ties in with a recent trend in graph streaming algorithms: Since many combinatorial objects such as matchings or independent sets may be of size $\Omega(n)$, streaming algorithms that output such objects require at least this amount of space. Consequently, many recent papers ask whether the task of approximating the output size is easier than outputting the object itself. As previously mentioned, this is not the case for the maximum independent set problem, where the space complexity of both computing a c-approximate independent set and finding a c-approximation to the size of a maximum independent set is $\tilde{\Theta}(\frac{n^2}{c^2})$ [11]. For the maximum matching problem, it is known that space $\Omega(n/c)$ is needed for computing a c-approximation, but space $\tilde{O}(n/c^2)$ is sufficient for outputting a c-approximation to the maximum matching size [12]. However, for graphs with arboricity c, the size of a maximum matching can even be approximated within a factor of $O(c)$ using $O(c \log^2 n)$ space [13]. Another example is a work by Cabello and Pérez-Lantero [14], which gives a polylogarithmic space streaming algorithm that approximates the maximum size of an independent set of intervals within a constant factor, while storing such a set would require $\Omega(n)$ space.

Starting Point: Frequency Moments. Approximating $\beta(G)$ is essentially the same as approximating the -1 (negative) frequency moment (or the harmonic mean) of a frequency vector derived from the graph stream. The pth frequency moment of a stream of n different items where item i appears f_i times is defined by $F_p = \sum_i |f_i|^p$. Approximating the frequency moments is one of the most studied problems in the data streaming literature, starting in 1996 with the seminal work of Alon, Matias and Szegedy [15]. It is known that all finite positive frequency moments can be approximated with sublinear space (see Woodruff's article [16] for an overview of the problem). Braverman and Chestnut [17] studied the problem of approximating the negative frequency moments, which turn out to be much harder to approximate: Computing a $(1 + \epsilon)$-aproximation to the harmonic mean in one pass requires $\Omega(n)$ space if the length of the input sequence

---

[1] We use the notation $\tilde{O}(.), \tilde{\Theta}(.)$ and $\tilde{\Omega}(.)$, which correspond to $O(.), \Theta(.)$ and $\Omega(.)$, respectively, where all polylogarithmic factors are ignored.

is $\Omega(n^2)$. While this lower bound is designed for arbitrary frequency vectors, it can be embedded into a graph with $\Theta(n^2)$ edges so that frequencies correspond to vertex degrees. This implies we cannot find an algorithm to approximate the Caro-Wei bound within a factor of $1 + \epsilon$ which guarantees that the space used will always be sublinear.

Our Results. Despite these lower bounds, we are able to provide upper and lower bounds that improve on those stated above. The key advance is that they incorporate a dependence on the target quantity, $\beta(G)$. This means when this quantity is suitably big (as is the case in many graphs of interest), we can in fact guarantee sublinear space. In more detail, we proceed as follows. Since in our setting the frequency vector is derived from the degrees of the vertices of the input graph, we can exploit the properties of the underlying graph. In our first result, we relate the space complexity of our algorithm to a given lower bound γ on $\beta(G)$. A meaningful lower bound γ is easy to obtain: It is easy to see that the Turán bound [18] for independent sets, which shows that $n/(\bar{d}+1)$ is a lower bound on the size of a maximum independent set, is also a lower bound on $\beta(G)$, where $\bar{d}$ is the average degree of the input graph. Our first result is then a one-pass randomized streaming algorithm with space $O(\frac{n \log n}{\gamma c^2})$ that approximates $\beta(G)$ within a factor of c with high probability (Theorem 1). Using $\gamma = \frac{n}{\bar{d}+1}$, the space becomes $O(\frac{\bar{d} \log n}{c^2})$, which is polylogarithmic for graphs of constant average degree such as planar graphs or bounded arboricity graphs. The algorithm can also give a $(1 + \epsilon)$-approximation using $O(\frac{n \log n}{\gamma \epsilon^2})$ space.

We prove that our algorithm is best possible (up to poly-log factors). Via a reduction from a hard problem in communication complexity, we show that every p-pass streaming algorithm for computing a c-approximation to $\beta(G)$ requires $\Omega(\frac{n}{\beta(G)c^2 p})$ space (Theorem 4). This lower bound also holds in the *vertex arrival order*, where vertices arrive one by one together with those incident edges that connect to vertices that have previously arrived (see Sect. 2 for a more precise definition). Our lower bound is more general than the lower bound from Braverman and Chestnut [17], since their lower bound only holds for $(1 + \epsilon)$-approximation algorithms and does not establish a dependency on the output quantity, i.e., the -1-negative frequency moment. Furthermore, their bound was not developed in the graphical setting where frequencies are derived from the vertex degrees.

Our lower bound shows that the promise that the input stream is in vertex arrival order is not helpful for approximating $\beta(G)$. However, if we regard the task of approximating $\beta(G)$ as obtaining a (hopefully large) lower bound on the size of a maximum independent set of the input graph, then any value sandwiched between $\beta(G)$ and the maximum independent set size would be equally suitable (or even superior). In the vertex arrival setting, we give a randomized one-pass streaming algorithm with space $O(\log^3 n)$, which outputs a value β' with $\beta' = \Omega(\beta(G)/\log n)$ and β' is at most the maximum independent set size (Theorem 2). Since the Caro-Wei bound is a polylogarithmic approximation to the maximum independent set size in polynomially bounded-independence graphs, a corollary of our result is that the maximum independent set size can be approximated within a polylogarithmic factor in polylogarithmic space in poly-nomially bounded-independence graphs (e.g., the approximation factor obtained on unit disc graphs is $O(\frac{\log^3 n}{(\log \log n)^2})$).

Our focus is on streaming models where edges only arrive. We briefly comment on when our results generalize to models which allow deletions following each algorithm.

Further Related Work. There has been substantial interest in the topic of streaming algorithms for graphs in the last two decades. Indeed, the introduction of the streaming model focused on graph problems [19]. McGregor provides a survey that outlines key results on well-studied problems such as finding sparsifiers, identifying connectivity structure, and building spanning trees and matchings [20].

Our work is the first to consider the graph frequency moments (or degree moments) in the data streaming model. They have previously been considered in the property testing literature [21–23], where the input graph can only be queried a sublinear number of times. There are important connections between the degree moments and network science and various other disciplines. For details we refer the reader to [22].

2   Preliminaries

The Independent Set problem is most naturally modeled as a problem over graphs $G = (V, E)$. A set $U \subseteq V$ is an independent set if for all pairs $u, w \in U$ we have $\{u, w\} \notin E$, i.e. there is no edge between u and w. Let $\alpha(G)$ be the *independence number* of graph G, i.e., the size of a maximum independent set in G.

We consider graphs defined by streams of edges. That is, we observe a sequence of unordered pairs $\{u, w\}$ which collectively define the (current) edge set E. We do not require V to be given explicitly, but take it to be defined implicitly as the union of all nodes observed in the stream. In the (arbitrary, possibly adversarial) edge arrival model, no further constraints are placed on the order in which the edges arrive. In the vertex arrival model, there is a total ordering on the vertices $\prec$ which is revealed incrementally. Given the final graph G, node v "arrives" so that all edges $\{u, v\} \in E$ such that $u \prec v$ are presented sequentially before the next vertex arrives. We do not assume that there is any further ordering among this group of edges.

3   Algorithm in the Edge-Arrival Model

In this section, we suppose that a lower bound $\gamma \leq \beta(G)$ is known. For example, $\gamma = \frac{n}{\bar{d}+1}$ is a suitable bound, where $\bar{d}$ is the average degree of the input graph. If no such bound is known, then the algorithm can be used with the trivial lower bound $\gamma = 1$.

We give an algorithm that computes an estimate B which approximates $\beta(G)$ within a factor of $1+\epsilon$ with probability at least $2/3$. By running $\Theta(\log n)$ copies of our algorithm and returning the median of the computed estimates, the success probability can be increased to $1 - \frac{1}{n^c}$, for any constant c.

The estimator B is computed as follows: First, take a uniform random sample $S \subseteq V$ such that every vertex $v \in V$ is included in S with probability $p = \frac{3}{\epsilon^2 \gamma}$. Then, while processing the stream, compute $\deg_G(v)$, for every vertex $v \in S$. Let $x_v \in \{0, 1\}$ be the indicator variable of the event $v \in S$. Then B is computed by $B = \frac{1}{p} \sum_{v \in V} a_v x_v$, where $a_v := \frac{1}{\deg_G(v)+1}$.

We first show that B is an unbiased estimator and we bound the variance of B.

Lemma 1. *Let B be the estimate computed as above. Then:*

$$\mathbb{E}[B] = \beta(G), \ and \ \mathbb{V}[B] < \frac{1}{p} \sum_{v \in V} a_v^2 \leq \frac{1}{p}\beta(G).$$

The proof is a fairly straightforward calculation of expectations, and is deferred to the full version of this paper.

Theorem 1. *Let $\gamma \leq \beta(G)$ be a given lower bound on $\beta(G)$. Then, there is a randomized one-pass approximation streaming algorithm in the edge arrival model with space $O\left(\frac{n \log n}{\gamma \epsilon^2}\right)$ that approximates $\beta(G)$ within a factor of $1 + \epsilon$, with high probability.*

Proof. By Chebyshev's inequality, the error probability of our estimate is at most $1/3$, since (recall that $p = \frac{3}{\epsilon^2 \gamma}$)

$$\mathbb{P}[|B - \beta(G)| \geq \epsilon\beta(G)] \leq \frac{\mathbb{V}[B]}{\epsilon^2 \beta(G)^2} < \frac{1}{p\epsilon^2 \beta(G)} \leq \frac{1}{3} .$$

By a standard Chernoff bounds argument, running $\Theta(\log n)$ copies of our algorithm and returning the median of the computed estimates allows us to obtain an error probability of $O(n^{-c})$, for any constant c. □

Remarks: Observe that the previous theorem also holds for large values of ϵ (e.g. $\epsilon = n^\delta$, for some $\delta > 0$). The core of our algorithm is to sample nodes with a fixed probability and to count their degree. This can easily be achieved in the model where edges are also deleted (the turnstile streaming model) without any further data structures, so our results hold in that stream model also.

4   Algorithm in the Vertex-Arrival Model

Let $v_1, \ldots, v_n$ be the order in which the vertices appear in the stream. Let $G_i = G[\{v_1, \ldots, v_i\}]$ be the subgraph induced by the first i vertices. Let $n_{d,i} := |\{v \in V(G_i) : \deg_{G_i}(v) \leq d\}|$ be the number of vertices of degree at most d in G_i, and let $n_d = \max_i n_{d,i}$.

Algorithm 1. Algorithm $\text{DEGTEST}(d, \epsilon)$

Require: Degree bound d, ϵ for a $1 + \epsilon$ approximation
1:   $p \leftarrow 1$, $S \leftarrow \varnothing$, $m \leftarrow 0$, $\epsilon' \leftarrow \epsilon/2$, $c \leftarrow \frac{28}{\epsilon'^2}$
2:   **while** stream not empty **do** {The current subgraph is G_i}
3:      $v \leftarrow$ next vertex in stream
4:      **if** $\text{COIN}(p)$ **then**
5:         $S \leftarrow S \cup \{v\}$ {Sample vertex with probability p}
6:         Update degrees of vertices in S, i.e., for every $u \in S$ adjacent to v, increment its degree {This ensures that for every $u \in S$ $\deg_{G_i}(u)$ is known}
7:         Remove every vertex $u \in S$ from S if $\deg_{G_i}(u) > d$
8:      **if** $p = 1$ **then**
9:         $m \leftarrow \max\{m, |S|\}$
10:    **if** $|S| = c\log(n)$ **then**
11:       $m \leftarrow c\log(n)/p$
12:       Remove each element from S with probability $\frac{1}{1+\epsilon'}$
13:       $p \leftarrow p/(1 + \epsilon')$
14: **return** m

We first give an algorithm, $\text{DEGTEST}(d, \epsilon)$, which with high probability returns a $(1 + \epsilon)$-approximation of n_d using $O(\frac{1}{\epsilon^2} \log^2 n)$ bits of space. In the description of the algorithm, we suppose that we have a random function COIN: $[0, 1] \rightarrow \{\texttt{false}, \texttt{true}\}$ such that $\text{COIN}(p) = \texttt{true}$ with probability p and $\text{COIN}(p)$ $= \texttt{false}$ with probability $1 - p$. Furthermore, the outputs of repeated invocations of COIN are independent.

Algorithm $\text{DEGTEST}(d, \epsilon)$ maintains a sample S of at most $c\log n$ vertices. It ensures that all vertices $v \in S$ have degree at most d in the current graph G_i (notice that $\deg_{G_i}(v) \leq \deg_{G_j}(v)$, for every $j \geq i$). Initially, $p = 1$, and all vertices of degree at most d are stored in S. Whenever S reaches the limiting size of $c\log n$, we downsample S by removing every element of S with probability $\frac{1}{1+\epsilon'}$ and update $p \leftarrow p/(1 + \epsilon')$. This guarantees that throughout the algorithm S constitutes a uniform random sample of all vertices of degree at most d in G_i.

The algorithm outputs $m \leftarrow c\log(n)/p$ as the estimate for n_d, where p is the smallest value of p that occurs during the course of the algorithm. It is updated whenever S reaches the size $c\log n$, since S is large enough at this moment to be used as an accurate predictor for $n_{d,i}$, and hence also for n_d.

Lemma 2. *Let* $0 < \epsilon \leq 1$. $\text{DEGTEST}(d, \epsilon)$ *(Algorithm 1) approximates* n_d *within a factor* $1 + \epsilon$ *with high probability, i.e.,*

$$\frac{n_d}{1 + \epsilon} \leq \text{DEGTEST}(d, \epsilon) \leq (1 + \epsilon)n_d,$$

and uses $O(\frac{1}{\epsilon^2} \log^2 n)$ *bits of space.*

For space reasons, we defer the proof of this Lemma to the full version of this paper and only give a brief outline here. We say that the algorithm is in phase i if the current value of p is $p = 1/(1 + \epsilon')^i$. We focus on the key

---

Algorithm 2. Algorithm in the Vertex-arrival Order

---

for every $i \in \{0, 1, \ldots, \lceil \log n \rceil\}$, run in parallel:

$\tilde{n}_{2^i} = \text{DegTest}(2^i, 1/2)$

end for

return $\max \left\{ \dfrac{\tilde{n}_{2^i}}{2(2^i + 1)} : i \in \{0, 1, \ldots, \lceil \log n \rceil\} \right\}$

---

moments $(j_i)_{i \geq 0}$ in the algorithm, where j_i is the smallest index j such that $n_{d,j} \geq c \log n (1 + \epsilon')^i (1 + \epsilon'/2)$. The core of our proof is to show that after iteration j_i, the algorithm is in phase $i + 1$ with high probability. For an intuitive justification of this claim, suppose that this is not true and the algorithm was in phase at most i after iteration j_i. Then, since S is a uniform sample, we expect the size of S to be at least $n_{d,j_i}/(1+\epsilon')^i \geq c \log n (1 + \epsilon'/2)$, which however would have triggered the downsampling step in Line 10 of the algorithm and would have transitioned the algorithm into the next phase. On the other hand, suppose that the algorithm was in phase at least $i + 2$ after iteration j_i. In iteration k when the algorithm transitioned into phase $i + 2$, the number of nodes $n_{d,k}$ of degree at most d was bounded by $n_{d,k} \leq n_{d,j_i}$. The transition from phase $i + 1$ to $i + 2$ would thus not have occurred, since the expected size of S in iteration k was at most $n_{d,j_i}/(1 + \epsilon')^{i+1} \leq c \log n (1 + \epsilon'/2)/(1 + \epsilon')$. In our proof, we make this intuition formal and conduct an induction over the phases. Let $j_{\tilde{i}}$ be the largest occurring value of j_i. Then $n_{d,j_{\tilde{i}}}$ is a good approximation of n_d and, as argued above, the algorithm is in phase $\tilde{i} + 1$ after iteration $j_{\tilde{i}}$. Using the largest occurring value of p, we can thus estimate n_d.

Next, we run multiple copies of DegTest in order to obtain our main algorithm, Algorithm 2. This consists of making multiple parallel guesses of the parameter d as powers of 2, and taking the guess which provides the maximum bound.

Theorem 2. *Let γ be the output of Algorithm 2. Then, with high probability:*

1. $\gamma = \Omega(\frac{\beta(G)}{\log n})$, *and*
2. $\gamma \leq \alpha(G)$.

Furthermore, the algorithm uses space $O(\log^3 n)$ bits.

Proof. For $0 \leq i < \lceil \log(n) \rceil$, let $V_i \subseteq V$ be the subset of vertices with $\deg_G(v) \in \{2^i, 2^{i+1} - 1\}$. Then,

$$\beta(G) = \sum_{v \in V} \frac{1}{\deg_G(v) + 1} = \sum_i \sum_{v \in V_i} \frac{1}{\deg_G(v) + 1} \leq \sum_i \frac{|V_i|}{2^i + 1}.$$

Let $i_{\max} := \arg\max_i \frac{|V_i|}{2^i + 1}$. Then, we further simplify the previous inequality:

$$\beta(G) \leq \cdots \leq \sum_i \frac{|V_i|}{2^i + 1} \leq \lceil \log(n) \rceil \cdot \frac{|V_{i_{\max}}|}{2^{i_{\max}} + 1} \leq \lceil \log(n) \rceil \cdot \frac{|V_{\leq i_{\max}}|}{2^{i_{\max}} + 1}. \quad (2)$$

where $V_{\leq i} = \cup_{j \leq i} V_j$. Let $d_{\max} = 2^{i_{\max}}$. Since $|V_{i_{\max}}| \leq n_{d_{\max}}$ and $\tilde{n}_{d_{\max}} =$ DEGTEST$(d_{\max}, \frac{1}{2})$ is a 1.5-approximation to $n_{d_{\max}}$, we obtain $\gamma = \Omega(\frac{\beta(G)}{\log n})$, which proves Item 1.

Concerning Item 2, notice that for every i and d, it holds

$$\alpha(G) \geq \alpha(G_i) \geq \beta(G_i) = \sum_{v \in V(G_i)} \frac{1}{\deg_{G_i}(v) + 1}$$

$$\geq \sum_{v \in V(G_i):\deg_{G_i}(v) \leq d} \frac{1}{\deg_{G_i}(v) + 1} \geq \frac{n_{i,d}}{d + 1},$$

and, in particular, the inequality holds for $n_{d_{\max}} = n_{i_{\max}, d_{\max}}$. Since the algorithm returns a value bounded by $\frac{\tilde{n}_{d_{\max}}}{2 \cdot (d_{\max} + 1)}$, and $\tilde{n}_{d_{\max}}$ constitutes a 1.5-approximation of $n_{d_{\max}}$, Item 2 follows.

Concerning the space requirements, the algorithm runs O$(\log n)$ copies of Algorithm 1 which itself requires O$(\log^2 n)$ bits of space. $\square$

Remark: On first glance, it may appear that our algorithm would translate to the turnstile model where edges can be deleted: the central step of sampling vertices at varying probabilities is reminiscent of steps from L_0 sampling algorithms [24]. However, there are a number of obstacles to achieving this. First, the algorithm computes a maximum over the estimate $\beta(G_i)$ for intermediate graphs G_i. This is correct when nodes and edges only arrive, but is not correct when a graph may be subject to deletions. We therefore leave the question of giving comparable bounds under the turnstile stream model as an open problem.

5   Space Lower Bound

Our lower bound follows from a reduction using a well-known hard problem from communication complexity. Let DISJ_n refer to the *two-party set disjointness problem* for inputs of size n. In this problem we have two parties, Alice and Bob. Alice knows $X \subset [n]$, while Bob knows $Y \subset [n]$. Alice and Bob must exchange messages until they both know whether $X \cap Y = \emptyset$ or $X \cap Y \neq \emptyset$.

Using $R(\text{DISJ}_n)$ to refer to the randomised (bounded error probability) communication complexity of DISJ_n, the following theorem is known.

Theorem 3 (Kalyanasundaram and Schnitger [25]). $R(\text{DISJ}_n) \in \Omega(n)$.

To get our lower bound, we will show a reduction from randomised set disjointness to randomised c-approximation of $\beta(G)$.

Theorem 4. *Every randomized constant error p-pass streaming algorithm that approximates $\beta(G)$ within a factor of c uses space $\Omega\left(\frac{n}{\beta(G)c^2 p}\right)$, even if the input stream is in vertex arrival order.*

Proof. Let $\text{ALG}_{c,n}$ be any streaming algorithm that performs p passes over a vertex arrival stream of an n-vertex graph G and returns a c-approximation of $\beta(G)$ with probability $\frac{2}{3}$. Suppose we are given an instance of DISJ_k. We will construct a graph G from X and Y which we can use to tell whether $X \cap Y = \emptyset$ by checking a c-approximation of $\beta(G)$.

Let $z \geq 2$ be an arbitrary integer. Set $q = 2zc^2$ and $a = kq$. Let $G = (V, E)$, where V is partitioned into disjoint subsets A, B, C, and U_i for $i \in [k]$. These are of size $|A| = |B| = a$, $|C| = z$, and $|U_i| = q$. So $n := |V| = kq + 2a + z = 3kq + z = z(6kc^2 + 1)$. Thus, $k \in \Theta(\frac{n}{zc^2})$ holds.

First consider the set of edges E_0 consisting of all $\{u, v\}$ with $u, v \in A \cup B$, $u \neq v$. Setting $E = E_0$ makes $A \cup B$ a clique, while all other vertices remain isolated.

Figure 1a shows this initial configuration. For clarity, we represent the structure using super-nodes and super-edges. A super-node is a subset of V (in this case we use A, B, C, and each U_i). Between the super-nodes, we have super-edges representing the existence of all possible edges between constituent vertices. So a super-edge between super-nodes Z_1 and Z_2 represents that $\{z_1, z_2\} \in E$ for every $z_1 \in Z_1$ and $z_2 \in Z_2$. The lack of a super-edge between Z_1 and Z_2 indicates that none of these $\{z_1, z_2\}$ are in E.

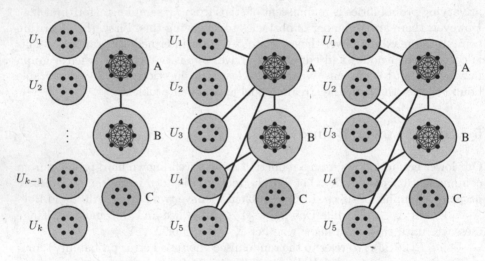

(a) Initial configuration.   (b) Example with $X = \{2, 4\}$ (c) Example with $X = \{2, 4\}$
and $Y = \{1, 2, 3\}$.   and $Y = \{1, 3\}$.

Fig. 1. Lower bound construction

Next we add dependence on X and Y. Let

$$E_X = \bigcup_{i \in [n] \setminus X} \left(\bigcup_{u \in U_i, v \in A} \{\{u, v\}\} \right) \text{ and } E_Y = \bigcup_{i \in [n] \setminus Y} \left(\bigcup_{u \in U_i, v \in B} \{\{u, v\}\} \right).$$

So E_X contains all edges from vertices in U_i to vertices in A exactly when index i is not in the set X. E_Y similarly contains all edges from U_i to B when $i \notin Y$.

Now let $E = E_0 \cup E_X \cup E_Y$. Adding these edge sets corresponds to adding a super-edge to Fig. 1a between U_i and A (or B) whenever i is not in X (or Y). Figures 1b and 1c illustrate this. In Fig. 1b, the intersection is non-empty, which creates a set of isolated nodes that push up the value of $\beta(G)$. Meanwhile, there is no intersection in Fig. 1c, so the only isolated nodes are those in C.

Now, consider $\beta(G)$. In the case where $X \cap Y = \emptyset$, we will have a super-edge connecting each U_i to at least one of A and B, so the degree of each vertex in each U_i is either a or $2a$. Similarly, $A \cup B$ is a clique, so each vertex has degree at least $(2a - 1)$. There are $2a$ such vertices, so they contribute at most $\frac{2a}{(2a-1)+1} = 1$ to β. Vertices in C are isolated and contribute exactly z to β. Therefore, $z \leq \beta(G) \leq \frac{kq}{a} + 1 + z = z + 2$.

Now consider the case where $X \cap Y \neq \emptyset$. This means that there exists some $i \in X \cap Y$, and so U_i will have no super-edges. So each vertex in U_i is isolated, and contributes exactly 1 to β. There are q such vertices, and also accounting for the contribution of vertices C, we obtain $\beta(G) \geq q + z = z(2c^2 + 1)$.

Since the minimum possible ratio of the β-values between graphs in the two cases is at least $\frac{z(2c^2+1)}{z+2} > c^2$ (using $z \geq 2$), a c-approximation algorithm for $\beta(G)$ would allow us to distinguish between the two cases.

Now, return to our instance of DISJ_k. We can have Alice initialise an instance of $\mathrm{ALG}_{c,n}$ and have all vertices in A, C, and each U_i arrive in any order. This only requires knowledge of X because only edges in E_0 and E_X are between these vertices and these are the only edges that will be added so far in the vertex arrival model. Alice then communicates the state of $\mathrm{ALG}_{c,n}$ to Bob. Bob can now have all vertices in B arrive in any order. This only requires knowledge of Y because only edges in E_0 and E_Y are still to be added. Bob then communicates the state of $\mathrm{ALG}_{c,n}$ back to Alice (if $p \geq 2$). This process continues until the p passes of the algorithm have been executed. Bob can then compute a c-approximation of $\beta(G)$ with probability at least $\frac{2}{3}$ from the final state of the algorithm, determining which case we are in and solving DISJ_k.

From Theorem 3, we know that Alice and Bob must have communicated at least $\Omega(k)$ bits. However, all they communicated was the state of $\mathrm{ALG}_{c,n}$. Therefore, $\Omega(k/p) = \Omega(\frac{n}{zc^2p})$ bits was being used by $\mathrm{ALG}_{c,n}$ at the time, since the algorithm runs in p passes.

Consider again the graph G. The above argument shows that in order to compute a c-approximation to $\beta(G)$, space $\Omega(\frac{n}{zc^2p})$ is needed. Since $\beta(G) \geq z$ in both cases, we obtain the space bound $\Omega(\frac{n}{\beta(G)c^2p})$. Last, recall that z and thus $\beta(G)$ can be chosen arbitrarily. The theorem hence holds for any value of $\beta(G)$. $\square$

Remark: The vertices of set C of the construction employed in the previous proof are isolated. This property may be considered undesirable – for example, it may be relatively easy to identify and separately count isolated vertices. However, this structure in the hard instances can be entirely circumvented by, for example, replacing each of these vertices $u \in C$ with a pair of nodes u_1, u_2, which are connected by an edge. We also note that, of course, the problem is no easier when deletions are allowed, and so the lower bound also holds for such models.

6   Conclusion

In this paper, we gave an optimal one-pass c-approximation streaming algorithm with space $O(\frac{n \operatorname{polylog} n}{c^2 \gamma})$ for approximating the Caro-Wei bound $\beta(G)$ in graph streams, where $\gamma \leq \beta(G)$ is a given lower bound. If the input stream is in vertex arrival order, then we showed that a quantity β' can be computed, which is at most a logarithmic factor below $\beta(G)$ and at most $\alpha(G)$, the maximum independent set size of the input graph.

From a technical perspective, we leverage this problem to advance the study of the degree moments in the streaming model. The fact that the frequencies are derived from the degrees of the input graph adds an additional dimension to the frequency moments problem, since, as illustrated by our two algorithms, the arrival order of edges can now be exploited. Furthermore, it seems plausible that exploiting additional graph structure could reduce the space complexity even further. For example, it is known that in claw-free graphs, it holds $\sum_{u \in \Gamma(v)} \frac{1}{\deg(u)} = O(1)$, for every vertex v [10]. It remains to be investigated whether such properties can give additional space improvements. Last, one of the objectives of this work was the popularization of the Caro-Wei bound, and we thus only addressed the -1-negative frequency moment. Our algorithm for the edge arrival model can in fact also be used for approximating any other negative degree moment $\sum_{v \in V} (\frac{1}{\deg_G(v)})^p$, for every $p < 0$, since the analysis only requires that the contribution of a vertex to the degree moment is at most 1, which is the case for all negative moments (it holds $(\frac{1}{\deg_G(v)})^p \leq 1$, for every $v \in V$ and $p < 0$). Generalizing our approach to positive frequency moments is left for future work.

Acknowledgements. We thank an anonymous reviewer whose comments helped us simplify Theorem 1. The work of GC is supported in part by European Research Council grant ERC-2014-CoG 647557; JD is supported by a Microsoft EMEA scholarship and the Alan Turing Institute under the EPSRC grant EP/N510129/1; CK is supported by EPSRC grant EP/N011163/1.

References

1. Halldórsson, B.V., Halldórsson, M.M., Losievskaja, E., Szegedy, M.: Streaming algorithms for independent sets in sparse hypergraphs. Algorithmica **76**(2), 490–501 (2016)
2. Karp, R.M.: Reducibility among combinatorial problems. In: Miller, R.E., Thatcher, J.W. (eds.) Complexity of Computer Computations. IRSS, pp. 85–103. Plenum Press, New York (1972). https://doi.org/10.1007/978-1-4684-2001-2_9
3. Håstad, J.: Clique is hard to approximate within $n^{1-\epsilon}$. Acta Math. **182**(1), 105–142 (1999)

4. Zuckerman, D.: Linear degree extractors and the inapproximability of max clique and chromatic number. Theory Comput. **3**(1), 103–128 (2007)
5. Feige, U.: Approximating maximum clique by removing subgraphs. SIAM J. Discret. Math. **18**(2), 219–225 (2005)
6. Halldórsson, M., Radhakrishnan, J.: Greed is good: approximating independent sets in sparse and bounded-degree graphs. In: STOC, pp. 439–448 (1994)
7. Wei, V.: A lower bound on the stability number of a simple graph. Technical report, Bell Labs (1981)
8. Griggs, J.R.: Lower bounds on the independence number in terms of the degrees. J. Comb. Theory Ser. B **34**(1), 22–39 (1983)
9. Caro, Y.: New results on the independence number. Technical report, Tel Aviv Univ (1979)
10. Halldórsson, M.M., Konrad, C.: Distributed large independent sets in one round on bounded-independence graphs. In: Moses, Y. (ed.) DISC 2015. LNCS, vol. 9363, pp. 559–572. Springer, Heidelberg (2015). https://doi.org/10.1007/978-3-662-48653-5_37
11. Halldórsson, M.M., Sun, X., Szegedy, M., Wang, C.: Streaming and communication complexity of clique approximation. In: Czumaj, A., Mehlhorn, K., Pitts, A., Wattenhofer, R. (eds.) ICALP 2012. LNCS, vol. 7391, pp. 449–460. Springer, Heidelberg (2012). https://doi.org/10.1007/978-3-642-31594-7_38
12. Assadi, S., Khanna, S., Li, Y.: On estimating maximum matching size in graph streams. In: ACM-SIAM Symposium on Discrete Algorithms, pp. 1723–1742 (2017)
13. Cormode, G., Jowhari, H., Monemizadeh, M., Muthukrishnan, S.: The sparse awakens: streaming algorithms for matching size estimation in sparse graphs. In: ESA (2017)
14. Cabello, S., Pérez-Lantero, P.: Interval selection in the streaming model. In: Dehne, F., Sack, J.-R., Stege, U. (eds.) WADS 2015. LNCS, vol. 9214, pp. 127–139. Springer, Cham (2015). https://doi.org/10.1007/978-3-319-21840-3_11
15. Alon, N., Matias, Y., Szegedy, M.: The space complexity of approximating the frequency moments. J. Comput. Syst. Sci. **58**(1), 137–147 (1999)
16. Woodruff, D.P.: Frequency moments. In: Liu, L., Özsu, M.T. (eds.) Encyclopedia of Database Systems, pp. 1169–1170. Springer, Boston (2009). https://doi.org/10.1007/978-0-387-39940-9
17. Braverman, V., Chestnut, S.R.: Universal sketches for the frequency negative moments and other decreasing streaming sums. In: APPROX/RANDOM, pp. 591–605 (2015)
18. Turán, P.: On an extremal problem in graph theory. Matematikai és Fizikai Lapok **48**(436–452), 137 (1941)
19. Henzinger, M., Raghavan, P., Rajagopalan, S.: Computing on data streams. Technical report SRC 1998–011, DEC Systems Research Centre (1998)
20. McGregor, A.: Graph stream algorithms: a survey. SIGMOD Rec. **43**(1), 9–20 (2014)
21. Gonen, M., Ron, D., Shavitt, Y.: Counting stars and other small subgraphs in sublinear-time. SIAM J. Discrete Math. **25**(3), 1365–1411 (2011)
22. Eden, T., Ron, D., Seshadhri, C.: Sublinear time estimation of degree distribution moments: the arboricity connection. CoRR abs/1604.03661 (2016)

23. Aliakbarpour, M., Biswas, A.S., Gouleakis, T., Peebles, J., Rubinfeld, R., Yodpinyanee, A.: Sublinear-time algorithms for counting star subgraphs with applications to join selectivity estimation. CoRR abs/1601.04233 (2016)
24. Jowhari, H., Sağlam, M., Tardos, G.: Tight bounds for Lp samplers, finding duplicates in streams, and related problems. In: ACM Principles of Database Systems (2011)
25. Kalyanasundaram, B., Schnitger, G.: The probabilistic communication complexity of set intersection. SIAM J. Discrete Math. 5(4), 545–557 (1992)

The Minimum Rooted-Cycle
Cover Problem

D. Cornaz[✉] and Y. Magnouche

LAMSADE, University Paris-Dauphine,
Place du Maréchal de Lattre de Tassigny, 75016 Paris, France
{denis.cornaz,youcef.magnouche}@dauphine.fr

Abstract. Given an undirected rooted graph, a cycle containing the
root vertex is called a rooted cycle. We study the combinatorial duality
between vertex-covers of rooted-cycles, which generalize classical vertex-
covers, and packing of disjoint rooted cycles, where two rooted cycles
are vertex-disjoint if their only common vertex is the root node. We
use Menger's theorem to provide a characterization of all rooted graphs
such that the maximum number of vertex-disjoint rooted cycles equals
the minimum size of a subset of non-root vertices intersecting all rooted
cycles, for all subgraphs.

Keywords: Kőnig's theorem · Menger's theorem

1 Introduction

Throughout $G = (V, E)$ is a simple undirected graph. The *minimum vertex-cover
problem* amounts to find a *vertex-cover* (that is, a set $T \subseteq V$ so that every edge
of G has at least one vertex in T) minimizing $|T|$. This is a very well studied
NP-hard problem, equivalent to finding a maximum *stable set* (equivalently, the
complement of a vertex-cover, or a *clique* in the complementary graph) [6]. In
this paper, we introduce the *minimum rooted-cycle cover problem* which contains
the vertex-cover problem, and which is, given a root vertex r of G, to remove a
minimum size subset of $V \setminus \{r\}$ so that r is contained in no cycle anymore. The
minimum vertex-cover problem is the particular case where r is adjacent with
all other vertices.

If we are given a set of terminal vertices of G, with at least two vertices, the
minimum multi-terminal vertex-cut problem is to remove a minimum number
of vertices, so that no path connects two terminal vertices anymore, see [1,2].
The weighted version of the minimum rooted-cycle cover problem contains the
minimum multi-terminal vertex-cut problem which is the particular case where
the neighborhood $N(r)$ of r is the set of terminal vertices with infinite weight.
In turn, if we replace r by $|N(r)|$ terminal vertices $t_1, \ldots, t_k$ where $N(r) =
\{v_1, \ldots, v_k\}$ and link t_i to v_i, then we obtain an instance of the minimum multi-
terminal vertex-cut problem the solution of which is a solution for the original
instance of the minimum rooted cycle cover problem.

© Springer International Publishing AG, part of Springer Nature 2018
J. Lee et al. (Eds.): ISCO 2018, LNCS 10856, pp. 115–120, 2018.
https://doi.org/10.1007/978-3-319-96151-4_10

Our main motivation to introduce the minimum rooted cycle cover problem is that it allows us to give short proofs of some min-max theorems, such results being fundamental in combinatorial optimization and linear programming [5]. Jost and Naves gave such results for the minimum multi-terminal vertex-cut problem in an unpublished manuscript [2] (actually we found independently this result).

The paper is organized as follows. In Sect. 2, we recall two classical theorems and give formal definitions. In Sect. 3, we give a characterization of all rooted graphs (G, r) so that the minimum number of non-root vertices intersecting all rooted cycles equals the maximum number of rooted cycles having only the root as common node, for all partial subgraphs. In Sect. 4, we revisit a result by Jost and Naves [2] in terms of rooted cycles. (We found the equivalent result independently.) This is a structural characterization in terms of excluded minors of pseudo-bipartite rooted graphs, that is, rooted graphs satisfying the min-max equality for all rooted minors.

2   Background

Let us recall two fundamental min-max theorems.

Given a graph, a *matching* is a subset of pairwise vertex-disjoint edges.

Kőnig's Theorem [3]. *Let G be a bipartite graph. The minimum size of a vertex-cover of G is equal to the maximum size of a matching of G.*

Take a graph G and fix distinct vertices s, t. A *st-path* of G is subset $P \subseteq V$ of vertices of G which can be ordered into a sequence $s = v_0, v_1, \ldots, v_k = t$ where $v_i v_{i+1}$ is an edge of G. The vertices v_0, v_k are the *extremities of P*, the other vertices are the *internal vertices of P*. Two *st*-paths P, Q are *internally vertex-disjoint* if $P \cap Q = \{s, t\}$. A subset D of vertices of G is an *st-vertex cut* if neither s nor t belongs to D, and D intersects every *st*-path.

Menger's Theorem [4]. *Let G be a graph and let s and t be two nonadjacent vertices of G. The minimum size of a st-vertex cut is equal to the maximum number of internally vertex-disjoint st-paths.*

A subset $C \subseteq V$ containing r and so that $C \setminus \{r\}$ is a path the extremities of which are adjacent with r is called a *rooted cycle of* (G, r). Two rooted cycles are *internally* vertex-disjoint if r is their only common vertex. A *rooted-cycle cover* of (G, r) is a subset of $T \subseteq V \setminus \{r\}$ of non-root vertices so that $C \cap T \neq \emptyset$ for all rooted cycle C. A rooted-cycle cover is minimum if $|T|$ is minimum. A *rooted-cycle packing* of (G, r) is a collection $C_1, \ldots, C_k$ of rooted cycles so that $C_i \cap C_j = \{r\}$ for all distinct $i, j = 1, \ldots, k$. A rooted-cycle packing is maximum if k is maximum. Clearly the minimum of the cover is at least the maximum of the packing.

We call G' a *subgraph of G* if it is obtained from G by deleting vertices, and G' is a *partial subgraph of G* if it is obtained by deleting vertices and/or edges.

3   Packing and Covering Rooted Cycles

The following consequence of Menger's theorem is useful to characterize rooted graphs for which the minimum equals the maximum.

Corollary 1. *Let G be a graph with a vertex t and a subset S, of at least k vertices, not containing t. If there are k internally vertex-disjoint vt-paths for every $v \in S$, then there are k distinct vertices $s_1, \ldots, s_k$ of S, with $s_i t$-paths P_i for each $i = 1, \ldots, k$, so that t is the only vertex belonging both to distinct P_i, P_j.*

Proof. Add a new vertex s to G and link it to every vertex in S. We only need to prove that there are k internally vertex-disjoint st-paths. If it is not the case, then, by Menger's theorem, there is a st-vertex cut D of size $|D| < k$. Let $v \in S \setminus D$. Clearly v is not adjacent with t. Thus D is a vt-vertex cut which is impossible since (again by Menger's theorem) there are k internally vertex-disjoint vt-paths.   □

Let K_4 be the complete graph on four vertices and r one of its vertices. The rooted graph $(\hat{G}, r)$ is a subdivision of (K_4, r) if it is obtained from (K_4, r) by inserting vertices in edges. Note that in any such subdivision, the vertex r has degree three. A *rooted partial subgraph of* (G, r) is a rooted graph (G', r) where G' is a partial subgraph of G.

Theorem 1. *The minimum size of a subset of non-root vertices intersecting all rooted cycles is equal to the maximum number of internally vertex-disjoint rooted cycles, for all partial rooted subgraphs of (G, r), if and only if no partial rooted subgraph of (G, r) is a subdivision of (K_4, r).*

Proof. ($\Rightarrow$) It suffices to see that, for any subdivision of (K_4, r), any two rooted cycles must have a non-root vertex in common while any rooted cycle cover needs at least two non-root vertices.
($\Leftarrow$) Let (G, r) be a minimum graph, that is, with a minimum number of edges, such that the minimum rooted cycle cover is strictly greater than the maximum packing of rooted cycles. Minimality implies that G has no vertices of degree < 3. It follows that the graph $G - r$ obtained from G by removing r has a cycle C with at least three distinct vertices s_1, s_2, s_3 (since G is a simple graph). Hence, by Corollary 1, it suffices to prove (1) below, since it implies that there are three internally vertex-disjoint $s_i r$-paths which form with C a subdivision of (K_4, r).

There are three internally vertex-disjoint vr-paths for every vertex $v \in V \setminus \{r\}$ (1)

Assume that (1) is not true, and let v be a counter-example. Remark that, for any vertex s nonadjacent with r, the minimality of G implies that no sr-vertex cut is a clique, in particular, it has at least two vertices (indeed otherwise v belongs to no inclusion wise minimal rooted cycle). Furthermore, no vertex-cut has only one vertex. We build a graph $\hat{G}$ from G as follows:

(a): If v is nonadjacent with r, then (by Menger's theorem) there is a vr-vertex cut with size 2, say $D = \{u, w\}$. Let $V' \ni v$ be the subset of vertices in the component, containing v, of the graph obtained from G by removing D. We let $\hat{G}$ be the graph obtained by removing V' and adding the edge $e = uw$.

(b): If v is adjacent with r, then, in the graph $G - vr$ (obtained from G by removing the edge vr), there is a vr-vertex cut $D = \{u\}$. Let $V' \ni v$ be the subset of vertices in the component, containing v, of the graph obtained by removing u from $G - vr$. We let $\hat{G}$ be the graph obtained by removing V' and adding the edge $e = ur$.

Observe now, that if there are ν vertex-disjoint rooted cycles in $\hat{G}$, then there are also ν vertex-disjoint rooted cycles in G. Indeed, if some rooted cycle of $\hat{G}$ contains the additional edge e, then e can be replaced by a path of G with all internal vertices in V'. Moreover, if there are τ vertices intersecting every rooted cycles of $\hat{G}$, then these τ vertices intersect also every rooted cycles of G. We have a contradiction, since the minimality of G implies $\tau = \nu$.   □

Finally, since series-parallel graphs are those with no minor K_4, one has:

Corollary 2. *Given a graph G, the minimum rooted cycle cover equals the maximum rooted cycle packing, for all partial subgraphs and every choice of a root r, if and only if G is series-parallel.*

4   Pseudo-bipartite Rooted Graphs

The *closed neighborhood of r* is the set of $N[r] = N(r) \cup \{r\}$.

A rooted graph is *pseudo-bipartite* if it is obtained from a bipartite graph $(V_1, V_2; E)$ by creating a root vertex linked to every vertex of $V_1 \cup V_2$, and then, by replacing some original edges $uv \in E$ by any graph G_{uv}, on new vertices, with edges between some new vertices and u (or v), more precisely:

Definition 1. *A rooted graph (G, r) is* pseudo-bipartite *if*

(a) *The subgraph $G[N(r)]$ induced by the neighbors of r is a bipartite graph G_B;*

(b) *There is a bipartition of G_B, so that every component of the graph $G \setminus N[r]$, that we obtain if we remove the closed neighborhood of r, has at most one neighbour in each side of G_B.*

Contracting a vertex v is to delete v, and to add edges so that its neighborhood $N(v)$ forms a clique. A *rooted minor of (G, r)* is a rooted graph $(\hat{G}, r)$ obtained from (G, r) by deleting vertices different from r, or by *contracting* vertices $v \in V \setminus N[r]$ outside the closed neighborhood of the root.

An *odd wheel* is a graph composed of an odd cycle together with one vertex, called the *center* of the wheel, to which all the vertices of the odd cycle are linked. A *rooted odd wheel* is a rooted graph (G, r) so that G is an odd wheel the center of which is the root r.

Theorem 2 [2]. *A rooted graph (G, r) is pseudo-bipartite if and only if it has no rooted minor which is a rooted odd wheel.*

Proof. Necessity holds since $G[N(r)] = G_B$ has no odd cycle and since every edge which appears in $G[N(r)]$ by contracting some vertex outside $N[r]$ necessarily links two vertices in different side of the bipartition of G_B.

To see sufficiency, suppose that no rooted-minor of (G, r) is a rooted odd-wheel. Condition (a) of Definition 1 is indeed satisfied since deleting all vertices but those of an odd cycle in the neighborhood of r leaves a rooted odd wheel. Assume that condition (b) is not satisfied. Let $U_1, \ldots, U_p$ be the components of the graph obtained by removing r and all its neighbors. If U_i has at least three neighbors x, y, z (in G_B), then contracting all the vertices in U_i and deleting all vertices but x, y, z (and r) leaves (K_4, r) which is a rooted odd wheel. It follows that U_i has at most two neighbors. Chose a bipartition V_1, V_2 of G_B so that the number of components $U_1, \ldots, U_p$ having two neighbors in the same side is minimum. There is a component U_i with two neighbors x, y in the same side, say $x, y \in V_1$ (otherwise the proof is done). Let V_1^x (resp. V_2^x) be the set of vertices in V_1 (resp. in V_2) reachable, from x, by a path of G_B. Similarly, define V_1^y and V_2^y. If either $V_1^x \cap V_1^y$ or $V_2^x \cap V_2^y$ is nonempty, then there exists a xy-path P in G_B. Yet contracting U_i and deleting all vertices but those of P (and r) leaves a rooted odd wheel. It follows that $V_1^x, V_2^x, V_1^y, V_2^y$ are pairwise disjoint, and hence $(V_1 \setminus V_1^x) \cup V_2^x, (V_2 \setminus V_2^x) \cup V_1^x$ is a possible bipartition for G_B. The way the bipartition V_1, V_2 was chosen implies that there is another component, say U_j, with either one neighbor in V_1^x and the other in V_2^y, or one neighbor in V_1^y and the other in V_2^x. Anyway, contracting both U_i and U_j creates an odd cycle in the neighborhood of r. Now deleting the vertices outside this odd cycle leaves a rooted odd wheel; contradiction. $\square$

Given a pseudo-bipartite rooted graph (G, r) with bipartite graph $G_B = G[N(v)]$, we let $\hat{G}$ be the graph obtained by removing r from (G, r), and then by creating two new vertices s and t, so that s is linked to every vertex of one side of G_B, and t is linked similarly to every vertex in the other side of G_B. Observe that:

A subset $P \subseteq V$ is a st-path of $\hat{G}$ if and only if $P \setminus \{s, t\} \cup \{r\}$ is a rooted cycle of (G, r). (2)

Since (2) holds, Menger's theorem implies that, for pseudo-bipartite rooted graphs, the minimum rooted cycle cover equals the maximum rooted cycle packing.

Observe, moreover, that for any rooted odd wheel with an odd cycle having $2k + 1$ vertices, the minimum size of a cover is $k + 1$ while the maximum packing is k. It follows that Theorem 2 implies Corollary 3 below, which is also a consequence of a result in [2] (namely Lemma 9).

Corollary 3 [2]. *The minimum size of a subset of non-root vertices intersecting all rooted cycles is equal to the maximum number of internally vertex-disjoint rooted cycles, for all rooted minor of (G, r), if and only if no rooted minor of (G, r) is a rooted odd wheel.* $\square$

Observe that, if every vertex is linked to r, then (G, r) is pseudo-bipartite if and only if the graph induced by $V \setminus \{r\}$ is bipartite. Hence by Theorem 2:

Remark 1. *Corollary 3 contains Kőnig's theorem as particular case.*

Note also that recognizing if (G, r) is pseudo-bipartite or not, it suffices to contract all vertices outside the closed neighborhood of r, to remove r, and to check if the remaining graph is bipartite or not. So one has:

Remark 2. *In can be checked in polynomial time if (G, r) is pseudo-bipartite or not.*

References

1. Garg, N., Vazirani, V.V., Yannakakis, M.: Multiway cuts in vertex weighted graphs. J. Algorithms **50**, 49–61 (2004)
2. Jost, V., Naves, G.: The graphs with the max-Mader-flow-min-multiway-cut property. CoRR abs/1101.2061 (2011)
3. Kőnig, D.: Graphok és alkalmazásuk a determinánsok és a halmazok elméletére [Hungarian]. Mathematikai és Természettudományi Értesítő **34**, 104–119 (1916)
4. Menger, K.: Zur allgemeinen Kurventheorie. Fundam. Math. **10**, 96–115 (1927)
5. Schrijver, A.: Theory of Linear and Integer Programming. Wiley, Chichester (1986)
6. Schrijver, A.: Combinatorial Optimization: Polyhedra and Efficiency. Springer, Heidelberg (2003)

Online Firefighting on Trees

Pierre Coupechoux[1], Marc Demange[2], David Ellison[2(⊠)], and Bertrand Jouve[3]

[1] LAAS-CNRS, Université de Toulouse, CNRS, Toulouse, France
pierre.coupechoux@laas.fr
[2] School of Science, RMIT University, Melbourne, Australia
{marc.demange,david.ellison2}@rmit.edu.au
[3] FRAMESPA and IMT, CNRS, Université de Toulouse, Toulouse, France
bertrand.jouve@cnrs.fr

Abstract. In the FIREFIGHTER problem, introduced by Hartnell in 1995, a fire spreads through a graph while a player chooses which vertices to protect in order to contain it. In this paper, we focus on the case of trees and we consider as well the FRACTIONAL FIREFIGHTER game where the amount of protection allocated to a vertex lies between 0 and 1. We introduce the online version of both FIREFIGHTER and FRACTIONAL FIREFIGHTER, in which the number of firefighters available at each turn is revealed over time. We show that the greedy algorithm on finite trees, which maximises at each turn the amount of vertices protected, is 1/2-competitive for both online versions; this was previously known only in special cases of FIREFIGHTER. We also show that, for FIREFIGHTER, the optimal competitive ratio of online algorithms ranges between 1/2 and the inverse of the golden ratio. The greedy algorithm is optimal if the number of firefighters is not bounded and we propose an optimal online algorithm which reaches the inverse of the golden ratio if at most 2 firefighters are available. Finally, we show that on infinite trees with linear growth, any firefighter sequence stronger than a non-zero periodic sequence is sufficient to contain the fire, even when revealed online.

1 Introduction and Definitions

Since it was formally introduced by B. Hartnell in 1995 ([1], cited in [2]) the firefighting problem - FIREFIGHTER - has raised the interest of many researchers. While this game started as a very simple model for fire spread and containment problems for wildfires, it can also represent any kind of threat able to spread sequentially in a network (diseases, viruses, rumours, flood ...).

It is a deterministic discrete-time one-player game defined on a graph. In the beginning, a fire breaks out on a vertex and at each step, if not blocked, the fire spreads to all adjacent vertices. In order to contain the fire, the player is given a number f_i of firefighters at each turn i and can use them to protect vertices which are neither burning nor already protected. The game terminates when the

B. Jouve—We acknowledge the support of GEO-SAFE, H2020-MSCA-RISE-2015 project # 691161.

fire cannot spread any further. In the case of finite graphs the aim is to save as many vertices as possible, while in the infinite case, the player wins if the game finishes, which means that the fire is contained.

This problem and its variants give rise to a generous literature; the reader is referred to [2] for a broad presentation of the main research directions. A significant amount of theoretical work deals with its complexity and approximability behaviour in various classes of graphs [3–6] and recently its parametrised complexity (e.g. [7]). It is known to be very hard, even in some restrictive cases. In particular, the case of trees was revealed to be very rich and a lot of research focuses on it. The problem is NP-hard on finite trees [3], even in more restricted cases [6]. Regarding approximability results on trees, it was first shown to be $\frac{1}{2}$-approximated by a greedy strategy [8], improved to $1 - \frac{1}{e}$ [5] and very recently to a polynomial time approximation scheme [9], which closes essentially the question of approximating firefighter problem in trees and motivates considering some generalisations. On general graphs the problem is hard to approximate within $n^{1-\varepsilon}$ [10]. A related research direction investigates integer linear programming models for the problem, especially on trees [9,11,12]. This line of research makes very natural a *relaxed* version where the amount of firefighters available at each turn is any non-negative number and the amount allocated to vertices lies between 0 and 1. A vertex with a protection less than 1 is *partially protected* and its unprotected part can burn partially and transmit only its fraction of fire to the adjacent vertices. Thus, the f_i may take any non-negative value. This defines a variant game called FRACTIONAL FIREFIGHTER which was introduced in [13].

Online optimisation [14] is a generalisation of approximation theory which represents situations where the information arrives over the time and one needs to make irrevocable decisions. We propose an online version of both firefighter problems and consider first results on trees. In our model, the graph is known and the sequence of available firefighters is revealed online. We then refer to the usual case where $(f_i)_{i \geq 1}$ is known in advance as *offline*. To our knowledge, this is the first attempt at analysing online firefighter problems. Even though our motivation is mainly theoretical, this paradigm is particularly natural in emergency management where one has to make quick decisions despite lack of information. Any progress in this direction, even on simplified models, contributes to understanding how lack of information impacts the quality of the solution.

Given a tree T rooted in r, $V(T)$ and $E(T)$ will denote the vertex set and the edge set of T, respectively. Given two vertices v and v', $v \prec v'$ denotes that v is an ancestor of v' (or v' is a descendant of v) and $v \preceq v'$ denotes that either $v = v'$ or $v \prec v'$.

For any vertex v, let $T[v]$ denote the sub-tree induced by v and its descendants. Let T_i denote the i-th level of T rooted in r, where $\{r\} = T_0$. The height $h(T)$ of T, rooted in r, is the maximum length of a path from r to a leaf. If $i > h(T)$, we have $T_i = \emptyset$.

If T is finite, the *weight* $w(v)$ of a vertex v is the number of vertices in its sub-tree (including v), i.e. $w(v) = |V(T[v])|$. When no ambiguity may occur, we will simply write $w_v = w(v)$. For any vertex $v \in T_i$ and any $i \leq j \leq i + h(T[v])$, we denote by v_j a vertex of maximum weight w_{v_j} in $T_j \cap V(T[v])$. While v_j is not defined if $i + h(T[v]) < j \leq h(T)$, we define $\bar{w}_{v_j}$ for all $j \leq h(T)$ and $v \in V(T)$ via: $\bar{w}_{v_j} = w_{v_j}$ if $j \leq i + h(T[v])$ and 0 otherwise. We denote by $B(T)$ the tree obtained from T by fusing all vertices from levels 0 and 1 into a new root vertex r_B: every edge $u_1u_2 \in E(T)$ with $u_1 \in T_1$ and $u_2 \in T_2$ gives rise to the edge $r_Bu_2 \in E(B(T))$. For $k \leq h(T)$, $B^k(T)$ will denote the kth iteration of B applied to T: all vertices from levels 0 to k are fused into a single vertex denoted by r_{B^k} which becomes the new root.

2   The Problems and Preliminary Results

2.1   FIREFIGHTER and FRACTIONAL FIREFIGHTER on Trees

In this paper we only play the game on finite or infinite trees. An instance of the FRACTIONAL FIREFIGHTER on trees is defined by a triple $(T, r, (f_i))$, where $T = (V(T), E(T))$ is a tree, $r \in V(T)$ is the root where the fire breaks out and $(f_i)_{i \geq 1}$ is the non-negative real *firefighter sequence*. Turn $i = 0$ is the initial state where r is burning and all other vertices are unprotected, and $i \geq 1$ corresponds to the different rounds of the game. At each turn $i \geq 1$ and for every vertex v, the player decides which amount $p(v)$ of protection to add to v. Throughout the game, for every vertex v the part of v which is burning is denoted by $b(v)$. Let us note that if T is finite, the game will end in at most $h(T)$ turns where $h(T)$ is the length of a longuest path form the root to the leaves.

Solutions on trees have a very specific structure: at each turn i, the amounts of fire are non-increasing along any root to leaf path, which means that the fire will spread only towards the leaves. Note that for any solution which allocates a positive amount of protection at turn i to a vertex $v \in T_k, k > i$, allocating the same amount of protection to v's father instead strictly improves the performance. So we may consider only algorithms that play in T_i at turn i. For an optimal offline algorithm, this property was emphasised in [8]. So, for any vertex $v \in T_i$, the amounts of fire $b(v)$ and protection $p(v)$ on v will not change after turn i.

A solution p is characterised by the values $p(v), v \in V(T)$. For any solution p, while $p(v)$ represents the amount of protection received directly, vertex v also receives protection through its ancestors, the amount of which is denoted by $P_p(v) = \sum_{v' \prec v} p(v')$ (used in Sect. 3). For any vertex v, we have the equality $p(v) + P_p(v) + b(v) = 1$.

Any solution p for FIREFIGHTER or FRACTIONAL FIREFIGHTER will satisfy the constraints:

$$\begin{cases} \sum_{v \in T_i} p(v) \leq f_i & (i) \\ \forall v, p(v) + P_p(v) \leq 1 & (ii) \end{cases}$$

In [12], a specific boolean linear model has been proposed for solving FIREFIGHTER on a tree T involving these constraints. Solving FRACTIONAL FIREFIGHTER on T corresponds to solving the relaxed version of this linear programme.

2.2   Online Version

We introduce online versions of FIREFIGHTER and FRACTIONAL FIREFIGHTER. The graph - a tree T in this work - and the ignition vertex - the root r - are known in advance but the firefighter sequence $(f_i)_{i \geq 1}$ is revealed over time by a second player called *adversary*. At each turn i, the adversary reveals f_i and then the player chooses where to allocate this resource.

Let us consider an online algorithm OA for one of the two problems and let us play the game on a finite tree T until the fire stops spreading. The value λ_{OA} achieved by the algorithm, defined as the amount of saved vertices, is measured against the best value performed by an algorithm knowing in advance the sequence (f_i). In the present case, it is simply the optimal value of the offline instance, referred to as the *offline optimal value*, denoted by β_I when considering the online FIREFIGHTER (I stands for "Integral") and β_F for the online FRACTIONAL FIREFIGHTER. We will call *Bob* such an algorithm able to see the future and guaranteeing the value β_I or β_F for online FIREFIGHTER and FRACTIONAL FIREFIGHTER.

OA is said to be γ-competitive, $\gamma \in]0,1]$ for the online FIREFIGHTER (resp. FRACTIONAL FIREFIGHTER) if for every instance, $\frac{\lambda_{OA}}{\beta_I} \geq \gamma$ (resp. $\frac{\lambda_{OA}}{\beta_F} \geq \gamma$). γ is also called the *competitive ratio* guaranteed by OA. An online algorithm will be called *optimal* if it guarantees the best possible competitive ratio.

Let us first note that one can reduce the problem to the case where $f_1 > 0$:

Proposition 1. *We can reduce online* (FRACTIONAL) FIREFIGHTER *on trees to instances where* $f_1 > 0$.

Proof. If $f_i = 0$ for all i such that $1 \leq i \leq k$, then the instance $(T, r, (f_i))$ is equivalent to the instance $(B^k(T), r_{B^k}, (f_{i+k}))$.

In the infinite case we do not define competitiveness but only ask whether the fire can be contained by an online algorithm. Sections 3 and 4 deal with the finite case while Sect. 5 deals with a class of infinite trees.

3   Competitive Analysis of a Greedy Algorithm

Greedy algorithms are usually very good candidates for online algorithms, sometimes the only known approach. Mainly two different greedy algorithms have been considered in the literature for FIREFIGHTER on a tree [2] and they are both possible online strategies in our set-up. The *degree greedy* strategy priorities saving vertices of large degree; it has·been shown in [6] that it cannot guarantee any approximation ratio on trees, even for a constant firefighter sequence. A second greedy algorithm was introduced in [8] for an integral sequence (f_i), maximising at each turn the weight of the vertices protected. We generalise it to any firefighter sequence for both the integral and the fractional problems: at each turn i, the algorithm Gr solves the linear programme $\mathcal{P}_i$ with variables $x(v), v \in T_i$ and constraints:

$$\mathcal{P}_i : \begin{cases} \max \sum_{v \in T_i} x(v)w(v) \\ \sum_{v \in T_i} x(v) \leq f_i & (i) \\ \forall v, x(v) + P_x(v) \leq 1 & (ii) \end{cases}$$

An optimal solution of $\mathcal{P}_i$ is obtained by ordering vertices $\{v_1, \ldots, v_{|T_i|}\}$ of level i by non-increasing weight and taking them one by one in this order and greedily assigning to vertex v_j the value $x(v_j) = \min(f_i - \sum_{k<j} x(v_k), 1 - P_x(v_j))$. Note that Gr is valid for both FIREFIGHTER and FRACTIONAL FIREFIGHTER.

It was shown in [8] that the greedy algorithm on trees gives a $\frac{1}{2}$-approximation of the restriction of FIREFIGHTER when a single firefighter is available at each turn. They claim that this approximation ratio remains valid for a fixed number $D \in \mathbb{N}$ of firefighters at each turn. We extend this result to any firefighter sequence $(f_i)_{i \geq 1}$, integral or not. Since Gr is an online algorithm, the performance can also be seen as a competitive ratio for the online version.

Theorem 1. *The greedy algorithm Gr is $\frac{1}{2}$-competitive for both online* FIREFIGHTER *and* FRACTIONAL FIREFIGHTER *on finite trees.*

Proof. Let us first consider the fractional case with an online instance $(T, r, (f_i))$ of FRACTIONAL FIREFIGHTER on a tree.

Let $x(v)$ and $y(v)$ be the amounts of firefighters placed on vertex v by Gr and *Bob*, respectively. We have $\lambda_{Gr} = \sum_v x(v)w(v)$ and $\beta_F = \sum_v y(v)w(v)$.

Recall that $P_x(v) = \sum_{v' \prec v} x(v')$ and $P_y(v) = \sum_{v' \prec v} y(v')$. We split $y(v)$ into two non-negative quantities, $y(v) = g(v) + h(v)$, with:
$g(v) = \min\{y(v), \max\{0, P_x(v) - P_y(v)\}\}$
and $h(v) = \max\{0, y(v) + \min\{0, P_y(v) - P_x(v)\}\}$.

We now claim that $\forall v' \in T$, $\sum_{v \preceq v'} g(v) \leq P_x(v')$ and prove it by induction.

Since $g(r) = 0$, it holds for the root r. Assuming that the inequality holds for a vertex v', let v'' be a child of v'. If $P_x(v'') - P_y(v'') \geq 0$, then we directly have:
$\sum_{v \preceq v''} g(v) = \sum_{v \prec v''} g(v) + g(v'') \leq \sum_{v \prec v''} y(v) + (P_x(v'') - P_y(v'')) = P_x(v'')$.

Else $g(v'') = 0$ and using $\sum_{v \preceq v'} g(v) \leq P_x(v')$ and $P_x(v'') \geq P_x(v')$, the inequality holds for v''; which completes the proof of the claim.

Thus: $\sum_{v'} \sum_{v \preceq v'} g(v) \leq \sum_{v'} P_x(v') = \sum_{v'} \sum_{v \preceq v'} x(v) \leq \sum_{v'} \sum_{v \preceq v'} x(v)$.
Since $w(v) = \sum_{v \preceq v'} 1$, by inverting the sums on both sides, we obtain:

$$\sum_v g(v)w(v) \leq \sum_v x(v)w(v) = \lambda_{Gr} \tag{1}$$

Let us now consider the coefficients $h(v)$. We claim that the coefficients $h(v)$ with $v \in T_i$ satisfy the constraints (i) and (ii) of $\mathcal{P}_i$: indeed for (i), we have $h(v) \leq y(v)$ and y satisfies constraint (i). For (ii) note that $h(v) + P_x(v) = \max\{P_x(v), y(v) + \min\{P_x(v), P_y(v)\}\} \leq \max\{P_x(v), y(v) + P_y(v)\} \leq 1$.

Hence, $\forall i, \sum_{v \in T_i} h(v)w(v) \leq \sum_{v \in T_i} x(v)w(v)$ and therefore:

$$\sum_{v \in T} h(v)w(v) \leq \sum_{v \in T} x(v)w(v) = \lambda_{Gr} \tag{2}$$

Finally, since $g(v) + h(v) = y(v)$, we conclude from Eqs. (1) and (2) that $\beta_F \leq 2\lambda_{Gr}$. Hence the Greedy algorithm is $\frac{1}{2}$-competitive for the online FRAC-TIONAL FIREFIGHTER problem. Since the greedy algorithm gives an integral solution if (f_i) has integral values and since $\beta_F \geq \beta_I$, it is also $\frac{1}{2}$-competitive for the FIREFIGHTER problem. This concludes the proof of theorem 1.

Conjecture 2.3 in [11] (which is also Conjecture 3.5 in [2]) claims that there is a constant ρ such that the optimal value of FRACTIONAL FIREFIGHTER is at most ρ times the optimal value of FIREFIGHTER. It was supported by extensive experimental tests [11], but finding such a constant and proving the ratio is one of the open problems proposed in [2] (Problem 7). It was shown in [15] that such a constant must be greater than $\frac{e}{e-1}$. Theorem 1 can be expressed by $\lambda_{Gr} \leq \beta_I \leq \beta_F \leq 2\lambda_{Gr}$, which shows that $\rho = 2$ is such a constant:

Corollary 1. *In* FRACTIONAL FIREFIGHTER, *the amount of vertices saved is at most twice the maximum number of vertices saved in* FIREFIGHTER.

4   Improved Competitive Algorithm for FIREFIGHTER

In this section, we investigate possible improvements for online strategies for FIREFIGHTER on finite trees. Let $\varphi = \frac{1+\sqrt{5}}{2}$ denote the golden ratio, satisfying $\varphi^2 = \varphi + 1$ and $\frac{1}{\varphi} = \varphi - 1$. For any integer $k \geq 2$, we denote

$$\alpha_{I,k} = \inf_{T \in \mathcal{T}} \max_{OA \in \mathcal{A}_L} \min_{(f_i) \in \mathbb{N}^{\mathbb{N}}, \sum_i f_i \leq k} \frac{\lambda_{OA}}{\beta_I},$$

where $\mathcal{T}$ denotes the set of finite rooted trees and $\mathcal{A}_L$ the set of online algorithms for FIREFIGHTER on finite trees, be the best possible competitive ratio for online FIREFIGHTER on finite trees if at most k firefighters are available. Note that the sequence $(\alpha_{I,k})_k$ is non-increasing. Then, $\alpha_I = \inf_{k \geq 2} \alpha_{I,k}$ is the best possible competitive ratio for online FIREFIGHTER on finite trees. The index I stands for *Integral* and refers to the problem FIREFIGHTER.

In what follows we give an optimal online algorithm for FIREFIGHTER on a finite tree in the case where at most two firefighters are available. Based on Proposition 1 we may assume $f_1 = 1$ and such an online instance is characterised by when the second firefighter is presented.

Lemma 1. *Let a and b be two vertices of maximum weights in T_1. If $\sum_i f_i \leq 2$, there is an optimal offline algorithm for* FIREFIGHTER *which places the first firefighter on either a or b.*

Proof. If the first firefighter is placed on $v \in T_1 \setminus \{a, b\}$ by an optimal offline algorithm and since at most two firefighters are available, $\exists u \in \{a, b\}$, $T[u]$ burns completely. Then replacing v by u when assigning the first firefighter would produce another optimal solution (necessarily $w_v = w_u$).

We suppose *Bob* has this property but note that even if $w_a > w_b$, he will not necessarily choose a; as illustrated by the graph $W_{1,10,20}$ (Fig. 1): if the firefighter sequence is $(1, 0, 1, 0, \ldots)$, then *Bob*'s first move needs to be on x. Note also that, when the root is of degree at least 3, the second firefighter is not necessarily in $V(T[a]) \cup V(T[b])$.

We now consider algorithm 1 and assume that the adversary will reveal at most two firefighters. The case where $f_1 = 2$ is trivial since an online algorithm can make the same decision as *Bob* by assigning both firefighters to two unburnt vertices of maximum weights. So, we consider a binary firefighter sequence.

The algorithm works on an updated version $\widetilde{T}$ of the tree: if one vertex is protected, then the corresponding sub-tree is removed and all the burnt vertices are fused into the new root $\tilde{r}$ so that the algorithm always considers vertices of level 1 in $\widetilde{T}$. Before starting the online process, the algorithm computes all weights of vertices. Weights of unburnt vertices will not change when updating $\widetilde{T}$. The value of $h(\widetilde{T})$, required in line 6 can be computed during the initial calculation of weights and easily updated with $\widetilde{T}$. For clarity, we do not detail all update in the algorithm. The notation $\bar{w}_{v_j}$ used at line 6 is defined in Sect. 1.

Algorithm 1

Require: A finite tree T with root r - An online adversary.
1: $(\widetilde{T}, \tilde{r}) \leftarrow (T, r)$; Compute w_v, $\forall v \in V(\widetilde{T})$
2: {Start of the online process}
3: At each turn, after fire spreads $\widetilde{T}$ is updated - burt vertices are fused to $\tilde{r}$;
4: **if** First Firefighter is presented **and** $\tilde{r}$ has some child **then**
5:     Let a and b denote two children of $\tilde{r}$ with maximum weight w_a, w_b and $w_a \geq w_b$
    ($a = b$ if $\tilde{r}$ has only one child);
6:     **if** $\min\limits_{2 \leq i \leq 1 + h(\widetilde{T})} \frac{w_a + \bar{w}_{b_i}}{w_b + \bar{w}_{a_i}} \geq \frac{1}{\varphi}$ **then**
7:         Place the first firefighter on a;
8:     **else**
9:         Place the first firefighter on b;
10: **if** Second Firefighter is presented **and** $\tilde{r}$ has some child **then**
11:     Place the firefighter on a maximum weight child v of $\tilde{r}$

Theorem 2. *Algorithm 1 is a $\frac{1}{\varphi}$-competitive online algorithm for online* FIRE-FIGHTER *with at most two available firefighters. It is optimal for this case.*

Proof. If the adversary does not present any firefighter before the turn $h(T)$, both algorithm 1 and *Bob* cannot save any vertex and by convention one considers then that the competitive ratio is 1.

Let us suppose that at least one firefighter is presented at some turn $k \leq h(T)$; the tree still has at least one unburnt vertex. During the $(k-1)$ first turns, the instance is updated into $(B^k(T), r_{B^k}, (f_{i+k}))$ with one firefighter presented during the first turn and the root has at least one child. Proposition 1 ensures that it is equivalent to the original instance.

If the root has only one child $a = b$ at line 5 and algorithm 1 selects a and saves all unburnt vertices inducing a competitive ratio of 1.

Else, $a \neq b$ with $w_a \geq w_b$ (line 5). Suppose first that the adversary presents a single firefighter during the whole process, then algorithm 1 places him on a or b while Bob places him on a, saving w_a. If $w_a \geq \varphi w_b$, then we have:

$$\forall i, 2 \leq i \leq 1 + h(T), \frac{w_a + \bar{w}_{b_i}}{w_b + \bar{w}_{a_i}} \geq \frac{w_a}{w_b + w_a} \geq \frac{\varphi w_b}{w_b + \varphi w_b} = \frac{1}{\varphi} \tag{3}$$

Therefore, (see line 6) the unique firefighter is placed on a by the algorithm, guaranteeing a competitive ratio of 1. Else we have $w_b > \frac{1}{\varphi} w_a$ and even placing the firefighter on b guarantees the ratio $\frac{1}{\varphi}$.

Suppose now that the adversary presents two firefighters. We consider two cases.

Case (i): If algorithm 1 places the first firefighter on a in line 7 and if the adversary presents the second firefighter at turn i, then the algorithm will save $w_a + \bar{w}_{x_i}$ for some $x \in T_1 \setminus \{a\}$ such that $\bar{w}_{x_i} = \max_{u \in T_1 \setminus \{a\}} \bar{w}_{u_i}$. For the same instance Bob will save $w_v + \bar{w}_{y_i}$ for some $v \in \{a, b\}$ and $y \in T_1 \setminus \{v\}$. If both solutions are not of the same value (the optimal one is strictly better), then necessarily $v = b$ and $y = a$. In this case the criterion of line 6 ensures that the related competitive ratio is $\frac{1}{\varphi}$.

Case (ii): Suppose now algorithm 1 places the first firefighter on b in line 9, and say the adversary presents the second firefighter at turn j. Line 5 ensures that:

$$\exists i, 2 \leq i \leq 1 + h(T), \frac{w_a + \bar{w}_{b_i}}{w_b + \bar{w}_{a_i}} < \frac{1}{\varphi} \tag{4}$$

It implies in particular that $w_a < \varphi w_b$ since in the opposite case Eq. (3) would hold. In case (ii), algorithm 1 saves $w_b + \bar{w}_{x_j}$ with $x \in T_1 \setminus \{b\}$ such that $\bar{w}_{x_j} = \max_{u \in T_1 \setminus \{b\}} \bar{w}_{u_j}$. Meanwhile, Bob selects $v \in \{a, b\}$ and, if it exists, y_j for some $y \in T_1 \setminus \{v\}$, for a total of $w_v + \bar{w}_{y_j}$ saved vertices. If $y \neq b$, then $\bar{w}_{y_j} \leq \bar{w}_{x_j}$, by definition of x, and thus:

$$\frac{w_b + \bar{w}_{x_j}}{w_a + \bar{w}_{y_j}} \geq \frac{w_b + \bar{w}_{x_j}}{w_a + \bar{w}_{x_j}} \geq \frac{w_b}{w_a} > \frac{1}{\varphi} \tag{5}$$

Finally, if $y = b$, then $v = a$ and the competitive ratio to evaluate is $\frac{w_b + \bar{w}_{x_j}}{w_a + \bar{w}_{b_j}}$. We claim that the following holds:

$$\frac{w_a + \bar{w}_{b_i}}{w_b + \bar{w}_{a_i}} \times \frac{w_b + \bar{w}_{x_j}}{w_a + \bar{w}_{b_j}} \geq \frac{1}{\varphi^2} \tag{6}$$

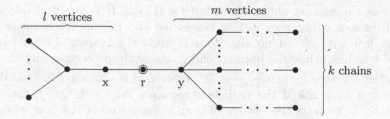

Fig. 1. The fire on graph $W_{k,l,m}$ starts at vertex r. Whether the firefighter should protect x or y on the first turn depends on the firefighter sequence.

If $i \geq j$ we have $\bar{w}_{a_i} \leq \bar{w}_{a_j}$ and since $a \neq b$, $\bar{w}_{a_j} \leq \bar{w}_{x_j}$. Hence:
$\frac{w_b + \bar{w}_{x_j}}{w_b + \bar{w}_{a_i}} \geq \frac{w_b + \bar{w}_{x_j}}{w_b + \bar{w}_{a_j}} \geq 1$ and therefore:

$$\frac{w_a + \bar{w}_{b_i}}{w_b + \bar{w}_{a_i}} \times \frac{w_b + \bar{w}_{x_j}}{w_a + \bar{w}_{b_j}} \geq \frac{w_a + \bar{w}_{b_i}}{w_a + \bar{w}_{b_j}} \geq \frac{w_a}{w_a + w_b}.$$

If now $i < j$, we get: $\frac{w_a + \bar{w}_{b_i}}{w_a + \bar{w}_{b_j}} \geq 1$ and therefore:

$$\frac{w_a + \bar{w}_{b_i}}{w_b + \bar{w}_{a_i}} \times \frac{w_b + \bar{w}_{x_j}}{w_a + \bar{w}_{b_j}} \geq \frac{w_b + \bar{w}_{x_j}}{w_b + \bar{w}_{a_i}} \geq \frac{w_b}{w_a + w_b}.$$

In both cases, since $\frac{w_a}{w_a + w_b} \geq \frac{w_b}{w_a + w_b} \geq \frac{1}{1+\varphi} = \frac{1}{\varphi^2}$ we deduc Eq. (6). Now, Eqs. (4) and (6) imply that in case (ii), when $y = b$, we also have $\frac{w_b + \bar{w}_{x_j}}{w_a + \bar{w}_{b_j}} \geq \frac{1}{\varphi}$. Together with Eq. (5), this concludes case (ii) and shows that algorithm 1 is $\frac{1}{\varphi}$-competitive.

Even though complexity analyses are not usually proposed for online algorithms, it is worth noting that line 6 only requires the weights of vertices in $V(T[a]) \cup V(T[b])$ and the maximum weight per level in $T[a]$ and $T[b]$ and consequently algorithm 1 requires $O(|V(T[a])| + |V(T[b])|)$ to choose the position of the first firefighter and $O(|V(T)|)$ in all.

We conclude this section with a hardness result justifying that the greedy algorithm Gr is optimal while algorithm 1 is optimal if at most two firefighters are available.

Proposition 2. *For all $k \geq 2$, $\frac{1}{2} \leq \alpha_{I,k} \leq \frac{1}{\varphi}$, more precisely:*

(i) $\alpha_I = \frac{1}{2}$, *meaning that the greedy algorithm is optimal for* FIREFIGHTER *in finite trees;*

(ii) $\alpha_{I,2} = \frac{1}{\varphi}$ *meaning that algorithm 1 is optimal if at most two firefighters are available;*

(iii) $\alpha_{I,4} < \frac{1}{\varphi}$.

Proof. Theorem 1 shows that $\alpha_I \geq \frac{1}{2}$. Given integers l, m, k such that $k|m - 1$, we define the graph $W_{k,l,m}$ as shown in Fig. 1. We will assume that $m > k^2$.

(i) Let us consider an online algorithm for $W_{k,l,m}$. If $f_1 = 1$, the algorithm will protect either x or y. If x is selected and the firefighter sequence is $(1, 1, 0, 0, \ldots)$, our online algorithm protects the branch of x and one of the k chains, while the optimal offline algorithm protects y and the star. Its performance is then $\frac{l+\frac{m-1}{k}}{l+m-1}$. If, however, y is protected instead during the first turn and if the firefighter sequence is $(1, 0, 1, 1, \ldots)$, the online algorithm protects the branch of y and one vertex of the star whilst the optimal algorithm protects the branch of x as well as the k chains, minus for $\frac{k(k+1)}{2}$ vertices. If $l = m$, for large values of m, the online algorithm which protects x is more performant and its competitive ratio tends to $\frac{1+\frac{1}{k}}{2}$. Considering, for instance, W_{k,k^3,k^3}, when $k \to +\infty$ shows that $\alpha \leq \frac{1}{2}$. Since the greedy algorithm Gr guarantees $\alpha_I \geq \frac{1}{2}$ we have $\alpha_I = \frac{1}{2}$.

(ii) Consider the sequence of graphs $W_{1,l,\lfloor \varphi l \rfloor}$. If the online algorithm protects x, the adversary selects the sequence $(1, 0, \ldots)$, whereas if the online algorithm protects y, $(1, 0, 1, 0, \ldots)$ is selected. In both cases, the performance tends to $\frac{1}{\varphi}$ when $l \to +\infty$.

(iii) If at most 4 firefighters are available, the graph $W_{4,901,1001}$ (Fig. 1) gives an example where $\frac{1}{\varphi}$ cannot be reached. Ideed, if $f_1 = 1$ and the online algorithm protects x, then the adversary will select the sequence $(1, 1, 0, \ldots)$, as in the proof (i), for a performance of $\frac{1151}{1901}$. If the online algorithm protects y, since firefighters are limited to 4, the adversary will select $(1, 0, 1, 1, 1, 0, \ldots)$, for a performance of $\frac{1002}{1645}$. This second choice is slightly better, but $\frac{1002}{1645} < \frac{1}{\varphi}$.

We have also proved that Theorem 2 holds if three firefighters are presented (i.e., $\alpha_{I,3} = \frac{1}{\varphi}$). However, the proof involves a much more technical case-by-case analysis and will not be detailed here.

5   Firefighting on Trees with Linear Growth

In this section, we consider infinite trees. We say that a rooted tree (T, r) has *linear growth* if the number of vertices per level increases linearly, i.e. $|T_i| = \mathcal{O}(i)$.

Remark 1. The linear growth property of T remains if we choose a different root r'. Indeed, if d is the distance between r and r', the set of vertices at distance i from r' is included in $\bigcup_{j=i-d}^{i+d} T_j$, the cardinal of which is a $\mathcal{O}(i)$.

Given two firefighter sequences (f_i) and (f_i'), we say that (f_i) is *stronger* than (f_i') if for all k, $\sum_{i=1}^{k} f_i \geq \sum_{i=1}^{k} f_i'$.

Lemma 2. *If the fire can be contained in an instance $(G, r, (f_i'))$ and if (f_i) is stronger than (f_i'), then the fire can also be contained in $(G, r, (f_i))$.*

Proof. Given a winning strategy in the instance $(G, r, (f_i'))$, if (f_i) firefighters are available, we contain the fire by protecting the same vertices, eventually earlier than in the initial strategy.

Theorem 3. *Let $(T, r, (f_i))$ be an online instance of* FRACTIONAL FIRE-FIGHTER. *If T has linear growth and (f_i) is stronger than some non-zero periodic sequence, then the fire can be contained.*

The proof of theorem 3 will use the following lemma:

Lemma 3. *For any $a > 0$, $\lim_{n \to +\infty} \prod_{j=1}^{n} \frac{aj-1}{aj} = 0$*

Proof. We have $ln \prod_{j=1}^{n} \frac{aj-1}{aj} = \sum_{j=1}^{n} ln(1 - \frac{1}{aj})$ and $\sum_{j=1}^{n} ln(1 - \frac{1}{aj}) \sim -\sum_{j=1}^{n} \frac{1}{aj}$. So $\sum_{j=1}^{n} ln(1 - \frac{1}{aj}) \to -\infty$ and $\prod_{j=1}^{n} \frac{aj-1}{aj} \to 0$.

We can now prove Theorem 3:

Proof. Since T has linear growth, let C be such that $\forall i, |T_i| \leq Ci$. For all n, let $(\delta_{n|i})$ denote the firefighter sequence where one firefighter is available every n turns; i.e. δ_i^n is equal to 1 if $n|i$ and 0 otherwise. That the sequence (f_i) is stronger than a non-zero periodic sequence means that (f_i) is stronger than $(\delta_{n|i})$, for all n greater than some m. First we will give an offline strategy to contain the fire with one firefighter every n turns. Then, we will show that online instances with (f_i) stronger than a $(\delta_{n|i})$ known to the player are winning. Finally, we will describe the winning strategy when such a $(\delta_{n|i})$ is unknown.

Given an integer n, let us first consider the instance $(T, r, (\delta_{n|i}))$. It follows from Lemma 3 that there exists an integer N such that $\prod_{j=1}^{N} \frac{Cnj-1}{Cnj} < \frac{1}{2Cn}$. Let $h(n) = 2nN$. A winning strategy is obtained by protecting at turn nj the unprotected vertex of T_{nj} with the highest number of descendants in level $h(n)$. Since $|T_{nj}| \leq Cnj$, the remaining number of unprotected vertices in $T_{h(n)}$ is reduced by at least $\frac{1}{Cnj}$ of its previous value. So the number of unprotected vertices of $T_{h(n)}$ remaining after nN turns is less than $|T_{h(n)}| \prod_{i=1}^{N} \frac{Cnj-1}{Cnj} \leq \frac{|T_{h(n)}|}{2Cn} \leq N$. Since N firefighters remain to be placed between turns N and $h(n)$, the strategy is winning in at most $h(n)$ turns.

If the player knows that (f_i) is stronger than $(\delta_{n|i})$, the above strategy can be adapted using Lemma 2.

In the general case, the player knows that (f_i) is stronger than $(\delta_{n|i})$ for some n, but he does not know which n. The strategy proceeds as follows: we initially play as though under the assumption that (f_i) is stronger than $(\delta_{n_0|i})$, with $n_0 = 100$. If the fire is not contained by turn $h(n_0)$, or later on by turn $h(n_k)$, we choose $n_{k+1} = h(n_k)(\lceil S_{h(n_k)} \rceil + 1)$, where $S_n = \sum_{i=1}^{n} f_i$. We now assume that (f_i) was stronger than $(\delta_{h(n_k)|i})$. It follows that after cancelling the first $h(n_k)$ terms of (f_i), the resulting sequence is stronger than $(\delta_{n_{k+1}|i})$. So we can consider that the first $h(n_k)$ turns were wasted and follow the strategy for n_{k+1} until turn $h(n_{k+1})$. Eventually, this strategy will win when n_k is large enough.

6   Final Remarks

In this paper, we introduce the online version of (FRACTIONAL) FIREFIGHTER and propose first results for both the finite and the infinite cases. To our knowledge, Theorem 1 is the first non trivial competitive (but also approximation) analysis for FRACTIONAL FIREFIGHTER and a first question would be to investigate whether a better competitive ratio can be obtained for FRACTIONAL FIRE-FIGHTER in finite trees. Even though the case of trees is already challenging despite allowing many simplifications, the main open question will be to study online (FRACTIONAL) FIREFIGHTER PROBLEM in other classes of finite graphs.

References

1. Hartnell, B.: Firefighter! an application of domination. Presented at the 10th Conference on Numerical Mathematics and Computing, University of Manitoba in Winnipeg, Canada (1995)
2. Finbow, S., MacGillivray, G.: The firefighter problem: a survey of results, directions and questions. Australas. J. Comb. **43**(6), 57–77 (2009)
3. Finbow, S., King, A., Macgillivray, G., Rizzi, R.: The firefighter problem for graphs of maximum degree three. Discrete Math. **307**(16), 2094–2105 (2007)
4. Fomin, F.V., Heggernes, P., van Leeuwen, E.J.: The firefighter problem on graph classes. Theor. Computut. Sci. **613**(C), 38–50 (2016)
5. Cai, L., Verbin, E., Yang, L.: Firefighting on trees: (1–1/e)-approximation, fixed parameter tractability and a subexponential algorithm. In: Proceedings of 19th International Symposium on Algorithms and Computation, ISAAC 2008, Gold Coast, Australia, 15–17 December 2008, pp. 258–269 (2008)
6. Bazgan, C., Chopin, M., Ries, B.: The firefighter problem with more than one firefighter on trees. Discrete Appl. Math. **161**(7–8), 899–908 (2013)
7. Bazgan, C., Chopin, M., Cygan, M., Fellows, M.R., Fomin, F.V., van Leeuwen, E.J.: Parameterized complexity of firefighting. J. Comput. Syst. Sci. **80**(7), 1285–1297 (2014)
8. Hartnell, B., Li, Q.: Firefighting on trees: how bad is the greedy algorithm? Congressus Numerantium **145**, 187–192 (2000)
9. Adjiashvili, D., Baggio, A., Zenklusen, R.: Firefighting on trees beyond integrality gaps. In: Proceedings of the Twenty-Eighth Annual ACM-SIAM Symposium on Discrete Algorithms, SODA 2017, Barcelona, Spain, Hotel Porta Fira, 16–19 January 2017, pp. 2364–2383 (2017)
10. Anshelevich, E., Chakrabarty, D., Hate, A., Swamy, C.: Approximability of the firefighter problem - computing cuts over time. Algorithmica **62**(1–2), 520–536 (2012)
11. Hartke, S.G.: Attempting to narrow the integrality gap for the firefighter problem on trees. In: Discrete Methods in Epidemiology, pp. 225–232 (2004)
12. MacGillivray, G., Wang, P.: On the firefighter problem. J. Comb. Math. Comb. Comput. **47**, 83–96 (2003)
13. Fogarty, P.: Catching the Fire on Grids, Ph.D. thesis. University of Vermont (2003)
14. Ausiello, G., Becchetti, L.: On-line algorithms. In: Paschos, V.T. (ed.) Paradigms of Combinatorial Optimization: Problems and New Approaches, vol. 2, pp. 473–509. ISTE - WILEY, London - Hoboken (2010)
15. Chalermsook, P., Vaz, D.: New integrality gap results for the firefighters problem on trees. In: Jansen, K., Mastrolilli, M. (eds.) WAOA 2016. LNCS, vol. 10138, pp. 65–77. Springer, Cham (2017). https://doi.org/10.1007/978-3-319-51741-4_6

A Multigraph Formulation for the Generalized Minimum Spanning Tree Problem

Ernando Gomes de Sousa[1], Rafael Castro de Andrade[2],
and Andréa Cynthia Santos[3(✉)]

[1] Federal Institute of Education, Science and Technology of Maranhão,
BR-230, Km 319, Zona Rural,
São Raimundo das Mangabeiras, Maranhão 65840-000, Brazil
ernando.sousa@ifma.edu.br
[2] Department of Statistics and Applied Mathematics, Federal University of Ceará,
Campus do Pici, Bloco 910., Fortaleza, Ceará 60455-900, Brazil
rca@lia.ufc.br
[3] Charles Delaunay Institute, LOSI, Technological University of Troyes,
12, rue Marie Curie, CS 42060, 10004 Troyes Cedex, France
andrea.duhamel@utt.fr

Abstract. Given a connected, undirected and m-partite complete graph
$G = (V_1 \cup V_2 \cup ... \cup V_m; E)$, the Generalized Minimum Spanning Tree
Problem (GMSTP) consists in finding a tree with exactly $m - 1$ edges,
connecting the m clusters $V_1, V_2, ..., V_m$ through the selection of a unique
vertex in each cluster. GMSTP finds applications in network design, irrigation agriculture, smart cities, data science, among others. This paper
presents a new multigraph mathematical formulation for GMSTP which
is compared to existing formulations from the literature. The proposed
model proves optimality for well-known GMSTP instances. In addition,
this work opens new directions for future research to the development
of sophisticated cutting plane and decomposition algorithms for related
problems.

Keywords: Generalized Minimum Spanning Tree
Trees of multigraph · Mixed integer linear programming formulations

1 Introduction

The Generalized Minimum Spanning Tree Problem (GMSTP) is defined in a
connected, undirected and m-partite complete graph $G = (V, E)$. Its vertex set
V is partitioned into m clusters, with $V = V_1 \cup V_2 \cup ... \cup V_m$ and $V_r \cap V_q = \emptyset$, for
all $r \neq q$ with $r, q \in M = \{1, ..., m\}$. Its edge set E is given by $E = \{\{i, j\} \mid i \in
V_r, j \in V_q\}$, for all $r \neq q$ with $r, q \in M$, for which $c_e \in R_+$ denotes the edge cost
of $e = \{i, j\} \in E$. GMSTP consists in finding a minimum cost tree, spanning a
unique vertex in each cluster, with exactly $m-1$ edges connecting the m clusters.

© Springer International Publishing AG, part of Springer Nature 2018
J. Lee et al. (Eds.): ISCO 2018, LNCS 10856, pp. 133–143, 2018.
https://doi.org/10.1007/978-3-319-96151-4_12

The GMSTP extends the so-called Minimum Spanning Tree (MST), which is the base of difficult problems and applications [20].

Classical optimization problems were generalized by means of partitioning the set of vertices V, as it is the case of GMSTP, the generalized traveling salesman problem [6] and the generalized vehicle routing problem [3]. The generalization counterpart of these problems belongs to the NP-hard class of problems. In particular, the GMSTP complexity was proved by Myung, Lee and Tcha [16] applying a reduction to the vertex cover problem. GMSTP has applications in network design, irrigation agriculture, smart cities, data science, among others.

In regarding mathematical models, Myung, Lee and Tcha [16] presented four formulations for GMSTP. One formulation has a polynomial number of restrictions and variables, while the others have an exponential number of restrictions. The authors also developed a Branch and Bound (B&B) using the linear relaxation of their polynomial formulation. In the study of Feremans, Labb and Laporte [9], four formulations were proposed and compared with the ones of [16]. Moreover, they investigate the linear relaxations polytopes of the proposed formulations and show that four polytopes are strictly contained in the remaining ones. One may note that three of the best formulations in the literature have an exponential number of constraints, while only one has a polynomial number of variables and constraints [16].

A mathematical formulation, an approximation algorithm and a Lagrangian relaxation for GMSTP were developed by Pop [18]. Given an instance for GMSTP, let us consider ω as the number of vertices in its biggest cluster and O^* as its optimal solution value. The approximation algorithm of [18] obtains solutions with rate up to $\omega(2 - 2/m) \times O^*$. It is important to highlight that their Lagrangian relaxation found better lower bounds than the linear relaxation of the eight mathematical formulations presented in [9]. Another approximation algorithm is found in [4], where the approximation rate is limited by $(1 + 4\sqrt{2} + 2\sqrt{2}/O^*)$ considering GMSTP grid graphs, where each 1×1 cell is a cluster.

The mathematical formulation proposed by [18] makes use of a graph G and a global simple graph G^* obtained as follows. Each vertex of G^* represents a cluster of G. Two clusters are connected by an edge in G^* if some edge in G has these clusters as its extremities. All edges in G^* have the same cost. The idea is to address the spanning tree constraints in G^*, while specific GMSTP constraints are handled using G. Tests were performed with Euclidean and non-Euclidean instances with up 240 vertices and 40 clusters. Other works use a different strategy to model the GMSTP such as transforming the problem into a Steiner tree problem. This is the case of Duin, Volgenant and Voss [7] and Golden, Raghavan and Stanojevic in [13].

Exact and heuristic algorithms are also available in the literature. For instance, a Branch-and-Cut (B&C) algorithm and several valid inequalities for GMSTP were proposed by Feremans, Labbé and Laporte [10]. The authors proved that many of the proposed inequalities are facets. The B&C uses a

Tabu Search as a local search and valid inequalities. Tests were performed with Euclidean and non-Euclidean instances with up to 200 vertices and 40 clusters.

A large number of metaheuristics were developed for GMSTP such as Tabu Search [8,12,17,22], Simulated Annealing [18], Variable Neighborhood Search [12,14], Greedy Randomized Adaptive Search Procedure (GRASP) [11] and Genetic Algorithm (GA) [5,13,21]. The work of Pop et al. [19] presents a two-level optimization approach for GMSTP, combining a GA and a dynamic programming algorithm. Among such heuristics, [5,21] produced the best known results.

In this work, we propose a novel mathematical formulation based on multigraph, which performs very well on known instances from the literature. Moreover, the proposed model can be easily extended to other generalized optimization problems and can be applied as a basis of cutting-plane methods.

The remainder of this work is organized as follows. In Sect. 2, the main GMSTP models from the literature are reviewed. The new GMSTP model is presented in Sect. 3. Numerical experiments are reported in Sect. 4. Concluding remarks and perspectives for future works are given in Sect. 5.

2 Models from the Literature

The model proposed by Myung et al. [16], uses a multicommodity flow idea to handle GMSTP. Initially, every edge $[i,j] \in E$ is replaced by two arcs (i,j) and (j,i) having same cost of edge $[i,j]$, resulting in a set of arcs A. Let us consider a cluster containing a root node, that is cluster $\{1\}$. The idea is to build an arborescence T, rooted at some node in V_1, spanning exactly one node of the remaining clusters $M \setminus \{1\}$. For this purpose, the root node sends $|M| - 1$ units of flow (commodities), one unit to exactly one node of every other cluster $k \in M \setminus \{1\}$. Only arcs defining T can transport these commodities. In model $(\mathcal{F}_{MLT})$, consider a node variable $y_i = 1$ if node i is in the solution, and $y_i = 0$ otherwise. Moreover, an arc variable $w_{ij} = 1$ if arc (i,j) is in the solution; and $w_{ij} = 0$, otherwise. Finally, let a non-negative variable f_{ij}^k be the flow from the root cluster to cluster k in arc (i,j). The model is given by

$$(\mathcal{F}_{MLT}) \quad \min \sum_{(i,j)\,\in\,A} c_{ij} w_{ij} \tag{1}$$

$$s.t. \quad \sum_{i \in V_k} y_i = 1, \qquad \forall k \in M \tag{2}$$

$$\sum_{(i,j)\in A:\, j\in V_k} w_{ij} \leq 1, \qquad \forall k \in M \setminus \{1\} \tag{3}$$

$$\sum_{j:(i,j)\in A} f_{ij}^k - \sum_{j:(j,i)\in A} f_{ji}^k = \left\{ \begin{array}{ll} y_i, & i \in V_1 \\ -y_i, & i \in V_k \\ 0, & i \notin V_1 \cup V_k \end{array} \right\} \forall k \in M \setminus \{1\},\ \forall i \in V \tag{4}$$

$$f_{ij}^k \leq w_{ij}, \qquad \forall (i,j) \in A;\quad \forall k \in M \setminus \{1\} \tag{5}$$

$$f_{ij}^k \geq 0, \qquad \forall (i,j) \in A;\quad \forall k \in M \setminus \{1\} \tag{6}$$

$$w_{ij} \in \{0,1\}, \forall (i,j) \in A \tag{7}$$
$$y_i \in \{0,1\}, \forall i \in V \tag{8}$$

Constraints (2) state that exactly one node of each cluster belongs to the solution T. Inequalities (3) define that at most one arc enters each node in T. The well known flow conservation constraints are provided in (4). Constraints (5) establish that if an arc is not in solution, then the flow in this arc is null. The domain of the variables are given from (6) to (8). Further details on this model can be found in [16].

The model of Pop [18] determines a spanning tree topology Z connecting all clusters in the global graph G^* previously defined. Moreover, if two clusters r_1 and r_2 are connected by an edge $[r_1, r_2] \in E^*$ in Z, then a decision is done among all the original edges connecting these two clusters, to obtain which original one $[i,j] \in E$ of G represents $[r_1, r_2]$ of Z. Pop's model makes use of the Martin's spanning tree polytope [15], defined in the following. Let us represent a spanning tree Z of G^*. In addition, a variable z_{rq} is associated with every edge $[r, q] \in E^*$, $r, q \in V^*$, where $z_{rq} = 1$ if $[r, q]$ is in Z; and $z_{rq} = 0$, otherwise. For every node $k \in V^*$ and every edge $[r, q] \in E^*$, variables λ_{krq} define if edge $[r, q]$ belongs to Z when $\lambda_{krq} = 1$, and in this case, it is oriented from q to r w.r.t. node k; and $\lambda_{krq} = 0$, otherwise. Any spanning tree Z of G^* is given by $(\mathcal{F}_{Martin})$ below. $N(i)$ denotes the set of nodes adjacent to node i in a given graph.

$$(\mathcal{F}_{Martin}) \qquad \sum_{[i,j] \in E^*} z_{ij} = m - 1 \tag{9}$$

$$z_{ij} - \lambda_{kij} - \lambda_{kji} = 0, \qquad \forall k \in V^*, \forall [i,j] \in E^* \tag{10}$$

$$\sum_{j \in N(i)} \lambda_{kij} = 1, \qquad \forall k \in V^*, \forall i \in V^* \setminus \{k\} \tag{11}$$

$$\lambda_{kkj} = 0, \qquad \forall k \in V^*, \forall j \in N(k) \tag{12}$$

$$z \geq 0 \tag{13}$$

$$\lambda \geq 0 \tag{14}$$

Following (9), the tree Z has $m - 1$ edges. Constraints (10) establish that if an edge $[i,j]$ is in Z, then an orientation, induced by the λ variables, is obtained w.r.t. every node $k \in V^*$. Considering Z rooted at k, constraints (11) determine that exactly one edge is oriented to every other node in $V^* \setminus \{k\}$. Domain of variables are defined from (13) to (14). Pop's model [18] has variables $x_{ij} = 1$ if edge $(i,j) \in E$, of the original graph G, is in the solution; and $x_{ij} = 0$, otherwise. It is given by

$$(\mathcal{F}_{Pop}) \qquad \min \sum_{[i,j] \in E} c_{ij} x_{ij} \tag{15}$$

$$s.t. \quad (2), (9)-(13), \text{ and}$$

$$\sum_{[i,j] \in E} x_{ij} = m - 1, \tag{16}$$

$$\sum_{i \in V_r} \sum_{j \in V_q} x_{ij} = z_{rq}, \qquad \forall [r, q] \in E^* \tag{17}$$

$$\sum_{j \in V_r} x_{ij} \leq y_i, \qquad \forall r \in M, \, \forall i \in V \setminus V_r \tag{18}$$

$$x_{ij} \in [0,1], \qquad \forall [i,j] \in E \tag{19}$$

$$y_i \in [0,1], \qquad \forall i \in V \tag{20}$$

$$z_{ij} \in \{0,1\}, \qquad \forall [i,j] \in E^* \tag{21}$$

Constraint (16) states that $m-1$ original edges of G belong to the solution. Equalities (17) define if an edge $[r,q]$ is in Z, then an original edge connecting clusters r and q is in the solution. Constraints (18) determine that a node $i \in V \setminus V_r$ can be adjacent to at most one edge of a given cluster r if i belongs to the solution. The domain of variables is defined from (19) to (21).

3   A Multigraph Formulation for GMSTP

The proposed GMSTP formulation is inspired by the formulation of Andrade [2], used for determining a forest of size f for any type of graph: complete or not, directed or not, and having one or many edges (or arcs) between any pair of nodes. Given a graph G, previously defined, the multigraph $H(G) = (V', E')$ is obtained as follow. Each cluster of G is considered as a vertex of V' and each edge $e \in E$ corresponds to exactly an edge $e' \in E'$ of same cost c_e, with $|E| = |E'|$. The general idea of the model is to simultaneously define arborescences, one for every vertex (cluster) $k \in V'$ distinguished as root, such that each root k determines an orientation of the edges in the solution according to the characterization theorem for forest in graphs (see Theorem 1 in Adasme et al. [1]).

Let us consider the following variables. First, let $\rho(v)$ denote the cluster containing vertex $v \in V$. Let an edge $e \in E$ be represented as an ordered pair $e = [i(e), j(e)]$, where the notations $i(e)$ and $j(e)$ indicate the first and the second coordinates of e, respectively. For every edge $e = [i(e), j(e)] \in E$ (or equivalently $e = [\rho(i(e)), \rho(j(e))] \in E'$), variables $x_e = 1$ define if e belongs to the solution; and $x_e = 0$, otherwise. Let every vertex (cluster) $k \in V'$ be an orientation for every edge e in the solution as follow. For every cluster $k \in V'$, $w_e^{k,0} = 1$ if $x_e = 1$ and in this case, edge $e = [\rho(i(e)), \rho(j(e))]$ is oriented from cluster $\rho(j(e))$ to cluster $\rho(i(e))$; and $w_e^{k,0} = 0$, otherwise. Following a similar idea, for every cluster $k \in V'$, $w_e^{k,1} = 1$ if $x_e = 1$ and edge $e = [\rho(i(e)), \rho(j(e))]$ is oriented from cluster $\rho(i(e))$ to cluster $\rho(j(e))$; and $w_e^{k,1} = 0$, otherwise. Finally, for every node $s \in V$ of the original graph G, a binary variable $y_s = 1$ if s belongs to the solution; and $y_s = 0$, otherwise. The proposed multigraph formulation for GMSTP is given from (22) to (32).

Constraint (23) defines that exactly $m-1$ edges are in the solution. Equalities (24) state that exactly a vertex of each cluster is selected. Constraints (25) determine that either variable $w_e^{k,0} = 1$ or variable $w_e^{k,1} = 1$ will define the orientation of edge $e \in E'$ w.r.t. every node $k \in V'$ if e belongs to the solution. Constraints (26) guarantee that for a given cluster $k \in V'$ and for every cluster $u \in V'$, with $u \neq k$, there is exactly one edge $e \in E'$ incident to u such that its orientation points to cluster u w.r.t. cluster k. Constraints (27) have the same idea of (26) while considering that node s is the chosen node of its cluster. These constraints guarantee that if node s does not belong to the solution, then all variables referred to the orientations of the edges incident to s must be zero. We can prove that constraints (27) are stronger than (18) because they consider all edges incident to s from all the others clusters while in (18) only edges incident to s from a given cluster are considered. Note that, by fixing at zero such variables, the corresponding x variables are also fixed at zero by (25). Constraints (28) and (29) state that there is no edge oriented from any other cluster to k. The domain of the variables are defined in (30) and (31). The model contains $(|V| + |E'| + 2|V'| \times |E'|)$ variables (only $|V|$ are required to be binary) and $(1 + |V'| + |V'||E'| + |V'|(|V'| - 1)(|V'| + 1))$ constraints.

$$(\mathcal{F}_{SAS}) \qquad \min \sum_{e \in E'} c_e x_e \tag{22}$$

$$s.t. \qquad \sum_{e \in E'} x_e = m - 1 \tag{23}$$

$$\sum_{s \in V_k} y_s = 1, \quad \forall k \in M \tag{24}$$

$$x_e - w_e^{k,0} - w_e^{k,1} = 0, \quad \forall e \in E', \ \forall k \in V' \tag{25}$$

$$\sum_{\substack{e \in E' \\ i(e)=u}} w_e^{k,0} + \sum_{\substack{e \in E' \\ j(e)=u}} w_e^{k,1} = 1, \quad \forall k \in V', \forall u \in V' \setminus \{k\} \tag{26}$$

$$\sum_{\substack{e \in E \\ i(e)=s}} w_{[\rho(i(e)),\rho(j(e))]}^{k,0} + \sum_{\substack{e \in E \\ j(e)=s}} w_{[\rho(i(e)),\rho(j(e))]}^{k,1} = y_s, \quad \forall k \in V', \forall u \in V' \setminus \{k\}, \forall s \in V_u \tag{27}$$

$$w_e^{k,0} = 0, \quad \forall k \in V' : k = i(e), \ \forall e \in E' \tag{28}$$

$$w_e^{k,1} = 0, \quad \forall k \in V' : k = j(e), \ \forall e \in E' \tag{29}$$

$$w_e^{k,0}, \ w_e^{k,1} \in [0,1], \forall k \in V', \ \forall e \in E' \tag{30}$$

$$y_s \in \{0,1\}, \ \forall s \in V \tag{31}$$

$$x_e \in [0,1], \ \forall e \in E' \tag{32}$$

A correctness proof of model $(\mathcal{F}_{SAS})$ is omitted. In fact, it is based on the fact that for a given $\bar{y} \in \{0,1\}^{|V|}$ satisfying (24), the resulting model is total dual integral. Thus, its corresponding optimal solution is a minimum spanning tree induced by the subgraph w.r.t. the nodes whose entries in $\bar{y}$ are equal to 1. Note that, excepting the bounds on w and x variables, all remaining constraints of the new model have to be satisfied at equality. Thus, an interesting question is to check if some of them are facets of the convex hull of integer feasible solutions.

4   Computational Results

Experiments were performed on an Intel Core i7-3770 processor 3.4 GHz, 8 cores, 16 GB RAM and Ubuntu 14.04 LTS Linux operating system. The software IBM

CPLEX solver 12.7.1, using C++, was used to run the models with a time limit of 3600 s. The goal of the experiments is to evaluate the performance of the proposed model ($\mathcal{F}_{SAS}$) compared to the following GMSTP formulations from literature: the polynomial multicommodity flow formulation of Myung, Lee and Tcha [16] and the formulation of Pop [18], refereed respectively here as ($\mathcal{F}_{MLT}$) and ($\mathcal{F}_{Pop}$). Tests with formulation of ($\mathcal{F}_{Pop}$) use all clusters as root nodes. This, because we observed in our numerical experiments that iteratively breaking cycles showed not to be more effective than using all nodes as root in model ($\mathcal{F}_{Pop}$).

A benchmark set of 40 instances is used in the experiments: 20 instances are original from Öncan, Cordeau and Laporte [17] and the 20 remaining were created by taking a subset of vertices of instances in [17]. We apply a preprocessing procedure [11] to eliminate some edges that will never belong to any optimal solution of the problem. This allowed to handle instances with up to 3 hundreds nodes. It is worth mentioning that without applying a preprocessing technique, only small instances can be solved to optimality by all models in a reasonable execution time. The preprocessing procedure allowed us to eliminate, on average, 85% of the original edges of the benchmark instances in an average running time of 1.37 s. The preprocessing works as follows. Let $P_{uv} = \{P_1, P_2, \ldots, P_Q\}$ be the set of all paths between vertices u and v not containing edge $[u, v]$ of cost c_{uv}. Let $\bar{c}_k$ be the biggest edge cost of each path $P_k \in P_{uv}$, $k = 1, \ldots, Q$. Moreover, consider $b_{uv} = \min_{k=1,\ldots,Q} \bar{c}_k$. In every optimal GMSTP solution T^*, if $b_{uv} < c_{uv}$, then $[u, v]$ does not belong to T^*. If $b_{uv} \le c_{uv}$ and T^* contains $[u, v]$, then we can show that there exists other optimal solution of same cost that does not contain this edge. In both cases, edge $[u, v]$ can be eliminated. The number of paths in P_{uv} can be exponential. Nevertheless, Ferreira et al. [11] show that it is sufficient to check if there exists a subset of clusters, distinct of the ones of u and v, where all paths from u to v passing by these clusters have their biggest edge cost smaller than c_{uv}. Thus, here only subsets formed by one cluster and by two clusters distinct of the ones of $[u, v]$ are considered.

Figure 1 illustrates the preprocessing procedure for the case where there exists one distinct cluster (cluster 2) in the three paths (excluding edge [2, 6]) from node

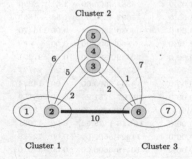

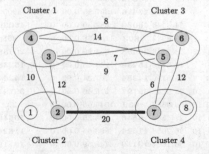

Fig. 1. Case of paths with one transit cluster from node 2 to 6.

Fig. 2. Case of paths with two transit clusters from node 2 to 7.

2 to 6. The smallest cost among the most costly edges of these paths is equal to $b_{2,6} = 2$. As $b_{2,6} < c_{2,6} = 10$, edge $[2, 6]$ can be eliminated. In Fig. 2, there are two distinct clusters (clusters 1 and 3) in the four paths (excluding edge $[2, 7]$) from node 2 to 7. The smallest cost among the most costly edges of these paths has cost equal to $b_{2,7} = 12$. As $b_{2,7} < c_{2,7} = 20$, edge $[2, 7]$ can be eliminated.

Table 1. Numerical results for the models

Intances				$(\mathcal{F}_{SAS})$ MIP				$(\mathcal{F}_{MLT})$ MIP				$(\mathcal{F}_{Pop})$ MIP			
Inst	μ	m	$|E|$	Cost	Time	GAP	Nodes	Cost	Time	GAP	Nodes	Cost	Time	GAP	Nodes
gr159	3	48	1649	36335	91.14	0.00	0	36335	198.48	0.00	0	36335	415.93	0.00	19667
gil140	3	78	578	1220	49.84	0.00	0	1220	37.17	0.00	0	1220	102.46	0.00	143
pr164	3	73	834	22692	119.41	0.00	0	22692	217.35	0.00	0	22692	3600.00	0.00	318230
pr199	3	72	1197	17241	227.19	0.00	0	17241	151.53	0.00	0	17241	3600.00	0.00	35736
lin218	3	76	1803	17739	437.20	0.00	0	17739	865.61	0.00	0	17739	928.56	0.00	5645
gr159	5	28	2916	28081	110.58	0.00	0	28081	223.33	0.00	0	28081	255.19	0.00	37554
gil140	5	59	698	1059	55.10	0.00	0	1059	50.87	0.00	0	1059	1502.29	0.00	15146
pr164	5	41	1346	18045	41.97	0.00	0	18045	90.67	0.00	0	18045	29.04	0.00	13778
pr199	5	49	2107	14431	233.82	0.00	0	14431	418.81	0.00	0	14431	3600.00	0.00	65927
lin218	5	50	2493	14407	321.12	0.00	0	14407	644.00	0.00	0	14407	2068.96	0.00	28023
gr159	7	21	3699	24198	52.25	0.00	0	24198	218.44	0.00	0	24198	107.92	0.00	29038
gil140	7	45	1094	872	47.09	0.00	0	872	95.43	0.00	0	872	591.07	0.00	6899
pr164	7	35	1533	17109	33.11	0.00	0	17109	64.03	0.00	0	17109	14.47	0.00	5976
pr199	7	37	2135	12585	103.51	0.00	0	12585	149.73	0.00	0	12585	3600.00	0.00	477100
lin218	7	36	3488	11870	229.78	0.00	0	11870	492.76	0.00	0	11870	3600.00	0.00	76804
gr159	10	14	4943	18813	36.42	0.00	0	18813	161.60	0.00	0	18813	51.16	0.00	17284
gil140	10	35	1659	726	50.39	0.00	0	726	111.50	0.00	0	726	3600.00	0.00	44272
pr164	10	21	2124	15889	9.94	0.00	0	15889	92.78	0.00	0	15889	0.31	0.00	0
pr199	10	27	3087	9826	108.70	0.00	0	9826	161.65	0.00	0	9826	751.00	0.00	246160
lin218	10	26	5135	8223	190.37	0.00	0	8223	503.25	0.00	0	8223	1410.25	0.00	138179
gr229	3	81	2778	74819	790.26	0.00	0	74819	3566.02	0.00	0	74819	3600.00	0.00	7655
gil262	3	95	2817	1255	3511.96	0.00	0	3763	3600.00	199.84	0	6007	3600.00	378.65	22
pr264	3	101	1526	29199	1425.47	0.00	0	29199	1330.97	0.00	0	29215	3600.00	0.05	3105
pr299	3	102	1960	23096	1221.02	0.00	0	23096	2356.20	0.00	0	23135	3600.00	0.17	2398
lin300	3	103	2633	23083	2620.33	0.00	0	51149	3600.00	121.59	0	23157	3600.00	0.32	2142
gr229	5	47	4569	54236	815.63	0.00	0	137333	3600.00	153.21	0	54236	3600.00	0.00	57393
gil262	5	63	5653	984	3272.10	0.00	0	3176	3600.00	222.76	0	1005	3600.00	2.13	742
pr264	5	55	2696	21351	416.20	0.00	0	21351	635.92	0.00	0	21351	3600.00	0.00	117217
pr299	5	69	3633	18582	1508.28	0.00	0	64042	3600.00	244.65	0	18881	3600.00	1.61	3890
lin300	5	62	4328	17416	1357.85	0.00	0	49053	3600.00	181.65	0	17692	3600.00	1.58	1865
gr229	7	34	6540	46030	587.30	0.00	0	46030	3091.37	0.00	0	46030	3600.00	0.00	115470
gil262	7	49	6583	802	1703.58	0.00	0	2986	3600.00	272.32	0	802	3600.00	0.00	689
pr264	7	43	3848	20438	552.99	0.00	0	20438	1192.56	0.00	0	20439	3600.00	0.00	197271
pr299	7	47	5446	15238	1254.45	0.00	0	15238	2488.00	0.00	0	15277	3600.00	0.26	24177
lin300	7	46	5199	14234	767.34	0.00	0	14234	1588.00	0.00	0	14234	3600.00	0.00	14834
gr229	10	23	9745	39797	516.46	0.00	0	39797	2853.96	0.00	0	39797	2281.24	0.00	266159
gil262	10	36	9553	639	1045.63	0.00	0	3221	3600.00	404.07	0	649	3600.00	1.56	3727
pr264	10	27	4369	16546	94.35	0.00	0	16546	400.01	0.00	0	16546	28.54	0.00	7051
pr299	10	35	7653	11624	803.65	0.00	0	11624	1920.75	0.00	0	11624	3600.00	0.00	72718
lin300	10	36	7667	10119	748.59	0.00	0	10119	1773.30	0.00	0	10119	3600.00	0.00	47010

Preliminary tests are shown in Table 1. The four first columns depict: the instance name in column 1, the value of parameter u in column 2 used to calculate the number of clusters for the corresponding instance, the number of clusters m in column 3, and the number of edges $|E|$ in column 4. Results obtained for each formulation is given in the remaining columns. For each experiments, four data are provided: the solution value in column "Cost" (optimal values are highlighted in bold), the running time in seconds in column "Time", the relative difference in percentage between the optimal value of each instance and the value obtained by each model in column "GAP", and the number of nodes generated in the B&B tree of IBM CPLEX in column "Nodes". Note that in formulation $(\mathcal{F}_{SAS})$ only the variables representing the vertices in V are binary. In model $(\mathcal{F}_{MLT})$ the variables representing the vertices V and the variables representing the arcs of the corresponding directed graph are binary. In model $(\mathcal{F}_{Pop})$, only the variables representing the edges of the global graph G' are binary. Bold values for instance cost indicate that CPLEX proved the optimality for that instances.

Considering the set of 40 instances tested, the proposed formulation $(\mathcal{F}_{SAS})$ proved optimality for all instances, while $(\mathcal{F}_{MLT})$ proved optimality for 32 instances and the model $(\mathcal{F}_{Pop})$ proved optimality for 17 instances. The formulation $(\mathcal{F}_{SAS})$ had an average runtime of 689.05 s, whereas $(\mathcal{F}_{MLT})$ had an average of 1423.65 s and $(\mathcal{F}_{Pop})$ of 2423.45 s. The formulations reached an average GAP of 0.00, 42.51 and 9.65 for the $(\mathcal{F}_{SAS})$, $(\mathcal{F}_{MLT})$, and $(\mathcal{F}_{Pop})$ models, respectively. The GAP of $(\mathcal{F}_{MLT})$ and $(\mathcal{F}_{Pop})$ was calculated based on result of model $(\mathcal{F}_{SAS})$, where all results have its optimality proved.

The model $(\mathcal{F}_{SAS})$ outperforms $(\mathcal{F}_{MLT})$ and $(\mathcal{F}_{Pop})$ models w.r.t. execution times. This probably happens due to a smaller number of integer variables among the three models. Finally, the numerical results indicate that $(\mathcal{F}_{SAS})$ and $(\mathcal{F}_{MLT})$ models obtained same linear relaxed cost solutions. Although this issue has not been formally investigated, we believe that this must always be true.

5   Conclusions

In this paper, a multigraph formulation $(\mathcal{F}_{SAS})$ is proposed for the GMSTP. A preprocessing has also been applied which allowed to reduce the problems size significantly. On average, 85% of the original edges were eliminated within 2 s. The proposed formulation presents a competitive performance compared to the models of $(\mathcal{F}_{MLT})$ and $(\mathcal{F}_{Pop})$. In particular, by founding better lower bounds and by proving optimality for larger instances than the other models. The originality of the proposed formulation by considering a multigraph opens new research directions in terms of adaptation for related problems, as well as for its use in cutting plane and decomposition approaches. We intend to develop a Benders decomposition and a B&C using the proposed model, and to develop valid inequalities to strengthen the model.

Acknowledgements. The authors are grateful to CNPq (grant 449254/2014-3) and FUNCAP (grant PNE-0112-00061.01.00/16) and to the anonymous referees for their helpful comments.

References

1. Adasme, P., Andrade, R., Letournel, M., Lisser, A.: Stochastic maximum weight forest problem. Networks **65**, 289–305 (2015)
2. Andrade, R.C.: Appointments on the spanning tree polytope. Annals of the ALIO/EURO - Montevideo (2014)
3. Afsar, H.M., Prins, C., Santos, A.C.: Exact and heuristic algorithms for solving the generalized vehicle routing problem with flexible fleet size. Int. Trans. Oper. Res. **21**(1), 153–175 (2014)
4. Bhattacharya, B., Ćustić, A., Rafiey, A., Rafiey, A., Sokol, V.: Approximation algorithms for generalized MST and TSP in grid clusters. In: Lu, Z., Kim, D., Wu, W., Li, W., Du, D.-Z. (eds.) COCOA 2015. LNCS, vol. 9486, pp. 110–125. Springer, Cham (2015). https://doi.org/10.1007/978-3-319-26626-8_9
5. Contreras-Bolton, C., Gatica, G., Barra, C.R., Parada, V.: A multi-operator genetic algorithm for the generalized minimum spanning tree problem. Expert Syst. Appl. **50**, 1–8 (2016)
6. Dror, M., Haouari, M.: Generalized Steiner problems and other variants. J. Comb. Optim. **4**, 415–436 (2000)
7. Duin, C.W., Volgenant, A., Voss, S.: Solving group Steiner problems as Steiner problems. Eur. J. Oper. Res. **154**, 323–329 (2004)
8. Feremans, C.: Generalized spanning tree and extensions. Ph.D. thesis, Universite Libre de Bruxelles (2001)
9. Feremans, C., Labbé, M., Laporte, G.: A comparative analysis of several formulations for the generalized minimum spanning tree problem. Networks **39**(1), 29–34 (2002)
10. Feremans, C., Labbé, M., Laporte, G.: The generalized minimum spanning tree problem: polyhedral analysis and branch-and-cut algorithm. Networks **43**(2), 71–86 (2004)
11. Ferreira, C.S., Ochi, L.S., Parada, V., Uchoa, E.: A GRASP-based approach to the generalized minimum spanning tree problem. Expert Syst. Appl. **39**(3), 3526–3536 (2012)
12. Ghosh, D.: Solving medium to large sized Euclidean generalized minimum spanning tree problems. Technical report NEP-CMP-2003-09-28 Indian Institute of Management, Research and Publication Department, India (2003)
13. Golden, B., Raghavan, S., Stanojevic, D.: Heuristic search for the generalized minimum spanning tree problem. INFORMS J. Comput. **17**(3), 290–304 (2005)
14. Hu, B., Leitner, M., Raidl, G.R.: Computing generalized minimum spanning trees with variable neighborhood search. In: Proceedings of the 18th Mini-Euro Conference on Variable Neighborhood Search (2005)
15. Martin, K.: Using separation algorithms to generate mixed integer model reformulations. Oper. Res. Lett. **10**(3), 119–128 (1991)
16. Myung, Y.S., Lee, C.H., Tcha, D.W.: On the generalized minimum spanning tree problem. Networks **26**(4), 231–241 (1995)
17. Öncan, T., Cordeau, J., Laporte, G.: A tabu search heuristic for the generalized minimum spanning tree problem. Eur. J. Oper. Res. **191**(2), 306–319 (2008)
18. Pop, P.C.: The generalized minimum spanning tree problem. Ph.D. thesis, University of Twente, Netherlands (2002)
19. Pop, P.C., Matei, O., Sabo, C., Petrovan, A.: A two-level solution approach for solving the generalized minimum spanning tree problem. Eur. J. Oper. Res. **170**, 1–10 (2017)

20. Santos, A.C., Duhamel, C., Andrade, R.: Trees and forests. In: Martí, R., Panos, P., Resende, M.G.C. (eds.) Handbook of Heuristics, pp. 1–27. Springer, Boston (2016)
21. Sousa, E.G., Andrade, R.C., Santos, A.C.: Algoritmo genético para o problema da árvore geradora generalizada de custo mínimo. SBPO **34**(4), 437–444 (2017)
22. Wang, Z., Che, C.H., Lim, A.: Tabu search for generalized minimum spanning tree problem. In: Yang, Q., Webb, G. (eds.) PRICAI 2006. LNCS, vol. 4099, pp. 918–922. Springer, Heidelberg (2006). https://doi.org/10.1007/978-3-540-36668-3_106

The Distance Polytope for the Vertex Coloring Problem

Bruno Dias[1], Rosiane de Freitas[1(✉)], Nelson Maculan[2], and Javier Marenco[3]

[1] Instituto de Computação, Universidade Federal do Amazonas, Manaus, Brazil
{bruno.dias,rosiane}@icomp.ufam.edu.br
[2] PESC/COPPE, Universidade Federal do Rio de Janeiro, Rio de Janeiro, Brazil
maculan@cos.ufrj.br
[3] Instituto de Ciencias, Universidad Nacional de General Sarmiento,
Buenos Aires, Argentina
jmarenco@dc.uba.br

Abstract. In this work we consider the distance model for the classical vertex coloring problem, introduced by Delle Donne in 2009. This formulation involves decision variables representing the distance between the colors assigned to every pair of distinct vertices, thus not explicitly representing the colors assigned to each vertex. We show close relations between this formulation and the so-called orientation model for graph coloring. In particular, we prove that we can translate many facet-inducing inequalities for the orientation model polytope into facet-inducing inequalities for the distance model polytope, and viceversa.

1 Introduction

Graph coloring is one of the most known problems in graph theory and combinatorial optimization. It has many applications, including channel assignment in wireless networks [1,2], scheduling [3], register allocation [4], school timetabling [5], and others. However, it is NP-hard (and its decision version is one of Karp's 21 NP-complete problems [6]), for which that are many approaches to reach optimal solutions. In this paper, we review some integer programming (IP) models for the *vertex coloring problem* (VCP) from the literature, and present a new IP formulation that involves decision variables representing the distance between the colors assigned to every pair of distinct vertices, thus not explicitly representing the colors assigned to each vertex. For this, the classical VCP can be formally described as follows: let $G = (V, E)$ be a simple undirected graph and, for each edge $(i, j) \in E$, the VCP asks for a mapping $c : V \to \mathbb{Z}_{\geq 0}$ such that $|c(i) - c(j)| \geq 1$ for each edge $(i, j) \in E$, i.e., the absolute difference between colors assigned to adjacent vertices must be greater than or equal to an uniform value equal to one [7,8].

We present an initial polyhedral study considering the distance IP formulation, for which we give two families of valid inequalities, show that they induce facets of the associated polytope. An overview of the IP formulations for the

© Springer International Publishing AG, part of Springer Nature 2018
J. Lee et al. (Eds.): ISCO 2018, LNCS 10856, pp. 144–156, 2018.
https://doi.org/10.1007/978-3-319-96151-4_13

VCP, such as our proposed IP formulation and polyhedral combinatorics results for it, can be seen in the following sections.

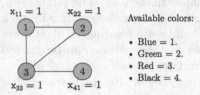

Available colors:

- Blue = 1.
- Green = 2.
- Red = 3.
- Black = 4.

Fig. 1. Example of VCP instance and its corresponding optimal solution with 3 colors. (Color figure online)

2   Known Integer Programming Models for VCP

2.1   Classic Formulation

In the classic formulation, two sets of binary variables are used: for each pair of vertex $i \in V$ and color k, we define a binary variable x_{ik} which has value 1 if color k is assigned to vertex i and 0 otherwise. For each color k, we also define a binary variable y_k which has value 1 if color k is assigned to any vertex and 0 otherwise. Note that the set of possible colors must be known in advance, since these variables are indexed not only by vertices (in the case of x variables), but also by colors. Instead of considering the set of possible colors as $\mathbb{N}$, we can limit it to the interval $[1, |V|]$, since, in the worst case, each vertex will have a different colors (which happens with complete graphs). The coloring function is then defined as $c : V \to [1, |V|]$. The IP formulation is then:

$$\text{Minimize} \quad \sum_{k=1}^{|V|} y_k \tag{1}$$

$$\text{Subject to} \quad \sum_{k=1}^{|V|} x_{ik} = 1 \qquad (\forall i \in V) \tag{2}$$

$$x_{ik} + x_{jk} \leq 1 \qquad (\forall (i,j) \in E; \ k = 1, 2, ..., |V|) \tag{3}$$

$$x_{ik} \leq y_k \qquad (\forall i \in V; \ k = 1, 2, ..., |V|) \tag{4}$$

$$x_{ik} \in \{0, 1\} \qquad (\forall i \in V; \ k = 1, 2, :.., |V|) \tag{5}$$

$$y_k \in \{0, 1\} \qquad (\forall k = 1, 2, ..., |V|) \tag{6}$$

The objective function (1) makes the number of variables with value 1 from the set y be the least possible, which equals to minimizing the number of used colors. Constraints (2) require that each vertex receive a color. The set of constraints (3) enforce that adjacent vertices do not share the same color. Constraints (4) ensure that if for a certain color k there is a variable x_{ik} with value 1 regardless of the vertex i, then the variable y_k must also be 1. Finally, sets

(5) and (6) are integrality constraints. In Fig. 1, we have an example of graph coloring problem and its solution encoded as variables of this formulation.

This formulation has polynomial dimension. There are $O(|V|^2)$ variables in the x set and $O(|V|)$ variables in the y set. The largest constraint set is (3), where there are $O(|V|^2|E|)$ constraints. It is also easily adaptable to other variations of vertex coloring, such as bandwidth coloring [1,9]. However, the formulation has symmetry problems, since another solution can be derived simply by exchanging the colors between all vertices of one color and all vertices of another color. Its lower bound is also very weak [10]. In spite of that, the formulation has been widely used in the literature to derive cutting planes and branch-and-cut algorithms for vertex coloring problems [11]

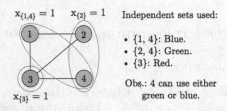

$x_{\{1,4\}} = 1$ $x_{\{2\}} = 1$ Independent sets used:

- $\{1, 4\}$: Blue.
- $\{2, 4\}$: Green.
- $\{3\}$: Red.

$x_{\{3\}} = 1$

Obs.: 4 can use either green or blue.

Fig. 2. Example of VCP instance and its corresponding optimal solution encoded as independent sets. (Color figure online)

2.2 Independent Set Formulation

We can note, by definition of VCP, that since vertices that share the same color are not adjacent to each other, then they form an independent set of the graph. Each one of them is called a color class. An IP formulation can be derived from this observation as follows [12,13]. Let $Q(G)$ be the set of all maximal independent sets of G. We define, for each independent set $S \in Q(G)$, a variable x_S which has value 1 if S will receive a unique color and 0 otherwise. The formulation is given by:

$$\text{Minimize} \quad \sum_{S \in Q(G)} x_S \tag{7}$$

$$\text{Subject to} \quad \sum_{S \in Q(G)\,:\,i \in S} x_S \geq 1 \qquad (\forall i \in V) \tag{8}$$

$$x_S \in \{0, 1\} \qquad (\forall S \in Q(G)) \tag{9}$$

The objective function (7) ensures that the number of maximal independent sets that will receive a unique color is minimized, which is equivalent to minimizing the number of used colors. Constraints (8) require that each vertex must be present in at least one of the independent sets that will be selected for coloring. Finally, the set (9) consists of integrality constraints. Figure 2 shows the solution of Fig. 1 encoded as variables of this formulation.

Although this formulation has $O(|V|)$ constraints, it has an exponential number of variables. Since there is one for each maximal independent set of the input graph, there are $O(3^{|V|/3})$ variables in the full model (following the same bound for the number of maximal independent sets [14]). However, this formulation can be used in a column generation algorithm, where the subproblem consists of finding a new and improving maximal independent sets [12,13]. This formulation does not share the symmetry problem of the classic one, since the colors are not directly assigned - rather, each independent set can use any color, as long as each one of them use a distinct color.

2.3   Asymmetric Representatives Formulation

This formulation is also based on the notion of color classes (that is, independent sets) and was proposed by Campêlo et al. [15]. Let W_k be the color class containing vertices which use color k. For each W_k, one vertex $i \in W_k$ is chosen to be the representative of such color class. Let also $N(i)$ be the set of vertices that are adjacent i and $\overline{N}(i)$ its complement (that is, $\overline{N}(i) = V \setminus (N(i) \cup \{i\})$). Define, for $i \in V$ and $j \in \overline{N}(i)$, variables x_{ij} with value 1 if vertex $i \in V$ is the representative of the color of $j \in \overline{N}(i)$ and 0 otherwise.

To avoid symmetry between vertices in the same color class, an order $\prec$ between vertices is defined such that if $i \prec j$, then j cannot be the representative of the color of i, and x_{ji} is fixed to 0. If we consider the complement $\overline{G}$ of G (where $V(\overline{G}) = V$ and $E(\overline{G}) = \{(i,j) \mid i,j \in V \text{ and } (i,j) \notin E\}$), then we can see that $\prec$ induces an orientation of $\overline{G}$, since vertices that are adjacent to each other in the complement can use the same color. Let $\overline{N}^+(i) = \{j \in \overline{N}(i) \mid i \prec j\}$, $\overline{N}^-(i) = \{j \in N(i) \mid j \prec i\}$, $G^+(i)$ be the induced subgraph containing only vertices from $\overline{N}^+(i)$ and $G^-(i)$ be the induced subgraph with vertices from $\overline{N}^-(i)$. The orientation induced by $\prec$ on $\overline{G}$ is acyclic, so there are two special sets of vertices: $S = \{s \mid \overline{N}^-(s) = \emptyset\}$ and $T = \{t \mid \overline{N}^+(t) = \emptyset\}$, whose induced subgraphs are cliques. Define variables y_i with value 1 if a vertex $i \in S$ and 0 otherwise. The formulation is defined as:

$$\text{Minimize} \quad \sum_{i \in V \setminus S} x_{ii} + |S| \tag{10}$$

$$\text{Subject to} \quad \sum_{j \in \overline{N}^s ps(i)} x_{ji} \geq 1 \quad (\forall i \in V \setminus S) \tag{11}$$

$$\sum_{j \in L} x_{ij} \leq y_i \quad (\forall i \in V \setminus T; \ L \subseteq \overline{N}^+(i) : \text{ind. clique has size 1 or 2}) \tag{12}$$

$$x_{ij} \in \{0,1\} \quad (\forall i \in V; \ j \in \overline{N}(i)) \tag{13}$$

Where $y_i = 1$ if $i \in S$ and $y_i = x_{ii}$ otherwise. The objective function (10) consists of minimizing the number of vertices outside S that are represented by themselves in the coloring plus the number of vertices in S (since these ones, called sources cannot be represented by any other vertex), which is equivalent to minimizing the number of used colors. Constraints (11) ensure that a vertex must be represented by itself or by some other vertex not adjacent to it. The set of constraints (12) require that vertices which are not adjacent to a certain vertex but are adjacent to each other be given different representatives. Figure 3 shows an example of instance using asymmetric representatives.

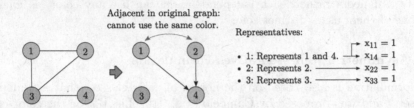

Fig. 3. Example of VCP instance and its corresponding optimal solution encoded using asymmetric representatives and its complement oriented graph.

This formulation has $O(|V|^2 - |E|)$ variables, since there is one variable for each edge in the complement graph. The largest constraint set is (12), having $O(|V|(|V|^2 - |E|)) = O(|V|^3 - |V||E|)$ elements. This means that the formulation has polynomial size, and it also avoids symmetry problems by defining the $\prec$ order. In [15], a polyhedral study and facet defining inequalities are provided.

2.4   Clique Based Formulation

A clique is a subset H of V such that its induced subgraph is itself a complete graph, that is, for all $i, j \in H$, then $(i, j) \in E$. Each clique of G corresponds to a independent set of its complement graph $\overline{G}$, and each member of the clique will have a different color (and, in the complement graph, each member would use the same color). The clique cover problem consists of partitioning V into k cliques. We can see that the least possible k for which the complement graph is partitioned into cliques is the chromatic number of the original graph.

Two vertices $i, j \in V$ of G are said to be indistinguishable if $(i, j) \in E$ and also $N(i) \cup \{i\} = N(j) \cup \{j\}$. A set of indistiguishable vertices form a supernode of G. We define the reversible clique partition R of G as a partition where each clique is a supernode r, that is, the vertices in the clique are indistiguishable. Consider, then, the graph G', where $V' = R$ and $E' = \{(r_1, r_2) \mid (i, j) \in E; r_1, r_2 \in R; r_1 \neq r_2; i \in r_1; j \in r_2\}$. By using multicoloring on G', where each supernode receives more than one color, equivalent to the number of elements in the supernode, we have a solution for VCP in the original graph. Based on this, a formulation is defined [5] which use, for each supernode $r \in R$ and color k, a variable x_{rk} which is valued 1 if k is assigned to r and 0 otherwise.

The formulation is defined as:

$$\text{Minimize} \quad \sum_{k=1}^{|V|} y_k \tag{14}$$

$$\text{Subject to} \quad \sum_{k=1}^{|V|} x_{rk} = |r| \qquad (\forall r \in R) \tag{15}$$

$$x_{r_1 k} + x_{r_2 k} \leq 1 \qquad (\forall (r_1, r_2) \in E'; \ k = 1, 2, ..., |V|) \tag{16}$$

$$x_{rk} \leq y_k \qquad (\forall r \in R; \ k = 1, 2, ..., |V|) \tag{17}$$

$$x_{rk} \in \{0, 1\} \qquad (\forall r \in R; \ k = 1, 2, ..., |V|) \tag{18}$$

$$y_k \in \{0, 1\} \qquad (\forall k = 1, 2, ..., |V|) \tag{19}$$

This model is the same one as the classic formulation applied on graph G' and considering the color set as $[1, |V|]$ instead of $[1, |V'|]$ (which is the same as $[1, |R|]$). However, as discussed earlier, the reversible clique partition must be multicolored instead of single colored, so the right-hand size in constraint set (15) is changed so that each supernode d receive $|d|$ colors, which ensure that each vertex in the original graph has one color assigned to it.

There are $O(|V||r|)$ variables in the x set and $O(|R|)$ variables in the y set, which follows a similar reasoning of the classic formulation. There are also $O(|R| + |V||E'|)$ constraints. This formulation also avoids symmetry issues: since colors assigned to supernodes can be distributed in any manner to the internal vertices, the problem of simply exchanging the value of the color between two vertices, resulting in a symmetric solution, is avoided. Figure 4 shows an example of instance using this formulation.

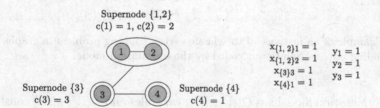

Fig. 4. Example of VCP instance and its corresponding optimal solution encoded by the clique formulation.

2.5 Orientation-Based Formulation

This formulation was proposed in [16] for the bandwidth coloring problem and can be readily used for the classic vertex coloring problem (since this one is a specific case of the latter). It is based on variables which induce an orientation on the input graph. Using this, we can directly use variables that contain the color of each vertex, instead of needing a variable for each pair of vertex and color. Following this, for each $i \in V$, we introduce the integer variable $z_i \in \mathbb{N}$

representing the color assigned to i and, for each edge $(i, j \in E)$ such that $i < j$, we define a binary variable y_{ij} which has value 1 if $z_i < z_j$ and 0 otherwise. Finally, the free variable z_{max} represents the maximum assigned color in the optimal solution. Fix $S \geq \chi(G)$. In this setting, the proposed formulation is the following.

$$\text{Minimize} \quad z_{max} \tag{20}$$
$$\text{Subject to} \quad z_i + 1 \leq z_j + s(1 - y_{ij}) \qquad \forall (i, j) \in E, \ i < j \tag{21}$$
$$z_j + 1 \leq z_j + s y_{ij} \qquad \forall (i, j) \in E, \ i < j \tag{22}$$
$$z_{max} \geq z_i \qquad \forall i \in V \tag{23}$$
$$z_i \in \mathbb{N} \qquad \forall i \in V \tag{24}$$
$$y_{ij} \in \{0, 1\} \qquad \forall (i, j) \in E, \ i < j \tag{25}$$

In the above formulation, constraint (21) ensures that $y_{ij} = 0$ if $x_i < x_j$, for $(i, j) \in E$. On the other hand, if $z_j < z_i$ then constraint (22) ensures that $y_{ij} = 1$. Both constraints also guarantee that $|z_i - z_j| \geq 1$, which ensures that adjacent vertices use different colors. Constraints (23) impose z_{max} to take a value greater than or equal to every used color, and in an optimal solution this bound will be tight. Finally, constraints (24)–(25) define integer variable bounds. Figure 5 shows the solution of Fig. 1 encoded as variables of this formulation.

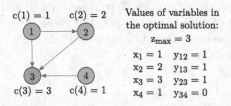

Fig. 5. Example of an instance of the classic vertex coloring problem in graphs and its corresponding optimal solution encoded by the orientation model.

The orientation model has $O(|V| + |E|)$ variables and $O(|V| + |E|)$ constraints. The constant s must be set to a sufficiently large value, in such a way that s is greater than or equal to the (unknown) parameter $\chi(G)$. Known upper bounds for graph coloring, such as $|V|$ and $\Delta(G) + 1$ (where $\Delta(G)$ is the maximum degree among all vertices of V), can be used in place of $\chi(G)$. Although the objective function is the minimization of the maximum used color instead of the number of used colors, these values are equivalent for classic vertex coloring [2]. The constant s is indeed needed in the formulation, since if we do not impose an upper bound to the x-variables, then the convex hull of feasible solutions is not a polytope in general, hence an IP formulation is not possible. Based on [16], we also remark that if $s \geq \chi(G) + 2$, then the corresponding polytope is full-dimensional.

3   Distance-Based Formulation for VCP

Given a graph $G = (V, E)$ and a set C of colors (which can be given as $C = \{1, 2, \ldots, |V|\}$ for VCP), the *distance model* employs an integer variable x_{ij} for every $i, j \in V$, $i < j$, denoting the difference between the colors assigned to i and j (i.e., if i takes color $c(i)$ and j takes color $c(j)$, then $x_{ij} = c(i) - c(j)$), and, for every $(i, j) \in E$, $i < j$, a binary variable y_{ij} which has value 1 if $x_{ij} < 0$ and 0 otherwise. A feasible solution is given by the following constraints.

$$x_{ik} = x_{ij} + x_{jk} \qquad \forall i, j, k \in V, \, i < j < k \qquad (26)$$

$$x_{ij} \geq 1 - |C|y_{ij} \qquad \forall (i, j) \in E, \, i < j \qquad (27)$$

$$x_{ij} \leq -1 + |C|(1 - y_{ij}) \qquad \forall (i, j) \in E, \, i < j \qquad (28)$$

$$x_{ij} \in \{-|C| + 1, \ldots, |V| - 1\} \qquad \forall i, j \in V, \, i < j \qquad (29)$$

$$y_{ij} \in \{0, 1\} \qquad \forall (i, j) \in E, \, i < j \qquad (30)$$

We note that, since this formulation does not provide the colors of the vertices, but only the differences between such colors for adjacent vertices, we cannot consider an objective function for determining the chromatic number, so our interest lies, primarily, in studying properties about the polyhedron defined by (26–30).

Figure 6 extends the solution of Fig. 5 for this formulation. The first set of constraints is composed by $O(|V|^3)$ equations, and may generate a large model in practice. The following result allows us to replace these contraints by $O(|V|^2)$ equations.

Theorem 1 ([17])**.** *If $V = \{1, \ldots, n\}$, then constraints (26) are equivalent to*

$$x_{i,i+1} + x_{i+1,i+2} = x_{i,i+2} \qquad \forall i \in V, i \leq n - 2 \qquad (31)$$

$$x_{ij} + x_{i+1,j-1} = x_{i,j-1} + x_{i+1,j} \qquad \forall i, j \in V, i \leq n - 3, i + 3 \leq j \qquad (32)$$

For $i, j \in V$, $i < j$, we define $x_{ji} = -x_{ij}$ as a notational convenience. Similarly, for $(i, j) \in E$, $i < j$, we define $y_{ji} = 1 - y_{ij}$. Let $PD(G, C)$ be the convex hull of the vectors (x, y) satisfying constraints (26)–(30). Theorem 1 allows us to calculate the dimension of this polytope. Call $n = |V|$ and $m = |E|$.

Theorem 2. *If $|C| > \chi(G) + 1$, then $\dim(PD(G, C)) = n + m - 1$.*

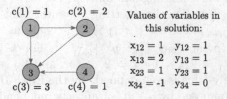

Fig. 6. Example of an instance of the classic vertex coloring problem in graphs and its corresponding optimal solution encoded by the distance model.

Idea of the proof. There are $n(n-1)/2$ x-variables and m y-variables. The family (31) is composed by $n-2$ equations, and the family (32) is composed by $(n-3)(n-2)/2$ equations. It is possible to show by induction on n that the equations in these two families are linearly independent (and this fact does not depend on the number of colors).

Take an optimal coloring $c : V \to \{1, \ldots, \chi(G)\}$ and let $v_1, \ldots, v_n$ be an ordering of V such that $c(v_i) \leq c(v_{i+1})$ for $i = 1, \ldots, n-1$. Construct the points $\{(x^k, y)\}_{k=0}^{n-1}$ in such a way that x^k represents the coloring c^k defined by $c^k(v_i) := c(v_i)$ for $i = 1, \ldots, n-k$ and $c^k(v_i) := c(v_i) + 1$ for $i = n-k+1, \ldots, n$. These points are affinely independent, showing that the projection of $PD(G, C)$ onto the x-variables has dimension $n-1$. This argument only requires $|C| > \chi(G)$.

Assume $\lambda x + \gamma y = \lambda_0$ for every $(x, y) \in PD(G, C)$. The previous construction shows that Eqs. (31) and (32) are linearly independent, so λ is a linear combination of them. For every $(i, j) \in E$, construct a coloring $c : V \to \{1, \ldots, \chi(G)+2\}$ with $\chi(G)+2$ colors such that $c(i) = \chi(G)+1$, $c(j) = \chi(G)+2$, and $c(k) \leq \chi(G)$ for $k \in V \setminus \{i, j\}$. Let (x, y) be the associated solution, and let (x', y') be the solution obtained by swapping the colors of i and j. The existence of these two solutions, together with the previous observation on λ, shows that $\gamma_{y_{ij}} = 0$. Then, the projection of $PD(G, C)$ onto the y-variables has dimension m, and so we have that $\dim(PD(G, C)) = n + m - 1$. $\qquad\square$

The formulation (26)–(30) is very similar to the *orientation model* for graph coloring [18]. In this model, we have an integer variable $z_i \in \{1, \ldots, |C|\}$ for every $i \in V$, and the y-variables as in the distance model. We also set the constant s from the orientation model to $|C|$, Define $PO(G, C)$ to be the convex hull of feasible solutions of the orientation model. The following result shows that we can translate many facet-inducing inequalities for $PO(G, C)$ into facet-inducing inequalities for $PD(G, C)$.

Theorem 3. *Let $\alpha z_i + \pi y \leq \alpha z_j + \pi_0$ be a valid (resp. facet-inducing) inequality for $PO(G, C)$, where $(i, j) \in E$. Then, $\alpha x_{ij} + \pi y \leq \pi_0$ is valid (resp. facet-inducing if $|C| \geq \chi(G) + 2$) for $PD(G, C)$.*

Idea of the Proof. Validity follows directly from the definition of the variable x_{ij}. For facetness, take $n + m$ affinely independent points $\{(z^k, y^k)\}_{k=1}^{n+m}$ in $PO(G, C)$ satisfying $\alpha z_i + \pi y \leq \alpha z_j + \pi_0$ with equality. For $k = 1, \ldots, n+m$, map the vector z^k into $\hat{z}^k = (z_1^k, z_2^k - z_1^k, z_3^k - z_2^k, z_4^k - z_3^k, \ldots, z_n^k - z_{n-1}^k)$. The vectors $\{(\hat{z}^k, y^k)\}_{k=1}^{n+m}$ are affinely independent since the mapping $z \to \hat{z}$ is a linear transformation with a nonsingular matrix. For $k = 1, \ldots, n+m$, remove the first component from $\hat{z}^k$, thus getting a vector that we shall call $\hat{x}^k = (x_{21}^k, x_{32}^k, x_{43}^k, \ldots, x_{n,n-1}^k)$. The set $\{(\hat{x}^k, y^k)\}_{k=1}^{n+m}$ contains, therefore, $n + m - 1$ affinely independent points, say $\{(\hat{x}^k, y^k)\}_{k=1}^{n+m-1}$. For $k = 1, \ldots, n+m-1$, define x^k to be $x_{t,t-1}^k = \hat{x}_{t,t-1}^k$ for $t = 2, \ldots, n$, and by setting the remaining x-variables accordingly (this is well-defined, by the Eqs. (31) and (32)). The set $\{(x^k, y^k)\}_{k=1}^{n+m-1}$ is composed by $n + m - 1$ affinely independent points satisfying $\alpha x_{ij} + \pi y \leq \pi_0$ with equality. $\qquad\square$

The construction in Theorem 3 can be generalized to the following setting. Let $\{(i_k, j_k)\}_{k=1}^p$, where $i_k, j_k \in V$, $i_k \neq j_k$, for $k = 1, \ldots, p$. If the inequality

$$\sum_{k=1}^p \alpha_k z_{i_k} + \pi y \leq \sum_{k=1}^p \alpha_k z_{j_k} + \pi_0,$$

is valid (resp. facet-inducing if $|C| \geq \chi(G) + 2$) for $PO(G, C)$, then the inequality

$$\sum_{k=1}^p \alpha_k x_{i_k, j_k} + \pi y \leq \pi_0$$

is valid (resp. facet-inducing) for $PD(G, C)$. Many inequalities presented in [19, 20] fit into this pattern, hence this result provides many facet-inducing inequalities for $PD(G, C)$.

Finally, the following theorem provides a converse result, thus showing that all facets of $PD(G, C)$ come from facets of $PO(G, C)$.

Theorem 4. *Assume $C = \{1, \ldots, |V|\}$. Let $\gamma x + \pi y \leq \pi_0$ be a valid (resp. facet-inducing) inequality for $PD(G, C)$. Then, $\sum_{i \neq j} \gamma_{ij}(z_i - z_j) + \pi y \leq \pi_0$ is valid (resp. facet-inducing if $|C| \geq \chi(G) + 2$) for $PO(G, C \cup \{|C| + 1\})$.*

Idea of the Proof. Validity follows directly from the variable definitions. For facetness, take $n + m - 1$ affinely independent points $\{(x^k, y^k)\}_{k=1}^{n+m-1}$ in $PD(G, C)$ satisfying $\gamma x + \pi y = \pi_0$. For $k = 1, \ldots, n + m - 1$, let $c^k : V \to C$ be a coloring compatible with (x^k, y^k) (note that there may be more than one such coloring, if $\max_{i \neq j} |x_{ij}^k| < |C| - 1$), and construct the point $(z^k, y^k) \in PO(G, C)$ representing the coloring c^k. Note that $(z^k, y^k) \in PO(G, C \cup \{|C| + 1\})$, since $PO(G, C) \subseteq PO(G, D)$ if $C \subseteq D$. Also note that the map $(z^k, y^k) \to (x^k, y^k)$ is a linear transformation $L(z, y) = A \times (z, y)$ for some nonsquare matrix A. Consider the points (z^1, y^1) and $(\bar{z}^1, y^1)$, where $\bar{z}_i^1 = z_i^1 + 1$ for $i \in V$. Both points belong to $PO(G, C \cup \{|C| + 1\})$ and are distinct, hence they are affinely independent. For $k = 2, \ldots, n + m - 1$, the point (z^k, y^k) is affinely independent w.r.t. the points $\{(\bar{z}^1, y^1)\} \cup \{(z^t, y^t)\}_{t < k}$, since otherwise the point $L(z^k, y^k)$ would be affinely dependent w.r.t. $\{L(\bar{z}^1, y^1)\} \cup \{L(z^t, y^t)\}_{t < k} = \{L(z^t, y^t)\}_{t < k}$, a contradiction. Hence, we construct the $n + m$ affinely independent points $\{(\bar{z}^1, y^1)\} \cup \{(z^t, y^t)\}_{t=1}^{n+m-1}$ satisfying the new inequality with equality. $\square$

4   Facet-Defining Inequalities for the Distance Model

One of the main implications of Theorem 3 is that we can transform most of the facet-defining inequalities for the orientation model into ones for the distance model. We show below three of these families of cuts and their transformation by means of such Theorem.

4.1   Clique Inequality

The clique inequality arises from the fact that all vertices in a clique must use different colors and has been used in other formulations [11,15]. Let $i \in V$ and $K \subseteq N(i)$ be a clique. In this setting, we define the following as the *clique inequality* associated with the vertex i and the clique K for the orientation model:

$$z_i \geq \sum_{k \in K} y_{ki} + 1$$

In order to set the clique inequality to an appropriate format to the distance model using Theorem 3, we have to add a z_j term (where $j \in N(i)$) to the RHS (right-hand side) of the expression so that we can obtain a $z_i - z_j$ expression on the LHS (left-hand side) to be replaced by x_{ij}. However, we must add another term to the LHS to compensate this addition in order to maintain the validity of the inequality. Since LHS $\geq$ RHS in the constraint, we can add a $\Delta(G) + 1$ term to the LHS (where $\Delta(G)$ is the maximum degree among all vertices in G), since it is an upper bound for the maximum used color in VCP, and the z_j variable is the color used by j (so it is bounded by $|C|$). The modified expression is then $z_i + \Delta(G) + 1 \geq z_j + \sum_{k \in K} y_{ki} + 1$ and the corresponding clique inequality for the distance model is:

$$x_{ij} + \Delta(G) \geq \sum_{f \in K} y_{fi}$$

4.2   Double Clique Inequality

The double clique inequality arises from cliques from the intersection of neighborhoods of two vertices, and has been used in some related problems, such as chromatic scheduling [20]. Let $(i, j) \in E$ and consider a clique $K \subseteq N(i) \cap N(j)$. Also, fix a vertex $p \in K$. In this setting, we define the following as the *double clique inequality* associated with the vertex i and the clique K for the orientation model:

$$z_i + 1 + \sum_{k \in K} (y_{ik} - y_{jk}) \leq z_j + (s - |K|)y_{ji}$$

By applying Theorem 3 and observing that the constant s from the orientation model is equivalent to $|C|$ in the distance model, we obtain the following inequality for the latter:

$$x_{ik} + 1 + \sum_{k \in K} (y_{ik} - y_{jk}) \leq (|C| - |K|)y_{ji}$$

In Table 1, a summary of informations about each formulation shown in this work is given.

Table 1. Summary of IP formulations for vertex coloring.

Formulation	Variables	Constraints	Reference
Classic	$O(|V^2|)$	$O(|V^2E|)$	–
Independent sets	$O(3^{|V|/3})$	$O(|V|)$	[12]
Asymmetric representatives	$O(|V|^2 - |E|)$	$O(|V|^3 - |V||E|)$	[15]
Clique cover	$O(|V||H|)$	$O(|H| + |V||E'|)$	[5]
Orientation-based	$O(|V| + |E|)$	$O(|V| + |E|)$	[16]
Distance-based	$O(|V|^2)$	$O(|V|^2)$	**This work**

5 Concluding Remarks

In this work, we presented a new IP formulation for the classical VCP, based on the orientation model for bandwidth coloring and the original distance model for chromatic scheduling. In this formulation, decision variables represent the distance between the colors assigned to every pair of distinct vertices, thus not explicitly representing the colors assigned to each vertex. We prove that we can translate many facet-inducing inequalities for the orientation model polytope into facet-inducing inequalities for the distance model polytope, and viceversa. Ongoing works include the implementation of a cut-and-branch method, such as to verify the behavior of the cuts proposed.

References

1. Dias, B.: Modelos teóricos e algoritmos para a otimização da alocação de canais em redes móveis sem fio. Master's thesis, Instituto de Computação - Universidade Federal do Amazonas (2014). in Portuguese
2. Koster, A.M.C.A.: Frequency assignment: models and algorithms. Ph.D. thesis, Universiteit Maastricht (1999)
3. de Freitas, R., Dourado, M., Szwarcfiter, J.: Graph coloring and scheduling problems. In: 4th Latin American Workshop on Cliques in Graphs (2010)
4. Chaitin, G.J.: Register allocation and spilling via graph coloring. In: Proceedings of the 1982 SIGPLAN Symposium on Compiler Construction, SIGPLAN 1982, pp. 98–105. ACM (1982)
5. Burke, E., Mareček, J., Parkes, A., Rudová, H.: A supernodal formulation of vertex colouring with applications in course timetabling. Ann. Oper. Res. **179**, 105–130 (2010)
6. Karp, R.: Reducibility among combinatorial problems. In: Miller, R.E., Thatcher, J.W. (eds.) Complexity of Computer Computations, pp. 85–103. Plenum Press (1972)
7. Dias, B., de Freitas, R., Maculan, N., Michelon, P.: Solving the bandwidth coloring problem applying constraint and integer programming techniques. Optimization Online (e-print) (2016)
8. Lai, X., Lü, Z.: Multistart iterated tabu search for bandwidth coloring problem. Comput. Oper. Res. **40**, 1401–1409 (2013)

9. Mak, V.: Polyhedral studies for minimum-span graph labelling with integer distance constraints. Int. Trans. Oper. Res. **14**(2), 105–121 (2007)
10. Malaguti, E., Toth, P.: A survey on vertex coloring problems. Int. Trans. Oper. Res. **17**(1), 1–34 (2010)
11. Méndez-Díaz, I., Zabala, P.: A branch-and-cut algorithm for graph coloring. Discrete Appl. Math. **154**(5), 826–847 (2006)
12. Mehrotra, A., Trick, M.: A column generation approach for graph coloring. INFORMS J. Comput. **8**(4), 344–354 (1996)
13. Mehrotra, A., Trick, M.A.: A branch-and-price approach for graph multi-coloring. In: Baker, E.K., Joseph, A., Mehrotra, A., Trick, M.A. (eds.) Extending the Horizons: Advances in Computing, Optimization, and Decision Technologies, vol. 37, pp. 15–29. Springer, Boston (2007). https://doi.org/10.1007/978-0-387-48793-9_2
14. Moon, J., Moser, L.: On cliques in graphs. Isr. J. Math. **3**, 23–28 (1965)
15. Campêlo, M., Campos, V., Corrêa, R.: On the asymmetric representatives formulation for the vertex coloring problem. Discrete Appl. Math. **156**, 1097–1111 (2008)
16. Dias, B., de Freitas, R., Marenco, J., Maculan, N.: Facet-inducing inequalities and a cut-and-branch for the bandwidth coloring polytope based on the orientation model. In: IX Latin and American Algorithms, Graphs and Optimization Symposium. Electronic Notes in Discrete Mathematics (2017, to appear)
17. Delle Donne, D.: Un algoritmo branch&cut para un problema de asignación de frecuencias en redes de telefonía celular. Ph.D. thesis, Departamento de Computación, FCEyN - Universidad de Buenos Aires (2009). In Spanish
18. Borndörfer, R., Eisenblätter, A., Grötschel, M., Martin, A.: The orientation model for frequency assignment problems. Technical report, ZIB Berlin (1998)
19. Marenco, J., Wagler, A.: Chromatic scheduling polytopes coming from the bandwidth allocation problem in point-to-multipoint radio access systems. Ann. Oper. Res. **150**, 159–175 (2007)
20. Marenco, J., Wagler, A.: Facets of chromatic scheduling polytopes based on covering cliques. Discrete Optim. **6**, 64–78 (2009)

A PTAS for the Time-Invariant Incremental Knapsack Problem

Yuri Faenza[1] and Igor Malinovic[2]([✉])

[1] Columbia University, New York City, USA
[2] École Polytechnique Fédéral de Lausanne, Lausanne, Switzerland
`igor.malinovic@epfl.ch`

Abstract. The Time-Invariant Incremental Knapsack problem (IIK) is a generalization of Maximum Knapsack to a discrete multi-period setting. At each time, capacity increases and items can be added, but not removed from the knapsack. The goal is to maximize the sum of profits over all times. IIK models various applications including specific financial markets and governmental decision processes. IIK is strongly NP-Hard [2] and there has been work [2,3,6,13,15] on giving approximation algorithms for some special cases. In this paper, we settle the complexity of IIK by designing a PTAS based on rounding a disjunctive formulation, and provide several extensions of the technique.

1 Introduction

Knapsack problems are among the most fundamental and well-studied in discrete optimization. Some variants forego the development of modern optimization theory, dating back to 1896 [11]. The best known representative is arguably *Maximum Knapsack* (MAX-K): given a set of items with specified profits and weights, and a threshold, find a most profitable subset of items whose total weight does not exceed the threshold. MAX-K is NP-complete [8], while admitting a *fully polynomial-time approximation scheme* (FPTAS) [7]. Many classical algorithmic techniques including greedy, dynamic programming, backtracking/branch-and-bound have been studied by means of solving this problem, e.g., see a survey by Kellerer et al. [9]. The algorithm of Martello and Toth [10] has been known to be the fastest in practice for exactly solving knapsack instances [1].

In order to model scenarios arising in real-world applications, more complex knapsack problems have been introduced (see the survey [9]) and recent works studied extensions of classical combinatorial optimization problems to multi-period settings, e.g., [6,13,14]. At the intersection of those two streams of research, Bienstock et al. [2] proposed a generalization of a MAX-K to a multi-period setting that they dubbed *Time-Invariant Incremental Knapsack* (IIK). In IIK, we are given a set of items $[n]$ with profits $p : [n] \to \mathbb{R}_{>0}$ and weights $w : [n] \to \mathbb{R}_{>0}$ and a knapsack with non decreasing capacity b_t over time $t \in [T]$. We can add items at each time as long as the capacity constraint is not violated, and once inserted, an item cannot be removed from the knapsack. The goal is to

© Springer International Publishing AG, part of Springer Nature 2018
J. Lee et al. (Eds.): ISCO 2018, LNCS 10856, pp. 157–169, 2018.
https://doi.org/10.1007/978-3-319-96151-4_14

maximize the total profit, which is defined to be the sum, over $t \in [T]$, of profits of items in the knapsack at time t.

IIK models a scenario where available resources (e.g., money, labour force) augment over time in a predictable way, allowing to grow our portfolio. E.g., take a bond market with an extremely low level of volatility, where all coupons render profit only at their common maturity time T (*zero-coupon* bonds) and an increasing budget over time that allows buying more and more (differently sized and priced) packages of those bonds. For variations of MAX-K that have been used to model financial problems, see the survey [9]. A different application arises in government-type decision processes, where items are assets of public utility (schools, parks, etc.) that can be built at a given cost and give a yearly benefit (both constant over the years), and the community will profit each year those assets are available.

Previous Works on IIK. Although the first publication on IIK appeared just very recently (Della Croce et al. [3]), the problem was previously studied by Bienstock et al. [2] and covered in several PhD theses [6,13,15]. Here we summarize all those results. The authors in [2] proved IIK to be strongly NP-hard and they gave an instance where the natural LP relaxation has unbounded integrality gap. In the same paper, a PTAS is designed for $T = O(\log n)$. This improves over Sharp [13], who gave a PTAS for the special case $p = w$ and T being constant. Again, when $p = w$, a 1/2-approximation algorithm for generic T is provided by Hartline [6]. Results by Ye [15] can be adapted to give an algorithm that solves IIK in time polynomial in n and of order $(\log T)^{O(\log T)}$ for a fixed approximation guarantee ε [12]. Della Croce et al. [3] provide an alternative PTAS for IIK with constant T, and a 1/2-approximation for arbitrary T under the assumption that every item alone fits into the knapsack at $t = 1$.

Our Contribution. In this paper, we provide, for any fixed ε, an algorithm that computes a $(1 - \varepsilon)$-approximate solution to IIK, and whose running time is polynomial in both the number of items n and the number of times T. In particular, our algorithm is a PTAS for IIK regardless of T.

Theorem 1. *There exists an algorithm that, when given as input $\varepsilon \in \mathbb{R}_{>0}$ and an instance $\mathcal{I}$ of IIK with n items and $T \geq 2$ times, produces a $(1 - \varepsilon)$-approximation to the optimum solution of $\mathcal{I}$ in time $O(T^{h(\varepsilon)} \cdot n f_{LP}(n))$. Here $f_{LP}(m)$ is the time required to solve a linear program with $O(m)$ variables and constraints, and $h : \mathbb{R}_{>0} \to \mathbb{R}_{\geq 1}$ is a function depending on ε only. In particular, there exists a PTAS for IIK.*

Theorem 1 dominates all previous results on IIK [2,3,6,13,15] and, due to the hardness results by Bienstock et al. [2], settles the complexity of the problem. Interestingly, it is based on designing a disjunctive formulation – a tool mostly common among integer programmers and practitioners[1] – and then rounding the solution to its linear relaxation with a greedy-like algorithm. We see Theorem 1 as an important step towards the understanding of the complexity landscape of

---

[1] See the full version of the paper [4] for a discussion on disjunctive programming.

knapsack problems over time. Theorem 1 is proved in Sect. 2: see the end of the current section for a sketch of the techniques we use and a detailed summary of Sect. 2. In Sect. 3, we show some extensions of Theorem 1 to more general problems.

Related Work on Other Knapsack Problems. Bienstock et al. [2] discuss the relation between IIK and the generalized assignment problem (GAP), highlighting the differences between those problems. In particular, there does not seem to be a direct way to apply, to IIK, the $(1 - 1/e - \varepsilon)$ approximation algorithm Fleischer et al. [5] for GAP. Other generalizations of MAX-K related to IIK, but whose current solving approaches do not seem to extend, are the multiple knapsack (MKP) and unsplittable flow on a path (UFP) problems. We discuss those problems and highlight the new ingredients introduced by our approach in the full version of the paper [4].

The Basic Techniques. In order to illustrate the ideas behind the proof of Theorem 1, let us first recall one of the PTAS for the classical MAX-K with capacity β, n items, profit and weight vector p and w respectively. Recall the greedy algorithm for knapsack:

1. Sort items so that $\frac{p_1}{w_1} \geq \frac{p_2}{w_2} \geq \cdots \geq \frac{p_n}{w_n}$.
2. Set $\bar{x}_i = 1$ for $i = 1, \ldots, \bar{\imath}$, where $\bar{\imath}$ is the maximum integer s.t. $\sum_{1 \leq i \leq \bar{\imath}} w_i \leq \beta$.

It is well-known that $p^T \bar{x} \geq p^T x^* - \max_{i \geq \bar{\imath}+1} p_i$, where x^* is the optimum solution to the linear relaxation. A PTAS for MAX-K can then be obtained as follows: "guess" a set S_0 of $\frac{1}{\varepsilon}$ items with $w(S_0) \leq \beta$ and consider the "residual" knapsack instance $\mathcal{I}$ obtained removing items in S_0 and items ℓ with $p_\ell > \min_{i \in S_0} p_i$, and setting the capacity to $\beta - w(S_0)$. Apply the greedy algorithm to $\mathcal{I}$ as to obtain solution S. Clearly $S_0 \cup S$ is a feasible solution to the original knapsack problem. The best solutions generated by all those guesses can be easily shown to be a $(1 - \varepsilon)$-approximation to the original problem.

IIK can be expressed in the form of the following integer program.

$$
\begin{aligned}
\max \quad & \sum_{t \in [T]} p^T x_t \\
\text{s.t.} \quad & w^T x_t \leq b_t \quad \forall t \in [T] \\
& x_t \leq x_{t+1} \quad \forall t \in [T-1] \\
& x_t \in \{0,1\}^n \quad \forall t \in [T].
\end{aligned}
\tag{1}
$$

By definition, $0 < b_t \leq b_{t+1}$ for $t \in [T-1]$. We also assume wlog that $1 = p_1 \geq p_2 \geq \ldots \geq p_n$.

When trying to extend the PTAS above for MAX-K to IIK, we face two problems. First, we have multiple times, and a standard guessing over all times will clearly be exponential in T. Second, when inserting an item into the knapsack at a specific time, we are clearly imposing this decision on all times that succeed it, and it is not clear a priori how to take this into account.

We solve these issues by proposing an algorithm that, in a sense, still follows the general scheme of the greedy algorithm sketched above: after some preprocessing, guess items (and insertion times) that give high profit, and then fill the remaining capacity with an LP-driven integral solution. However, the way of achieving this is different from the PTAS above. In particular, some of the techniques we introduced are specific for IIK and not to be found in methods for solving non-incremental knapsack problems.

An Overview of the Algorithm:

(i) *Sparsification and other simplifying assumptions.* We first show that by losing at most a 2ε fraction of the profit, we can assume the following (see Sect. 2.1): item 1, which has the maximum profit, is inserted into the knapsack at some time; the capacity of the knapsack only increases and hence the insertion of items can only happen at $J = O(\frac{1}{\varepsilon} \log T)$ times (we call them *significant*); and the profit of each item is either much smaller than $p_1 = 1$ or it takes one of $K = O(\frac{1}{\varepsilon} \log \frac{T}{\varepsilon})$ possible values (we call them *profit classes*).

(ii) *Guessing of a stairway.* The operations in the previous step give a $J \times K$ grid of "significant times" vs "profit classes" with $O(\frac{1}{\varepsilon^2} \log^2 \frac{T}{\varepsilon})$ entries in total. One could think of the following strategy: for each entry (j, k) of the grid, guess how many items of profit class k are inserted in the knapsack at time t_j. However, those entries are still too many to perform guessing over all of them. Instead, we proceed as follows: we guess, for each significant time t_j, *which is the class k* of maximum profit that has an element in the knapsack at time t_j. Then, for profit class k and carefully selected profit classes "close" to k, we either *guess exactly how many items* are in the knapsack at time t_j or if these are *at least* $\frac{1}{\varepsilon}$. Each of the guesses leads to a natural IP. The optimal solution to one of the IPs is an optimal solution to our original problem. Clearly, the number of possible guesses affects the number of the IPs, hence the overall complexity. We introduce the concept of "stairway" to show that these guesses are polynomially many for fixed ϵ. See Sect. 2.2 for details. We remark that, from this step on, we substantially differ from the approach of [2], which is also based on a disjunctive formulation.

(iii) *Solving the linear relaxations and rounding.* Fix an IP generated at the previous step, and let x^* be the optimal solution of its linear relaxation. A classical rounding argument relies on LP solutions having a small number of fractional components. Unfortunately, x^* is not as simple as that. However, we show that, after some massaging, we can control the entries of x^* where "most" fractional components appear, and conclude that the profit of $\lfloor x^* \rfloor$ is close to that of x^*. See Sect. 2.3 for details. Hence, looping over all guessed IPs and outputting vector $\lfloor x^* \rfloor$ of maximum profit concludes the algorithm.

Assumption: We assume that expressions $\frac{1}{\varepsilon}$, $(1+\varepsilon)^j$, $\log_{1+\varepsilon} \frac{T}{\varepsilon}$ and similar are to be rounded up to the closest integer. This is just done for simplicity of notation and can be achieved by replacing ϵ with an appropriate constant fraction of it, which will not affect the asymptotic running time.

2   A PTAS for IIK

2.1   Reducing IIK to Special Instances and Solutions

Our first step will be to show that we can reduce IIK, without loss of generality, to solutions and instances with a special structure. The first reduction is immediate: we restrict to solutions where the highest profit item is inserted in the knapsack at some time. We call these 1-*in solutions*. This can be assumed by guessing which is the highest profit item that is inserted in the knapsack, and reducing to the instance where all higher profit items have been excluded. Since we have n possible guesses, the running time is scaled by a factor $O(n)$.

Observation 1. *Suppose there exists a function* $f : \mathbb{N} \times \mathbb{N} \times \mathbb{R}_{>0} \to \mathbb{N}$ *such that, for each* $n, T \in \mathbb{N}$, $\varepsilon > 0$, *and any instance of* IIK *with* n *items and* T *times, we can find a* $(1 - \varepsilon)$-*approximation to a 1-in solution of highest profit in time* $f(n, T, \varepsilon)$. *Then we can find a* $(1 - \varepsilon)$-*approximation to any instance of* IIK *with* n *items and* T *times in time* $O(n) \cdot f(n, T, \varepsilon)$.

Now, let $\mathcal{I}$ be an instance of IIK with n items, let $\varepsilon > 0$. We say that $\mathcal{I}$ is ε-*well-behaved* if it satisfies the following properties.

($\varepsilon 1$) For all $i \in [n]$, one has $p_i = (1 + \varepsilon)^{-j}$ for some $j \in \{0, 1, \ldots, \log_{1+\varepsilon} \frac{T}{\varepsilon}\}$, or $p_i \leq \frac{\varepsilon}{T}$.
($\varepsilon 2$) $b_t = b_{t-1}$ for all $t \in [T]$ such that $(1+\varepsilon)^{j-1} < T - t + 1 < (1+\varepsilon)^j$ for some $j \in \{0, 1, \ldots, \log_{1+\varepsilon} T\}$, where we set $b_0 = 0$.

See Fig. 1 for an example. Note that condition ($\varepsilon 2$) implies that the capacity can change only during the set of times $\mathcal{T} := \{t \in [T] : t = T + 1 - (1 + \varepsilon)^j$ for some $j \in \mathbb{N}\}$, with $|\mathcal{T}| = O(\log_{1+\varepsilon} T)$. $\mathcal{T}$ clearly gets sparser as t becames smaller. Note that for T not being a degree of $(1 + \varepsilon)$ there will be a small fraction of times t at the beginning with capacity 0; see Fig. 1.

Next theorem implies that we can, wlog, assume that our instances are ε-well-behaved (and our solutions are 1-in).

Theorem 2. *Suppose there exists a function* $g : \mathbb{N} \times \mathbb{N} \times \mathbb{R}_{>0} \to \mathbb{N}$ *such that, for each* $n, T \in \mathbb{N}$, $\varepsilon > 0$, *and any* ε-*well-behaved instance of* IIK *with* n *items and* T *times, we can find a* $(1 - 2\varepsilon)$-*approximation to a 1-in solution of highest profit in time* $g(n, T, \varepsilon)$. *Then we can find a* $(1 - 4\varepsilon)$-*approximation to any instance of* IIK *with* n *items and* T *times in time* $O(T + n(n + g(n, T, \varepsilon)))$.

Proof. Fix an IIK instance $\mathcal{I}$. The reason why we can restrict ourselves to finding a 1-in solution is Observation 1. Consider instance $\mathcal{I}'$ with n items having the same weights as in $\mathcal{I}$, T times, and the other parameters defined as follows:

- For $i \in [n]$, if $(1+\varepsilon)^{-j} \leq p_i < (1+\varepsilon)^{-j+1}$ for some $j \in \{0, 1, \ldots, \log_{1+\varepsilon} \frac{T}{\varepsilon}\}$, set $p_i' := (1+\varepsilon)^{-j}$; otherwise, set $p_i' := p_i$. Note that we have $1 = p_1' \geq p_2' \geq \ldots \geq p_n'$.
- For $t \in [T]$ and $(1 + \varepsilon)^{j-1} < T - t + 1 \leq (1 + \varepsilon)^j$ for some $j \in \{0, 1, \ldots, \log_{1+\varepsilon} T\}$, set $b_t' := b_{T-(1+\varepsilon)^j+1}$, with $b_0' := 0$.

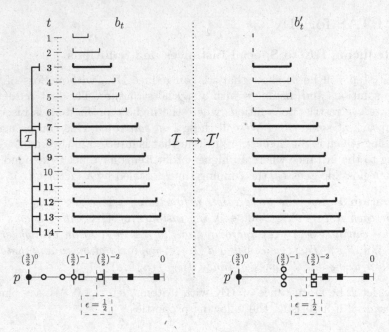

Fig. 1. An example of obtaining an ε-well-behaved instance for $\varepsilon = \frac{1}{2}$ and $T = 14$.

One easily verifies that $\mathcal{I}'$ is ε-well-behaved. Moreover, $b'_t \leq b_t$ for all $t \in [T]$ and $\frac{p_i}{1+\varepsilon} \leq p'_i \leq p_i$ for $i \in [n]$, so we deduce:

Claim 3. *Any solution $\bar{x}$ feasible for $\mathcal{I}'$ is also feasible for $\mathcal{I}$, and $p(\bar{x}) \geq p'(\bar{x})$.*

Proof of the next claim is given in the full version [4].

Claim 4. *Let x^* be a 1-in feasible solution of highest profit for $\mathcal{I}$. There exists a 1-in feasible solution x' for $\mathcal{I}'$ such that $p'(x') \geq (1-\varepsilon)^2 p(x^*)$.*

Let $\hat{x}$ be a 1-in solution of highest profit for $\mathcal{I}'$ and $\bar{x}$ is a solution to $\mathcal{I}'$ that is a $(1-\varepsilon)$-approximation to $\hat{x}$. Claims 3 and 4 imply that $\bar{x}$ is feasible for $\mathcal{I}$ and we deduce:

$$p(\bar{x}) \geq p'(\bar{x}) \geq (1-2\varepsilon)p'(\hat{x}) \geq (1-2\varepsilon)p'(x') \geq (1-2\varepsilon)(1-\varepsilon)^2 p(x^*) \geq (1-4\varepsilon)p(x^*).$$

In order to compute the running time, it is enough to bound the time required to produce $\mathcal{I}'$. Vector p' can be produced in time $O(n)$, while vector b' in time T. Moreover, the construction of the latter can be performed before fixing the highest profit object that belongs to the knapsack (see Observation 1). The thesis follows. □

2.2 A Disjunctive Relaxation

Fix $\varepsilon > 0$. Because of Theorem 2, we can assume that the input instance $\mathcal{I}$ is ε-well-behaved. We call all times from $\mathcal{T}$ *significant*. Note that a solution over the latter times can be naturally extended to a global solution by setting $x_t = x_{t-1}$ for all non-significant times t. We denote significant times by $t(1) < t(2) < \cdots < t(|\mathcal{T}|)$. In this section, we describe an IP over feasible 1-in solutions of an ε-well-behaved instance of IIK. The feasible region of this IP is the union of different regions, each corresponding to a partial assignment of items to significant times. In Sect. 2.3 we give a strategy to round an optimal solution of the LP relaxation of the IP to a feasible integral solution with a $(1 - 2\varepsilon)$-approximation guarantee. Together with Theorem 2 (taking $\varepsilon' = \frac{\varepsilon}{4}$), this implies Theorem 1.

In order to describe those partial assignments, we introduce some additional notation. We say that items having profit $(1 + \varepsilon)^{-k}$ for $k \in [\log_{1+\varepsilon} \frac{T}{\varepsilon}]$, belong to *profit class* k. Hence bigger profit classes correspond to items with smaller profit. All other items are said to belong to the *small* profit class. Note that there are $O(\frac{1}{\varepsilon} \log \frac{T}{\varepsilon})$ profit classes (some of which could be empty). Our partial assignments will be induced by special sets of vertices of a related graph called *grid*.

Definition 1. *Let $J \in \mathbb{Z}_{>0}, K \in \mathbb{Z}_{\geq 0}$, a grid of dimension $J \times (K + 1)$ is the graph $G_{J,K} = ([J] \times [K]_0, E)$, where*

$$E := \{\{u, v\} : \ u, v \in [J] \times [K]_0, \ u = (j, k)$$
$$\text{and either } v = (j + 1, k) \text{ or } v = (j, k + 1)\}.$$

Definition 2. *Given a grid $G_{J,K}$, we say that*

$$S := \{(j_1, k_1), (j_2, k_2), \ldots, (j_{|S|}, k_{|S|})\} \subseteq V(G_{J,K})$$

is a stairway *if $j_h > j_{h+1}$ and $k_h < k_{h+1}$ for all $h \in [|S| - 1]$.*

Lemma 1. *There are at most 2^{K+J+1} distinct stairways in $G_{J,K}$.*

Proof. The first coordinate of any entry of a stairway can be chosen among J values, the second coordinate from $K + 1$ values. By Definiton 2, each stairway correspond to exactly one choice of sets $J_1 \subseteq [J]$ for the first coordinates and $K_1 \subseteq [K]_0$ for the second, with $|K_1| = |J_1|$. $\square$

Now consider the grid graph with $J := |\mathcal{T}| = \theta(\frac{1}{\varepsilon} \log T), K = \log_{1+\varepsilon} \frac{T}{\varepsilon}$, and a stairway S with $k_1 = 0$. See Fig. 2 for an example. This corresponds to a partial assignment that can be informally described as follows. Let $(j_h, k_h) \in S$ and $t_h := t(j_h)$. In the corresponding partial assignment no item belonging to profit classes $k_h \leq k < k_{h+1}$ is inside the knapsack at any time $t < t_h$, while the first time an item from profit class k_h is inserted into the knapsack is at time t_h (if $j_{|S|} > 1$ then the only items that the knapsack can contain at times $1, \ldots, t_{|S|} - 1$ are the items from the small profit class). Moreover, for each $h \in [|S|]$, we focus on the family of profit classes $\mathcal{K}_h := \{k \in [K] : \ k_h \leq k \leq k_h + C_\varepsilon\}$ with $C_\varepsilon = \log_{1+\varepsilon} \frac{1}{\varepsilon}$. For each $k \in \mathcal{K}_h$ and every (significant) time t in the set $\mathcal{T}_h := \{t \in \mathcal{T} : \ t_{h-1} < t \leq t_h\}$, we will either specify exactly the number of items taken from profit class k at time t, or impose that there are at least $\frac{1}{\varepsilon} + 1$ of those items (this is established by map ρ_h below). Note that we can assume that the items taken within a profit class are those with minimum weight: this may exclude some feasible 1-in solutions, but it will always keep at least a feasible 1-in solution of maximum profit. No other constraint is imposed.

More formally, set $k_{|S|+1} = K + 1$ and for each $h = 1, \ldots, |S|$:

(i) Set $x_{t,i} = 0$ for all $t \in [t_h - 1]$ and each item i in a profit class $k \in [k_{h+1} - 1]$.

(ii) Fix a map $\rho_h : \mathcal{T}_h \times \mathcal{K}_h \to \{0, 1, \ldots, \frac{1}{\varepsilon} + 1\}$ such that for all $t \in \mathcal{T}_h$ one has $\rho_h(t, k_h) \geq 1$ and $\rho_h(\bar{t}, k) \geq \rho_h(t, k)$, $\forall(\bar{t}, k) \in \mathcal{T}_h \times \mathcal{K}_h$, $\bar{t} \geq t$.

Additionally, we require $\rho_h(\bar{t}, k) \geq \rho_{h+1}(t, k)$ for all $h \in [|S| - 1]$, $k \in \mathcal{K}_h \cap \mathcal{K}_{h+1}$, $\bar{t} \in \mathcal{T}_h$, $t \in \mathcal{T}_{h+1}$. Thus, we can merge all ρ_h into a function $\rho : \cup_{h \in [|S|]}(\mathcal{T}_h \times \mathcal{K}_h) \to \{0, 1, \ldots, \frac{1}{\varepsilon} + 1\}$. For each profit class $k \in [K]$ we assume that items from this class are $I_k = \{1(k), \ldots, |I_k|(k)\}$, so that $w_{1(k)} \leq w_{2(k)} \leq \cdots \leq w_{|I_k|(k)}$. Based on our choice (S, ρ) we define the polytope:

$$P(S, \rho) = \{x \in \mathbb{R}^{Tn} : w^T x_t \leq b_t \ \forall t \in [T]$$
$$x_t \leq x_{t+1} \ \forall t \in [T-1]$$
$$0 \leq x_t \leq 1 \ \forall t \in [T]$$
$$\forall h \in [|S|] :$$
$$x_{t,i(k)} = 0, \ \forall t < t_h, \ \forall k < k_{h+1}, \ \forall i(k) \in I_k$$
$$x_{t,i(k)} = 1, \ \forall t \in \mathcal{T}_h, \ \forall k \in \mathcal{K}_h, \ \forall i(k) : i \leq \rho(t, k)$$
$$x_{t,i(k)} = 0, \ \forall t \in \mathcal{T}_h, \ \forall k \in \mathcal{K}_h : \rho(t, k) \leq \frac{1}{\varepsilon},$$
$$\forall i(k) : i > \rho(t, k)\}.$$

The linear inequalities are those from the IIK formulation. The first set of equations impose that, at each time t, we do not take any object from a profit class k, if we guessed that the highest profit object in the solution at time t belongs to a profit class $k' > k$ (those are entries corresponding to the dark grey area in Fig. 2). The second set of equations impose that for each time t and class k for which a guess $\rho(t, k)$ was made (light grey area in Fig. 2), we take the $\rho(t, k)$ items of smallest weight. As mentioned above, this is done without loss of generality: since profits of objects from a given profit class are the same, we can assume that the optimal solution insert first those of smallest weight. The last set of equations imply that no other object of class k is inserted in time t if $\rho(t, k) \leq \frac{1}{\varepsilon}$.

Note that some choices of S, ρ may lead to empty polytopes. Fix S, ρ, an item i and some time t. If, for some $t' \leq t$, $x_{t',i} = 1$ explicitly appears in the definition of $P(S, \rho)$ above, then we say that i is t-included. Conversely, if $x_{\bar{t},i} = 0$ explicitly appears for some $\bar{t} \geq t$, then we say that i is t-excluded.

Theorem 5. *Any optimum solution of*

$$\max \sum_{t \in [T]} p_t^T x_t \quad s.t. \quad x \in (\cup_{S,\rho} P(S, \rho)) \cap \{0, 1\}^{Tn}$$

is a 1-in solution of maximum profit for $\mathcal{I}$. Moreover, the number of constraints of the associated LP relaxation is at most $nT^{f(\varepsilon)}$ for some function $f : \mathbb{R}_{>0} \to \mathbb{R}_{>0}$ depending on ε only.

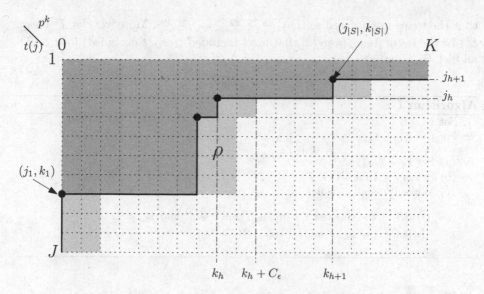

Fig. 2. An example of a stairway S, given by thick black dots. Entries (j, k) lying in the light grey area are those for which a value ρ is specified. No item corresponding to the entries in the dark grey area is taken, except on the boundary in bold.

Proof. Note that one of the choices of (S, ρ) will be the correct one, i.e., it will predict the stairway S associated to an optimal 1-in solution, as well as the number of items that this solution takes for each entry of the grid it guessed. Then there exists an optimal solution that takes, for each time t and class k for which a guess $\rho(t, k)$ was made, the $\rho(t, k)$ items of smallest weight from this class, and no other object if $\rho(t, k) \leq \frac{1}{\epsilon}$. These are exactly the constraints imposed in $P(S, \rho)$. The second part of the statement follows from the fact that the possible choices of (S, ρ) are

$$(\# \text{ stairways}) \cdot (\# \text{ possible values in each entry of } \rho)^{(\# \text{ entries of a vector } \rho)}$$

$$= 2^{O(\frac{1}{\epsilon} \log \frac{T}{\epsilon})} \cdot O\left(\frac{1}{\epsilon}\right)^{O(\frac{1}{\epsilon} \log \frac{T}{\epsilon})C_\epsilon} = \left(\frac{T}{\epsilon}\right)^{O(\frac{1}{\epsilon^2} \log^2 \frac{1}{\epsilon})},$$

and each (S, ρ) has $g(\epsilon)O(Tn)$ constraints, where g depends on ϵ only.   □

2.3 Rounding

By convexity, there is a choice of S and ρ as in the previous section such that any optimum solution of

$$\max \sum_{t \in [T]} p^T x_t \quad \text{s.t.} \quad x \in P(S, \rho) \tag{2}$$

is also an optimum solution to $\max\{\sum_{t \in [T]} p^T x_t : x \in \text{conv}(\cup_{S, \rho} P(S, \rho))\}$. Hence, we can focus on rounding an optimum solution x^* of (2). We assume that the items are ordered so that $\frac{p_1}{w_1} \geq \frac{p_2}{w_2} \geq \cdots \geq \frac{p_n}{w_n}$. Moreover, let $\mathcal{I}^t$ (resp. $\mathcal{E}^t$) be the set of items from $[n]$ that are t-included (resp. t-excluded) for $t \in [T]$, and let $W_t := w^T x_t^*$.

Algorithm 1

1: Set $\bar{x}_0 = \mathbf{0}$.
2: For $t = 1, \ldots, T$:

    (a) Set $\bar{x}_t = \bar{x}_{t-1}$.
    (b) Set $\bar{x}_{t,i} = 1$ for all $i \in \mathcal{I}^t$.
    (c) While $W_t - w^T \bar{x}_t > 0$:

        (i) Select the smallest $i \in [n]$ such that $i \notin \mathcal{E}^t$ and $\bar{x}_{t,i} < 1$.
        (ii) Set $\bar{x}_{t,i} = \bar{x}_{t,i} + \min\{1 - \bar{x}_{t,i}, \frac{W_t - w^T \bar{x}_t}{w_i}\}$.

Respecting the choices of S and ρ (i.e., included/excluded items), at each time t, Algorithm 1 greedily adds objects into the knapsack, until the total weight is equal to W_t. Recall that in MAX-K one obtains a rounded solution which differs from the fractional optimum by the profit of at most one item. Here the fractionality pattern is more complex, but still under control. In fact, as we show below, $\bar{x}$ is such that $\sum_{t \in [T]} p^T \bar{x}_t = \sum_{t \in [T]} p^T x_t^*$ and, for each $h \in [|S|]$ and $t \in [T]$ such that $t_h \leq t < t_{h-1}$, vector $\bar{x}_t$ has at most $|S| - h + 1$ fractional components that do not correspond to items in profit classes $k \in K$ with at least $\frac{1}{\epsilon} + 1$ t-included items. We use this fact to show that $\lfloor \bar{x} \rfloor$ is an integral solution that is $(1 - 2\epsilon)$-optimal.

Theorem 6. *Let x^* be an optimum solution to (2). Algorithm 1 produces, in time $O(T + n)$, a vector $\bar{x} \in P(S, \rho)$ such that $\sum_{t \in [T]} p^T \lfloor \bar{x}_t \rfloor \geq (1 - 2\varepsilon) \sum_{t \in [T]} p^T x_t^*$.*

Theorem 6 will be proved in a series of intermediate steps. Define $\mathcal{F}_t := \{i \in [n] : 0 < \bar{x}_{t,i} < 1\}$ to be the set of fractional components of $\bar{x}_t$ for $t \in [T]$. Recall that Algorithm 1 sorts items by monotonically decreasing profit/weight ratio. For items from a given profit class $k \in [K]$, this induces the order $i(1) < i(2) < \ldots$ – i.e., by monotonically increasing weight – since all $i(k) \in I_k$ have the same profit.

The following claim (proved in the full version [4]) shows that $\bar{x}$ is in fact an optimal solution to $\max\{x : x \in P(S, \rho)\}$.

Claim 7. *For each $t \in [T]$, one has $w^T \bar{x}_t = w^T x_t^*$ and $p^T \bar{x}_t = p^T x_t^*$.*

For $t \in [T]$ define $\mathcal{L}_t := \{k \in [K] : |I_k \cap \mathcal{I}^t| \geq \frac{1}{\epsilon} + 1\}$ to be the set of classes with a large number of t-included items. Furthermore, for $h = 1, 2, \ldots, |S|$:

- Recall that $\mathcal{K}_h = \{k \in [K] : k_h \le k \le k_h + C_\varepsilon\}$ are the classes of most profitable items present in the knapsack at times $t \in [T] : t_h \le t < t_{h-1}$, since by definition no item is taken from a class $k < k_h$ at those times. Also by definition $\rho(t_h, k_h) \ge 1$, so the largest profit item present in the knapsack at any time $t \in [T] : t_h \le t < t_{h-1}$ is item $1(k_h)$. Denote its profit by p_{max}^h.

- Define $\bar{\mathcal{K}}_h := \{k \in [K] : k_h + C_\varepsilon < k\}$, i.e., it is the family of the other classes for which an object may be present in the knapsack at time $t \in [T] : t_h \le t < t_{h-1}$.

Proofs of next two claims are given in the full version [4].

Claim 8. *Fix $t \in [T]$, $t_h \le t < t_{h-1}$. Then $|I_k \cap \mathcal{F}_t| \le 1$ for all $k \in [K] \cup \{\infty\}$. Moreover, $|((\cup_{k \in \bar{\mathcal{K}}_h} I_k) \cap \mathcal{F}_t) \setminus \mathcal{F}_{t_{h-1}}| \le 1$.*

Claim 9. *Let $h \in [|S|]$, then: $p((\cup_{k \in \bar{\mathcal{K}}_h \setminus \mathcal{L}_t} I_k) \cap \mathcal{F}_t) \le \epsilon \sum_{\bar{h}=h}^{|S|} p_{max}^{\bar{h}}$, $\forall t : t_h \le t < t_{h-1}$.*

Proof of Theorem 6. We focus on showing that, $\forall t \in [T]$:

$$\sum_{i \in [n] \setminus I_\infty} p_i \lfloor \bar{x}_{t,i} \rfloor \ge \sum_{i \in [n] \setminus I_\infty} p_i \bar{x}_{t,i} - \sum_{i \in ([n] \setminus I_\infty) \cap \mathcal{F}_t} p_i \ge (1 - \epsilon) \sum_{i \in [n] \setminus I_\infty} p_i \bar{x}_{t,i}. \quad (3)$$

The first inequality is trivial and, if $t < t_{|S|}$, so is the second, since in this case $\bar{x}_{t,i} = 0$ for all $i \in [n] \setminus I_\infty$. Otherwise, t is such that $t_h \le t < t_{h-1}$ for some $h \in [|S|]$ with $t_0 = T + 1$. Observe that:

$$([n] \setminus I_\infty) \cap \mathcal{F}_t = ((\cup_{k \in (\mathcal{K}_h \cup \bar{\mathcal{K}}_h) \setminus \mathcal{L}_t} I_k) \cap \mathcal{F}_t) \cup ((\cup_{k \in (\mathcal{K}_h \cup \bar{\mathcal{K}}_h) \cap \mathcal{L}_t} I_k) \cap \mathcal{F}_t)$$
$$= ((\cup_{k \in \bar{\mathcal{K}}_h \setminus \mathcal{L}_t} I_k) \cap \mathcal{F}_t) \cup ((\cup_{k \in \mathcal{L}_t} I_k) \cap \mathcal{F}_t)$$

For $k \in [K]$ denote the profit of $i \in I_k$ with p^k. We have:

$$\begin{aligned}\sum_{i \in ([n] \setminus I_\infty) \cap \mathcal{F}_t} p_i \bar{x}_{t,i} &= p((\cup_{k \in \bar{\mathcal{K}}_h \setminus \mathcal{L}_t} I_k) \cap \mathcal{F}_t) + p((\cup_{k \in \mathcal{L}_t} I_k) \cap \mathcal{F}_t) \\ \text{(By Claim 9 and Claim 8)} &\le \epsilon \sum_{\bar{h}=h}^{|S|} p_{max}^{\bar{h}} + \sum_{k \in \mathcal{L}_t} p^k.\end{aligned} \quad (4)$$

If $k = k_{\bar{h}} \in \mathcal{L}_t$ for $\bar{h} \in [|S|]$ then $\sum_{i \in I_k} p_i \bar{x}_{t,i} \ge (\frac{1}{\epsilon} + 1) p^k = p_{max}^{\bar{h}} + \frac{1}{\epsilon} p^k$. Together with $\rho(k_h, t_h) \ge 1$ $\forall h \in [|S|]$ and the definition of $\mathcal{L}_t$ this gives:

$$\sum_{i \in [n] \setminus I_\infty} p_i \bar{x}_{t,i} \ge \sum_{\bar{h}=h}^{|S|} p_{max}^{\bar{h}} + \frac{1}{\epsilon} \sum_{k \in \mathcal{L}_t} p^k. \quad (5)$$

Put together, (4) and (5) imply (3). Morever, by Claim 8, $|I_\infty \cap \mathcal{F}_t| \le 1$ for all $t \in [T]$ and since we are working with an ϵ-well-behaved instance $p_i \le \frac{\epsilon}{T} = \frac{\epsilon}{T} p_{max}^1$ so $\sum_{t \in [T]} \sum_{i \in I_\infty \cap \mathcal{F}_t} p_i \le \epsilon p_{max}^1$. The last fact with (3) and Claim 7 gives the statement of the theorem. $\quad \square$

Theorem 1 now easily follows from Theorems 2, 5, and 6.

Proof of Theorem 1. Since we will need items to be sorted by profit/weight ratio, we can do this once and for all before any guessing is performed. Classical algorithms implement this in $O(n \log n)$. By Theorem 2, we know we can assume that the input instance is ε-well-behaved, and it is enough to find a solution of profit at least $(1 - 2\varepsilon)$ the profit of a 1-in solution of maximum profit – by Theorem 6, this is exactly vector $\lfloor \bar{x} \rfloor$. In order to produce $\lfloor \bar{x} \rfloor$, as we already sorted items by profit/weight ratio, we only need to solve the LPs associated with each choice of S and ρ, and then run Algorithm 1. The number of choices of S and ρ are $T^{f(\varepsilon)}$, and each LP has $g(\varepsilon)O(nT)$ constraints, for appropriate functions f and g (see the proof of Theorem 5). Algorithm 1 runs in time $O(\frac{T}{\varepsilon} \log \frac{T}{\varepsilon} + n)$. The overall running time is:

$$O(n \log n + n(n + T + T^{f(\varepsilon)}(f_{LP}(g(\varepsilon)O(nT)) + \frac{T}{\varepsilon} \log \frac{T}{\varepsilon}))) = O(nT^{h(\varepsilon)}f_{LP}(n)),$$

where $f_{LP}(m)$ is the time required to solve an LP with $O(m)$ variables and constraints, and $h : \mathbb{R} \to \mathbb{N}_{\geq 1}$ is an appropriate function.   □

3 Generalizations

Following Theorem 1, one could ask for a PTAS for the general incremental knapsack (IK) problem. This is the modification of IIK (introduced in [2]) where the objective function is $p_\Delta(x) := \sum_{t \in [T]} \Delta_t \cdot p^T x_t$, where $\Delta_t \in \mathbb{Z}_{>0}$ for $t \in [T]$ can be seen as time-dependent discounts. We show here some partial results. All proofs from this section are given in the full version [4].

Corollary 1. *There exists a PTAS-preserving reduction from IK to IIK, assuming $\Delta_t \leq \Delta_{t+1}$ for $t \in [T-1]$. Hence, the hypothesis above, IK has a PTAS.*

In [4], we show an auxiliary claim that there is a PTAS for IK when $\|\Delta\|_\infty$ is polynomially bounded in n and T. The direct corollary is that IK has a PTAS when every item fits into the knapsack at time $t = 1$. Additionally, our technique gives a PTAS for IK when T is constant.

Of independent interest is the fact that there is a PTAS for the modified version of IIK when each item can be taken multiple times. Unlike Corollary 1, this is not based on a reduction between problems, but on a modification on our algorithm.

Corollary 2. *There is a PTAS for the following modification of IIK: in (1), replace $x_t \in \{0,1\}^n$ with: $x_t \in \mathbb{Z}_{>0}^n$ for $t \in [T]$; and $0 \leq x_t \leq d$ for $t \in [T]$, where we let $d \in (\mathbb{Z}_{>0} \cup \{\infty\})^n$ be part of the input.*

Acknowledgements. We thank Andrey Kupavskii for valuable combinatorial insights on the topic. Yuri Faenza's research was partially supported by the SNSF *Ambizione* fellowship PZ00P2_154779 *Tight formulations of 0/1 problems*. Some of the work was done when Igor Malinović visited Columbia University partially funded by a gift from the SNSF.

References

1. Andonov, R., Poirriez, V., Rajopadhye, S.: Unbounded knapsack problem: dynamic programming revisited. Eur. J. Oper. Res. **123**(2), 394–407 (2000)
2. Bienstock, D., Sethuraman, J., Ye, C.: Approximation algorithms for the incremental knapsack problem via disjunctive programming. arXiv preprint: arXiv:1311.4563 (2013)
3. Della Croce, F., Pferschy, U., Scatamacchia, R.: Approximation results for the incremental knapsack problem. In: Brankovic, L., Ryan, J., Smyth, W.F. (eds.) IWOCA 2017. LNCS, vol. 10765, pp. 75–87. Springer, Cham (2018). https://doi.org/10.1007/978-3-319-78825-8_7
4. Faenza, Y., Malinovic, I.: A PTAS for the time-invariant incremental knapsack problem. arXiv preprint: arXiv:1701.07299v4 (2018)
5. Fleischer, L., Goemans, M.X., Mirrokni, V.S., Sviridenko, M.: Tight approximation algorithms for maximum general assignment problems. In: Proceedings of SODA 2006, pp. 611–620 (2006)
6. Hartline, J.: Incremental optimization. Ph.D. thesis, Cornell University (2008)
7. Ibarra, O.H., Kim, C.E.: Fast approximation algorithms for the knapsack and sum of subset problems. J. ACM **22**(4), 463–468 (1975)
8. Karp, R.M.: Reducibility among combinatorial problems. In: Miller, R.E., Thatcher, J.W., Bohlinger, J.D. (eds.) Complexity of Computer Computations, pp. 85–103. Springer, Boston (1972). https://doi.org/10.1007/978-1-4684-2001-2_9
9. Kellerer, H., Pferschy, U., Pisinger, D.: Knapsack Problems. Springer, Heidelberg (2004). https://doi.org/10.1007/978-3-540-24777-7
10. Martello, S., Toth, P.: Knapsack Problems: Algorithms and Computer Implementations. Wiley, New York (1990)
11. Mathews, G.B.: On the partition of numbers. Proc. Lond. Math. Soc. **28**(1), 486–490 (1896)
12. Sethuraman, J., Ye, C.: Personal communication (2016)
13. Sharp, A.: Incremental algorithms: solving problems in a changing world. Ph.D. thesis, Cornell University (2007)
14. Skutella, M.: An introduction to network flows over time. In: Cook, W., Lovász, L., Vygen, J. (eds.) Research Trends in Combinatorial Optimization, pp. 451–482. Springer, Heidelberg (2008). https://doi.org/10.1007/978-3-540-76796-1_21
15. Ye, C.: On the trade-offs between modeling power and algorithmic complexity. Ph.D. thesis, Columbia University (2016)

On Bounded Pitch Inequalities
for the Min-Knapsack Polytope

Yuri Faenza[1], Igor Malinović[2(✉)], Monaldo Mastrolilli[3], and Ola Svensson[2]

[1] Columbia University, New York City, USA
[2] École Polytechnique Fédérale de Lausanne, Lausanne, Switzerland
igor.malinovic@epfl.ch
[3] IDSIA Lugano, Manno, Switzerland

Abstract. In the min-knapsack problem one aims at choosing a set of objects with minimum total cost and total profit above a given threshold. In this paper, we study a class of valid inequalities for min-knapsack known as *bounded pitch inequalities*, which generalize the well-known unweighted cover inequalities. While separating over pitch-1 inequalities is NP-Hard, we show that approximate separation over the set of pitch-1 and pitch-2 inequalities can be done in polynomial time. We also investigate integrality gaps of linear relaxations for min-knapsack when these inequalities are added. Among other results, we show that, for any fixed t, the t-th CG closure of the natural linear relaxation has the unbounded integrality gap.

1 Introduction

The min-knapsack problem (MinKnap)[1]

$$\min c^T x \quad \text{s.t.} \quad p^T x \geq 1, \ x \in \{0,1\}^n \tag{1}$$

is the variant of the max knapsack problem (MaxKnap) where, given a cost vector c and a profit vector p, we want to minimize the total cost given a lower bound on the total profit. MinKnap is known to be NP-Complete, even when $p = c$. Moreover, it is easy to see that the classical FPTAS for MaxKnap [11,14] can be adapted to work for MinKnap, thus completely settling the complexity of MinKnap.

However, *pure* knapsack problems rarely appear in applications. Hence, one aims at developing techniques that remain valid when less structured constraints are added on top of the original knapsack one. This can be achieved by providing *strong* linear relaxations for the problem: then, any additional linear constraint can be added to the formulation, providing a good starting point for any branch-and-bound procedure. The most common way to measure the strength of a linear relaxation is by measuring its *integrality gap*, i.e. the maximum ratio between

---

[1] Note that $c \in \mathbb{R}_+^n$, $p \in \mathbb{R}_+^n$ and the constraint is scaled so that the right-hand side is 1.

© Springer International Publishing AG, part of Springer Nature 2018
J. Lee et al. (Eds.): ISCO 2018, LNCS 10856, pp. 170–182, 2018.
https://doi.org/10.1007/978-3-319-96151-4_15

the optimal solutions of the linear and the integer programs (or of its inverse if the problem is in minimization form) over all the objective functions.

Surprisingly, if we aim at obtaining linear relaxations with few inequalities and bounded integrality gap, MINKNAP and MAXKNAP seem to be very different. Indeed, the standard linear relaxation for MAXKNAP has integrality gap 2, and this can be boosted to $(1 + \epsilon)$ by an extended formulation with $n^{O(1/\epsilon)}$ many variables and constraints, for $\epsilon > 0$ [2]. Conversely, the standard linear relaxation for MINKNAP has unbounded integrality gap, and this remains true even after $\Theta(n)$ rounds of the Lasserre hierarchy [13]. No linear relaxation for MINKNAP with polynomially many constraints and constant integrality gap can be obtained in the original space [6]. It is an open problem whether an extended relaxation with this properties exists. Recent results showed the existence [1] and gave an explicit construction [9] of a linear relaxation for MINKNAP of quasi-polynomial size with integrality gap $2 + \epsilon$. This is obtained by giving an approximate formulation for *Knapsack Cover inequalities* (KC) (see [4] and the references therein). Adding those exponentially many inequalities, that can be approximately separated [4], gives an integrality gap of 2. The bound on the integrality gap is tight, even in the simpler case when $p = c$. One can then look for other classes of well-behaved inequalities that can be added to further reduce the integrality gap. A prominent family is given by the so called *bounded pitch* inequalities [3] defined in Sect. 2. Here, we remark that the *pitch* is a parameter measuring the complexity of an inequality, and the associated separation problem is NP-Hard already for pitch-1. The pitch-1 inequalities are often known in the literature as *unweighted cover inequalities* (see e.g. [1]).

In this paper, we study structural properties and separability of bounded pitch inequalities for MINKNAP, and the strength of linear relaxations for MINKNAP when they are added. Let $\mathcal{F}$ be the set given by pitch-1, pitch-2, and inequalities from the linear relaxation of (1). We first show that, for any arbitrarily small precision, we can solve in polynomial time the *weak separation problem* for the set $\mathcal{F}$. Even better, our algorithm either certifies that the given point x^* violates an inequality from $\mathcal{F}$, or outputs a point that satisfies all inequalities from $\mathcal{F}$ and whose objective function value is arbitrarily close to that of x^*. We define such an algorithm as a $(1 + \epsilon)$-*oracle* in Sect. 2; see Sect. 3 for the construction. A major step of our procedure is showing that non-redundant pitch-2 inequalities have a simple structure.

It is then a natural question whether bounded pitch inequalities can help to reduce the integrality gap below 2. We show that, when $p = c$, if we add to the linear relaxation of (1) pitch-1 and pitch-2 inequalities, the integrality gap is bounded by $3/2$; see Sect. 4.1. However, this is false in general. Indeed, we also prove that KC plus bounded pitch inequalities do not improve upon the integrality gap of 2; see Sect. 4.3. Moreover, bounded pitch alone can be much weaker than KC: we show that, for each fixed k, the integrality gap may be unbounded even if all pitch-k inequalities are added. Using the relation between bounded pitch and Chvátal-Gomory (CG) closures established in [3], this implies that, for each fixed t, the integrality gap of the t-th CG closure can be unbounded;

see Sect. 4.2. For an alternative proof that having all KC inequalities bounds the integrality gap to 2 see the full version of the paper [7].

2   Basics

A MinKnap instance is a binary optimization problem of the form (1), where $p, c \in \mathbb{Q}^n$ and we assume $0 \leq p_1 \leq p_2 \leq \cdots \leq p_n \leq 1, 0 < c_i \leq 1, \forall i \in [n]$. We will often deal with its *natural linear relaxation*

$$\min c^T x \quad \text{s.t.} \quad p^T x \geq 1, \ x \in [0, 1]^n. \tag{2}$$

The NP-Hardness of MinKnap immediately follows from the fact that MaxKnap is NP-Hard [12], and that a MaxKnap instance

$$\max v^T x \quad \text{s.t.} \quad w^T x \leq 1, \ x \in \{0, 1\}^n. \tag{3}$$

can be reduced into a MinKnap instance (1) as follows: each $x \in \{0, 1\}^n$ is mapped via $\pi : \mathbb{R}^n \to \mathbb{R}^n$ with $\pi(x) = \mathbf{1} - x$; $p_i = \frac{w_i}{\sum_{j=1}^n w_j - 1}$ and $c_i = v_i$ for $i \in [n]$. Note that the reduction is not approximation-preserving.

We say that an inequality $w^T x \geq \beta$ with $w \geq 0$ is *dominated* by a set of inequalities $\mathcal{F}$ if $w'^T x \geq \beta'$ can be written as a conic combination of inequalities in $\mathcal{F}$ for some $\beta' \geq \beta$ and $w' \leq w$. $w^T x \geq \beta$ is *undominated* if any set of valid inequalities dominating $w^T x \geq \beta$ contains a positive multiple of it.

Consider a family $\mathcal{F}$ of inequalities valid for (1). We refer to [10] for the definition of *weak separation oracle*, which is not used in this paper. We say that $\mathcal{F}$ admits a $(1 + \epsilon)$-*oracle* if, for each fixed $\epsilon > 0$, there exists an algorithm that takes as input a point $\bar{x}$ and, in time polynomial in n, either outputs an inequality from $\mathcal{F}$ that is violated by $\bar{x}$, or outputs a point $\bar{y}$, $\bar{x} \leq \bar{y} \leq (1 + \epsilon)\bar{x}$ that satisfies all inequalities in $\mathcal{F}$. In particular, if $\mathcal{F}$ contains the linear relaxation of (1), $0 \leq \bar{y} \leq 1$.

Let $\sum_{i \in T} w_i x_i \geq \beta$ be a valid inequality for (1), with $w_i > 0$ for all $i \in T$. Its *pitch* is the minimum k such that, for each $I \subseteq T$ with $|I| = k$, we have $\sum_{i \in I} w_i \geq \beta$. Undominated pitch-1 inequalities are of the form $\sum_{i \in T} x_i \geq 1$. Note that the map from MaxKnap to MinKnap instances defined above gives a bijection between *minimal cover inequalities*

$$\sum_{i \in I} x_i \leq |I| - 1$$

for MaxKnap and undominated pitch-1 inequalities for the corresponding MinKnap instance. Since, given a MaxKnap instance, it is NP-Hard to separate minimal cover inequalities [8], we conclude the following.

Theorem 1. *It is NP-Hard to decide whether a given point satisfies all valid pitch-1 inequalities for a given MinKnap instance.*

Given a set $S \subseteq [n]$, such that $\beta := 1 - \sum_{i \in S} p_i > 0$, the *Knapsack cover inequality* associated to S is given by

$$\sum_{i \in [n] \setminus S} \min\{p_i, \beta\} x_i \geq \beta \tag{4}$$

and it is valid for (1).

For a set $S \subseteq [n]$, we denote by χ^S its characteristic vector. An ϵ-*approximate solution* for a minimization integer programming problem is a solution $\bar{x}$ that is feasible, and whose value is at most $(1 + \epsilon)$ times the value of the optimal solution. An algorithm is called a *polynomial time approximation scheme (PTAS)* if for each $\epsilon > 0$ and any instance of the given problem it returns an ϵ-approximate solution in time polynomial in the size of the input. If in addition the running time is polynomial in $1/\epsilon$, then the algorithm is a *fully polynomial time approximation scheme (FPTAS)*.

Given a rational polyhedron $P = \{x \in \mathbb{R}^n : Ax \geq b\}$ with $A \in \mathbb{Z}^{m \times n}$ and $b \in \mathbb{Z}^m$, the first *Chvátal-Gomory (CG) closure* [5] of P is defined as follows:

$$P^{(1)} = \{x \in \mathbb{R}^n : \lceil \lambda^\top A \rceil x \geq \lceil \lambda^\top b \rceil, \ \forall \lambda \in \mathbb{R}^m\}.$$

Equivalently, one can consider all $\lambda \in [0, 1]^m$ such that $\lambda^\top A \in \mathbb{Z}^n$. For $t \in \mathbb{Z}_{\geq 2}$, the t-th CG closure of P is recursively defined as $P^{(t)} = (P^{(t-1)})^{(1)}$. The CG closure is an important tool for solving integer programs, see again [5].

3   A $(1 + \epsilon)$-oracle for Pitch-1 and Pitch-2 Inequalities

In this section, we show the following:

Theorem 2. *Given a* MinKnap *instance* (1), *there exists a* $(1+\epsilon)$-*oracle for the set* $\mathcal{F}$ *containing: all pitch-1 inequalities, all pitch-2 inequalities and all inequalities from the natural linear relaxation of* (1).

We start with a characterization of inequalities of interest for Theorem 2.

Lemma 1. *Let* K *be the set of feasible solutions of a* MinKnap *instance* (1). *All pitch-2 inequalities valid for* K *are implied by the set composed of:*

 (i) *Non-negativity constraints* $x_i \geq 0$ *for* $i \in [n]$;
 (ii) *All valid pitch-1 inequalities;*
 (iii) *All inequalities of the form*

$$\sum_{i \in I_1} x_i + 2 \sum_{i \in I_2} x_i \geq 2 \tag{5}$$

 where $I \subseteq [n]$, $|I| \geq 2$, $\beta(I) := 1 - \sum_{i \in [n] \setminus I} p_i$, $I_1 := \{i \in I : p_i < \beta(I)\} \neq \emptyset$ *and* $I_2 := I \setminus I_1$.

The inequalities in (iii) are pitch-2 and valid.

Proofs of Lemma 1 and Theorem 2 are given in Sects. 3.1 and 3.2, respectively.

3.1   Restricting the Set of Valid Pitch-2 Inequalities

We will build on two auxiliary statements in order to prove Lemma 1.

Claim 3. *If $w^T x \geq \beta$ and $u^T x \geq \beta$ are distinct valid inequalities and $u \geq w$, then the latter inequality is dominated by the former.*

Proof. $u^T x \geq \beta$ can be obtained summing nonnegative multiples of $w^T x \geq \beta$ and $x_i \geq 0$ for $i \in [n]$, which are all valid inequalities.   □

Claim 4. *Let*

$$\sum_{i \in T_1} x_i + 2 \sum_{i \in T_2} x_i \geq 2 \tag{6}$$

be a valid inequality for MINKNAP, *with $T_1 \cap T_2 = \emptyset$ and $T_1, T_2 \subseteq [n]$. Then, (6) is dominated by the inequality in (iii) with $I = T_1 \cup T_2$.*

Proof. One readily verifies that Inequality (5) with I as above is valid. Suppose now that $i \in T_1 \setminus I_1$. Then the integer solution that takes all elements in $([n] \setminus I) \cup \{i\}$ is feasible for MINKNAP, but it does not satisfy (6), a contradiction. Hence $T_1 \subseteq I_1$. Since $T_2 = I \setminus T_i \supseteq I \setminus I_1 = I_2$, (5) dominates (6) componentwise, and the thesis follows by Claim 3.   □

Proof of Lemma 1. The fact that an inequality of the form (5) is pitch-2 and valid is immediate. Because of Claim 4, it is enough to show the thesis with (5) replaced by (6). Consider a pitch-2 inequality valid for K:

$$\sum_{i \in T} w_i x_i \geq 1, \tag{7}$$

where $T \subseteq [n]$ is the support of the inequality, $w \in \mathbb{R}_+^{|T|}$. Without loss of generality one can assume that $T = [h]$ for some $h \leq n$ and $w_1 \leq w_2 \leq \cdots \leq w_h$. Since (7) is pitch-2 we have that $w_1 + w_i \geq 1$ for all $i \in [h] \setminus \{1\}$. We can also assume $w_h \leq 1$, since otherwise $\sum_{i \in [h-1]} w_i x_i + x_h \geq 1$ is valid and dominates (7) by Claim 3.

Let $j \in [h]$ be the maximum index such that $w_j < 1$. Note that such j exists, since, if $w_1 \geq 1$, then (7) is a pitch-1 inequality. If $1 - w_1 \leq 1/2$, then, by Claim 3, (7) is dominated by the valid pitch-2 inequality

$$\sum_{i \in [j]} x_i + 2 \sum_{i=j+1}^{h} x_i \geq 2, \tag{8}$$

which again is of the type (6). Hence $1 - w_1 > 1/2$ and again via Claim 3, (7) is dominated by

$$w_1 x_1 + \sum_{i=2}^{j} (1 - w_1) x_i + \sum_{i=j+1}^{h} x_i \geq 1, \tag{9}$$

since $w_i + w_1 \geq 1$ for all $i \neq 1$, so one has $w_i \geq 1 - w_1 > 1/2$. Thus, we can assume that (7) has the form (9). Note that inequality

$$\sum_{i=2}^{h} x_i \geq 1 \tag{10}$$

is a valid pitch-1 inequality, since we observed $w_1 < 1$. Therefore, (7) is implied by (8) and (10), taken with the coefficients w_1 and $1 - 2w_1$ respectively. Recalling that (8) is a valid pitch-2 inequality of the form (6) concludes the proof. ∎

3.2   A $(1 + \epsilon)$-oracle

We will prove Theorem 2 in a sequence of intermediate steps. Our argument extends the weak separation of KC inequalities in [4].

Let $\bar{x}$ be the point we want to separate. Note that it suffices to show how to separate over inequalities (i)-(ii)-(iii) from Lemma 1. Separating over (i) is trivial. We first show how to separate over (iii).

Claim 5. *For $\alpha \in]0,1]$, let z^α be the optimal solution to the following IP P_α, and $v(z^\alpha)$ its value:*

$$\min \sum_{i \in [n]:\ p_i < \alpha} \bar{x}_i z_i + 2 \sum_{i \in [n]:\ p_i \geq \alpha} \bar{x}_i z_i \quad s.t. \quad \sum_{i \in [n]} p_i(1 - z_i) \leq 1 - \alpha, \ z \in \{0,1\}^n. \tag{11}$$

If $v(z^\alpha) < 2$, then $\bar{x}$ violates Inequality (5) with $I := \{i \in [n] : z_i^\alpha = 1\}$, otherwise $\bar{x}$ does not violate any Inequality (5) with $\beta(I) = \alpha$.

Proof. Fix a feasible solution $\bar{z}$ to (11), and let $I := \{i \in [n] : \bar{z}_i = 1\}$. Then:

$$\beta := \beta(I) = 1 - \sum_{i \in [n] \setminus I} p_i = 1 - \sum_{i \in [n]} p_i(1 - \bar{z}_i) \geq \alpha.$$

Hence:

$$\sum_{i \in I:\ p_i < \beta} \bar{x}_i + 2 \sum_{i \in I:\ p_i \geq \beta} \bar{x}_i = \sum_{i \in [n]:\ p_i < \beta} \bar{x}_i \bar{z}_i + 2 \sum_{i \in [n]:\ p_i \geq \beta} \bar{x}_i \bar{z}_i$$
$$\leq \sum_{i \in [n]:\ p_i < \alpha} \bar{x}_i \bar{z}_i + 2 \sum_{i \in [n]:\ p_i \geq \alpha} \bar{x}_i \bar{z}_i = v(\bar{z}),$$

where the central inequality holds at equality if $\alpha = \beta$. Hence, if $v(z^\alpha) < 2$, the inequality with $I := \{i \in [n] : z_i^\alpha = 1\}$ from (11) is violated by $\bar{x}$. Else, all inequalities from (11) with $\beta(I) = \alpha$ are satisfied. □

Note that P_α is a MINKNAP instance, hence we can use the appropriate FPTAS to find, for each $\epsilon > 0$, an ϵ-approximate solution for it.

Since all data are rationals, we can assume there exists $q \in \mathbb{N}$ such that, for each $i \in [n]$, $p_i = r_i/q$ for some $r_i \in \mathbb{N}$.

Claim 6. *Let $r \in \{r_i + 1 :\ i \in [n]\}$ and, for $\alpha = r/q$, let $\bar{z}^\alpha$ be the solution output by the FPTAS for problem P_α and $\bar{v}^\alpha$ its objective function value. If $\bar{v}^\alpha < 2$ for some α, then $\bar{x}$ violates Inequality (5) with $I = \{i \in [n] : \bar{z}^\alpha = 1\}$. Else, $(1 + \epsilon)\bar{x}$ satisfies all inequalities in Lemma 1 (iii).*

Proof. Let $r = r_i + 1$ for some $i \in [n]$. If $\bar{v}^\alpha < 2$ then by Claim 5 $\bar{x}$ violates the corresponding inequality (5). Otherwise, $v^\alpha \geq 2/(1 + \epsilon)$, and $(1 + \epsilon)\bar{x}$ is feasible for any pitch-2 Inequality (5) induced by I with $\beta(I) = \alpha$.

Now let $I^* \subseteq [n]$ with $\beta(I^*) = \frac{r^*}{q} < 1$. There exists $i^* \in [n]$ such that $r_{i^*} < r^* \leq r_{i^*+1} \leq q$ (with $r_{n+1} = q$). Let $\alpha := \frac{r_{i^*+1}}{q} \leq \alpha^* := \beta(I^*)$. The set of feasible solutions of P_α contains that of P_{α^*}, and $\{i \in [n] : p_i < \alpha\} = \{i \in [n] : p_i < \alpha^*\}$. Hence, $v^{\alpha^*} \geq v^\alpha$ and consequently $v^{\alpha^*} < 2$ implies $v^\alpha < 2$. Thus, for separating all inequalities in Lemma 1 (iii), it suffices to check (11) for all $\alpha = \frac{r}{q}$ as in the statement of the claim.   □

The following claim follows in a similar fashion to the previous one by observing that, for $\beta(I^*) = \frac{1}{q}$, (11) separates over undominated pitch-1 inequalities.

Claim 7. *Let* $\alpha = 1/q$, *and* $\bar{z}^\alpha$ *be the solution output by the FPTAS for problem* P_α, *and* $\bar{v}^\alpha$ *its objective function value. If* $\bar{v}^\alpha < 2$, *then* $\bar{x}$ *violates the pitch-1 inequality with support* $I = \{i \in [n] : \bar{z}^\alpha = 1\}$. *Else,* $(1 + \epsilon)\bar{x}$ *satisfies all valid pitch-1 inequalities.*

Next claim shows how to round a point in the unit cube that almost satisfies all pitch-1 and pitch-2 inequalities, to one that satisfies them and is still contained in the unit cube.

Claim 8. *Let* $\bar{x} \in [0,1]^n$ *be such that* $(1 + \epsilon)\bar{x}$ *satisfies all inequalities from Lemma 1, and define* $\bar{y} \in \mathbb{R}^n$ *as follows:* $\bar{y}_i = \min\{1, \frac{1+\epsilon}{1-\epsilon}\bar{x}_i\}$ *for* $i \in [n]$. *Then* $\bar{y} \in [0,1]^n$ *and* $\bar{y}$ *satisfies all inequalities from Lemma 1.*

Proof. Clearly $\bar{y} \in [0,1]^n$. Let $J = \{i \in [n] : \bar{y}_i = 1\}$. If $J = \emptyset$, $(1+\epsilon)\bar{x} < \bar{y} \leq 1$, hence $\bar{y}$ satisfies all pitch-2 inequalities. Thus, $J \neq \emptyset$. Consider a pitch-2 inequality of the form (5), and note that the left-hand side of the inequality computed in $\bar{y}$ is lower bounded by $\sum_{i \in J} \alpha_i$, where α_i is the coefficient of x_i. First assume there exists $j \in J \cap I_2$. Then $\sum_{i \in J} \alpha_i \geq \alpha_j = 2$. Similarly, if $j, j' \in J$, then $\sum_{i \in J} \alpha_i \geq \alpha_j + \alpha_{j'} \geq 2$. In both cases, $\bar{y}$ satisfies the pitch-2 inequality. Hence, we can assume $J = \{j\} \subseteq I_1$. Then:

$$\sum_{i \in I} \alpha_i \bar{x}_i \geq \frac{2}{1+\epsilon}, \text{ from which we deduce}$$

$$\sum_{i \in I \setminus \{j\}} \alpha_i \bar{x}_i \geq \frac{2}{1+\epsilon} - \bar{x}_j \geq \frac{2}{1+\epsilon} - 1 = \frac{1-\epsilon}{1+\epsilon} \text{ and}$$

$$\sum_{i \in I} \alpha_i \bar{y}_i = \sum_{i \in I \setminus \{j\}} \alpha_i \bar{y}_i + 1 = \frac{1+\epsilon}{1-\epsilon} \sum_{i \in I \setminus \{j\}} \alpha_i \bar{x}_i + 1 \geq 2,$$

as required. A similar (simpler) argument shows that $\bar{y}$ also satisfies all pitch-1 inequalities $\sum_{i \in I} x_i \geq 1$.   □

Algorithm 1

1: Let $\epsilon' = \frac{\epsilon}{2+\epsilon}$.
2: For $r \in \{r_i + 1 : i \in [n]\}$ and for $\alpha = r/q$, run the FPTAS for P_α with approximation factor ϵ'. If any of the output solution $\bar{z}^\alpha$ has value $\bar{v}^\alpha < 2$, output inequality (5) with $I = \{i \in [n] : \bar{z}^\alpha = 1\}$ and stop.
3: For $\alpha = 1/q$, run the FPTAS for P_α with approximation factor ϵ'. If the output solution $\bar{z}^\alpha$ has value $\bar{v}^\alpha < 2$, output inequality $\sum_{i:\bar{z}^\alpha=1} x_i \geq 1$ and stop.
4: Output point $\bar{y}$ constructed as in Claim 8 with ϵ' and stop. Note that $\bar{x} \leq \bar{y} \leq \frac{1+\epsilon'}{1-\epsilon'}\bar{x} = (1+\epsilon)\bar{x}$.

Proof of Theorem 2. We can now sum up our $(1+\epsilon)$-oracle, see Algorithm 1. Correctness and polynomiality follow from the discussion above. ∎

In the full version [7] we give an example showing that inequalities of pitch-3 and higher do not have the nice structure of pitch-2. For later use (in Sect. 4.3), we observe here that when $I \subseteq [n]$ is fixed, we can *efficiently* and *exactly* solve the separation problem over bounded pitch inequalities with support I just by solving an LP. This can be seen as an application of polarity (see, e.g., Nemhauser and Wolsey [15]), while being restricted to a specific support. Clearly, we are interested in valid inequalities $\alpha^T x \geq 1$ with $\alpha \geq 0$ and points $0 \leq x^* \leq 1$. Let $\beta = 1 - p([n] \setminus I)$. We can assume $\beta > 0$, otherwise there is no valid inequality as above with support I. Call $J \subseteq I$ *massive* if $\sum_{i \in J} p_i \geq \beta$. Consider the following LP:

$$\min \sum_{i \in I} \alpha_i x_i^*$$
$$\text{s.t.} \quad \sum_{i \in J} \alpha_i \geq 1 \text{ for all massive } J \subseteq I \tag{12}$$
$$\alpha \geq 0$$

Note that, for each feasible solution $\bar{\alpha}$ to the previous LP, we have that $\bar{\alpha}^T x \geq 1$ is a valid inequality for the original MINKNAP instance, and conversely that all inequalities with support I can be obtained in this way. Hence, let α^* be the optimal solution to the previous LP. If $(\alpha^*)^T x^* < 1$, we obtain an inequality whose support is contained in I, that is violated by x^*. The support of the inequality can be extended to I by setting $\alpha_i = \epsilon$ for all $i \in I$ with $\alpha_i = 0$. On the other hand, if $(\alpha^*)^T x^* \geq 1$, x^* satisfies all inequalities with support I.

4 Integrality Gap of IPs for MinKnap with Bounded Pitch Inequalities

4.1 When $p = c$

Theorem 9. *Consider an instance of* MINKNAP *(1) with $p = c$. Denote by K the linear relaxation of (1) to which all pitch-1 and pitch-2 inequalities have been added. The integrality gap of K is at most 3/2.*

Proof. Let $p = c$, and let $\bar{x}$ be the optimal integer solution to (1). We can assume $p^T \bar{x} > 1.5$, else we are done.

Claim 10. *The support of $\bar{x}$ has size 2.*

Proof. Let k be the size of the support of $\bar{x}$. If $k = 1$, then $\bar{x}$ is also the optimal fractional solution. Now assume $k \geq 3$. Remove from $\bar{x}$ the cheapest item as to obtain $\bar{x}'$. We have

$$p^T x' \geq \left(1 - \frac{1}{k}\right) p^T \bar{x} > \frac{2}{3} \cdot 1.5 = 1,$$

contradicting the fact that $\bar{x}$ is the optimal integral solution. $\square$

Hence, we can assume that the support of $\bar{x}$ is given by $\{i, j\}$, with $0 < p_i \leq p_j \leq 1$. Since $p_i + p_j > 1.5$, we deduce $p_j > .75$. Since $p_j \leq 1$, we deduce $p_i > .5$.

Claim 11. *Let $\ell < j$ and $\ell \neq i$. Then $p_\ell < .25$.*

Proof. Recall that for $S \subseteq [n]$ we denote its characteristic vector with χ^S. If $0.25 \leq p_\ell < p_i$, then $\chi^{\{\ell,j\}}$ is a feasible integral solution of cost strictly less than $\bar{x}$. Else if $0.5 < p_i \leq p_\ell < p_j$, then $\chi^{\{\ell,i\}}$ is a feasible integral solution of cost strictly less than $\bar{x}$. In both cases we obtain a contradiction. $\square$

Because of the previous claim, we can assume w.l.o.g. $j = i + 1$.

Claim 12. $p_n + \sum_{\ell=1}^{i-1} p_\ell < 1$.

Proof. Suppose $p_n + \sum_{\ell=1}^{i-1} p_\ell \geq 1$. Since $p_\ell < .25$ for all $\ell = 1, \ldots, i - 1$, there exists $k \leq i - 1$ such that $x_n + \sum_{\ell=1}^{k} p_k \in [1, 1.25[$. Hence $x^{\{1,\ldots,k,n\}}$ is a feasible integer solution of cost at most 1.25, a contradiction. $\square$

Because of the previous claim, the pitch-2 inequality $\sum_{k=i}^{n} x_k \geq 2$ is valid. The fractional solution of minimum cost that satisfies this inequality is the one that sets $x_i = x_j = 1$ (since $j = i + 1$) and all other variables to 0. This is exactly $\bar{x}$.   ∎

4.2   CG Closures of Bounded Rank of the Natural MinKnap Relaxation

For $t \in \mathbb{N}$, let K^t be the linear relaxation of (1) given by: the original knapsack inequality; non-negativity constraints; all pitch-k inequalities, for $k \leq t$.

Lemma 2. *For $t \geq 2$, the integrality gap of K^t is at least $\max\{\frac{1}{2}, \frac{t-2}{t-1}\}$ times the integrality gap of K^{t-1}.*

Proof. Fix $t \geq 2$, and let C be the cost of the optimal integral solution to (1). Let C/v' be the integrality gap of K^t. Since v' is the optimal value of K^t, by the strong duality theorem (and Caratheodory's theorem), there exist nonnegative multipliers $\alpha, \alpha_1, \ldots, \alpha_n, \gamma_1, \ldots, \gamma_{n+1}$ such that the inequality $c^T x \geq v'$ can be obtained as a conic combination of the original knapsack inequality (with multiplier α), non-negativity constraints (with multipliers $\alpha_1, \ldots, \alpha_n$), and at most $n + 1$ inequalities of pitch at most t (with multipliers $\gamma_1, \ldots, \gamma_{n+1}$). By scaling, we can assume that the rhs of the latter inequalities is 1. Hence $v' = \alpha + \sum_{i=1}^{r} \gamma_i$. Proof of Claim 13 is given in the full version [7].

Claim 13. *Let $d^T x \geq 1$ be a valid pitch-t inequality for* (1), *and assume w.l.o.g. that $d_1 \leq d_2 \leq \cdots \leq d_n$. Then inequality $\sum_{i=2}^n d_i x_i \geq \max\{\frac{1}{2}, \frac{t-2}{t-1}\}$ is a valid inequality of pitch at most $t-1$ for* (1).

Now consider the conic combination with multipliers $\alpha, \alpha_1, \ldots, \alpha_n, \gamma_1, \ldots, \gamma_{n+1}$ given above, where each inequality of pitch-t is replaced with the inequality of pitch at most $t-1$ obtained using Claim 13. We obtain an inequality $(c')^T x \geq v''$, where one immediately checks that $c' \leq c$ and

$$v'' \geq \alpha + \sum_{i=1}^{n+1} \gamma_i \max\left\{\tfrac{1}{2}, \tfrac{t-2}{t-1}\right\} \geq \max\left\{\tfrac{1}{2}, \tfrac{t-2}{t-1}\right\}\left(\alpha + \sum_{i=1}^{n+1} \gamma_i\right)$$
$$= \max\left\{\tfrac{1}{2}, \tfrac{t-2}{t-1}\right\} v'.$$

Hence the integrality gap of K^t is

$$\frac{C}{v'} \geq \frac{C}{v''} \max\left\{\frac{1}{2}, \frac{t-2}{t-1}\right\}$$

and the thesis follows since the integrality gap of K^{t-1} is at most C/v''.   ∎

Lemma 3. *For a fixed $\epsilon > 0$ and square integers $n \geq 4$, consider the MINKNAP instance K defined as follows:*

$$\min \quad \epsilon y + \sqrt{n}z + \sum_{i=1}^n x_i$$
$$\text{st}$$
$$(n - \sqrt{n})y + \tfrac{n}{2}z + \sum_{i=1}^n x_i \geq n$$
$$y, \quad z, \quad x \in \{0,1\}.$$

For every fixed $t \in \mathbb{N}$, the integrality gap of K^t is $\Omega(\sqrt{n})$.

Proof. Because of Lemma 2, it is enough to show that the integrality gap of K^1 is $\Omega(\sqrt{n})$. Clearly, the value of the integral optimal solution of the instance is $\sqrt{n} + \epsilon$. We claim that the fractional solution

$$(\bar{y}, \bar{x}, \bar{z}) = \left(1, \underbrace{\frac{1}{n - \sqrt{n} + 1}, \ldots, \frac{1}{n - \sqrt{n} + 1}}_{n \text{ times}}, \frac{2}{\sqrt{n}}\right)$$

is a feasible point of K^1. Since $\epsilon \bar{y} + \sqrt{n}\bar{z} + \sum_{i=1}^n \bar{x}_i = \epsilon + 2 + \frac{n}{n - \sqrt{n} + 1}$, the thesis follows.

Observe that $(n - \sqrt{n})\bar{y} + \frac{n}{2}\bar{z} + \sum_{i=1}^n \bar{x}_i = (n - \sqrt{n}) + \frac{n}{2}\frac{2}{\sqrt{n}} + \frac{n}{n - \sqrt{n} + 1} > n$, hence $(\bar{y}, \bar{x}, \bar{z})$ satisfies the original knapsack inequality.

Now consider a valid pitch-1 inequality whose support contains y. Since $\bar{y} = 1$, $(\bar{y}, \bar{x}, \bar{z})$ satisfies this inequality. Hence, the only pitch-1 inequalities of interest do not have y in the support. Note that such inequalities must have z in the support, and some of the x_i. Hence, all those inequalities are dominated by the valid pitch-1 inequalities

$$z + \sum_{i \in I} x_i \geq 1 \; \forall I \subseteq [n], |I| = n - \sqrt{n} + 1,$$

which are clearly satisfied by $(\bar{y}, \bar{x}, \bar{z})$.   ∎

Theorem 14. *For a fixed $q \in \mathbb{N}$, let $CG^q(K)$ be the q–th CG closure of the* MINKNAP *instance K as defined in Lemma 3. The integrality gap of $CG^q(K)$ is $\Omega(\sqrt{n})$.*

Proof. We will use the following fact, proved (for a generic covering problem) in [3]. Let $t, q \in \mathbb{N}$ and suppose $(\bar{y}, \bar{z}, \bar{x}) \in K^t$. Define point (y', x', z'), where each component is the minimum between 1 and $(\frac{t+1}{t})^q$ times the corresponding component of $(\bar{y}, \bar{z}, \bar{x})$. Then $(y', x', z') \in CG^q(K)$. Now fix t, q. We have therefore that

$$\epsilon y' + \sqrt{n} z' + \sum_{i=1}^{n} x_i' \leq \left(\frac{t+1}{t}\right)^q \left(\epsilon \bar{y} + \sqrt{n} \bar{z} + \sum_{i=1}^{n} \bar{x}_i\right)$$
$$= \left(\frac{t+1}{t}\right)^q \left(\epsilon + 2 + \frac{n}{n - \sqrt{n} + 1}\right)$$

and the claim follows in a similar fashion to the proof of Lemma 3.   ∎

4.3   When All Bounded Pitch and Knapsack Cover Inequalities Are Added

Consider the following MINKNAP instance with $\epsilon_n = \frac{1}{\sqrt{n}}$:

$$\min \quad \sum_{i \in [n]} x_i + \frac{1}{\sqrt{n}} \sum_{j \in [n]} z_j$$

$$\text{s.t.} \quad \sum_{i \in [n]} x_i + \frac{1}{n} \sum_{j \in [n]} z_j \geq 1 + \epsilon_n \tag{13}$$

$$x, z \in \{0, 1\}^n.$$

Lemma 4. *For any fixed $k \in \mathbb{N}$ and $n \in \mathbb{N}$ sufficiently large, point $(\bar{x}, \bar{z}) \in \mathbb{R}^{2n}$ with $\bar{x}_i = \frac{1 + \epsilon_n}{n}, \bar{z}_i = \frac{k}{n}$ satisfies the natural linear relaxation, all KC and all inequalities of pitch at most k valid for (13). Observing that the optimal integral solution is 2, this gives an IG of $\frac{2}{1 + \frac{k}{n}} \approx 2$.*

Proof. We prove the statement by induction. Fix $k \in \mathbb{N}$. Note that $(\bar{x}, \bar{z})$ dominates componentwise the point generated at step $k - 1$, and the latter by induction hypothesis satisfies all inequalities of pitch at most $k - 1$. Let

$$\sum_{i \in I} w_i x_i + \sum_{j \in J} w_j z_j \geq \beta \tag{14}$$

be a valid KC or pitch-k inequality with support $I \cup J$, which gives that $w_i, w_j \in \mathbb{R}_{>0}, \; \forall i \in I, \; \forall j \in J$ and $\beta > 0$. Observe the following.

Claim 15. $|I| \geq n - 1$. In addition, $|I| = n - 1$ or $|J| \leq n(1 - \epsilon_n)$ implies $w_i \geq \beta$, $\forall i \in I$.

Proof. Since all coefficients in (14) are strictly positive and $\beta > 0$, $|I| \leq n - 2$ gives that the feasible solution $(\chi^{[n] \setminus I}, \mathbf{0})$ for (13) is cut off by (14), a contradiction.

Furthermore, if $|I| = n-1$ and $w_{i^*} < \beta$ for some $i^* \in I$, then $(\chi^{([n] \setminus I) \cup \{i^*\}}, \mathbf{0})$ is cut off, again a contradiction.

Finally, if $|I| = n$, $|J| \leq n(1 - \epsilon_n)$ and $w_{i^*} < \beta$ for some $i^* \in I$, then $(\chi^{\{i^*\}}, \chi^{\{[n] \setminus J\}})$ does not satisfy (14), but it is feasible in (13). $\square$

We first show the statement for (14) being a KC. By the definition of KC: $\beta = 1 + \epsilon_n - |[n] \setminus I| - \frac{|[n] \setminus J|}{n}$, $w_i = \min\{1, \beta\}$, $\forall i \in I$ and $w_j = \min\{\frac{1}{n}, \beta\}$, $\forall j \in J$. If $I = [n]$, then $\sum_{i \in I} w_i \bar{x}_i = \min\{1, \beta\} \cdot (1 + \epsilon_n) \geq \beta$ since $\beta \leq 1 + \epsilon_n$. Otherwise, $|[n] \setminus I| = 1$ so $w_i = \beta$, $\forall i \in I$ and $\sum_{i \in I} w_i \bar{x}_i = \beta \frac{(n-1)(1+\epsilon_n)}{n} > \beta$ for sufficiently large n.

Conversely, let (14) be a valid pitch-k inequality. By Claim 15, if $|I| = n - 1$ or $|J| \leq n(1 - \epsilon_n)$ then $w_i \geq \beta$, $\forall i \in I$ so the proof is analogous to the one for KC. Otherwise, $|I| = n$ and $|J| > n(1 - \epsilon_n)$. Consider the LP (12) in Sect. 3 specialized for our case – that is, we want to detect if $(\bar{x}, \bar{z})$ can be separated via an inequality with support $I \cup J$. Since $(\chi^{i, \bar{\imath}}, \mathbf{0})$ is feasible in (13) for $i, \bar{\imath} \in I$, then

$$\alpha_i + \alpha_{\bar{\imath}} \geq 1. \tag{15}$$

Furthermore, for n large enough one has $|J| > n(1 - \epsilon_n) \geq k$ so

$$\sum_{j \in K} \alpha_j \geq 1 \tag{16}$$

for any k-subset K of J. We claim that the minimum in (12) is attained at $\bar{\alpha}_i = 1/2$, $\forall i \in I$ and $\bar{\alpha}_j = 1/k$, $\forall j \in J$. Indeed, the objective function of (12) computed in $\bar{\alpha}$ is given by

$$|I| \cdot \frac{1}{2} \cdot \frac{1 + \epsilon}{n} + |J| \cdot \frac{1}{k} \cdot \frac{k}{n} = \frac{1 + \epsilon_n}{2} + \frac{|J|}{n}.$$

On the other hand, by summing (15) for all possible pairs with multipliers $\frac{1+\epsilon_n}{2(n-1)}$ and (16) for all subsets of J of size k with multipliers $\binom{|J|-1}{k-1}^{-1} \cdot \frac{k}{n}$, simple linear algebra calculations lead to

$$\sum_{i \in I} \bar{x}_i \alpha_i + \sum_{j \in J} \bar{z}_j \alpha_j \geq \frac{1 + \epsilon_n}{2} + \frac{|J|}{n},$$

showing the optimality of $\bar{\alpha}$. Recalling $|J| > n(1 - \epsilon_n)$, we conclude that

$$\sum_{i \in I} \bar{x}_i \bar{\alpha}_i + \sum_{j \in J} \bar{z}_j \bar{\alpha}_j = \frac{1 + \epsilon_n}{2} + \frac{|J|}{n} > \frac{1 + \epsilon_n}{2} + 1 - \epsilon_n > 1,$$

hence $(\bar{x}, \bar{z})$ satisfies all inequalities with support $I \cup J$. ∎

Acknowledgments. Supported by the Swiss National Science Foundation (SNSF) project 200020-169022 "Lift and Project Methods for Machine Scheduling Through Theory and Experiments". Some of the work was done when the second and the third author visited the IEOR department of Columbia University, partially funded by a gift of the SNSF.

References

1. Bazzi, A., Fiorini, S., Huang, S., Svensson, O.: Small extended formulation for knapsack cover inequalities from monotone circuits. In: Proceedings of SODA 2017, pp. 2326–2341 (2017)
2. Bienstock, D.: Approximate formulations for 0–1 knapsack sets. Oper. Res. Lett. **36**(3), 317–320 (2008)
3. Bienstock, D., Zuckerberg, M.: Approximate fixed-rank closures of covering problems. Math. Program. **105**(1), 9–27 (2006)
4. Carr, R.D., Fleischer, L.K., Leung, V.J., Phillips, C.A.: Strengthening integrality gaps for capacitated network design and covering problems. In: Proceedings of SODA 2000, pp. 106–115 (2000)
5. Conforti, M., Cornuejols, G., Zambelli, G.: Integer Programming. Springer, Heidelberg (2014). https://doi.org/10.1007/978-3-319-11008-0
6. Dudycz, S., Moldenhauer, C.: Approximated extended formulations for the knapsack cover problem. Technical report, EPFL (2015)
7. Faenza, Y., Malinović, I., Mastrolilli, M., Svensson, O.: On bounded pitch inequalities for the min-knapsack polytope. arXiv preprint arXiv:1801.08850 (2018)
8. Ferreira, C.E.: On combinatorial optimization problems arising in computer system design. Ph.D. thesis, Technical University of Berlin, Germany (1994)
9. Fiorini, S., Huynh, T., Weltge, S.: Strengthening convex relaxations of 0/1-sets using boolean formulas. arXiv preprint arXiv:1711.01358 (2017)
10. Grötschel, M., Lovász, L., Schrijver, A.: Geometric Algorithms and Combinatorial Optimization. Springer, New York (1988). https://doi.org/10.1007/978-3-642-78240-4
11. Ibarra, O.H., Kim, C.E.: Fast approximation algorithms for the knapsack and sum of subset problems. J. ACM **22**(4), 463–468 (1975)
12. Karp, R.M.: Reducibility among combinatorial problems. In: Miller, R.E., Thatcher, J.W., Bohlinger, J.D. (eds.) Complexity of Computer Computations. The IBM Research Symposia Series, pp. 85–103. Springer, Boston (1972). https://doi.org/10.1007/978-1-4684-2001-2_9
13. Kurpisz, A., Leppänen, S., Mastrolilli, M.: On the hardest problem formulations for the 0/1 lasserre hierarchy. Math. Oper. Res. **42**(1), 135–143 (2017)
14. Lawler, E.L.: Fast approximation algorithms for knapsack problems. Math. Oper. Res. **4**(4), 339–356 (1979)
15. Nemhauser, G.L., Wolsey, L.A.: Integer and Combinatorial Optimization. Wiley Interscience Series in Discrete Mathematics and Optimization. Wiley, New York (1988)

Efficient Algorithms for Measuring
the Funnel-Likeness of DAGs

Marcelo Garlet Millani[1], Hendrik Molter[1(✉)], Rolf Niedermeier[1],
and Manuel Sorge[1,2]

[1] Institut für Softwaretechnik und Theoretische Informatik,
TU Berlin, Berlin, Germany
{m.garletmillani,h.molter,rolf.niedermeier}@tu-berlin.de
[2] Department of Industrial Engineering and Management,
Ben-Gurion University of the Negev, Beer Sheva, Israel
sorge@post.bgu.ac.il

Abstract. Funnels are a new natural subclass of DAGs. Intuitively, a
DAG is a funnel if every source-sink path can be uniquely identified by
one of its arcs. Funnels are an analog to trees for directed graphs that
is more restrictive than DAGs but more expressive than in-/out-trees.
Computational problems such as finding vertex-disjoint paths or tracking
the origin of memes remain NP-hard on DAGs while on funnels they
become solvable in polynomial time. Our main focus is the algorithmic
complexity of finding out how funnel-like a given DAG is. To this end,
we study the NP-hard problem of computing the arc-deletion distance to
a funnel of a given DAG. We develop efficient exact and approximation
algorithms for the problem and test them on synthetic random graphs
and real-world graphs.

1 Introduction

Directed acyclic graphs (DAGs) are finite directed graphs (digraphs) without
directed cycles and appear in many applications, including the representation
of precedence constraints in scheduling, data processing networks, causal structures, or inference in proofs. From a more graph-theoretic point of view, DAGs
can be seen as a directed analog of trees; however, their combinatorial structure
is much richer. Thus a number of directed graph problems remain NP-hard even
when restricted to DAGs. This motivates the study of subclasses of DAGs. We
study *funnels* which are DAGs where each source-sink path has at least one
private arc, that is, no other source-sink path contains this arc. In independent
work, Lehmann [13] studied essentially the same graph class.

Funnels are both of combinatorial and graph-theoretic as well as of practical
interest: First, funnels are a natural compromise between DAGs and trees as,

M. G. Millani—Partially supported by the DFG, project FPTinP (NI 369/16).

H. Molter—Partially supported by the DFG, project MATE (NI 369/17).

M. Sorge—Supported by the People Programme (Marie Curie Actions) of the European Union's Seventh Framework Programme (FP7/2007-2013) under REA grant
agreement number 631163.11 and Israel Science Foundation (grant no. 551145/14).

© Springer International Publishing AG, part of Springer Nature 2018
J. Lee et al. (Eds.): ISCO 2018, LNCS 10856, pp. 183–195, 2018.
https://doi.org/10.1007/978-3-319-96151-4_16

similarly to in- or out-trees, the private-arc property guarantees that the overall number of source-sink paths is upper-bounded linearly by its number of arcs, yet multiple paths connecting two vertices are possible. Second, in Sect. 2 we show that funnels, in a divide & conquer spirit, allow for a vertex partition into a set of *forking* vertices with indegree one and possibly large outdegree and a set of *merging* vertices with outdegree one and possibly large indegree. This partitioning helps in designing our algorithms. Third, in terms of applications, due to the simpler structure of funnels, problems such as DAG PARTITIONING [4,14] or VERTEX DISJOINT PATHS, (also known as k-LINKAGE) [2,7] become tractable on funnels while they are NP-hard on DAGs. Lehmann [13] showed that a variation of the problem NETWORK INHIBITION, which is NP-hard on DAGs, can be solved in polynomial time on funnels. Altogether, we feel that funnels are one of so far few natural subclasses of DAGs.

The focus of this paper is on investigating the complexity of turning a given DAG into a funnel by a minimum number of arc deletions. The motivation for this is twofold. First, due to the noisy nature of real-world data, we expect that DAGs from practice are not pure funnels, even though they may adhere to some form of funnel-like structure. To test this hypothesis we need efficient algorithms to determine funnel-likeness. Second, as mentioned above, natural computational problems become tractable on funnels (e.g., k-LINKAGE [15]). Thus it is promising to try and develop fixed-parameter algorithms for such NP-hard DAG problems with respect to distance parameters to funnels. This approach is known as exploiting the "distance from triviality" [5,10,18]. A natural way to measure the distance of a given DAG D to a funnel is the *arc-deletion distance to a funnel*, the minimum number of arcs that need to be deleted from D to obtain a funnel. The problem of computing this distance parallels the well-studied NP-hard FEEDBACK ARC SET problem where the task is to turn a given digraph into a DAG by a minimum number of arc deletions. Even FEEDBACK ARC SET on tournaments is NP-hard and it received considerable interest over the last years [1,3,6,11].

Formally, we study the ARC-DELETION DISTANCE TO A FUNNEL (ADDF) problem, where, given a DAG D, we want to find its arc-deletion distance d to a funnel. We show that ADDF is NP-hard and that it admits a linear-time factor-two approximation algorithm and a fixed-parameter algorithm with linear running time for constant d.[3] In experiments we demonstrate that our algorithms are useful in practice.

Due to the lack of space, proofs of results marked with ($\star$) are omitted.[4]

2   Funnels: Definition and Properties

In this section we formally define funnels. We provide several equivalent characterizations, summarized in Theorem 1, and analyze some basic properties of funnels. We use standard terminology from graph theory.

---

[3] There is also a simple $\mathcal{O}(5^d \cdot |V| \cdot |A|)$-time algorithm for general digraphs [15].

[4] A full version is available on arXiv [17].

Fig. 1. Example of a funnel (left) and a DAG which is not a funnel (right). Private arcs are marked as dashed lines. The DAG on the right is not a funnel because all arcs in an (s_1, t)-path are shared. Removing one arc from it turns it into a funnel. A forbidden subgraph for funnels is marked in bold.

To define funnels as a proper subclass of DAGs, we limit the number of paths that may exist between two vertices (which can be exponential in DAGs but is one in trees). Requiring every path between two vertices to be unique would possibly be too restrictive, and in the case of a single source such DAGs would simply be so-called out-trees. Instead, we require each path going from a source to a sink to be uniquely identified by one of its *private* arcs. We say that an arc is private if there is only one source-sink path which goes through that arc. An example of a funnel can be seen in Fig. 1.

Definition 1 (Funnel). *A DAG is a* funnel *if every source-sink path has at least one private arc.*

From this definition it is clear that the number of source-sink paths in a funnel is linearly upper-bounded in its number of arcs.

Different characterizations of funnels reveal certain interesting properties which these digraphs have, and are used in subsequent proofs and algorithms. We summarize these characterizations in the theorem below. In the following, $\mathsf{out}^*(v)$ denotes the set of vertices that can be reached from v in a given DAG, $\mathsf{out}(v)$ denotes the set of outneighbors of v and $\mathsf{outdeg}(v)$ denotes v's outdegree; $\mathsf{in}^*(v), \mathsf{in}(v)$ and $\mathsf{indeg}(v)$ are defined analogously.

Theorem 1 ($\star$). *Let D be a DAG. The following statements are equivalent:*

1. *D is a funnel.*
2. *For each vertex $v \in V$: $\mathsf{indeg}(v) > 1 \Rightarrow \forall u \in \mathsf{out}^*(v) : \mathsf{outdeg}(u) \leq 1$.*
3. *No subgraph of D is contained in $\mathcal{F} = \{D_i\}_{i=0}^{\infty}$, where*
   - *$D_k = (V_k, A_k)$,*
   - *$V_k = \{u_1, u_2, v_0, w_1, w_2\} \cup \{v_i\}_{i=1}^{k}$, and*
   - *$A_k = \{(u_1, v_0), (u_2, v_0), (v_k, w_1), (v_k, w_2)\} \cup \{(v_i, v_{i+1})\}_{i=1}^{k-1}$.*
4. *D does not contain D_0 or D_1 (defined above) as a topological minor.[5]*

Definition 1 does not give us a very efficient way of checking whether a given DAG is a funnel or not. A simple algorithm which counts how many paths go

---

[5] A graph H is called a *topological minor* of a graph G if a subgraph of G can be obtained from H by subdividing edges (that is, replacing arcs by directed paths).

through each arc would take $\mathcal{O}(|A|^2)$ time. Using the characterization in Theorem 1(2) we can follow some topological ordering of the vertices of a DAG and check in linear time whether it is a funnel.

The *degree characterization* in Theorem 1(2) provides some additional insight about the structure of a funnel. We can see that a funnel can be partitioned into two induced subgraphs: One is an out-forest and the other is an in-forest. Note that this partition is not necessarily unique. For use below, an FM-*labeling* for given a DAG with vertex set V is a function $L : V \rightarrow \{\text{FORK}, \text{MERGE}\}$ which gives a *label* to each vertex. An FM-labeling for a funnel is called *funnel labeling* if the vertices in the out-forest of the funnel are assigned the label FORK and vertices in the in-forest are assigned the label MERGE. The following holds.

Observation 1. *Let $D = (V, A)$ be a funnel and L be a funnel labeling for D. Then there is no $(v, u) \in A$ with $L(v) = \text{MERGE}$ and $L(u) = \text{FORK}$.*

With a simple counting argument it is also possible to give an upper bound on the number of arcs in a funnel. This bound is sharp.

Observation 2 ($\star$). *Let $D = (V, A)$ be a funnel. Then $|A| \leq |V|^2/4 + |V| - 2$.*

Considering that a DAG has at most $|V|(|V| - 1)/2$ arcs, Observation 2 implies that a funnel can have at most roughly half as many arcs as a DAG. This means that funnels are not necessarily sparse (unlike forests). While the degree characterization is useful for algorithms, the characterizations by forbidden subgraphs and minors (Theorem 1(3 and 4)) help us to understand the local structure of a funnel and of graphs that are not funnels. These characterizations also imply that being a funnel is a *hereditary* graph property, that is, deleting vertices does not destroy the funnel property.

3   Computing the Arc-Deletion Distance to a Funnel

In this section we show ADDF is NP-hard, and present a linear-time factor-2 approximation algorithm and an exact fixed-parameter algorithm. Our algorithms also compute the set of arcs to be deleted. We remark that the corresponding vertex-deletion distance minimization problem is also NP-hard and that it can be solved in $\mathcal{O}(6^d \cdot |V| \cdot |A|)$ time, where d is the number of vertices to delete [15]. The following result can be shown by a reduction from 3-SAT.

Theorem 2 ($\star$).   ADDF *is NP-hard.*

A Factor-2 Approximation Algorithm. We now give a linear-time factor-2 approximation algorithm for ADDF. We mention in passing that on tournament DAGs the algorithm always finds an optimal solution and on real-world DAGs, the approximation factor is typically close to one (see Sect. 4). The approximation algorithm works in three phases and makes extensive use of FM-labelings (defined in Sect. 2). First, we greedily compute an FM-labeling which we call L_a

Algorithm 1. Satisfying an FM-labeling.

```
1: function ArcDeletionSet(DAG D = (V, A), L : V → {FORK, MERGE})
2:     B := ∅
3:     for all v ∈ V do
4:         if L(v) = MERGE then
5:             Choose an arbitrary u ∈ out(v) with L(u) = MERGE (if it exists)
6:             B := B ∪ {(v, w) | w ≠ u ∧ w ∈ out(v)}
7:         else if L(v) = FORK then
8:             Choose an arbitrary u ∈ in(v) with L(u) = FORK (if it exists)
9:             B := B ∪ {(w, v) | w ≠ u ∧ w ∈ in(v)}
10:    return B
```

for the input graph (assigning each vertex v a FORK or a MERGE label). The labeling will be a funnel labeling of the output funnel indicating for each vertex whether it can have indegree or outdegree greater than one. To construct L_a, we try to minimize the number of arcs to be removed when only considering v. This strategy guarantees that, if the approximation algorithm assigns the wrong label to v, in the optimal solution many arcs incident to v need to be removed. This allows us to derive the approximation factor. Formally, we assign a label to a vertex v using the following rule.

$$
L_a(v) := \begin{cases} \text{FORK}, & \text{if outdeg}_D(v) > \text{indeg}_D(v), \\ \text{FORK}, & \text{if outdeg}_D(v) = \text{indeg}_D(v) \land \\ & \exists u \in \text{in}(v) : L_a(u) = \text{FORK}, \\ \text{MERGE}, & \text{otherwise.} \end{cases}
$$

Since we can assign a label whenever we know the labels of all incoming neighbors, the label of each vertex can be computed, in linear time, by following a topological ordering of the DAG.

In the second phase, after assigning labels to all vertices, we *satisfy* the labels by removing arcs. That is, for each FORK vertex v, we choose an arbitrary inneighbor u with $L(u) = $ FORK (if it exists) and remove all arcs incoming to v from vertices other than u. Similarly, for each MERGE vertex v we choose an arbitrary outneighbor u with $L(u) = $ MERGE (if it exists) and remove all arcs outgoing from v to vertices other than u. See Algorithm 1 for the pseudocode of the second phase. For use below we call the second-phase algorithm `ArcDeletionSet`.

In the third phase, we *greedily relabel* vertices, that is, we iterate over each vertex v (in an arbitrary order), changing v's label if the change immediately leads to an improvement in the solution size. To check if there is an improvement, we only need to consider the incident arcs of v and the labels of its endpoints. This completes the description of our approximation algorithm.

To argue about optimal solutions and for use in a search-tree algorithm below, we now show that if the input FM-labeling L corresponds to an optimal solution, then `ArcDeletionSet` outputs an optimal arc set: Say that an FM-labeling L

of a DAG D is *optimal* if it is a funnel labeling for some funnel $D - A'$, $A' \subseteq A$, such that A' has minimum size among all arc sets whose deletion makes D a funnel.

Proposition 1 ($\star$). *Let $D = (V, A)$ be a DAG, let $A' \subseteq A$ be a minimum arc set such that $D' = D - A'$ is a funnel, and let L^* be an optimal labeling for D'. Then $|\mathtt{ArcDeletionSet}(D, L^*)| = |A'|$.*

We now give a guarantee of the approximation factor. Due to space constraints we provide only a proof sketch. A full proof and an example for the tightness of the approximation factor can be found in [17].

Theorem 3 ($\star$). *There is a linear-time factor-two approximation for ADDF.*

Proof (Sketch). Let A' be a minimum arc set such that $D - A'$ is a funnel and $B = \mathtt{ArcDeletionSet}(D, L_a)$, where L_a is computed by the above described procedure. Let L^* be an optimal FM-labeling for the input DAG $D = (V, A)$. We define two functions $b : V \to \mathcal{P}(B)$ and $a : V \to \mathcal{P}(A')$ such that $\biguplus_{v \in V} b(v) = B$ and $\biguplus_{v \in V} a(v) = A'$, where $\uplus$ is a disjoint union and $\mathcal{P}(X)$ denotes the family of all subsets of a set X. Our goal is to assign each arc in A' or B to one of its endpoints via a or b, respectively, such that $|b(v)| \leq 2|a(v)|$ for every $v \in V$. We say that a vertex v has *type* $T(v) = \mathrm{FM}$ if $L_a(v) = \mathrm{FORK}$ and $L^*(v) = \mathrm{MERGE}$. The types FF, MM and MF are defined analogously. A vertex v is *correctly labeled* if $L_a(v) = L^*(v)$.

We define a and b in such a way that $|b(v)| = |a(v)|$ if v is correctly labeled. To this end, we only assign a removed arc to a correctly labeled vertex v if both endpoints are correctly labeled. For an incorrectly labeled vertex, we assign the arcs which are potentially removed by $\mathtt{ArcDeletionSet}$ when considering v, together with those of correctly labeled vertices. We additionally need to define a and b in such a way that no arc is assigned to both endpoints.

By construction, it is easy to show that $|b(v)| = |a(v)|$ if v is correctly labeled. We now consider an incorrectly labeled vertex v.

If $\mathsf{indeg}(v) = 1 = \mathsf{outdeg}(v)$, then the approximation removes at most one of the incident arcs of v. If both are removed, then the algorithm changes the label in the third phase, which implies that v was correctly labeled either before or after the change. As only one arc of v is removed, we can treat (u, v) and (v, w) as a single arc (u, w) and assign it to the same vertex that (u, w) would be assigned to if it would be removed.

For the remaining cases, we use a counting argument based on the amount of neighbors of v with each type. We additionally need an exchange argument, that is, whenever we have an arc $(v, u) \in A'$ where $T(v) = \mathrm{FM}$ and $T(u) = \mathrm{MF}$, some arc in $b(u)$ needs to be assigned to $b(v)$ instead. The exchange is possible because such arcs are always in A' but never in B, meaning that the approximation has an "advantage" over the optimal solution with respect to these arcs.

Because the functions a and b partition A' and B, respectively, we obtain that $\left|\biguplus_{v \in V} b(v)\right| = |B| \leq 2|A'| = 2\left|\biguplus_{v \in V} a(v)\right|$. $\qquad\square$

A Fixed-Parameter Algorithm. Using the forbidden subgraph characterization (Theorem 1(3)), we can compute a digraph's arc-deletion distance d to a funnel in $\mathcal{O}(5^d \cdot (|V|^2 + |V| \cdot |A|))$ time: After contracting the arcs on each vertex with in- and outdegree one into a single arc, it is enough to destroy all subgraphs D_0 or D_1 as in Theorem 1(3). The optimal arc-deletion set to destroy all these subgraphs can be found by branching into the at most five possibilities for each subgraph D_0 or D_1.

In this section, we show that, if the input is a DAG, we can solve ADDF in $\mathcal{O}(3^d \cdot (|V| + |A|))$ time instead; thus, in particular, we have linear running time if $d \in \mathcal{O}(1)$. Moreover, the resulting algorithm has also better running time in practice. As in the approximation algorithm, we again label the vertices. Proposition 1 shows that, after the vertices are correctly labeled with either MERGE or FORK, solving ADDF can be done in linear time on DAGs. Hence, the complicated part of the problem lies in finding such a labeling.

In the following, we describe a search-tree algorithm that receives a DAG $D = (V, A)$ and an upper bound $d \in \mathbb{N}$ on the size of the solution as input, and it maintains a partial labeling $L \colon V \to \{\text{FORK}, \text{MERGE}\}$ of the vertices and a partial arc-deletion set A' that will constitute the solution in the end. Initially, $A' = \emptyset$ and $L(v)$ is undefined for each $v \in V$, denoted by $L(v) = \bot$. The algorithm exhaustively and alternately applies the data reduction and branching rules described below and aborts if $|A'| > d$. The rules either determine a label of a vertex (based on preexisting labels and on the degree of the vertex) or put some arcs into the solution A'. Herein, when we say that an arc is put into the solution, we mean that it is deleted from D and put into A'. To show that the algorithm finds a size-d arc deletion set to a funnel if there is one, we ensure that the rules are *correct*, meaning that, if there is a solution of size d that respects the labeling L and contains A' before applying a data reduction rule or branching rule, then there is also such a solution in at least one of the resulting instances.

Reduction Rule 1 labels vertices of indegree (outdegree) at most one in a greedy fashion, based on the label of the single predecessor (successor) if it exists.

Reduction Rule 1 (Set Label) ($\star$). *Let $v \in V$ be an unlabeled vertex.*

Set $L(v) := $ FORK if at least one of the following is true: I) $\text{indeg}(v) = 0$; II) $\text{indeg}(v) = 1$ and $\exists u \in \text{in}(v) : L(u) = $ FORK; III) $\text{outdeg}(v) > 1$, $\text{indeg}(v) = 1$ and $\forall u \in \text{out}(v) : L(u) \neq \bot$.

Set $L(v) := $ MERGE if at least one of the following is true: I) $\text{outdeg}(v) = 0$; II) $\text{outdeg}(v) = 1$ and $\exists u \in \text{out}(v) : L(u) = $ MERGE; III) $\text{outdeg} v = 1$, $\text{indeg}(v) > 1$ and $\forall u \in \text{in}(v) : L(u) \neq \bot$.

Having labeled some vertices—whose labels will be as in an optimal labeling in some branch of the search tree—we simulate in Satisfy Label the behavior of `ArcDeletionSet` and remove arcs from labeled vertices.

Reduction Rule 2 (Satisfy Label) ($\star$). *Let v be some vertex where $L(v) =$ FORK and outdeg$(v) > 1$. If $\exists u \in in(v) : L(u) =$ FORK, then put the arcs $\{(x, v) \mid x \in in(v) \land x \neq u\}$ into the solution. Otherwise, put $\{(x, v) \mid x \in in(v) \land L(x) =$ MERGE$\}$ into the solution.*

Let v be some vertex where $L(v) =$ MERGE and outdeg$(v) > 1$. If $\exists u \in$ out$(v) : L(u) =$ MERGE, then put the arcs $\{(v, x) \mid x \in$ out$(v) \land x \neq u\}$ into the solution. Otherwise, put $\{(v, x) \mid x \in$ out$(v) \land L(x) =$ FORK$\}$ into the solution.

To assign a label to each remaining vertex, we branch into assigning one of the two possible labels. Key to an efficient running time is the observation that there is always a vertex which, regardless of the label set, has some incident arc which then has to be in the solution. This observation is exploited in Branching Rule 1.

Branching Rule 1 (Label Branch). *If there is some vertex v such that $\forall w \in$ in$(v) : L(w) \neq \perp$ or $\exists w \in$ in$(v) : L(w) =$ FORK, then branch into two possibilities: Set $L(v) :=$ FORK; Set $L(v) :=$ MERGE.*

If there is some vertex v such that $\forall w \in$ out$(v) : L(w) \neq \perp$ or $\exists w \in$ out$(v) : L(w) =$ MERGE, then branch into two possibilities: Set $L(v) :=$ FORK; Set $L(v) :=$ MERGE.

The final Branching Rule 2 tries all possibilities of satisfying a label of a vertex.

Branching Rule 2 (Arc Branch). *If there is a vertex v with $L(v) =$ FORK and* indeg$(v) > 1$, *then branch into all possibilities of removing all but one incoming arc of v. If there is a vertex v with $L(v) =$ MERGE and* outdeg$(v) > 1$, *then branch into all possibilities of removing all but one outgoing arc of v.*

The correctness of Arc Branch follows from Proposition 1. To show the algorithm's correctness, it remains to show the following central lemma.

Lemma 1. *Let D be a DAG. If Label Branch, Arc Branch, Set Label, and Satisfy Label are not applicable, then D is a funnel and all vertices have a label.*

Proof. First, note that if the label of a vertex has been set, it will be satisfied by either applying Satisfy Label or by branching with Arc Branch. Since satisfying all labels turns D into a funnel (Theorem 1(2)), it is enough to show that all vertices have a label if Label Branch, Set Label, and Satisfy Label are not applicable.

We first show that if there is some forbidden subgraph $D' = (V', A') \subseteq D$, that is, D' is isomorphic to some D_i from Theorem 1(3), and if additionally Set Label and Satisfy Label are not applicable, then Label Branch is applicable. Let D' be the forbidden subgraph in D with the smallest number of vertices. Let $v, u \in V'$ be two (not necessarily distinct) vertices in D' such that indeg$_{D'}(v) > 1$, outdeg$_{D'}(u) > 1$. Observe that all vertices between v and u in D' (if any) have in- and outdegree one in D, because D' has the smallest number of vertices. We distinguish two cases.

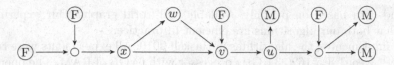

Fig. 2. A DAG where Satisfy Label and Set Label are not applicable. The letter F stands for a FORK label and M stands for MERGE. Label Branch cannot be applied to v since u does not have a label, yet it can be applied to $x \in \text{in}^*(w)$.

Case 1: $\forall w \in \text{in}_D(v) : L(w) \neq \bot$. Then either $\text{outdeg}_D(v) > 1$, meaning that we can apply Label Branch (as required), or $L(v) = \text{MERGE}$ due to Set Label. Since all vertices between v and u have in- and outdegree one, we also know from the latter case that there is some arc (x, y) in the (uniquely defined) (v, u)-path such that $L(x) = \text{MERGE}$ and $L(y) = \bot$. Note that it cannot happen that $L(y) = \text{FORK}$ since Satisfy Label is not applicable. We also know that $\text{outdeg}_D(y) > 1$ since Set Label is not applicable. This implies Label Branch is applicable on y.

Case 2: $\exists w \in \text{in}_D(v) : L(w) = \bot$. This case is illustrated in Fig. 2.

We show that we can find some vertex in $\text{in}^*(w)$ to which we can apply Label Branch. Consider the longest (x, w)-path that only contains vertices in $\text{in}^*(w)$ which do not have a label. Clearly, $\forall y \in \text{in}(x) : L(y) \neq \bot$ and $\text{indeg}(x) > 0$ since all sources have a label. Thus, we can apply Label Branch on x.

Since only these two cases are possible, and in both we can apply Label Branch, it follows, by contraposition, that D is a funnel and all vertices have a label if Label Branch, Set Label, and Satisfy Label are not applicable. □

By combining the previous data reduction and branching rules, we obtain a search-tree algorithm for ADDF on DAGs:

Theorem 4 (⋆). ADDF *can be solved in* $\mathcal{O}(3^d \cdot (|V| + |A|))$ *time, where d is the arc-deletion distance to a funnel of a given DAG* $D = (V, A)$.

To improve the running time of the search-tree algorithm in practice, we compute a lower bound of the arc-deletion distance to a funnel of the input and we stop expanding a branch of the search tree when the lower bound exceeds the available budget. A simple method for computing a lower bound is to find arc-disjoint forbidden subgraphs. Clearly, the sum of the arc-deletion distances to a funnel of the subgraphs found is not larger than the distance of the input DAG. To find such subgraphs, we first look for vertices with both in- and out-degree greater than one, which are not allowed in funnels. Then we search for paths $v_1, v_2, \ldots, v_k$ such that $\text{indeg}(v_1) > 1$ and $\text{outdeg}(v_k) > 1$. With some bookkeeping we can find a maximal set of arc-disjoint forbidden subgraphs in linear time.

4   Empirical Evaluation of the Developed Algorithms

In this section, we empirically evaluate the approximation algorithm and the fixed-parameter algorithm for ADDF described in Sect. 3. We used artificial data sets and data based on publicly available real-world graphs. Our experiments show that both our algorithms are efficient in practice.

We implemented the algorithms in Haskell 2010. All experiments were run on an Intel® Xeon® E5-1620 3.6 GHz processor with 64 GB of RAM. The operating system was GNU/Linux, with kernel version 4.4.0-67. For compiling the code, we used GHC version 7.10.3. The code is released as free software [16].

Experiments on Synthetic Funnel-like DAGs. We generated random funnel-like DAGs through the following steps. (1) Choose the number of vertices, arc density $p \in [0,1]$, and some $s \in \mathbb{N}$. (2) Fix a topological ordering of the vertices. (3) Uniformly at random assign a label FORK or MERGE to each vertex. (4) Create an out-forest with FORK vertices, and an in-forest with MERGE vertices. (5) Add random arcs from FORK to MERGE vertices until a density of p (relative to the maximum number of arcs allowed by the labeling) is achieved. (6) Add s random arcs which respect the topological ordering. Steps (1) through (5) result in a funnel which we call *planted* funnel below.

For a fixed labeling, the algorithm above generates funnels uniformly at random from the input parameters. The labeling, however, is drawn uniformly at random from all $2^{|V|}$ possible labelings, without considering how many different funnels exist with a given labeling. Hence, funnels with fewer arcs have a larger chance of being generated than funnels with many arcs (when compared to the chances in a uniform distribution). We consider this bias to be harmless for the experiments since, for the exact algorithm, the number of arcs is not decisive for the running time, and for the approximation algorithm the number of arcs should not have a big impact on the solution quality.

For $n \in \{250, 300, 500, 1000\}$, $p \in \{0.15, 0.5, 0.85\}$ and $s \in \{125, 150, 175\}$ we generated 30 funnels with n vertices and density p, and then added s random arcs as described above. This gives us a total of 1080 DAGs.

Our fixed-parameter algorithm computed the arc-deletion distance to funnel of 1059 instances (98%) within 10 min. The approximation algorithm finished on average in less than 72 ms. A cumulative curve with the percentage of instances solved within a certain time range is depicted in Fig. 3a. Most instances were solved fairly quickly: 932 (86%) instances were solved optimally within 15 s. We can also observe that there were essentially two types of instances: Easy ones which were solved within few seconds, and harder ones which often were not solved within 10 min. That is, if we limit the running time to five seconds, then we can solve 856 (79%) instances, and if we increase it to sixty seconds, we can solve only 141 additional instances.

Figure 3b shows the relation between the error of the approximation algorithm with the density of the planted funnel. The approximation algorithm found an optimal solution in 574 (54%) instances, and in 260 (25%) it removed only one more arc than necessary. As the arc-deletion distance to a funnel of most instances was greater than 100, this means that the approximation ratio is very close to one. Since the DAGs used here are already close to funnels, most decisions of the approximation algorithm are correct. Intuitively, having correct local information helps the approximation make a globally optimal decision, and so it

(a) Percentage of instances solved exactly within a time-range.

(b) Percentage of instances with an approximation error below a certain value.

Fig. 3. Running time and approximation error.

is unsurprising that the approximation factor in funnel-like DAGs is much better than the theoretical bound. This is supported also by the fact that the approximation performed worse on sparse planted funnels than on dense ones, since the proportion of "wrong" information regarding the arcs is larger on sparse funnels (when adding the same number of random arcs).

Experiments on DAGs Based on Real-World Data Sets. We obtained ten digraphs from the Konect database [12], containing food-chains, interactions between animals, and source-code dependencies. We also downloaded the dependency network of all packages in Arch Linux.[6] Since most of the gathered digraphs contain cycles, we performed a pre-processing step turning them into DAGs: we merged cycles into a single vertex, and then removed self-loops. For each of the eleven DAGs we computed a lower bound and an approximation of its arc-deletion distance to funnel. We also attempted to compute the real distance, stopping the algorithm if no solution was found within four hours.

The dataset was divided into six small DAGs ($\leq$ 156 vertices and $\leq$ 1197 arcs) and five larger ones ($\geq$ 5730 vertices and $\geq$ 26218 arcs). In the small ones, our fixed-parameter algorithm solved ADDF within one second, and our approximation algorithm found the correct distance in $\leq$ 2 ms. In two of the six small DAGs the distance was 60 and 129, which means that the exact algorithm is in practice much faster than what the worst-case upper bound predicts.

On the larger DAGs the fixed-parameter algorithm could not solve ADDF within four hours. By computing a lower bound for the distance, we managed to give an upper bound for the approximation factor, which was at most 1.16. This means that the approximation algorithm is practical since it is fast ($\leq$ 228 ms on average) and yields a near-optimal solution. Relative to the number of arcs, the arc-deletion distance to a funnel parameter was small (9% on average).

---

[6] Listed at https://www.archlinux.org/packages/ and obtained using `pacman`.

5   Conclusion

Our results add to the relatively small list of fixed-parameter tractability results for directed graphs and introduce a novel interesting structural parameter for directed (acyclic) graphs. In particular, our approximation and fixed-parameter algorithms could help to establish the arc-deletion distance to a funnel as a useful "distance-to-triviality measure" [5,10,18] for designing fixed-parameter algorithms for NP-hard problems on DAGs. We consider finding lower bounds for the approximation factor and for the running time of an exact algorithm to be an interesting task for future work. Finally, funnels might provide a basis for defining some useful digraph width or depth measures [8,9,15].

References

1. Ailon, N., Alon, N.: Hardness of fully dense problems. Inf. Comput. **205**(8), 1117–1129 (2007)
2. Bang-Jensen, J., Gutin, G.Z.: Digraphs: Theory, Algorithms and Applications. Springer Monographs in Mathematics. Springer, London (2008). https://doi.org/10.1007/978-1-84800-998-1
3. Bessy, S., Fomin, F.V., Gaspers, S., Paul, C., Perez, A., Saurabh, S., Thomassé, S.: Kernels for feedback arc set in tournaments. J. Comput. Syst. Sci. **77**(6), 1071–1078 (2011)
4. van Bevern, R., Bredereck, R., Chopin, M., Hartung, S., Hüffner, F., Nichterlein, A., Suchý, O.: Fixed-parameter algorithms for DAG partitioning. Discrete Appl. Math. **220**, 134–160 (2017)
5. Cai, L.: Parameterized complexity of vertex colouring. Discrete Appl. Math. **127**(3), 415–429 (2003)
6. Charbit, P., Thomassé, S., Yeo, A.: The minimum feedback arc set problem is NP-hard for tournaments. Comb. Probab. Comput. **16**(1), 1–4 (2007)
7. Fortune, S., Hopcroft, J., Wyllie, J.: The directed subgraph homeomorphism problem. Theor. Comput. Sci. **10**(2), 111–121 (1980)
8. Ganian, R., Hlinený, P., Kneis, J., Langer, A., Obdržálek, J., Rossmanith, P.: Digraph width measures in parameterized algorithmics. Discrete Appl. Math. **168**, 88–107 (2014)
9. Ganian, R., Hlinený, P., Kneis, J., Meister, D., Obdržálek, J., Rossmanith, P., Sikdar, S.: Are there any good digraph width measures? J. Comb. Theory Ser. B **116**, 250–286 (2016)
10. Guo, J., Hüffner, F., Niedermeier, R.: A structural view on parameterizing problems: distance from triviality. In: Downey, R., Fellows, M., Dehne, F. (eds.) IWPEC 2004. LNCS, vol. 3162, pp. 162–173. Springer, Heidelberg (2004). https://doi.org/10.1007/978-3-540-28639-4_15
11. Kenyon-Mathieu, C., Schudy, W.: How to rank with few errors. In: Proceedings of the 39th STOC, pp. 95–103. ACM (2007)
12. Kunegis, J.: KONECT - The Koblenz network collection. In: Proceedings of the 22nd WWW, pp. 1343–1350. ACM (2013)
13. Lehmann, J.: The computational complexity of worst case flows in unreliable flow networks. Bachelor thesis, Institut für Theoretische Informatik, Universität zu Lübeck, October 2017

14. Leskovec, J., Backstrom, L., Kleinberg, J.: Meme-tracking and the dynamics of the news cycle. In: Proceedings of 15th ACM SIGKDD, pp. 497–506. ACM (2009)
15. Millani, M.G.: Funnels–algorithmic complexity of problems on special directed acyclic graphs. Master thesis, Department of Electrical Engineering and Computer Science, TU Berlin, August 2017. http://fpt.akt.tu-berlin.de/publications/theses/MA-marcelo-millani.pdf
16. Millani, M.G.: Parfunn - Parameters for Funnels, August 2017. https://gitlab.tubit.tu-berlin.de/mgmillani1/parfunn
17. Millani, M.G., Molter, H., Niedermeier, R., Sorge, M.: Efficient algorithms for measuring the funnel-likeness of DAGs. CoRR abs/1801.10401 (2018)
18. Niedermeier, R.: Reflections on multivariate algorithmics and problem parameterization. In: Proceedings of the 27th STACS, pp. 17–32. Schloss Dagstuhl-Leibniz-Zentrum für Informatik (2010)

Jointly Optimizing Replica Placement, Requests Distribution and Server Storage Capacity on Content Distribution Networks

Raquel Gerhardt[(✉)], Tiago Neves, and Luis Rangel

Industrial Engineer School of Volta Redonda,
Federal Fluminense University, Volta Redonda, RJ, Brazil
quel_gerhardt@hotmail.com

Abstract. A Content Distribution Network includes dedicated servers creating an architecture that moves the content to servers that are closer to the user, reducing delays and traffic. In this structure several problems are studied, including the Problem of Allocation of Storage Capacity (SCAP) and the Replica Placement and Request Distribution Problem (RPRDP). This work analyzes these problems in an integrated way and proposes the creation of a new problem named Capacities, Replicas and Requests Distribution Problem (CRRDP), which enables the dynamic allocation of disk space on the servers and distribution of replicas and requests. The main contributions of this work are the creation of a new problem and a new formulation which associates variables and restrictions presents in mathematical formulations for this problems. The Mathematical formulation was analyzed and computational results shows that operational costs can be reduced and that it is possible to disable unused servers over the network.

Keywords: Content Distribution Networks
Combinatorial optimization · Mathematical formulation
Storage capacity optimization

1 Introduction

The demand for fast and reliable Internet services and the augment of information traffic require, from content providers, innovative ways to deliver such services in the most efficient manner to the customers. In this scenario it is possible to observe the increasing use of Content Distribution Networks (CDNs). The CDNs are able to continuously improve service performance. Such improvement is achieved through replication of the contents in strategically positioned servers, closer to the users, and the requests are handled by these surrogate servers, that can deliver the contents faster and at lower cost [1,2].

© Springer International Publishing AG, part of Springer Nature 2018
J. Lee et al. (Eds.): ISCO 2018, LNCS 10856, pp. 196–207, 2018.
https://doi.org/10.1007/978-3-319-96151-4_17

In the CDNs context, there are several optimization problems that have already been addressed in multiple ways [1–6]. This work proposes a new optimization problem, called the Capacities, Replicas and Requests Distribution Problem (CRRDP). This new problem involves the simultaneous optimization of servers disk capacities, replica positioning and the distribution of requests through CDN servers, solving two related optimization problems in a jointly way, the Storage Capacity Allocation Problem (SCAP) and the Replication Replica Placement and Request Distribution Problem (RPRDP). With data volume increasing and content popularity, *CloudCDN* [7] structures bring new insights that utilize network virtualization to facilitate its operations. Such insights are also used in the this paper and will be better explored latter.

The remaining of this paper is organized as follows: Section 2 presents the problem definition. Section 3 shows the mathematical formulations analyzed. Section 4 exposes the computational result and Sect. 5 concludes the work.

2   Problem Definition

Some contents are more popular in determined regions than others. In addition, it is not always possible to allocate servers with sufficient capacity to serve all customers in all regions. Therefore, sometimes it is necessary to spend a higher cost to deliver these contents and assume the risk of not providing the service in the Quality of Service (QoS) expected by the user. In this way, it is possible to define a quality indicator for a CDN service, adding the following costs: total cost = request reply cost + content replication cost + backlog penalties.

During the CDN operation, if servers storage capacities are not totally used, the available resources may be underutilized. To avoid this kind of inefficiency, one option is to redistribute total capacity across servers according to the volume of requests, increasing or reducing the server capacity to better serve customers [3].

There are a number of papers in the literature [7] that handle multiple requests in jointly way (that deal with requests from different users as they were equal). Such approach may not be so advantageous for the clients, since network problems may occur and QoS specifications may be different. In order to reduce the network load without violating the requests' quality constraints, it is important to optimize the positioning of the replicas throughout a CDN network and the distribution of the requests over the servers, dealing with requests individually. It is also important to have the possibility of a request being handled by multiple servers simultaneously, as long as they have a copy of the desired content. The RPRDP, proposed by [1], addresses all this realistic characteristics and is one of the most general problems in the CDNs context.

Besides the mentioned aspects, the RPRDP also considers during the optimization process: servers' bandwidth and disk space limits; the existence of multiple contents; changes in the network over time; content submission; content removal; and the appearance of new requests. This problem tries to fulfill clients' quality criteria, using its maximum limits whenever possible [1]. However, this problem still does not treat the problem of capacity allocation of servers and considers that the location of servers is already known.

The SCAP [3] appears as an alternative to treat this issue, since its objective is to determine the proportion of the total storage capacity available in the network that should be allocated on each CDN server. However, this problem was not proposed in a multi period context, and thus, it does not consider the changes that may occur in the network and in the servers during the entire horizon. Besides, the request handling is not considered by SCAP although its results increase the chances of a request being serviced by a nearby server, thereby reducing the delay and even loading the network links.

Based on the gaps identified in each of the mentioned optimization problems and in the face of a increasing demand scenario, we claim that it is necessary to analyze these problems together. In this way, it is possible to describe a new optimization problem in the CDNs context, called the Capacity, Replica and Request Distribution Problem, or CRRDP. The CRRDP treats in an integrated way the problem of the positioning of the replicas in the servers and the distribution of the requests, and it also deals with the dynamic allocation of the storage capacity of the servers, which can be optimized to meet the variations of the demand over time periods. The characteristics of the CRRDP addressed in this work are: (1) Requests are treated individually and can be handled by multiple servers simultaneously; (2) Storage capacity of servers can be changed according to demand fluctuation; (3) New requests and contents may come up and contents can be removed; (4) Network delays can change over the time horizon; (5) Bandwidth constraints are considered for clients and servers; (6) clients' QoS requirements are fulfilled whenever possible; (7) The problem is Offline, meaning that all changes are known in advance.

3   Mathematical Formulations

The model of CRRDP should contain characteristics of SCAP and RPRDP. The mathematical formulations proposed by Uderman [3] and Neves [1] are analyzed separately. In the Sect. 3.3 a new formulation to solve the CRRDP is presented.

3.1   Storage Capacity Allocation Problem

Uderman [3] has studied a series of mathematical problems that deal with storage capacity of the servers optimization. The author proposed a mathematical formulation for the SCAP solution, suitable for CDNs without a hierarchical topology and capable of supporting contents of any size, reducing the number of variables needed to describe the model and reducing complexity. Uderman defined the problem as follows: Let ψ be the set of contents, where each content k has size L_k; let E be the total storage capacity of a CDN; let V be the set of servers; let J be the set of clients, where each client j has a request rate λ_j, a demand (for contents) distribution p_j and an source server $o_j \in V$, that is the closest CDN server to j (and thus the server that should preferably handle j); The distance between a client $j \in J$ and a server $v \in V$ is given by $d_{j,v} : J \times V$. The SCAP aims to minimize the distance between the client and the requested content, given the demand rates for the content, defining the optimal allocation of the total available storage space E through the servers set V. $X_{j,v}(k)$ is a binary variable equals to 1 if a client j receive the content k from the server v and equals to zero in the other cases. The variable $\delta_v(k)$ is auxiliary that is equal to one if the content k is positioned on the server v and equal to zero in the other cases. Using a static approach in which there is a single time period, the optimal allocation of the total available capacity in a CDN can be obtained by the formulation described by Uderman [3], here transcribed for didactic reasons. The formulation can be expressed as follows:

$$Max \sum_{j \in J} \lambda_j \sum_{k \in \psi} p_j(k) \sum_{v \in V} (d_{j,o_j} - d_{j,v}) X_{j,v}(k) \tag{1}$$

$S.t.$

$$\sum_{j \in J} X_{j,v}(k) \leq U \cdot \delta_v(k), \ \forall v \in V, \ \forall k \in \psi, \ U \geq |J|, \tag{2}$$

$$\sum_{v \in V} \sum_{k \in \psi} \delta_v(k) \cdot L_k \leq E, \tag{3}$$

$$X_{j,v}(k) \in [0,1], \ \forall v \in V, \ \forall k \in \psi, \ \forall j \in J, \tag{4}$$

$$\delta_v(k) \in [0,1], \ \forall v \in V, \ \forall k \in \psi, \tag{5}$$

The objective of the problem (1) is to minimize the distance between the client and the content to be requested, since the distance between the client and its source server will always be less than or equal to the distance between client j and v that satisfies the request, the term $d_{j,o_j} - d_{j,v}$ in the objective function will always be less than or equal to zero. The constraints (2) represent that a k content can only be delivered to the client if it is present on the server v, and the constraints (3) limit the capacity allocated on all servers to the total available storage capacity.

3.2   Replica Placement and Request Distribution Problem

The server storage capacity, determined by the SCAP, has a direct influence on the RPRDP. The RPRDP consists of determining the optimal positioning of content replicas in the network and distributing the requests through the servers, in order to reduce the operational cost of the CDN.

In order to distribute the requests, the RPRDP model, presented in [1], considers that a request can be redirected only to servers that have a copy of the requested content, checking the QoS parameters. If such parameters are violated, a penalization is added in the Objective Function of the model.

Thus, in a simplified way, given a set R of requests from customers; a set of S of CDN servers and a set C of contents, one must determine the best locations (servers in S) to place the replicas of each contents of C and the distribution of each request in R through servers in S in order to minimize the penalties applied for delaying requests handling. The bandwidth limits of servers and clients must be ensured, as well as the servers disk space.

The formulation presented in [1] considers the capacity of each server as a constant, thus, it does not contemplate the optimization of the disk space used by each server. The inclusion of this characteristics will be proposed later.

3.3   Capacity, Replica and Request Distribution Problem

Due to the interdependence between the SCAP and the RPRDP, both problems were studied by Uderman [3], so that the results of the first one were directly used as input data for the second. In this approach, SCAP is used to find an optimal distribution of storage space for each CDN server. In his work, Uderman uses two approaches. In the first, the SCAP is solved using average rates of the requests, but the results obtained by the SCAP solution are used only once, at the beginning of the optimization horizon. In the second, the SCAP is solved for each period of the planning horizon, considering as input data the current state of the CDN. However, this second approach continues to treat the two problems (SCAP and RPRDP) separately, which means that the changes that occur over time are not observed with care.

Modifying the RPRDP formulation proposed by Neves [1], and later revised by Gerhardt [8], it is possible to make a formulation that also can deal with the concept of storage space optimization. This formulation can be used to solve the CRRDP. The main objective of this problem is to optimally allocate the available storage capacity, distribute the content replicas and customer requests in order to improve resource utilization and reduce operational costs, always observing QoS standards. In this way, the CRRDP can be seen as a more general problem, which solves RPRDP and SCAP in a jointly approach.

The formulation for CRRDP, proposed in this work, can be described as follows: Considering S the set of CDN servers, C the set of replicated contents, R the set of requests to be met and E the total space available for distribution between servers. Let c_{ijt} be the cost of servicing the request i on the server j, in the period t, q_{it} punishment for backlogging the request i in the period t, h_{kjlt} the cost of replicating the contents k on server j from server l in period t. In addition, L_k the content size k (in bytes), MB_j maximum server band j (in bytes/second), BR_i request bandwidth requirement i (in bytes/second), BX_i is the maximum bandwidth of client i (in bytes/second), δ is the length of the periods in seconds, $G(i)$ the content required by the request i and N a constant that represents the sum of the size of all contents. The following sets of variables are also defined:

- x_{ijt} = fraction of content requested by request i delivered by server j in period t
- $y_{kjt} = \begin{cases} 1, \text{ if and only if the content } k \text{ is replicated on the server } j \text{ in the period } t \\ 0, \text{ otherwise} \end{cases}$
- $b_{it} = backlog$ of the request i in the period t
- $w_{kjlt} = \begin{cases} 1, \text{ if and only if the content } k \text{ is copied by the server } j \text{ from} \\ \quad \text{server } l \text{ in period } t \\ 0, \text{ otherwise} \end{cases}$
- r_{jt} = space allocated on server j during period t

$$Min \sum_{i \in R} \sum_{j \in S} \sum_{t \in T} c_{ijt} x_{ijt} + \sum_{i \in R} \sum_{t \in T} q_{it} b_{it} + \sum_{k \in C} \sum_{j \in S} \sum_{l \in S} \sum_{t \in T} h_{kjlt} w_{kjlt} \qquad (6)$$

$S.t.$

$$\sum_{j \in S} L_{G(i)} x_{ijt} - b_{i(t-1)} + b_{it} = D_{it}, \ \forall i \in R, \forall t \in [B_{G(i)}, E_{G(i)}], \qquad (7)$$

$$\sum_{i \in R} L_{G(i)} x_{ijt} \leq \delta M B_j, \ \forall j \in S, \ \forall t \in T, \qquad (8)$$

$$\sum_{j \in S} L_{G(i)} x_{ijt} \leq \delta B X_i, \ \forall i \in R, \ \forall t \in T, \qquad (9)$$

$$y_{G(i)jt} \geq x_{ijt}, \ \forall i \in R, \ \forall j \in S, \ \forall t \in T, \qquad (10)$$

$$\sum_{j \in S} y_{kjt} \geq 1, \ \forall k \in C, \forall t \in [B_k, E_k], \qquad (11)$$

$$y_{kjt} = 0, \ \forall k \in C, \ \forall j \in S, \ \forall t \notin [B_k, E_k], \qquad (12)$$

$$y_{kjB_k} = 0, \ \forall k \in C, \ \forall j \in \{S | j \neq O_k\}, \qquad (13)$$

$$y_{kj(t+1)} \leq \sum_{l \in S} w_{kjlt}, \ \forall k \in C, \ \forall j \in S, \ \forall t \in T, \qquad (14)$$

$$y_{kjt} \geq w_{kljt}, \ \forall k \in C, \forall j, \ \forall l \in S, \ \forall t \in T, \qquad (15)$$

$$\sum_{k \in C} L_k y_{kjt} \leq r_{jt}, \ \forall j \in S, \ \forall t \in T, \qquad (16)$$

$$\sum_{j \in S} r_{jt} \leq E, \ \forall t \in T, \qquad (17)$$

$$x_{ijt} \in [0, 1], \ \forall i \in R, \forall j \in S, \forall t \in T, \qquad (18)$$

$$y_{kjt} \in \{0, 1\}, \ \forall k \in C, \forall j \in S, \forall t \in T, \qquad (19)$$

$$b_{it} \geq 0, \ \forall i \in R, \forall t \in T, \qquad (20)$$

$$w_{kjlt} \in \{0, 1\}, \ \forall k \in C, \forall j \in S, \forall l \in S, \forall t \in T, \qquad (21)$$

$$r_{jt} \in [0, N], \ \forall j \in S, \forall t \in T. \qquad (22)$$

This formulation objective (6) is to minimize the cost of delivering content to customers as well as the punishment for delaying requests' handling and also the cost of replicating contents on the servers. The constraints (7) guarantee that the demand will be fully satisfied. The constraints (8) and (9) ensure the bandwidth limits of servers and clients. Restrictions (10) assure that a request can only be served by a server that has a copy of the requested content. The constraints (11) and (12) control the replicas, ensuring that there is at least one copy of the content during its lifetime and that there is no replication outside this period. Restrictions (13) ensure that during the appearance of the content, all other servers except the source server do not contain any replica of the content out of its lifetime. Restrictions (14) require that a new replica to be created for each replication (no partial replication is allowed), and constraints (15) require replication can occur only from servers that has a copy of the content.

Restrictions (16) indicate that the sum of the space occupied on a server must not exceed the space allocated to this server. Restrictions (17) ensure that on each period, the sum of disk spaces allocated to servers must be less than or equal to the total space available. The remaining restrictions are integrality and non-negativity restrictions.

4   Formulation Analysis Throughout Computational Tests

In order to validate the model, we performed tests using two mathematical formulations: one the RPRDP proposed in [1]; and one for the CRRDP, proposed in this work. Both formulations were solved using the CPLEX [9] software, that is a well known software for solving Integer Programming Problems. We choose the standard CPLEX configuration for all instances, meaning that a Branch-and-Bound like algorithm was used. The objective of the tests is to verify if the dynamic allocation of space can reduce costs without compromising the quality of service.

The instances used are the same ones used in several works [1,3,8] and are available at LABIC [10] website. This set of instances is the first set that simultaneously considers several characteristics close to reality, such as QoS requirements, different server capacity and dynamic content. As mentioned in [1], these instances are divided into four classes, A, B, C, and D. The instances of class A are small scale instances, used for testing and, therefore, virtually all values used for this class are chosen arbitrarily. Instances of class B are instances constructed based on values found in the literature and based on market equipment available when these instances were created (2008). The instances of class C are instances similar to the instances of class B, however, in instances of class C the servers have less storage capacity. Class D instances are instances with more severe restrictions in terms of storage capacity and bandwidth on the servers. For each possible size of 10, 20, 30 or 50 servers, 5 instances of each class were used. For all instances, it was considered that each period lasts 60 s.

4.1   CRRDP Dynamic

The solution status reported by the CPLEX [9] on CRRDP instances can be classified as optimal (*Optimal*), optimal within the established tolerance of 4 decimal places (*Tol. Opt.*) or best solution found within the timeout of 10800 s (*Time*).

Figure 1 shows the solution status presented by the CPLEX. The percentage difference (gap) between the two approaches is obtained by the following equation: $\left(\frac{F_1 - F_2}{F_1}\right) \cdot 100$, where F_1 is the result of the objective function for the RPRDP and F_2 is the result of the objective function for the CRRDP. The positive gap shows that the solution of the CRRDP presented a smaller objective function value when compared to the RPRDP solutions, indicating reduction of operational costs.

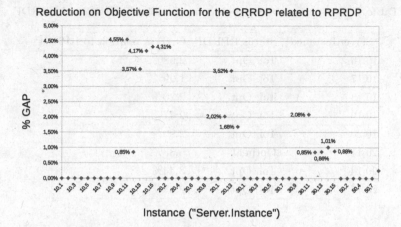

Fig. 1. Results for instances with status **optimal** or **optimal within tolerance**

Figure 1 shows the results for the instances that obtained the status optimal or optimal within the tolerance. In this way, the gaps indicate a reduction of up to 4.55% of an optimal solution over RPRDP when using the dynamic storage capacity allocation.

The data shown in Fig. 1 represent that the better use of servers can reduce the distribution costs of the contents in the network, because the allocation of the space is made according to the volume of requests. The dynamic allocation of space allows the reduction of the distances between the clients and the servers that handle them, especially when the volume of requests is high. This is possible because the dynamic allocation provides more storage capacity to the source servers with higher request rate, allowing these servers to handle more requests, avoiding unnecessary forwarding, reducing the delays and ensuring the quality of service.

Some results presented a gap of 0% in the best solution found. These results are among the instances belonging to classes A and B, which are the simplest and mainly used for testing [1,2]. In these cases, both approaches easily find the best possible solutions.

The Table 1 shows the results for eight hard to solve instances with status Optimal or Tol. Opt for the RPRDP and with status time for the CRRDP. Although the optimal solution were not found for the CRRDP the final gaps presented by CPLEX for these instances were less than 0.5% meaning that the values are to close to the optimal. The first column exposes the instances analyzes. The second column presents the solution status for the RPRDP. The third column depicts the relative difference between the final objective functions found by the CPLEX. The relative difference is calculated in the same manner used in Fig. 1. It is possible to note that even though the optimal solution for these instances were not found for, the gaps calculated for all of them show positive values. This means that even if we are not able to reach the optimal result found

Table 1. Comparison between objective function RPRDP and CRRDP

"Server.Instance"	Status RPRDP	GAP in relation to RPRDP
10.16	Tol. Opt.	2,08%
10.17	Tol. Opt.	1,99%
10.18	Tol. Opt.	2,20%
10.19	Tol. Opt.	2,47%
10.20	Tol. Opt.	2,99%
20.11	Optimal	5,33%
20.14	Tol. Opt.	3,44%
20.17	Tol. Opt.	2,31%

for CRRDP in the time limit is already better than the optimal solution of the RPRDP, thus, it can be concluded that there was a reduction of the operational costs for these instances.

For these instances were not found, the gaps calculated for all of them show positive values. This means that even if we are not able to reach the optimal result, the best result found for CRRDP in the time limit is already better than the optimal solution for the RPRDP, thus, it can be concluded that, for these instances, there was also reduction of the operational costs. The CPLEX gaps founded for these instances were less than 0.5% and represents that the values are to close to the optimal.

4.2   Instances that Do Not Consider All the Servers as Source

In a scenario with no disk space optimization on the servers, resources may be waste due to underutilization on some servers. This means that after optimizing the disk space allocation, some servers may not be used to handle requests and, thus, be deactivated, leading to further reduction of costs. In order to prove this concept, some instances were created based on instances of class D, that are instances closer to reality and that present a scarcity of resources. For this, we used instances of class D with 20 servers as baseline and, during the creation of the new instances, it was considered that only 15 out of the 20 servers could be used as source. The number of requests of each instance was preserved, but the sources of the requests were redistributed among 15 randomly chosen servers following a Uniform Distribution.

The results for these cases show that, no disk space was allocated for some servers on some periods, meaning that they were not needed for the problem solution. The Fig. 2 graphically shows a comparison between the percentage of servers occupied per period and also the total space allocated on the servers.

Figure 2 indicate that the use of dynamic allocation may determine that some servers can be disabled or that their resources can be transfered to another application, in the case of the use of shared servers. Some servers, like servers 2,

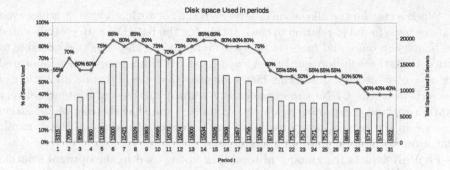

Fig. 2. Distribution of the allocated space on the optimization horizon

11, 12, and 19, have no disk space allocated in more than 90% of periods. The Fig. 2 also shows that in none of the periods we had 100% of the servers used, and, in some cases only 40% of the servers available on the network were used.

For all these instances, the solution status obtained was *Time*, that is, even in intermediate solutions, it is possible to reduce the number of servers in the network without reducing the quality of service.

4.3 CRRDP with Allocation Cost

In order to make the model even closer to reality, a modification was proposed in the objective function of the CRRDP, adding costs for the allocation of disk space in the servers. Without a cost associated with allocation of disk space on servers the use of a server becomes indifferent to the objective function, but, such allocation generates costs on real CDN environments. In addition, the inclusion of these costs helps to differentiate solutions that were previously considered equal by the model, favoring solutions that use a smaller fraction of the available space. Thus, considering F the cost of allocating a unit of disk space on a server, the objective function described in Eq. (6), should be rewritten as:

$$Min \sum_{i \in R}\sum_{j \in S}\sum_{t \in T} c_{ijt}x_{ijt} + \sum_{i \in R}\sum_{t \in T} q_{it}b_{it} + \sum_{k \in C}\sum_{j \in S}\sum_{l \in S}\sum_{t \in T} h_{kjlt}w_{kjlt} + F\sum_{j \in S}\sum_{t \in T} r_{jt} \quad (23)$$

Through the results obtained with the new objective function, it is possible to notice that, in some cases, the results are inconclusive, but they strongly indicate that the use of these associated costs helps the search process through more robust cuts, making possible to solve instances closer to the reality. In 80% of cases it was possible to observe a positive gap between the objective functions of the CRRDP and of the RPRDP. This indicates that the dynamic allocation reduced the operational costs. In all cases it was possible observe that it was not necessary to use all the servers to solve the problem since some servers were not used during the planning horizon. This shows that dynamic allocation of server space can reduce the replication and content delivery costs.

When a cost for the allocation of servers in the network is added, it is observed that the gaps found in relation to the solution for the RPRDP with cost allocated are increasing more and more as the cost increases. In the tests it is possible to notice that the solution gap increases almost 10 times more when allocated the cost of at least 1 unit. This fact indicates that the dynamic allocation of the servers in the CDNs can considerably reduce the cost of content delivery solution without compromising quality of service and that the cost of allocation of space in the servers can potentialize the necessity of the use of the dynamic allocation of the total space of a CDN.

Figure 3 depicts the amount of total disk space used in the optimal solution as the value of F increases for instance 30.11. It is possible to notice that when the value of F is augmented from 0 to 1, there is already a reduction of the total space allocated, and, as this value increases, the total space allocated is reduced until a certain limit.

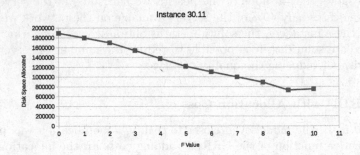

Fig. 3. Analysis Sensitivity of the constant F for instance 30.11

5   Conclusions and Future Work

We showed that when the content replication, request distribution and storage capacity are jointly optimized it is possible to reduce the operational costs of a CDN. We were able to obtain optimal results for almost all instances belonging to Classes A, B and C and the results presented a reduction of up to 4.55% on the costs. However, there were some instances, where the results are inconclusive. On such instances, it was observed that CPLEX could not even solve the linear relaxation of the problem in the proposed time limit due to the great number of variables and restrictions.

Through the analysis of occupation of the servers on each period, it was possible to identify that some servers did not have disk space allocated during some moments of the optimization horizon. This fact indicates that the use of the CRRDP may determine that some servers can be disabled, reducing operational cost even further without affecting quality of service and reliability.

It was identified the possibility of including a cost for disk space allocation in the model. A sensitivity analysis of such inclusion in the objective function identified that optimizing content replication, request distribution and storage capacity together become essential on scenarios where this cost is relevant.

The model used was able to solve two optimization problems jointly (SCAP and RPRDP), obtaining success for most of the instances used. The model was able to prove that the dynamic allocation of servers' disk space was able to indeed reduce the operational cost of a CDN without violating the required quality standards.

As future work we intend to verify if some specialized techniques, such as cut or column generation, can improve the solvability of the problem.

References

1. Neves, T., Drummond, L., Ochi, L., Albuquerque, C., Uchoa, E.: Solving replica placement and request distribution in content distribution networks. Electron. Notes Discrete Math. **36**, 89–96 (2010)
2. Neves, T., Ochi, L., Albuquerque, C.: A new hibrid heuristic dor replica placement and request distribution in content distribution networks. Optim. Lett. **9**(4) (2015)
3. Uderman, F., Neves, T., Albuquerque, C.: Optimizing server storage capacity on content distribution networks. In: Anais do Simpósio Brasileiro de Redes de Computadores - SBRC2011 (2011)
4. Tang, X., Xu, J.: On replica placement for QOS-aware content distribution. In: Proceedings of Infocon 2004, pp. 806–815 (2004)
5. Aioffi, W., Mateus, G., Almeida, J., Loureiro, A.: Dynamic content distribution for mobile enterprise networks. IEEE J. Sel. Areas Commun. **23**(10), 2022–2031 (2005)
6. Li, W.E.A.: Analysis and Performance Study for Coordinated Hierarchical Cache Placement Strategies, 1st edn. Elsevier (2010)
7. Hu, H., Wen, Y.: Joint content replication and request routing for social video distribution over cloud CDN: a community clustering method. IEEE Trans. Circuits Syst. Video Technol. **26**(7), 1320–1333 (2016)
8. Gerhardt, R., Neves, T., Albuquerque, C.: Análise de redundância em uma formulação matemática para o problema de posicionamento de réplicas e distribuição de requisições em redes de distribuição de conteúdos. SBPO - Simpósio Brasileiro de Pesquisa Operacional 8 (Agosto 2015)
9. IBM ILOG CPLEX Optimization Studio (COS): ILOG S.A., CPLEX 10 user's manual (2006)
10. Neves, T.: Synthetic instances for a management problem in content distribution networks. Technical report, UFF. http://labic.ic.uff.br/ (url world wide web (2011) 02/2012)

An Exact Column Generation-Based Algorithm for Bi-objective Vehicle Routing Problems

Estèle Glize[1(✉)], Nicolas Jozefowiez[2], and Sandra Ulrich Ngueveu[1]

[1] CNRS, LAAS, INSA, INP Toulouse, Toulouse, France
{glize,ngueveu}@laas.fr
[2] LCOMS, Université de Lorraine, Metz, France
nicolas.jozefowiez@univ-lorraine.fr

Abstract. We propose a new exact method for bi-objective vehicle routing problems where edges are associated with two costs. The method generates the minimum complete Pareto front of the problem by combining the scalarization of the objective function and the column generation technique. The aggregated objective allows to apply the exact algorithm for the mono-objective vehicle routing problem of Baldacci et al. (2008). The algorithm is applied to a bi-objective VRP with time-windows. Computational results are compared with a classical bi-objective technique. The results show the pertinence of the new method, especially for clustered instances.

Keywords: Combinatorial MOP · Vehicle routing problem
Exact method · Column generation

1 Introduction

This paper proposes a competitive method for solving to optimality a variant of the *vehicle routing problem* (*VRP*) [1]: the *bi-objective VRP* (*BOVRP*) where edges are associated with two costs. These objectives can be conflictive: in motor vehicle, the travel time differs from the distance. In this case, solving to optimality means finding the non-dominated set. Many industrials are interested in finding a good compromise.

Multi-objective *VRPs* (*MOVRPs*) are more and more studied. A complete survey of *MOVRP* can be found in *Jozefowiez et al.* [2]. In addition to the minimization of travel distance, most *MOVRPs* aim to minimize the number of vehicles or to maximize the fairness of routes.

In the larger scope of multi-objective integer programming (*MOIP*), exact methods are divided into two classes: methods working on the feasible solution space [3] and those working on the objective space [4]. These last methods solve a sequence of mono-objective problems and so, rely on the efficiency of single-objective integer programming solvers. The ϵ-constraint method is the most

© Springer International Publishing AG, part of Springer Nature 2018
J. Lee et al. (Eds.): ISCO 2018, LNCS 10856, pp. 208–218, 2018.
https://doi.org/10.1007/978-3-319-96151-4_18

commonly used objective space search algorithm [5–7] as its efficiency is verified. The balanced box method of *Boland et al.* [8] shows significant improvements for solution of *MOIPs* by exploring the decision space smartly. Recently, the efficient method of *Dai and Charkhgard* [9] which combines the balanced box method and the ϵ-constraint technique, has been applied to a *2-Dimensional Knapsack Problem* and to the *bi-objective Assignment Problem*.

Section 2 gives preliminaries about bi-objective optimization and an introduction of an algorithm to solve the mono-objective *VRP*. It also defines a formulation of the *BOVRP*. The branch-and-price method is presented in Sect. 3. Then, Sect. 4 introduces a classical bi-objective technique and computationally compares the two methods. Finally, Sect. 5 concludes about this new method.

2 Preliminaries

2.1 Problem Definition

Let $G = (V, E)$ be a non-oriented graph. A node $i \in V \backslash v_0$ is called a customer and has a demand q_i. These demands are satisfied by a fleet of K vehicles of capacity Q. A vehicle k starts and returns at a node v_0 called the depot and performs a route r_k by passing through a set of customers. The route r_k is said to be feasible if the total capacity of a vehicle is not exceeded by the demands.

An edge $e \in E$ of the graph has two costs c_e^1 and c_e^2. Each route r_k provides two costs c_k^1 and c_k^2 representing the sum of the two costs on the used edges. The aim of the studied *BOVRP* is to minimize the sum of each cost of the routes used.

Let Ω be the set of feasible routes r_k and a_{ik} be equal to 1 if the customer i belongs to the route r_k. The Set Partitioning formulation of the *BOVRPR* [10] is stated in the model (1).

$$\begin{cases} minimize \; (\sum_{r_k \in \Omega} c_k^1 \theta_k, \; \sum_{r_k \in \Omega} c_k^2 \theta_k) \\ \sum_{r_k \in \Omega} a_{ik} \theta_k \geq 1 \; (v_i \in V \backslash \{v_0\}), \\ \sum_{r_k \in \Omega} \theta_k \leq K, \\ \theta_k \in \{0, 1\} \; (r_k \in \Omega). \end{cases} \quad (1)$$

where θ_k is a variable that indicates if the route $r_k \in \Omega$ is selected in the solution ($\theta_k = 1$) or not ($\theta_k = 0$).

2.2 Single Objective Algorithm for the *VRP*

The method presented in this paper works on the objective space and is based on the exact algorithm for the single objective *VRP* of Baldacci et al. [11].

This state-of-the-art method considers the set partitioning formulation of the *VRP* as a master problem *MP*. As this formulation contains an exponential number of variables θ_k, $r_k \in \Omega$, *MP* needs to be solved optimally on a reduced set of columns. This set is obtained with a three steps algorithm:

1. Compute a good lower bound LB by column generation algorithm and a good upper bound UB. Compute the gap $\gamma = UB - LB$.
2. Generate all routes with a reduced cost lower than γ. Indeed, it can be proven that routes with a reduced cost higher than γ cannot be in the optimal integer solution. Let $\overline{\Omega}$ be this reduced set of routes.
3. Solve the initial integer problem on $\overline{\Omega}$ to obtain the optimal (integer) solution.

The second step uses dynamic programming to generate the lower bounds on reduced cost necessary to go from the depot to the node i with a load lower than q in non-elementary paths. Then, it solves an *elementary shortest path problem with resource constraints* (*ESPPRC*) with a bi-directional labeling algorithm to produce interesting paths. Finally, feasible routes are produced by combining pairs of these paths.

As previously mentioned, this method will be used to solve the *BOVRP*. In the following, we will refer to the second step by *GENROUTE(UB,LB)* with *UB* and *LB* the upper bound and the lower bound previously computed.

2.3   Multi-objective Optimization

The main purpose of this work is to obtain, in a-posteriori fashion, the minimum complete Pareto front of the *BOVRP*. All concepts of multi-objective optimization are detailed in [12], but an introduction is given in the following of the paper.

Let denote Θ the set of combinations of θ_k, $r_k \in \Omega$, which lead to a feasible solution. An element $\theta \in \Theta$ is a binary vector of size $card(\Omega)$ with $card(\Omega)$ the size of the set Ω indicating which routes are in the solution θ. For $\theta \in \Theta$, let $F(\theta) = (c_1(\theta), c_2(\theta)) = (\sum_{r_k \in \Omega} c_k^1 \theta_k, \sum_{r_k \in \Omega} c_k^2 \theta_k)$ be the function vector to minimize. $\mathcal{Y} = F(\Theta)$ represents the objective space and $y = F(\theta) \in \mathcal{Y}$ a point in the objective space. The following definitions are only valid for a minimization problem and specify the output of the method.

Definition 1. $(a, b) \in \Theta^2$.

$$a \text{ is Pareto dominant with respect to } b \Leftrightarrow \begin{cases} f_i(a) \leq f_i(b) \ \forall i \in \{1, 2\} \\ f_i(a) < f_i(b) \ \exists i \in \{1, 2\} \end{cases}$$

Definition 2. *A solution $a \in \Theta$ is said to be an efficient (or a Pareto-optimal) solution if $\nexists b \in \Theta$, $b \neq a$, such that b is Pareto dominant with respect to a.*

Definition 3. *A point $y \in \mathcal{Y}$ is said to be a non-dominated point if the solution $a \in \Theta \backslash F(a) = y$ is an efficient solution.*

Definition 4. *A non-dominated point $y \in \mathcal{Y}$ is said to be supported if it is located on the boundary of the convex hull of $\mathcal{Y}$. A non-dominated point $y \in \mathcal{Y}$ is said to be non-supported if it is located on the interior of the convex hull of $\mathcal{Y}$.*

A complete Pareto front is the set of all non-dominated points of the problem. Furthermore, as the same point $y \in \mathcal{Y}$ can be associated with several different solutions in Θ, the number of efficient solutions can be larger than the number of non-dominated points. In this paper, the method provides the minimum Pareto front that-is-to-say that only one efficient solution per non-dominated point is provided.

To lighten the notation, we will refer to the costs of a point $y \in \mathcal{Y}$ associated to a solution $\theta \in \Theta$, by $c^1(y)$ and $c^2(y)$ instead of the cost of the solution $c^1(\theta)$ and $c^2(\theta)$.

3   *Two-Step* Method

3.1   Global Algorithm

The *two-step* method is an objective space search method, that is to say that the algorithm determines areas in the objective space in which non-dominated points could be present. Once the areas are delimited, the method scalarizes the objectives to apply a single objective method to go through them. The scalarization we use is a weighted-sum of the two objectives [13]. The set partitioning formulation in this case is Model (2), where all notations are the same for the formulation (1). It is called the master problem MP_λ and its linear relaxation is denoted LMP_λ.

$$
\begin{cases}
minimize \; \lambda \sum_{r_k \in \Omega} c_k^1 \theta_k + (1 - \lambda) \sum_{r_k \in \Omega} c_k^2 \theta_k \\
\sum_{r_k \in \Omega} a_{ik} \theta_k \geq 1 \; (v_i \in V \backslash \{v_0\}), \\
\sum_{r_k \in \Omega} \theta_k \leq K, \\
\theta_k \in \{0, 1\} \; (r_k \in \Omega).
\end{cases}
\tag{2}
$$

Let introduce the call of the second step of *Baldacci et al.* method for the weight λ: $GENROUTE \; (UB, LB, \lambda)$. In this algorithm, the gap γ is computed with respect to λ: $\gamma = \lambda(c^1(UB) - c^1(LB)) + (1 - \lambda)(c^2(UB) - c^2(LB))$. We also denote $c(S)_\lambda$ as the weighted cost of a point S for the weight λ such that $c(S)_\lambda = \lambda c^1(S) + (1 - \lambda)c^2(S)$.

The algorithm of the *two-step* method returns the set Θ of all non-dominated points of the *BOVRP* and is described in Algorithm 1. It is decomposed in two steps as the algorithm of Ulungu and Teghem [13]. First, the supported points are computed thanks to the function *findSupportedPoint* described in Sect. 3.2. Let S_1 and S_2 denote the optimal solutions that minimize the costs c^1 and c^2 respectively. The aim of the first step is to generate all routes that can conduct to a non-dominated point in the triangle defined by S_1, S_2 and their ideal point $I = (c^1(S_1), c^2(S_2))$. These set of routes are returned as $\overline{\Omega_{supp}}$. All supported points are also returned as they are optimal solutions of MP_λ, for some λ [14]. Figure 1 represents a Pareto front with only supported points S_1 to

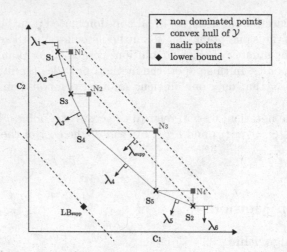

Fig. 1. Example front with supported points and their nadir points.

S_5. S_1 minimizes the first cost c_1 ($\lambda_1 = 1$) and S_2 minimizes the second cost c_2 ($\lambda_6 = 0$). For instance, $\forall \lambda \in [\lambda_2; \lambda_3]$, S_3 is the optimal solution of MP_λ.

Then, the non-supported points are found in areas not explored yet. These areas are triangles defined by two consecutive supported points and their nadir point $N = (c^1(S_2), c^2(S_1))$. The search in a triangle is performed by the function *findAllPoint* detailed in Sect. 3.3.

The intermediate functions *getOptimalSolution* and *gradient* are used in the global algorithm and are described in Algorithms 2 and 3 respectively. The first one takes a direction λ in input and aims to return the optimal solution of MP_λ using the *GENROUTE* algorithm. The other gives the gradient of the line between the two points it receives in input.

3.2   First Step

The first step is defined by the function *findSupportedPoint*(S_1, S_2, λ_{supp}) in Algorithm 4. It needs as input the non-dominated points S_1 and S_2 and a direction λ_{supp}. It generates all routes that can conduct to a non-dominated point situated below the line ($S_1 S_2$).

To do so, we compute the optimal solution LB_{supp} of $LMP_{\lambda_{supp}}$. Then, we apply *GENROUTE* ($UB = S_1$, LB_{supp}, λ_{supp}) to have a reduced set of routes $\overline{\Omega_{supp}}$. By taking the point S_1 as upper bound and the direction λ_{supp}, we ensure that all non-dominated points situated below the line ($S_1\ S_2$) is represented by a combination of routes in $\overline{\Omega_{supp}}$.

Finally, we search all supported points in $\overline{\Omega_{supp}}$ in a dichotomic approach summarized in Algorithm 5.

Algorithm 1. *Two-step* method

Input: A graph G representing the *BOVRP*
Output: A set Θ of all non-dominated points
1: Set $\Theta = \emptyset$;
2: $S_1 \leftarrow getOptimalSolution(1)$; // S_1 the optimal solution minimizing c_1
3: $S_2 \leftarrow getOptimalSolution(0)$; // S_2 the optimal solution minimizing c_2
4: $\Theta = \Theta \cup \{S_1\} \cup \{S_2\}$;
5: $\lambda_{supp} \leftarrow gradient(S_1, S_2)$;
6: $(\Theta, \overline{\Omega_{supp}}) \leftarrow findSupportedPoint(S_1, S_2, \lambda_{supp})$; // return all supported points and a set of routes
7: Compute $c(S_1)_{\lambda_{supp}} = \lambda_{supp} c^1(S_1) + (1 - \lambda_{supp}) c^2(S_2)$;
8: **for** S_i and S_j two consecutive points in Θ such that $c^1(S_i) < c^1(S_j)$ **do**
9:    $\Theta \leftarrow findAllPoint(S_i, S_j, c(S_1)_{\lambda_{supp}}, \lambda_{supp}, \overline{\Omega_{supp}})$;
10: **end for**
11: **return** Θ

Algorithm 2. *getOptimalSolution*(λ)

Input: A value λ
Output: The optimal solution S of MP_λ
1: Solve LMP_λ to obtain a lower bound LB;
2: Solve MP_λ to obtain an upper bound UB;
3: $\overline{\Omega} \leftarrow GENROUTE(UB, LB, \lambda)$;
4: Solve MP_λ on $\overline{\Omega}$ to obtain the optimal solution S;
5: **return** S;

Algorithm 3. *gradient*(S_i, S_j)

Input: Two points S_i and S_j
Output: The gradient of $(S_i S_j)$
1: $\lambda = \frac{abs(c^1(S_j) - c^1(S_i))}{abs(c^2(S_i) - c^1(S_i) - c^2(S_j) + c^1(S_j))}$;
2: **return** λ;

Algorithm 4. *findSupportedPoint*$(S_1, S_2, \lambda_{supp})$

Input: Points S_1 and S_2 and the direction λ_{supp}
Output: Set Θ_{supp} of supported points. Set of routes $\overline{\Omega_{supp}}$
1: Set $\Theta_{supp} = \emptyset$ and $\overline{\Omega_{supp}} = \emptyset$;
2: Solve $LMP_{\lambda_{supp}}$ to obtain LB_{supp};
3: $\overline{\Omega_{supp}} \leftarrow GENROUTE(UB=S_1, LB_{supp}, \lambda_{supp})$;
4: $dichotomicSearch(\Theta_{supp}, S_1, S_2, \overline{\Omega_{supp}})$;
5: **return** $(\Theta_{supp}, \overline{\Omega_{supp}})$;

Algorithm 5. $dichotomicSearch(\Theta_{supp}, S_i, S_j, \overline{\Omega_{supp}})$

Input: Set Θ_{supp}. 2 points S_i and S_j. Set of routes $\overline{\Omega_{supp}}$.

1: $\lambda_i \leftarrow gradient(S_i, S_j)$;
2: Solve MP_{λ_i} on $\overline{\Omega_{supp}}$ to obtain the optimal solution S_k;
3: **if** $S_k \neq S_i$ and $S_k \neq S_j$ **then**
4:     $\Theta_{supp} = \Theta_{supp} \cup \{S_k\}$;
5:     **if** $c^1(S_i) + 1 < c^1(S_k)$ and $c^2(S_i) - 1 > c^2(S_k)$ **then**
6:         $dichotomicSearch(\Theta_{supp}, S_i, S_k, \overline{\Omega_{supp}})$;
7:     **end if**
8:     **if** $c^2(S_j) + 1 < c^2(S_k)$ and $c^1(S_k) - 1 > c^1(S_k)$ **then**
9:         $dichotomicSearch(\Theta_{supp}, S_k, S_j, \overline{\Omega_{supp}})$;
10:    **end if**
11: **end if**

After the first step, all supported non-dominated points are already found. It calls the algorithm *GENROUTE* only once and solves the integer problem for each supported points. However, the set of routes $\overline{\Omega_{supp}}$ contains important information and has to be returned to be used in the second step.

3.3 Second Step

The second step aims to explore each upper right triangle defined by two consecutive non-dominated points computed in the first step and their nadir point. In practice, input data are integers, therefore we substract one to each coordinate of the nadir because non-dominated points having a same coordinate than the nadir are weakly dominated. For instance, in Fig. 1, if a point is located on the segment $[N_1 S_3]$ (resp. $[S_1 N_1]$), it is strictly dominated by S_3 (resp. S_1). The search is performed as explained in Algorithm 6. It requires in input two non-dominated points S_i and S_j such that $c^1(S_i) < c^1(S_j)$. First, a condition has to be checked: if the nadir $N = (c^1(S_j) - 1, c^2(S_i) - 1)$ is such that $c(N)_{\lambda_{supp}} \leq c(S_1)_{\lambda_{supp}}$, then directly apply the function *findInTriangle* to search on the set of known routes $\overline{\Omega_{supp}}$. Indeed, if $c(N)_{\lambda_1} \leq c(S_1)_{\lambda_1}$, we already have generated all routes that can conduct to a non-dominated point in this triangle in the first step. Otherwise, some routes has to be generated before applying *findInTriangle* because the nadir point is upside the line $(S_1 S_2)$.

Figure 1 represents a partial front obtained after the first step. The second step explores the triangles defined by two consecutive points S and their nadir N like $S_3 S_4 N_2$ or $S_4 S_5 N_3$. We already have generated in the first phase all routes that can conduct to a non-dominated point below the dotted line passing through S_1 and S_2. So, all interesting routes for non-dominated points in $S_3 S_4 N_2$ are already generated because the nadir N_2 is under this line. The condition $c(N_2)_{\lambda_{supp}} \leq c(S_1)_{\lambda_{supp}}$ is satisfied. On the contrary, we don't have all the routes that can conduct to a non-dominated point in $S_4 S_5 N_3$ as the dotted line passing through N_3 is above the dotted line passing through S_1 and S_2. So, we have the condition $c(N_3)_{\lambda_{supp}} > c(S_1)_{\lambda_{supp}}$.

Algorithm 6. $findAllPoint(S_i, S_j, c(S_1)_{\lambda_{supp}}, \lambda_{supp}, \overline{\Omega_{supp}})$

Input: Points S_i and S_j. The cost $c(S_1)_{\lambda_{supp}}$, the direction λ_{supp} and set $\overline{\Omega_{supp}}$ of the first step

Output: The set Θ_{nd} of non-dominated points

  Set $N = (c^1(S_j) - 1, c^2(S_i) - 1)$; Set $\Theta_{nd} = \emptyset$

  if $c(N)_{\lambda_{supp}} \leq c(S_1)_{\lambda_{supp}}$ **then**

   $findInTriangle(\Theta_{nd}, S_i, S_j, \overline{\Omega_{supp}})$;

  else

   $\lambda_i \leftarrow Gradient(S_i, S_j)$;

   Solve LMP_{λ_i} to obtain LB_i;

   $\overline{\Omega_i} \leftarrow GENROUTE(N, LB_i, \lambda_i)$;

   $findInTriangle(\Theta_{nd}, S_i, S_j, \overline{\Omega_i})$

  end if

  return Θ_{nd};

The algorithm $findInTriangle(\Theta_{nd}, S_i, S_j, \overline{\Omega})$ is similar to $DichotomicSearch$. It works on the computed set $\overline{\Omega}$ given in input and requires two non-dominated points S_i and S_j and their gradient λ_i. It solves the integer problem MP_{λ_i} on $\overline{\Omega}$ with two additional constraints: $\sum_{r_k \in \Omega} c_k^1 \theta_k \leq c^1(N)$ and $\sum_{r_k \in \Omega} c_k^2 \theta_k \leq c^2(N)$. If there is no optimal solution, it means that there is no non-dominated points in the area and the function stops. Otherwise, if there is an optimal solution S_k, S_k is added to Θ_{nd} and the function $findInTriangle$ is called again for (S_i, S_k) and (S_k, S_j).

4   Computational Experiments

To compare proposed the *two-step* method to the state-of-the-art, we have implemented a *reference* method. It is the more direct way to use the *Baldacci et al.* method in an ϵ-constraint technique as it is classically done in the literature. The algorithm, summarized in Algorithm 7, is based on the ϵ-constraint formulation that aims to minimize the first cost c^1 under the constraint that the second cost has to be lower than a certain value ϵ.

At the beginning, ϵ is set to $+\infty$. At each iteration, the reference method consists in finding a lower bound LB and an upper bound UB. Then, it applies the mono-objective algorithm $GENROUTE(UB, LB)$ to obtain $\overline{\Omega}$, the restricted set of routes with (i) their reduced cost within the gap $\gamma = UB - LB$ and (ii) below the constraints that the second cost is lower than ϵ. The algorithm optimally solves the integer problem restricted to $\overline{\Omega}$. If a new integer solution is found, ϵ is set to the value of the second objective of the solution minus one and the process is repeated. If no new optimal solution is found, the algorithm stops.

Algorithm 7. Algorithm of the reference method

$\epsilon \leftarrow +\infty$
while $\exists$ a solution **do**
   Solve the linear relaxation of the problem for ϵ to obtain LB
   Find a feasible solution UB of the integer problem for ϵ
   $\overline{\Omega} \leftarrow GENROUTE(UB, LB)$
   Solve the integer problem on $\overline{\Omega}$ to obtain S_{OPT}
   if $\exists S_{OPT}$ **then**
      Set $\epsilon \leftarrow S_{OPT}{}^{2} - 1$
   end if
end while

Results. To the best of our knowledge, there is no benchmark for multi-objective vehicle routing problems with different costs on edges. Therefore we propose new instances for the BOVRP with time windows ($BOVRPTW$) which minimizes two different route costs. Each instance is a combination of two Solomon's instances. The first instance provides the first edge costs, the time windows, the charges and the capacities. The second instance only provides the second edge costs. We have tested the algorithm on 20 instances of 25 customers and 20 instances of 50 customers, which correspond to the 25 and 50 first customers of Solomon's instances. Each method returns the minimum complete Pareto Front of the $BOVRPTW$.

The experiments have been conducted on a Xeon E5-2695 processor with a 2.30 GHz CPU and 3.5Go in a single thread. The implementation is in C++ and the linear problems and the integer problems are solved with Gurobi 7.1. The time limit for all experiments is 6 hours. Results are reported in Table 1 for the 27 instances for which at least one of the method tested converged.

Table 1 presents the features of the instances like the number of customers, if it is a clustered instance or not, the number of strict non-dominated points and the number of non-supported non-dominated points in the final Pareto front. It also shows the mean CPU time in seconds on 10 executions as well as their standard deviations for each method. We can notice that the *reference* method dominates the *two-step* method on instances easily solved - inferior to 49 s for the two methods. On the contrary, the *two-step* method dominates the other on harder instances. Furthermore, the instances noted 9, 21, 22 and 23 of 25 customers are only solved by the *two-step* method. 3 of them are clustered instances with very few non-dominated points in the final exact Pareto front. It suggests that the *two-step* method outperforms the state-of-the-art method for graphs with clustered structure.

Table 1 also exhibits the execution time in seconds for the methods in graphs with 50 customers (instances from 24 to 30). The number of non-dominated points is, as expected, larger than for the graphs with 25 clients. The execution of the algorithms is more time consuming and only converges for 2 instances in the *reference* method and for 7 in the *two-step* methods over 20 tested instances.

Table 1. Execution time of the methods for some graphs

#	Instance	Instance features				Reference method		The two-step method	
		Number of customers	Clustered?	Number of non-dominated points	Number of non-supported points	Mean time (s)	σ (s)	Mean time (s)	σ (s)
1	R105_C1	25	no	33	28	31	1	32	1
2	R105_C2	25	no	32	25	24	1	49	2
3	R109_C1	25	no	26	14	1055	130	289	30
4	R109_C2	25	no	28	17	1028	120	464	41
5	R102_RC1	25	no	68	55	-	-	2113	152
6	R105_RC1	25	no	34	28	24	1	44	2
7	R109_RC1	25	no	45	34	1794	187	660	51
8	RC101_R1	25	no	27	16	18	1	42	1
9	RC105_R1	25	no	19	13	193	19	139	10
10	RC106_R1	25	no	23	14	445	57	253	42
11	RC101_C1	25	no	9	3	8	0	8	0
12	RC102_C1	25	no	18	12	3823	382	473	37
13	RC105_C1	25	no	21	14	191	14	49	2
14	RC105_C2	25	no	22	13	204	16	38	2
15	RC106_C1	25	no	14	8	233	18	121	12
16	RC106_C2	25	no	28	19	456	54	198	10
17	C101_C2	25	yes	5	1	10059	782	6321	856
18	C106_R1	25	yes	22	14	-	-	16108	2227
19	C106_RC1	25	yes	6	4	-	-	3850	488
20	C106_C2	25	yes	5	1	-	-	9389	1403
21	R101_RC2	50	no	176	155	762	96	424	28
22	R105_RC2	50	no	139	118	-	-	11654	1220
23	R101_C2	50	no	114	93	476	57	301	21
24	R105_C1	50	no	154	134	-	-	12278	1226
25	R105_C2	50	no	169	145	-	-	11897	1340
26	RC101_R1	50	no	132	116	3651	370	1982	65
27	RC105_C2	50	no	79	58	-	-	17519	2105

5 Conclusion

In this paper, we proposed an exact method to solve the bi-objective vehicle routing problem: the *two-step* method. It scalarizes the objective function and uses column generation to find all non-dominated points, supported and non-supported points. The method is also generic for all classes of *BOVRP* as it doesn't exploit any specific property. To show the efficiency of the scheme proposed we have implemented the ϵ-constraint method combined with the *GENROUTE* component. Computational experiments showed that the *two-step* method outperforms this reference method, especially for clustered graphs.

References

1. Dantzig, G.B., Ramser, J.H.: The truck dispatching problem. Manage. Sci. **6**(1), 80–91 (1959)
2. Jozefowiez, N., Semet, F., Talbi, E.G.: Multi-objective vehicle routing problems. Eur. J. Oper. Res. **189**(2), 293–309 (2008)
3. Parragh, S.N., Tricoire, F.: Branch-and-bound for bi-objective integer programming. Optimization Online (2015)
4. Boland, N., Charkhgard, H., Savelsbergh, M.: The triangle splitting method for biobjective mixed integer programming. In: Lee, J., Vygen, J. (eds.) IPCO 2014. LNCS, vol. 8494, pp. 162–173. Springer, Cham (2014). https://doi.org/10.1007/978-3-319-07557-0_14
5. Moradi, S., Raith, A., Ehrgott, M.: A bi-objective column generation algorithm for the multi-commodity minimum cost flow problem. Eur. J. Oper. Res. **244**(2), 369–378 (2015)
6. Bérubé, J.F., Gendreau, M., Potvin, J.Y.: An exact ϵ-constraint method for bi-objective combinatorial optimization problems: application to the traveling salesman problem with profits. Eur. J. Oper. Res. **194**(1), 39–50 (2009)
7. Özlen, M., Azizoğlu, M.: Multi-objective integer programming: a general approach for generating all non-dominated solutions. Eur. J. Oper. Res. **199**(1), 25–35 (2009)
8. Boland, N., Charkhgard, H., Savelsbergh, M.: A criterion space search algorithm for biobjective integer programming: the balanced box method. INFORMS J. Comput. **27**(4), 735–754 (2015)
9. Dai, R., Charkhgard, H.: A two-stage approach for bi-objective integer linear programming. Oper. Res. Lett. **46**(1), 81–87 (2018)
10. Balinski, M.L., Quandt, R.E.: On an integer program for a delivery problem. Oper. Res. **12**(2), 300–304 (1964)
11. Baldacci, R., Christofides, N., Mingozzi, A.: An exact algorithm for the vehicle routing problem based on the set partitioning formulation with additional cuts. Math. Program. **115**(2), 351–385 (2008)
12. Ehrgott, M.: Multicriteria Optimization, vol. 491. Springer Science & Business Media, Heidelberg (2005)
13. Ulungu, E., Teghem, J.: The two phases method: an efficient procedure to solve bi-objective combinatorial optimization problems. Found. Comput. Dec. Sci. **20**(2), 149–165 (1995)
14. Geoffrion, A.M.: Proper efficiency and the theory of vector maximization. J. Math. Anal. Appl. **22**(3), 618–630 (1968)

Multi-start Local Search Procedure for the Maximum Fire Risk Insured Capital Problem

Maria Isabel Gomes[1]([✉]), Lourdes B. Afonso[1], Nelson Chibeles-Martins[1], and Joana M. Fradinho[2]

[1] Centre of Mathematics and Applications, Faculty of Science and Technology, Nova University of Lisbon, Caparica, Portugal
mirg@fct.unl.pt
[2] Department of Mathematics, Faculty of Science and Technology, Nova University of Lisbon, Caparica, Portugal

Abstract. A recently European Commission regulation requires insurance companies to determine the maximum value of insured fire risk policies of all buildings that are partly or fully located within circle of a radius of 200 m. In this work, we present the multi-start local search meta-heuristics that has been developed to solve the real case of an insurance company having more than 400 thousand insured buildings in mainland Portugal. A random sample of the data set was used and the solutions of the meta-heuristic were compared with the optimal solution of a MILP model based on the Maximal Covering Location Problem. The results show the proposed approach to be very efficient and effective in solving the problem.

Keywords: Meta-heuristics · Local search · Solvency II
Continuous location problem

1 Introduction

Recently the European Union (EU) has published a new legistative programme - Solvency II - aiming at the harmonization of insurance industry across the European market and defining a policyholders protection framework that is risk-sensitive [1]. Among other aspects Solvency II comprises risk-based capital requirements that need to be allocate in order to ensure the financial stability of insurance companies with assets and liabilities valued on a market consistent basis. More precisely the Solvency Capital Requirement (SRC) should reflect a level of eligible own funds that enables the insurance undertakings to absorb significant losses without compromising the fulfilling of its obligations. As a risk-sensitive and prudential regime, Solvency II wants to take into account all possible outcomes, including events of major magnitude. Therefore, the capital requirement for catastrophe risk should assess all possible catastrophes, as

J. Lee et al. (Eds.): ISCO 2018, LNCS 10856, pp. 219–227, 2018.
https://doi.org/10.1007/978-3-319-96151-4_19

natural catastrophes and man-made catastrophes, and establish how these risks should be quantified to integrate the whole.

In this work we will focus on the **man-made catastrophe risk** which comprises extreme events directly accountable to men (as motor vehicle liability risk; marine risk; aviation risk; fire risk; liability risk; credit and suretyship risk). Specifically we will address the capital requirement for **fire risk** (as fire, explosion and acts of terrorism) that should "(...) be equal to the loss in basic own funds of insurance and reinsurance undertakings that would result from an instantaneous loss of an amount that (...) is equal to the sum insured by the insurance or reinsurance undertaking with respect to the largest fire risk concentration". The fire risk assumes 100% damage on the total sum of the capital insured for each building located partly or fully within a 200 m radius [2,3]. Until now, and to best of our knowledge, the choice of 200 m as the radius for the concentration was based on statistics and expert judgment.

This problem can be stated as: "find the centre coordinates of a circle with a fixed radius that maximizes the coverage of total fire risk insured". This problem can be viewed as a particular instance of the Maximal Covering Location Problem (MCLP) with fixed radius [4]. Church and Revelle, in 1974, were the first authors to address the MCLP under the assumption that both demand and possible site locations are discrete points [5]. Mehrez extended this seminal work by proposing a zero-one integer linear formulation considering as possible site locations the entire plan [6]. A widely used approach to the continuous space optimization has been to discretize the demand region, transforming the problem into a discrete location model [7]. Under the assumption that demand point is either covered or not by the facility, it has been proven that a discrete and finite set contains an optimal coverage solution [5].

This work has been motivated by the real case of an insurance company that, having more than 400 thousand buildings in Portugal covered by a fire insurance policy, needs to determine the maximum accumulated risk within a circle with 200 m radius. Each building can be viewed as a "demand point" of the MCLP. Such a number of points leads to "enormous" MILP model which only "super" computers might be able to cope with. Although few meta-heuristic approaches already have been proposed to solve the MCLP, to the best of our knowledge, non of them fits our problem. For instance, Bruno et al. have developed a agent-based framework that could suit this problem. However no computational experiments were performed to access the scalability of the proposed framework to large problem instances [8]. Maximo et al. developed a meta-heuristic, named by the authors Intelligent-Guided Adaptive Search, to deal with large-scale instances where both demand points and possible site locations are discrete points [9]. The assumption of having discrete possible site locations may lead to non-optimal solutions, in our problem context.

An algorithm had to be designed so that the insurance company could use it at least once a year. Therefore, we have developed a meta-heuristic - the **Fire Risk Meta-heuristic** - inspired by the pattern search method proposed by Custódio and Vicente [10] that can be run in an ordinary desktop computer.

This paper will develop as follows. In the next section the meta-heuristic will be described in detail. Then test results will be reported to assess the quality of the proposed approach. Lastly, some conclusions and future work are given.

2   Fire Risk Meta-Heuristic

The **Fire Risk Meta-heuristic** is a multi-start local search procedure where intensification and exploration strategies have been defined. In a nutshell, this procedure can be stated as: given an initial coordinate point (randomly selected) for the circle centre, determine the total fire risk within a k meters radius; generate and evaluate four neighbourhood points by increasing/decreasing each coordinate by a Δ value; make the best neighbourhood point as the new center.

Initial Solution: (x_0, y_0) is randomly selected from the search space where the maximal risk is to be determined.

Objective Function: For a given solution s, the objective function value $f(s)$ is the fire risk covered by the circle with centre in s and radius k.

Neighbourhood Structure: At iteration i and considering a given Δ_i value, compute four new centre coordinates as shown in Fig. 1.

The step size Δ_i varies according to the quality of the neighbour solutions.

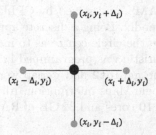

Fig. 1. Neighbourhood structure of a given point (x_i, y_i) [11]

Stopping Rules: Two stopping rules have been defined: one for the local search procedure and one for the multi-start algorithm. For the local search procedure one assumes as stopping criteria a small value for Δ_i, Δ_{min}. The meta-heuristic stopping criteria has two components: minimum number of iterations, i_{min}, and maximum fire risk of a single building, *Best*. The i_{min} is set empirically so that an adequate exploration of the search space is performed. The second component assures that the optimal circle must have a total fire risk (the objective function value) greater than the largest fire risk associated to a single building.

Algorithm Description: Algorithm 1 presents the pseudo-code of the Fire Risk Meta-heuristic. Parameters initialization is performed in lines 1 to 5. The local search procedure is presented between in lines 11 to 26. Given an initial solution s_i and the step size Δ_i, a set of four neighbourhood solutions are generated (line 12). The best of the four (in term of the objective function value) is compared with the value of s_i. If the best neighbourhood has a better value (a success), the current solution moves to neighbouring solution (lines 14 and 15). When the algorithm produces two consecutive successes, the step size is increased; the new step size doubles the previous one (lines 16 and 17). With a larger step size, we aim at broadening the search procedure in "promising" areas of the search space. If none of the four neighbours presents better values (line 21) the current solution does not change and the step size decreases - the idea is to intensify the search near the current solution, since it might be a local optimum (lines 22 and 23). The local search procedure stops when the step size is small enough (line 26) and the most promising solution found at the moment is update (or not) - line 27. This local search algorithm is embedded in the multi-start procedure. In the restart step, the random initial solution of iteration i (s_i) is generated and the step size is reset to Δ_0 (lines 7 to 10). The multi-start algorithm ends when both stopping criteria are met.

3   Results

In order to assess the Fire Risk Meta-heuristic solution, the MILP formulation for the maximal covering location problem (MCLP) was adapted from Farahani et al. [12], formulated in GAMS, and solved by CPLEX to optimality [13]. The formulation is given in Appendix. Being a discrete approach, it provides a lower bound to our problem, since the circle centre as to match one of the buildings.

The Fire Risk Meta-heuristic was programmed in R software, a requisite of the insurance company. The ggplot2 and ggmap were the packages chosen to plot all the maps using Google Maps information. All the results were obtained by a PC with Intel® Xeon 10 cores and 32 GB of RAM.

Data

Given the volume of data for a national study and being this work a first step towards the development of an optimization approach, we confined the study to the Lisbon area. The Insurance Company provided a data set which encompasses the chosen geographical area and has 46 843 buildings (points). Each points is defined by the two geographical coordinates (longitude and latitude) and the fire capital insured.

The MILP model is unable to solve real instance with 46 thousand points since it leads to out of memory issues. Our approach was then to select two random samples of tractable size (1000 points): samples A and B. Figure 2 shows the distribution of the insured capitals over the samples areas. Notice that no extreme point exists on sample A (by extreme we mean a building with an insured capital so large that it will obviously belong the optimal circle), while in sample B there are a few of such points. Figure 3 depicts the geographical location of the sample A.

Algorithm 1. Fire Risk Meta-heuristic

1: *Best   Find maximum risk of a single building*
2: Δ_0 *Set step size*
3: s_0 *Generate initial solution*
4: $s^* = s_0$
5: $i = 0$ *Iteration counter*
6: **While** $i < i_{min} \vee f(s^*) < Best$ **do**
7:    **If** $i > 0$ **then**
8:       s_i *Generate iteration i initial solution*
9:       $\Delta_i = \Delta_0$ *Set step size*
10:    **EndIf**
11:    **Do**
12:       $\mathcal{N}(s_i)$ *Generation of candidate neighbours*
13:       s' *Set the best neighbour $s' \in \mathcal{N}(s_i)$*
14:       **If** $f(s') > f(s_i)$ is a better neighbour (success) **then**
15:          $s_{i+1} = s'$
16:          **If** *this is the second consecutive success* **then**
17:             $\Delta_{i+1} = 2 \cdot \Delta_i$
18:          **else**
19:             $\Delta_{i+1} = \Delta_i$
20:          **EndIf**
21:       **else** *(unsuccess)*
22:          $s_{i+1} = s_i$
23:          $\Delta_{i+1} = \dfrac{\Delta_i}{2}$
24:       **EndIf**
25:       $i = i + 1$
26:    **Until** $\Delta_{i-1} < \Delta_{min}$
27: **If** $f(s_i) > f(s^*)$ **then** $s^* = s_i$
28: **EndWhile**
29: **Return** $s^*, f(s^*)$

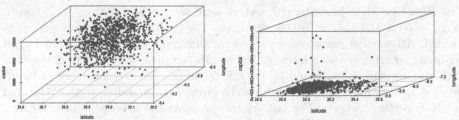

Fig. 2. Distribution of the 1000 insured capitals for samples A (*left*) and B (*right*)

In terms of descriptive statistics both samples also present differences. Table 1 shows the maximum and minimum fire insured capital, the amplitude, the average and standard deviation within each sample. With these two sets of points, we aimed at assessing the capability of the algorithm for overcoming the trap of local optimality.

Fig. 3. Geographical location of the 1000 points

Table 1. Samples descriptive statistics (€)

Sample	Maximum	Minimum	Amplitude	Average	Stand. dev.
A	14490,6	239,2	14251,4	10342,7	2785,9
B	5112903,0	5833,9	5107069,1	123052,0	335729,9

Meta-heuristic validation

The Risk Fire Meta-heuristic parameters were set to $\Delta_0 = 200 \cdot 2^7 = 25\,600$ m, $\Delta_{min} = 50$ m and $i_{min} = 50\,000$ iterations. For each sample 20 runs were performed. All starting point coordinates were randomly chosen while assuring an adequate covering the search space. The results are presented in Table 2 together with the optimal values found by GAMS/CPLEX. The columns "no. of buildings" show how many builds lie within the circle corresponding to the "Optimal value" (or "Best value"). The last column presents the average number of iterations needed to find the "Best value".

For sample A, the optimal fire risk found by GMAS/CPLEX is of € 28 857.46 and two buildings are covered. The algorithm was able to find a better value for the covered fire risk (€ 33 323, 56 *vs.* € 28 857, 46). In fact, the meta-heuristic only found two different values (not shown): the best value was found in 19 out of the 20 runs, and in one single run the same value as the one proposed by GAMS/CPLEX was found. As mentioned above, the MILP model assumes that the circle centre as being one of the building, while the meta-heuristic allows it to be placed in any point of the plane. The highest fire risk value improves in 15%

Table 2. Results (€)

GAMS/CPLEX			Fire-risk Meta-heuristic		
Sample	Optimal value	No. of buildings	Best value	No. of buildings	Average no. of iterations
A	28857,46	2	33233,56	3	2785,9
B	5112903,0	1	5112903,0	1	30431

the optimal value found by the MILP model. Figure 4 shows the two buildings covered of the MILP optimal solution (on the left). On the right side of Fig. 4, one depicts the three building covered by the 19 runs that provided the best risk value. Lastly, we should refer that the meta-heuristic took on average 54 s per run and GMAS/CPLEX needed about 0.03 s to reach the optimal solution.

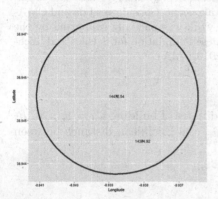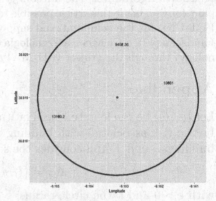

Fig. 4. Sample A: optimal solution from the MILP model (*left*) and meta-heuristic best solution (*right*)

Regarding sample B, one should emphasise that the optimal value was found by the meta-heuristic in all the 20 runs performed. These results suggest the proposed algorithm is able to escape from local optima. Additional tests are nonetheless needed to support this suggestion.

4 Final Remarks and Future Work

In this work we proposed a new meta-heuristic approach to determine the coordinates of a k radius circle that covers the maximum fire risk according to the EU legislative programme Solvency II. Until last year, insurance companies reported only the largest capital that covers one single insured building. The new legislative programme demands companies to solve a much harder problem, which can only be done using computational tools.

In the two tested samples, the Risk Fire Meta-heuristic found optimal value or even a better value than the best one computed by a MILP model. Notice, this MILP model provides a lower-bound for the problem since it is a discrete based model and, therefore, not allowing the circle centre coordinates to be defined outside the existing set of points.

The meta-heuristic has been applied to a data set of more than 46 thousand points [14] and it is currently being used by the insurance company.

As future work we will pursue three main directions. First, we will extended the computational experiments and assess the meta-heuristic solution quality with regard to the optimal solution provided by 01 integer formulation proposed by Mehrez, in 1983 [6]. Second, additional computational experiments will be performed so that statistical tests can be used to validate the usefulness of this method. Third, the computational performance of the algorithm will also be investigated to overcome the long times it is currently taking.

Acknowledgements. The authors gratefully acknowledge Liberty Insurance Company for who made this article possible to Magentakoncept–Consultores Lda and CMA-FCT-UNL for the computational support. This work was partially supported by the Fundação para a Ciência e a Tecnologia (Portuguese Foundation for Science and Technology) through the project UID/MAT/00297/2013(CMA).

Appendix

Let (a_i, b_i) be the longitude and latitude coordinates of building i, $i = 1, ..., n$, R_i be the risk associated with building i, and D_{ij} the Euclidean distance between buildings i and j, and consider the set

$$\Delta_{ij} = \{(i,j) : D_{ij} \leq 2k + \epsilon\},$$

with $\epsilon > 0$ and k the circle radius.

Let two binary variables be defined as: $x_{ij} = 1$ if building i is covered by the circle centred at j, 0 otherwise; and $y_j = 1$ if j is the centre of the circle.

$$\max \quad \sum_{ij \in \Delta_{ij}} R_{ij} x_{ij}$$

$$\text{s.t.} \quad D_{ij} x_{ij} \leq k y_i \quad (i,j) \in \Delta_{ij}$$

$$\sum_{i:=1}^{n} y_i = 1$$

$$x_{ii} \leq y_i \quad i \in \{1, ..., n\}$$

$$x_{ij}, y_i \in \{0, 1\}$$

The objective function is defined by the sum of the fire risks insured of the buildings covered by the circle. The first constraint assures that only the buildings distancing less that k from the circle centre y_j will be considered. The second constraint assures that only one circle is determined. The last constraint is needed since one has a maximization model. Notice that $D_{ii} = 0$, $i = 1, ..., n$. Therefore, all $x_{ii} = 1$ will verify the first constraint, whatever the value of y_i.

References

1. Loyds, L.: What is solvency II? (2016). https://www.lloyds.com/the-market/operating-at-lloyds/solvency-ii/about/what-is-solvency-ii
2. European Insurance and Occupational Pensions Authority. The underlying assumptions in the standard formula for the Solvency Capital Requirement calculation, 25 July 2014 (2014)
3. Regulations commission delegated regulation (eu) 2015/ec of 10 October 2014 supplementing Directive 2009/138/EC of the European Parliament and of the Council of 25 November 2009 on the taking-up and pursuit of the business of Insurance and Reinsurance (Solvency II)
4. Plastria, F.: Continuous covering location problems. In: Drezner, Z., Hamacher, H.W. (eds.) Facility Location: Applications and Theory, pp. 37–79. Springer, Berlin (2012)
5. Church, R.L., ReVelle, C.S.: The maximal covering location problem. Pap. Reg. Sci. Assoc. **32**(1), 101–118 (1974)
6. Mehrez, A.: A note on the linear integer formulation of the maximal covering location problem with facility placement on the entire plane. J. Reg. Sci. **23**, 553–555 (1983)
7. Wei, R., Murray, A.T.: Continuous space maximal coverage: insights, advances and challenges. Comput. Oper. Res. **62**, 325–336 (2015)
8. Bruno, G., Genovese, A., Sgalambro, A.: An agent-based framework for modeling and solving location problems. Topics **18**, 81–96 (2010)
9. Máximo, V.R., Nascimento, M.C.V., Carvalho, A.C.P.L.F.: Intelligent-guided adaptive search for the maximum covering location problem. Comput. Oper. Res. **78**, 129–137 (2017)
10. Custódio, A.L., Vicente, L.N.: Using sampling and simplex derivatives in pattern search methods. SIAM J. Optim. **18**(2), 537–555 (2007)
11. Fradinho, J.M.: Metaheuristic method to evaluate the fire risk sub-module in Solvency II. Master thesis on Applied Mathematics (Actuarial Science, Statistics and Operations Research). Faculty of Science and Technology, Nova University of Lisbon (2018)
12. Farahani, R.Z., Asgari, N., Heidari, N., Hosseininia, M., Goh, M.: Covering problems in facility location: a review. Comput. Ind. Eng. **62**, 368–407 (2012)
13. GAMS Development Corporation. General Algebraic Modeling System (GAMS) Release 24.2.3. Washington, DC, USA (2014)
14. Afonso, L.B., Fradinho, J.M. Chibeles-Martins, N., Gomes, M.I.: Metaheuristic method to evaluate the fire risk sub-module in Solvency II (2018)

A Branch-and-Bound Procedure
for the Robust Cyclic Job Shop Problem

Idir Hamaz[1](✉), Laurent Houssin[1](✉), and Sonia Cafieri[2](✉)

[1] LAAS-CNRS, Universite de Toulouse, CNRS, UPS, Toulouse, France
{ihamaz,lhoussin}@laas.fr
[2] ENAC, Universite de Toulouse, 31055 Toulouse, France
sonia.cafieri@enac.fr

Abstract. This paper deals with the cyclic job shop problem where
the task durations are uncertain and belong to a polyhedral uncertainty
set. We formulate the cyclic job shop problem as a two-stage robust
optimization model. The cycle time and the execution order of tasks
executed on the same machines correspond to the *here-and-now* decisions
and have to be decided before the realization of the uncertainty. The
starting times of tasks corresponding to the *wait-and-see* decisions are
delayed and can be adjusted after the uncertain parameters are known.
In the last decades, different solution approaches have been developed
for two-stage robust optimization problems. Among them, the use of
affine policies, column generation algorithms, row and row-and-column
generation algorithms. In this paper, we propose a Branch-and-Bound
algorithm to tackle the robust cyclic job shop problem with cycle time
minimization. The algorithm uses, at each node of the search tree, a
robust version of the Howard's algorithm to derive a lower bound on the
optimal cycle time. We also develop a heuristic method that permits to
compute an initial upper bound for the cycle time. Finally, encouraging
preliminary results on numerical experiments performed on randomly
generated instances are presented.

Keywords: Cyclic job shop problem · Robust optimization
Branch-and-Bound algorithm

1 Introduction

Most models for scheduling problems assume deterministic parameters. In contrast, real world scheduling problems are often subject to many sources of uncertainty. For instance, activity durations can decrease or increase, machines can
break down, new activities can be incorporated, *etc.* In this paper, we focus on
scheduling problems that are cyclic and where activity durations are affected by
uncertainty. Indeed, the best solution for a deterministic problem can quickly
become the worst one in the presence of uncertainties.

In this work, we focus on the *Cyclic Job Shop Problem* (CJSP) where processing times are affected by uncertainty. Several studies have been conducted on the

© Springer International Publishing AG, part of Springer Nature 2018
J. Lee et al. (Eds.): ISCO 2018, LNCS 10856, pp. 228–240, 2018.
https://doi.org/10.1007/978-3-319-96151-4_20

CJSP in its deterministic setting. The CJSP with identical jobs is studied in [1] and the author shows that the problem is $\mathcal{NP}$-hard and proposes a Branch-and-Bound algorithm to solve the problem. The more general CJSP is investigated in [2], where the author proposes a mixed linear integer programming formulation and presents a Branch-and-Bound procedure to tackle the problem. A general framework for modeling and solving cyclic scheduling problems is presented in [3]. The authors present different models for cyclic versions of CJSP. However, a few works consider cyclic scheduling problems under uncertainty. The cyclic hoist scheduling problem with processing time window constraints where the hoist transportation times are uncertain has been investigated by Che et al. [4]. The authors define a robustness measure for cyclic hoist schedule and present a bi-objective mixed integer linear programming model to optimize both the cycle time and the robustness.

Two general frameworks have been introduced to tackle optimization problems under uncertainty. These frameworks are Stochastic Programming (SP) and Robust Optimization (RO). The main difference between the two approaches is that the Stochastic Programming requires the probability description of the uncertain parameters while RO does not. In this paper, we focus on the RO paradigm. More precisely, we model the robust CJSP as a two-stage RO problem. The cycle time and the execution order of tasks on the machines corresponding to the *here-and-now* decisions have to be decided before the realization of the uncertainty, while the starting times of tasks corresponding to the *wait-and see* decisions are delayed and can be adjusted after the uncertain parameters are known. In recent years there has been a growing interest in the two-stage RO and in the multi-stage RO in general. The two-stage RO is introduced in [5], referred to as *adjustable optimization*, to address the over-conservatism of single stage RO models. Unfortunately, the two-stage RO problems tend to be intractable [5]. In order to deal with this issue, the use of affine policies ([5]) and decomposition algorithms ([6–8]) have been proposed.

This paper deals with the CJSP where the task durations are uncertain and belong to a polyhedral uncertainty set. The objective is to find a minimum cycle time and an execution order of tasks executed on the same machines such that a schedule exists for each possible scenario in the uncertainty set. To tackle the problem we design a Branch-and-Bound algorithm. More precisely, at each node of the search tree, we solve a robust Basic Cyclic Scheduling Problem (BCSP), which corresponds to the CJSP without resource constraints, using a robust version of Howard's algorithm to get a lower bound. We also propose a heuristic algorithm to find an initial upper bound on the cycle time. Finally, we provide results on numerical experiments performed on randomly generated instances.

This paper is structured as follows. In Sect. 2, we present both the Basic Cyclic Scheduling Problem and the Cyclic Job Shop Problem in their deterministic case and introduce the polyhedral uncertainty set considered in this study. Section 3 describes a Branch-and-Bound ($B\&B$) procedure to solve the robust CJSP. Numerical experiments performed on randomly generated instances are reported and discussed in Sect. 4. Finally, some concluding remarks and perspectives are drawn in Sect. 5.

2   The Cyclic Scheduling Problems

In this section, we first introduce the Basic Cyclic Scheduling Problem which corresponds to the CJSP without resource constraints. This problem will represent a basis for the Branch-and-Bound method designed for the robust CJSP solving. Next, we present the CJSP in its deterministic case. Finally, we present the uncertainty set that we consider in this paper and we formulate the CJSP with uncertain processing times as a two-stage robust optimization problem.

2.1   Basic Cyclic Scheduling Problem (BCSP)

Let $\mathcal{T} = \{1, ..., n\}$ be a set of n generic operations. Each operation $i \in \mathcal{T}$ has a processing time p_i and must be performed infinitely often. We denote by $< i, k >$ The kth occurrence of the operation i and by $t(i, k)$ the starting time of $< i, k >$.

The operations are linked by a set $\mathcal{P}$ of *precedence constraints* (uniform constraints) given by

$$t(i, k) + p_i \leqslant t(j, k + H_{ij}), \quad \forall (i, j) \in \mathcal{P}, \ \forall k \geq 1, \tag{1}$$

where i and j are two generic tasks and H_{ij} is an integer representing the depth of recurrence, usually referred to as *height*.

Furthermore, two successive occurrences of the same task i are not allowed to overlap. This constraint corresponds to the non-reentrance constraint and can be modeled as a uniform constraint with a height $H_{ii} = 1$.

A schedule S is an assignment of starting time $t(i, k)$ for each occurrence $< i, k >$ of tasks $i \in \mathcal{T}$ such that the precedence constraints are met. A schedule S is called *periodic* with cycle time α if it satisfies

$$t(i, k) = t(i, 0) + \alpha k, \quad \forall i \in \mathcal{T}, \ \forall k \geq 1. \tag{2}$$

For the sake of simplicity, we denote by t_i the starting time of the occurrence $< i, 0 >$. Since the schedule is periodic, a schedule can be completely defined by the vector of the starting times $(t_i)_{i \in \mathcal{T}}$ and the cycle time α.

The objective of the BCSP is to find a schedule that minimizes the cycle time α while satisfying precedence constraints. Note that other objective functions can be considered, such as work-in-process minimization [2].

A directed graph $G = (\mathcal{T}, \mathcal{P})$, called *uniform graph*, can be associated with a BCSP such that each node $v \in \mathcal{T}$ (resp. arc $(i, j) \in \mathcal{P}$) corresponds to a generic task (resp. uniform constraint) in the BCSP. Each arc $(i, j) \in \mathcal{P}$ is labeled with two values, a *length* $L_{ij} = p_i$ and a *height* H_{ij}.

We denote by $L(c)$ (resp. $H(c)$) the length (resp. height) of a circuit c in graph G, representing the sum of lengths (resp. heights) of the arcs composing the circuit c.

Let us recall the necessary and sufficient condition for the existence of a feasible schedule.

Theorem 1 (Hanen C. [2]). *There exists a feasible schedule if and only if any circuit of G has a positive height.*

A graph that satisfies the condition of Theorem 1 is called consistent. In the following, we assume that the graph G is always consistent. In other words, a feasible schedule always exists.

The minimum cycle time is given by the maximum circuit ratio of the graph G that is defined by

$$\alpha = \max_{c \in \mathcal{C}} \frac{L(c)}{H(c)}$$

where $\mathcal{C}$ is the set of all circuits in G. The circuit c with the maximum circuit ratio is called a *critical circuit*. Thus, the identification of the critical circuit in graph G allows one to compute the minimum cycle time.

Many algorithms for the computation of the cycle time and the critical circuit can be found in the literature. A binary search algorithm with time complexity $\mathcal{O}(nm \left(log(n) + log(\max_{(i,j) \in E}(L_{ij}, H_{ij}))\right))$ has been proposed in [9]. Experimental study about maximum circuit ratio algorithms has been presented in [10]. This study shows that the Howard's algorithm is the most efficient among the tested algorithms.

Once the optimal cycle time α is determined by one of the algorithms cited above, the optimal periodic schedule can obtained by computing the longest path in the graph $G = (\mathcal{T}, \mathcal{P})$ where each arc $(i, j) \in \mathcal{P}$ is weighted by $p_i - \alpha H_{ij}$.

The BCSP can also be solved by using the following linear program:

$$\min \quad \alpha \tag{3}$$
$$s.t. \ t_j - t_i + \alpha H_{ij} \geq p_i \ \forall (i, j) \in \mathcal{P} \tag{4}$$

where t_i represents $t(i, 0)$, i.e., the starting time of the first occurrence of the task i. Note that the precedence constraints (4) are obtained by replacing in (1) the expression of $t(i, k)$ given in (2).

2.2 Cyclic Job Shop Problem (CJSP)

In the present work, we focus on the cyclic job shop problem (CJSP). Contrary to the BCSP, in the CJSP, the number of machines is lower than the number of tasks. As a result, an execution order of the operations executed on the same machine have to be determined.

Each occurrence of an operation $i \in \mathcal{T} = \{1, ..., n\}$ has to be executed, without preemption, on the machine $M_{(i)} \in \mathcal{M} = \{1, ..., m\}$. Operations are grouped on a set of jobs $\mathcal{J}$, where a job $j \in \mathcal{J}$ represents a sequence of generic operations that must be executed in a given order. To avoid overlapping between the tasks executed on the same machine, for each pair of operations i and j where $M_{(i)} = M_{(j)}$, the following *disjunctive constraint* holds

$$\forall i, j \ s.t. \ M_{(i)} = M_{(j)}, \ \forall k, l \in \mathbb{N} : t(i, k) \leq t(j, l) \Rightarrow t(i, k) + p_i \leq t(j, l) \tag{5}$$

To summarize, the CJSP is defined by

- a set $\mathcal{T} = \{1, ..., n\}$ of n generic tasks,
- a set $\mathcal{M} = \{1, ..., m\}$ of m machines,
- each task $i \in \mathcal{T}$ has a processing time p_i and has to be executed on the machine $M_{(i)} \in \mathcal{M}$,
- a set $\mathcal{P}$ of precedence constraints,
- a set $\mathcal{D}$ of disjunctive constraints that occur when two tasks are mapped on the same machine,
- a set $\mathcal{J}$ of jobs corresponding to a sequence of elementary tasks. More precisely, a job J_j defines a sequence $J_j = O_{j,1} ... O_{j,k}$ of operations that have to be executed in that order.

The CJSP can be represented by directed graph $G = (\mathcal{T}, \mathcal{P} \cup \mathcal{D})$, called *disjunctive graph*. The sequence of operations that belong to the same job are linked by uniform arcs in $\mathcal{P}$ where the heights are equal to 0. Additionally, for each pair of operations i and j executed on the same machine, a disjunctive pair of arcs (i, j) and (j, i) occurs. These arcs are labeled respectively with $L_{ij} = p_i$ and $H_{ij} = K_{ij}$, and $L_{ji} = p_j$ and $H_{ji} = K_{ji}$ where K_{ij} is an occurrence shift variable to determine that satisfy $K_{ij} + K_{ji} = 1$ (see [2] for further details). Note that the K_{ij} variables are integer variables and not binary variables as is the case for the non-cyclic job shop problem. Two dummy nodes s and e representing respectively the start and the end of the execution are added to the graph. An additional arc (e,s) with $L_{es} = 0$ and $H_{ij} = WIP$ is considered. The WIP parameter is an integer, called a work-in-process, and represents the number of occurrences of a job concurrently executed in the system.

A lower bound on each occurrence shift value K_{ij} that makes the graph G consistent can be obtained as follows (see [2, 11]):

$$K_{ij}^- = 1 - min\{H(\mu) \,|\, \mu \text{ is a path from } j \text{ to } i \text{ in } G = (\mathcal{T}, \mathcal{P} \cup \emptyset)\}. \qquad (6)$$

Since $K_{ij} + K_{ji} = 1$, one can deduce an upper bound:

$$K_{ij}^- \le K_{ij} \le 1 - K_{ji}^-. \qquad (7)$$

The objective of the problem is to find an assignment of all the occurrence shifts, in other words, determining an order on the execution of operations mapped to the same machine such that the cycle time is minimum. Note that, once the occurrence shifts are determined, the minimum cycle time can be obtained by computing the critical circuit of the associated graph G.

2.3 CJSP Problem with Uncertain Processing Times ($\mathcal{U}^\Gamma$-CJSP)

We define the uncertainty set through the *budget of uncertainty* concept introduced in [12]. The processing times $(p_i)_{i \in \mathcal{T}}$ are uncertain and each processing time p_i belongs to the interval $[\bar{p}_i, \bar{p}_i + \hat{p}_i]$, where $\bar{p}_i$ is the nominal value and $\hat{p}_i$ the deviation of the processing time p_i from its nominal value. We associate a binary variable ξ_i to each operation $i \in \mathcal{T}$. The variable ξ_i is equal to 1 if the processing time of the operation i takes its worst-case value, 0 otherwise. For a given budget of uncertainty Γ, that is a positive integer representing the maximum number of tasks allowed to take their worst-case values, the processing time deviations can be modeled trough the following uncertainty set:

$$\mathcal{U}^\Gamma = \left\{ (p_i)_{i \in \mathcal{T}} \in \mathbb{R}^n : p_i = \bar{p}_i + \hat{p}_i \xi_i, \, \forall i \in \mathcal{T}; \, \xi_i \in \{0,1\}; \, \sum_{i \in \mathcal{T}} \xi_i \leq \Gamma \right\}.$$

The BCSP problem under the uncertainty set $\mathcal{U}^\Gamma$ is studied in [13]. Three exact algorithms are proposed to solve the problem. Two of them use a negative circuit detection algorithm as a subroutine and the last one is a Howard's algorithm adaptation. Results of numerical experiments show that the Howard algorithm adaptation yields efficient results.

The problem we want to solve in this study can be casted as follows:

$$\min \quad \alpha \tag{8}$$

$$s.t. \, \forall p \in \mathcal{U}^\Gamma : \exists t \geq 0 \begin{cases} t_j - t_i + \alpha H_{ij} \geq p_i & \forall (i,j) \in \mathcal{P} \\ t_j - t_i + \alpha K_{ij} \geq p_i & \forall (i,j) \in \mathcal{D} \end{cases} \tag{9}$$

$$K_{ij} + K_{ji} = 1 \qquad \forall (i,j) \in \mathcal{D} \tag{10}$$

$$K_{ij}^- \leq K_{ij} \leq 1 - K_{ji}^- \qquad \forall (i,j) \in \mathcal{D} \tag{11}$$

$$K_{ij} \in \mathbb{Z} \qquad \forall (i,j) \in \mathcal{D} \tag{12}$$

$$\alpha \geq 0 \tag{13}$$

In other words, we aim to find a cycle α and occurrence shifts $(K_{ij})_{(i,j) \in \mathcal{D}}$ such that, for each possible value of the processing times $p \in \mathcal{U}^\Gamma$, there always exists a feasible vector of starting time $(t_i)_{i \in \mathcal{T}}$.

Note that, once the occurrence shifts are fixed, the problem can be solved as a robust BCSP by using the algorithms described in [13]. The following theorem characterizes the value of the optimal cycle time for $\mathcal{U}^\Gamma$-CJSP:

Theorem 2 ([13]). *The optimal cycle time α of the $\mathcal{U}^\Gamma$-CJSP is characterized by*

$$\alpha = \max_{c \in \mathcal{C}} \left\{ \frac{\sum_{(i,j) \in c} \bar{L}_{ij}}{\sum_{(i,j) \in c} H_{ij}} + \max_{\xi : \sum_{i \in \mathcal{T}} \xi_i \leq \Gamma} \left\{ \frac{\sum_{(i,j) \in c} \hat{L}_{ij} \xi_i}{\sum_{(i,j) \in c} H_{ij}} \right\} \right\},$$

where $\bar{L}_{ij} = \bar{p}_i$, $\hat{L}_{ij} = \hat{p}_i$ and $\mathcal{C}$ is the set of all circuits in G.

3   Branch-and-Bound Method

We develop a Branch-and-Bound algorithm for solving $\mathcal{U}^\Gamma$-CJSP. Each node of the Branch-and-Bound corresponds to a subproblem defined by the subgraph $G_s = (\mathcal{T}, \mathcal{P} \cup \mathcal{D}_s)$, where $\mathcal{D}_s \subseteq \mathcal{D}$ is a subset of occurrence shifts already fixed.

The algorithm starts with a root node G_{root} where $\mathcal{D}_{root} = \emptyset$, in other words, no occurrence shifts are fixed. The branching is performed by fixing an undetermined occurrence shift K_{ij} and creates a child node for each possible value of K_{ij} in $[K_{ij}^-, 1 - K_{ji}^-]$. Each of these nodes is evaluated by computing the associated cycle time, such that a schedule exists for each $p \in \mathcal{U}^\Gamma$. This evaluation is made by means of the robust version of Howard's algorithm. Our method explores the search tree in best-first search (BeFS) manner, and, in order to branch, it chooses the node having the smallest lower bound. This search strategy can lead to a good feasible solution. A feasible solution is reached when all occurrence shifts are determined. Note that the nominal starting times (i.e. the starting times when no deviation occurs) can be determined by computing the longest path in the graph G where each arc (i, j) is valued by $p_i - \alpha H_{ij}$, and the adjustment is accomplished by shifting the starting of the following tasks by the value of the deviation. More details are provided in the next subsections.

3.1   Computation of an Initial Upper Bound of the Cycle Time

In order to compute an initial upper bound, we design a heuristic that combines a greedy algorithm with a local search. The greedy algorithm assigns randomly a value to a given occurrence shift K_{ij} in the interval $[K_{ij}^-, 1 - K_{ji}^-]$, and updates the bounds on the rest of the occurrences shifts such that the graph remains consistent. These two operations are repeated until all occurrence shifts are determined. Once all occurrence shifts are determined, a feasible schedule is obtained, consequently the associated optimal cycle time represents an upper bound of the global optimal cycle time. The local search algorithm consists in improving the cycle time by adjusting the values of the occurrence shifts that belong to the critical circuit. The idea behind these improvements is justified by the following proposition:

Proposition 1. *Let $(K_{ij})_{(i,j) \in \mathcal{D}}$ be a vector of feasible occurrence shifts and $\bar{\alpha}$ the associated cycle time given by the critical circuit c. Let $(u, v) \in \mathcal{D}$ be a disjunctive arc such that $(u, v) \in c$. If the following relation holds:*

$$\max_{l \in P^{uv}} \max_{p \in \mathcal{U}^\Gamma} \sum_{(i,j) \in l} p_i - \bar{\alpha} H_{ij} + p_v - \bar{\alpha}(K_{vu} - 1) \leq 0, \tag{14}$$

where P^{uv} is the set of paths from u to v, then the solution $(K'_{ij})_{(i,j) \in \mathcal{D}}$ where $K'_{uv} = K_{uv} + 1$ and $K'_{vu} = K_{vu} - 1$ has a cycle time less or equal to $\bar{\alpha}$.

Proof. Let $(K_{ij})_{(i,j) \in \mathcal{D}}$ be a vector of feasible occurrence shifts, $\bar{\alpha}$ the associated cycle time given by the critical circuit c and $(u, v) \in \mathcal{D}$ a disjunctive arc that belongs to c. Let us assume that relation (14) is verified. It is easily seen that putting $K'_{uv} = K_{uv} + 1$ makes the height of the circuit c increase by one and consequently makes decrease the value of its circuit ratio. In order to maintain the condition $K'_{uv} + K'_{vu} = 1$ verified, increasing the value of K_{uv} by one involve decreasing the value of K_{vu} by one. Now, it follows that decreasing the value of K_{vu} by one must ensure that the values of the circuits passing through the disjunctive arc (u, v) do not exceed $\bar{\alpha}$. This condition is verified, because by (14) we have:

$$\max_{l \in P^{uv}} \frac{\max_{p \in \mathcal{U}^{\Gamma}} \sum_{(i,j) \in l} p_i + p_v}{\sum_{(i,j) \in l} H_{ij} + (K_{vu} - 1)} \leq \bar{\alpha}$$

In other words, the maximum circuit ratio passing by the disjunctive arc (j, i) has a value less or equal to $\bar{\alpha}$. Moreover, since the value of $\bar{\alpha}$ and the values of the processing times are positives, then $\sum_{(i,j) \in l} H_{ij} + (K_{vu} - 1) > 1$. This ensure that the associated graph to the robust CJSP is still consistent and the solution $(K'_{ij})_{(i,j) \in \mathcal{D}}$ is feasible. $\qquad\square$

The pseudo-code of the proposed heuristic is given in Algorithm 1.

Algorithm 1. Initial upper bound computation

1: Compute a lower bounds on the occurrences shifts K_{ij};
2: **for all** $(i, j) \in \mathcal{D}$ **do**
3:     Update bounds on the occurrence shifts;
4:     Affect randomly value to K_{ij} on the interval $[K_{ij}^-, 1 - K_{ji}^-]$;
5: **end for**
6: Compute the associated cycle time $\bar{\alpha}$ and the critical circuit c.
7: **while** $it < it_{max}$ **do**
8:     Let $(u, v) \in \{(u, v) \in \mathcal{D}$ such that $(u, v) \in c\}$;
9:     $l_{uv} \leftarrow \max_{l \in P^{uv}} \max_{p \in \mathcal{U}^{\Gamma}} \sum_{(i,j) \in l} p_i - \bar{\alpha} H_{ij}$;
10:     **if** $l_{uv} + p_v - \bar{\alpha}(K_{vu} - 1) \leq 0$ **then**
11:         $K_{uv} \leftarrow K_{uv} + 1$;
12:         $K_{vu} \leftarrow K_{vu} - 1$;
13:     **end if**
14:     Compute the associated cycle time $\bar{\alpha}$ and the critical circuit c;
15:     it$\leftarrow$ it+1;
16: **end while**

3.2 Lower Bound

In the Branch-and-Bound algorithm, an initial lower bound is derived and compared to the incumbent. If the value of the initial lower bound and the value of the incumbent are equal, then an optimal solution is obtained and the Branch-and-Bound is stopped. It is easily seen that the problem where the disjunctive arcs are ignored is a relaxation of the initial problem. Consequently, the associated cycle time, α_{basic}, is a lower bound on the optimal cycle time. Furthermore, an other lower bound can be computed by reasoning on the machine charges. Let $M_{(i)} \in \mathcal{M}$ be a given machine and $S \subseteq \mathcal{T}$ the set of operations mapped on the machine $M_{(i)}$, then the optimal cycle time $\alpha_{opt} \geq \sum_{i \in S} p_i$, for each $p \in \mathcal{U}^{\Gamma}$.

Since this relation is verified for each machine, one can deduce the following lower bound:

$$\alpha_{machine} = \max_{m \in \mathcal{M}, p \in \mathcal{U}^\Gamma} \left\{ \sum_{i \in \mathcal{T} : M_{(i)} = m} p_i \right\}.$$

In the Branch-and-Bound procedure, we set the initial lower bound LB to the maximum value between $\alpha_{machine}$ and α_{basic}.

3.3 Node Evaluation

In the Branch-and-Bound algorithm, we aim to find a feasible vector $(K_{ij})_{(i,j) \in \mathcal{D}}$ of occurrence shifts such that the value of the associated cycle time that ensure, for each $p \in \mathcal{U}^\Gamma$, the existence of schedule is minimum. In order to fathom nodes with partial solution in the search tree, it has to be evaluated by computing an associated lower bound. Let us consider a given node of the search tree defined by the subgraph $G_s = (\mathcal{T}, \mathcal{P} \cup \mathcal{D}_s)$, where $\mathcal{D}_s \subseteq \mathcal{D}$ is the set of fixed occurrence shifts. This subgraph represents a relaxation of the initial problem since only a subset a disjunctive arcs is considered. Consequently, the associate cycle time is a lower bound on the optimal cycle time.

3.4 Branching Scheme and Branching Rule

To our knowledge, two branching schemes have been proposed for the cyclic job shop problem. In both of the branching schemes, the branching is performed on the unfixed occurrence shifts. The first one is introduced in [2]. Based on the interval of possibles values $[K_{ij}^-, 1 - K_{ji}^-]$ for the occurrence shift K_{ij} such that $(i,j) \in \mathcal{D}$, the author uses a dichotomic branching. In the first generated node, the interval of possible values of the occurrence shifts K_{ij} is restricted to $[K_{ij}^-, c_{ij}]$ and in the second one it is restricted to $[c_{ij} + 1, 1 - K_{ji}^-]$, where c_{ij} is the middle of the initial interval. The second branching scheme is introduced in [11]. The branching consists in selecting an unfixed disjunction and generate a child node for each possible value of the occurrence shift K_{ij} in the interval $[K_{ij}^-, 1 - K_{ji}^-]$. In each node, the algorithm assigns the corresponding possible value to the occurrence shift K_{ij}. In this paper, we follow the same branching scheme introduced in [11]. This branching scheme allows us to have, at each node, a subproblem which corresponds to a robust BCSP. Consequently, we can use the existing robust version of the Howard's algorithm to find the cycle time ensuring, for each $p \in \mathcal{U}^\Gamma$, the existence of a schedule. Different branching rules have been tested and numerical tests show that branching on occurrence shifts K_{ij} where $K_{ij}^- + K_{ji}^-$ is maximum yields best running times. This performance can be explained by the fact that this branching rule generates a small number of child nodes, which limits the size of the search tree.

4   Numerical Experiments

We implemented the Branch-and-Bound algorithm in C++ and conducted the numerical experiments on an Intel Xeon E5-2695 processor running at 2.30 GHz CPU. The time limit for each instance is set up to 900 s.

Since there are no existing benchmarks for the CJSP, even in its deterministic setting, we generate randomly 20 instances for each configuration as follows. We consider instances where the number of tasks n varies $\{10, 20, 30, 40, 50, 60, 80, 100\}$, the number of jobs j in $\{2, 3, 4, 5, 6, 10, 16\}$ and the number of machines m in $\{5, 6, 8, 10\}$. Each nominal duration $\bar{p}_i$ of task i is generated with uniform distribution in $[1, 10]$ and its deviation value $\hat{p}_i$ in $[0, 0.5\bar{p}_i]$.

Table 1 reports average solution times for the instances having from 10 to 40 tasks and with a budget of uncertainty varying from 0% to 100%. All these instances are solved before reaching the time limit. The average running times show that the Branch-and-Bound algorithm is not very sensitive to the variation of the budget of the uncertainty, but there is still a small difference. This can be explained by the number of the nodes explored in the Branch-and-Bound tree which can differ from an instance with a given value of Γ to another one. Table 1 also displays the percentage of deviation, for a given budget of uncertainty Γ, of the optimal cycle time α_Γ from the nominal optimal cycle time α_{nom}, where all tasks take their nominal values. This percentage of deviation is computed as $\text{Dev}_\alpha = \frac{\alpha_\Gamma - \alpha_{nom}}{\alpha_{nom}}$. The table shows that the percentage of deviations varies from 25.41% to 56.43%. In other words, these deviations represent the percentage of the nominal cycle time that has to be increased in order to protect a schedule against the uncertainty. We remark that the deviations stabilize when the budget of the uncertainty is greater than 20 or 30 percent. This situation occurs probably when the number of arcs of the circuit having the maximum number of arcs is less than Γ. In this case, increasing Γ does not influence the optimal cycle time. The second situation occurs when heights of other circuits than the actual critical circuit c have greater value than the height of c. In this case, increasing the budget of the uncertainty does not make the value of c lower than the others.

Table 2 shows the number of the instances that are solved before reaching the time limit. These results concern instances having from 50 to 100 tasks. The table shows that the Branch-and-Bound is not able to solve all these instances in less then 900 s. For example, among instances with 80 tasks, 16 jobs and 5 machines, only three instances have been solved.

Table 1. Average solution times in seconds for the Branch-and-Bound algorithm and percentage value of the deviation of the cycle time from the nominal cycle.

# Tasks	# Jobs	# Machines	$\Gamma(\%)$	$Dev_\alpha(\%)$	Time (s)
10	2	5	0	0	0.012
			10	25.41	0.0123
			20	41.89	0.0137
			30	48.95	0.0157
			40	51.67	0.0171
			50	53.18	0.0185
			70	53.49	0.0214
			90	53.49	0.0251
			100	53.49	0.0256
20	3	6	0	0	0.2980
			10	33.61	0.2136
			20	49.49	0.2432
			30	55.44	0.5685
			40	56.43	0.2994
			50	56.43	0.3258
			70	56.43	0.3783
			90	56.43	0.3996
			100	56.43	0.4695
30	5	8	0	0	22.6434
			10	38.25	12.0085
			20	49.03	15.4099
			30	50.91	66.2474
			40	50.91	16.3096
			50	50.91	17.1160
			70	50.91	13.8331
			90	50.91	15.7524
			100	50.91	15.8249
40	4	8	0	0	138.9174
			10	37.92	55.9220
			20	54.46	92.1455
			30	54.91	96.7587
			40	54.91	134.4442
			50	54.91	155.2327
			70	54.91	187.2088
			90	54.91	204.4813
			100	54.91	177.2455

Table 2. Number of solved instances in less than 900 s among 20 instances.

# tasks	# jobs	# machines	$\Gamma(\%)$								
			0	10	20	30	40	50	70	90	100
50	5	10	11	10	11	14	13	13	12	12	12
60	6	10	7	5	4	4	4	3	3	1	1
80	16	5	3	0	0	0	0	0	0	0	0
100	10	10	3	1	3	1	0	0	0	0	0

5 Concluding Remarks and Perspectives

In this paper, we consider the cyclic job shop problem where the task durations are subject to uncertainty and belong to a polyhedral uncertainty set. We model the problem as two-stage robust optimization problem where the cycle time and the execution order of tasks mapped on the same machine have to be decided before knowing the uncertainty, and the starting times of tasks have to be determined after the uncertainty is revealed. We propose a Branch-and-Bound method that solves instances with at most 40 tasks but starts to have difficulties with bigger instances. The next step is to investigate other techniques such as decomposition algorithms.

References

1. Roundy, R.: Cyclic schedules for job shops with identical jobs. Math. Oper. Res. **17**(4), 842–865 (1992)
2. Hanen, C.: Study of a np-hard cyclic scheduling problem: the recurrent job-shop. Eur. J. Oper. Res. **72**(1), 82–101 (1994)
3. Brucker, P., Kampmeyer, T.: A general model for cyclic machine scheduling problems. Discrete Appl. Math. **156**(13), 2561–2572 (2008)
4. Che, A., Feng, J., Chen, H., Chu, C.: Robust optimization for the cyclic hoist scheduling problem. Eur. J. Oper. Res. **240**(3), 627–636 (2015)
5. Ben-Tal, A., Goryashko, A., Guslitzer, E., Nemirovski, A.: Adjustable robust solutions of uncertain linear programs. Math. Program. **99**(2), 351–376 (2004)
6. Thiele, A., Terry, T., Epelman, M.: Robust linear optimization with recourse. Rapport Technique, 4–37 (2009)
7. Zeng, B., Zhao, L.: Solving two-stage robust optimization problems using a column-and-constraint generation method. Oper. Res. Lett. **41**(5), 457–461 (2013)
8. Ayoub, J., Poss, M.: Decomposition for adjustable robust linear optimization subject to uncertainty polytope. Comput. Manage. Sci. **13**(2), 219–239 (2016)
9. Gondran, M., Minoux, M., Vajda, S.: Graphs and Algorithms. Wiley, New York (1984)
10. Dasdan, A.: Experimental analysis of the fastest optimum cycle ratio and mean algorithms. ACM Trans. Des. Autom. Electron. Syst. (TODAES) **9**(4), 385–418 (2004)

11. Fink, M., Rahhou, T.B., Houssin, L.: A new procedure for the cyclic job shop problem. IFAC Proc. Vol. **45**(6), 69–74 (2012)
12. Bertsimas, D., Sim, M.: The price of robustness. Oper. Res. **52**(1), 35–53 (2004)
13. Hamaz, I., Houssin, L., Cafieri, S.: Robust basic cyclic scheduling problem. Technical report (2017)

An Exact Algorithm for the Split-Demand One-Commodity Pickup-and-delivery Travelling Salesman Problem

Hipólito Hernández-Pérez and Juan José Salazar-González[⊠]

DMEIO, Facultad de Ciencias, Universidad de La Laguna,
38200 La Laguna, Tenerife, Spain
hhperez@ull.edu.es, jjsalaza@ull.es

Abstract. This paper concerns the problem of designing a route of minimum cost for a capacitated vehicle moving a single commodity between a set of customers. The route must allow two characteristics uncommon in the literature. One characteristic is that a customer may be visited several times. The other characteristic is that a customer may be used as intermediate location to temporarily collect and deliver part of the load of the vehicle. Routes with these characteristics may lead to cost reductions when compared to routes without them. The paper describes a branch-and-cut algorithm based on a relaxation of a model of Mixed Integer Programming. Preliminary computational results on benchmark instances demonstrate the good performance of the algorithm compared with the original model.

1 Introduction

The *Split-Demand One-Commodity Pickup-and-Delivery Travelling Salesman Problem* (SD1PDTSP) is a generalization of the One-Commodity Pickup-and-Delivery Traveling Salesman Problem (1-PDTSP) addressed in e.g. [5]. The 1-PDTSP looks for a route for a capacitated vehicle to move a single commodity between customers, visiting each customer exactly once. The SD1PDTSP extends the 1-PDTSP by allowing more than one visit to each customer, and therefore may find routes with smaller cost. The SD1PDTSP is also a generalization of the *Capacitated Vehicle Routing Problem* (CVRP), aimed at designing the routes for a fleet of identical vehicles to deliver a commodity from the depot to a set of customers. In the CVRP, each route starts and ends at the depot, and the load of a vehicle through a route should never exceed the vehicle capacity. The CVRP can be seen as a pickup-and-delivery single-vehicle problem with only one pickup location (the depot) which can be visited by the vehicle a number of times at most the fleet size, and several delivery locations (the customers) that can be visited at most once. Even more, the SD1PDTSP can be seen as a generalization of the *Split Delivery Vehicle Routing Problem* (SDVRP).

This work has been partially supported by MTM2015-63680-R (MINECO/FEDER), ProID2017010132 (Gobierno Canarias) and 2016TUR11 (Fundación CajaCanarias).

© Springer International Publishing AG, part of Springer Nature 2018
J. Lee et al. (Eds.): ISCO 2018, LNCS 10856, pp. 241–252, 2018.
https://doi.org/10.1007/978-3-319-96151-4_21

The SD1PDTSP is defined as follows. Let us consider a finite set of locations. Each location is related to a customer, with a known positive or negative demand of a commodity. For example, the commodity can be bicycles of identical type, the locations can represent bike stations in a city, and the demand can be the difference between the number of bicycles at the beginning of a day and at the end of the previous day in each station. We assume that the sum of all demands is equal to zero. Customers with negative demands correspond to pickup locations, and customers with positive demands correspond to delivery locations. The travel distances (or costs) between the locations are assumed to be known. There is *one* vehicle with a given capacity that must visit each location *at least* once through a route to move the commodity between the customers as they require. Each visit may partially satisfy the demand of a customer, and all the visits to that customer must end up with exactly its complete demand. The SD1PDTSP consists of finding a minimum-cost route for the vehicle such that it satisfies the demand of all customers without violating the vehicle capacity. Although a customer may be visited several times, a maximum number of allowed visits is assumed on each customer. When this parameter is 1 for each customer, we refer to the *split-forbidden variant* of the problem, and the SD1PDSTP coincides with the 1-PDTSP. As for the 1-PDTSP [4, 6, 7], checking whether a feasible solution exists for the SD1PDTSP is a NP-complete problem.

The vehicle is not required to leave any location with an a-priori known load (neither empty nor full). If a location is considered the starting (ending) point of the route, the initial (final) load of the vehicle in the SD1PDTSP is a decision that must be determined within the optimization problem. Although our results can be adapted to the variant with a fixed initial load of the vehicle in a location, we do not consider it in this paper.

Since several visits to a location are allowed, the vehicle could deliver some units of the commodity in a location and collect them later in another visit. Similarly, the vehicle can collect some units of the commodity in a location and deliver them later in another visit. The SD1PDTSP allows these solutions and therefore it can be seen as an inventory-routing problem where each customer has an a-priori stock of the commodity, requires to have an a-posteriori stock, has a capacity, and the demand is the difference between the a-priori and a-posteriori stocks. In other words, a customer in the SD1PDTSP may be used to *temporarily* deliver or collect units of commodity. This characteristic is called *preemption* and it may provide routes with smaller costs respect to the non-preemption variant.

The SD1PDTSP was introduced in [9], together with a flow formulation and a branch-and-cut algorithm to solve it. When the vehicle capacity is large enough and the maximum number of visits to a location is one, the SD1PDTSP coincides with the *Travelling Salesman Problem* (TSP), thus the SD1PDTSP is NP-hard too. [1–3] study a very similar problem where the vehicle must start the route with empty load, and the number of visits to each customer is unlimited. [1] propose a branch-and-cut algorithm for computing a lower bound of the optimal solution value, and a tabu search heuristic algorithm to compute an upper bound on instances with up to 100 customers. [3] propose an exact approach, analyzed on instances with up to 60 customers. A hybrid iterative local search heuristic is designed in [2], and analyzed on instances with up to 100 customers. A math-heuristic approach is described in [8] to solve instances with up to 500 customers.

This paper is organized as follows. Section 2 presents two models for the SD1PDTSP. The first model is taken from the literature while the second model is new. Section 3 describes a branch-and-cut algorithm to solve the second formulation. Preliminary computational results in Sect. 4 show the good performance of the algorithm on benchmark instances.

2 Mathematical Models

This section shows a formulation introduced by [9] (corrected in [8]), and introduces a new formulation inspired by [3]. We need some notation before.

2.1 Problem Definition

Let $I = \{1, \ldots, n\}$ be the set of locations, all representing *customers*. Let p_i be the units of product in i before starting the service at location i, and p'_i the units of product desired in i at the end of the service. Let $d_i = p'_i - p_i$ be the demand of customer i. When $d_i > 0$ then the location i requires d_i units of product to be delivered by the vehicle. When $d_i < 0$ then the location i provides $-d_i$ units of product that must be collected by the vehicle. We assume that $\sum_{i \in I} d_i = 0$, so the number of units of the product in the system remains equal before and after performing the vehicle service. We also assume that all locations must be served by the vehicle, including those with zero demand (if any). In addition, there is a capacity q_i associated with each location $i \in I$, meaning that this location can store between 0 and q_i units of the commodity, and satisfying that $0 \leq p_i \leq q_i$ and $0 \leq p'_i \leq q_i$. The capacity of the vehicle is given a-priori, and it is denoted by Q. We denote by c_{ij} the travel cost for the vehicle to go from i to j, with $i, j \in I$. Let m be a known integer value representing the maximum number of visits allowed to a customer.

Although SD1PDTSP assumes that also zero-demand customers must be visited by the vehicle, it is possible to adapt the contributions of this paper to deal with the variants where, either zero-demand customers must be discarded, or zero-demand customers are optionally visited if convenient.

2.2 First Model

Let V_i be an ordered set of m nodes representing potential visits to location i. Since all nodes in V_i are identical, we intend that the sequence of visits of the vehicle to i is represented by consecutive nodes in V_i with i_1 representing the first visit. The set $V = \cup_{i \in I} V_i$ is the node set of a directed graph $G = (V, A)$ where A is the arc set connecting nodes associated with different locations. For a given subset S of nodes, we write $\delta_A^+(S) = \{(v, w) \in A : v \in S, w \notin S\}$ and $\delta_A^-(S) = \{(v, w) \in A : v \notin S, w \in S\}$. Given an arc $a = (v, w)$ we also denote the cost c_a from v to w as the travel cost c_{ij} from the location i associated with v to the location j associated with w.

We consider the following mathematical variables. For each arc $a \in A$, a binary variable x_a assumes value 1 if and only if the route includes a, and a continuous variable f_a is the load of the vehicle when traversing a. For each node $v \in V$, a binary variable y_v assumes value 1 if and only if the route includes v, and a continuous variable g_v determines the number $|g_v|$ of units delivered (if $g_v > 0$) or collected (if $g_v < 0$) when performing the visit v. Then, the SD1PDTSP can be formulated as:

$$\min \sum_{a \in A} c_a x_a \tag{1}$$

subject to:

$$y_{i_1} = 1 \qquad \text{for all } i \in I \tag{2}$$

$$\sum_{a \in \delta_A^+(v)} x_a = \sum_{a \in \delta_A^-(v)} x_a = y_v \qquad \text{for all } v \in V \tag{3}$$

$$\sum_{a \in \delta_A^+(S)} x_a \geq y_v + y_w - 1 \qquad \text{for all } S \subseteq V , \, v \in S , \, w \in V \setminus S \tag{4}$$

$$\sum_{a \in \delta_A^+(S) \setminus \delta_A^+(1_1)} x_a \geq y_{i_{l+1}} \qquad \text{for all } i \in I , \, l = 1, \dots, m-1 \, (i_l \neq 1_1),$$

$$S \subseteq V : \, 1_1, i_l \in S , \, i_{l+1} \in V \setminus S \tag{5}$$

$$\sum_{a \in \delta_A^-(v)} f_a - \sum_{a \in \delta_A^+(v)} f_a = g_v \qquad \text{for all } v \in V \tag{6}$$

$$0 \leq f_a \leq Q x_a \qquad \text{for all } a \in A \tag{7}$$

$$\sum_{1 \leq l \leq m} g_{i_l} = d_i \qquad \text{for all } i \in I \tag{8}$$

$$0 \leq p_i + \sum_{1 \leq k \leq l} g_{i_k} \leq q_i \qquad \text{for all } i \in I , \, l = 1, \dots, m-1 \tag{9}$$

$$-q_i y_{i_l} \leq g_{i_l} \leq q_i y_{i_l} \qquad \text{for all } i \in I , \, l = 2, \dots, m \tag{10}$$

$$y_v, x_a \in \{0, 1\} \qquad \text{for all } v \in V , \, a \in A. \tag{11}$$

Equations (2) ensure that each customer is visited by the vehicle. Equations (3) force the vehicle to enter and leave once each node v with $y_v = 1$. Inequalities (4) ensure a connected route. Inequalities (5) ensure that the vehicle visits i_l before i_{l+1} if $1 \leq l < m$ and $y_{i_{l+1}} = 1$. Constraints (6)–(8) ensure that the load of the vehicle is able to satisfy the demand decided at each visit. Constraints (9) guarantee that the storage of product in a customer i is always between 0 and its capacity q_i. Inequalities (10) impose that product can be delivered to or collected from a location in each visit, i.e., the preemption characteristic. Note that this characteristic can be easily forbidden by replacing (10) with:

$$g_{i_l} \geq 0 \qquad \text{for all } i \in I : d_i \geq 0 \,, \; l = 1, \ldots, m$$
$$g_{i_l} \leq 0 \qquad \text{for all } i \in I : d_i < 0 \,, \; l = 1, \ldots, m.$$

In such case, when preemption is not desired in the problem, then the parameters p_i, p_i' and q_i are useless, and parameters d_i are enough for validating a non-preemptive route.

To better understand the validity of model (1)–(11), observe that (2)–(4) and (11) ensure that (x, y) represents a cycle visiting each customer. When a customer is visited several times (say, $y_{i_l} = y_{i_{l+1}} = 1$) then constraints (5) ensure that the cycle goes from i_l to i_{l+1} with 1_1 not in between. In other words, starting from 1_1, constraints (5) ensure that the cycle will continue to i_l, then to i_{l+1}, and then to 1_1. Constraints (6)–(8) guarantee a flow of product collected or delivered within the cycle. Inequalities (9) use the fact that the visits i_l and i_{l+1} are done in this order, and then ensure that the cumulative product at the customer i is always within the limits 0 and q_i. Inequalities (10) guarantee that collections and deliveries are only done on visits.

2.3   Second Model

We first start presenting a relaxed (invalid) formulation based on an smaller graph. Let $G' = (I, B)$ be the oriented graph where each node represents a location (instead of a visit) and each arc $b \in B$ represents an arc between two locations. For a given subset $S \subseteq I$, we write $\delta_B^+(S) = \{(i, j) \in B : i \in S, j \notin S\}$ and $\delta_B^-(S) = \{(i, j) \in B : i \notin S, j \in S\}$. Then we define the following decision variables. For each arc $b \in B$, let be x_b an integer variable representing the number of times the edge b is traversed by the vehicle. Observe that we are using the notation x for variables in both models; however, this should not confuse the reader as the subindex identifies the model univocally. For example, x_a is a binary variable and x_b can be greater than one. For each $i \in I$, let z_i be an integer variable representing the number of visits to location i. For each $b = (i, j) \in B$ let c_b be the travel cost c_{ij} from i to j. Consider now the following mathematical formulation:

$$\min \sum_{b \in B} c_b x_b \tag{12}$$

subject to:

$$\sum_{b \in \delta_B^-(i)} x_b = \sum_{b \in \delta_B^+(i)} x_b = z_i \quad \text{for all } i \in I \tag{13}$$

$$1 \leq z_i \leq m \quad \text{for all } i \in I \tag{14}$$

$$\sum_{b \in \delta_B^+(S)} x_b \geq \max \left\{ 1, \left\lceil \frac{|\sum_{i \in S} d_i|}{Q} \right\rceil \right\} \quad \text{for all } S \subset I \tag{15}$$

$$x_b \geq 0 \text{ and } x_b \in \mathbb{Z} \quad \text{for all } b \in B. \tag{16}$$

Equations (13) forces the same number of arc in the route going in and out of at each location. Inequations (14) limit the number of visits to each location. Inequations (15) ensure the connectivity of the route and the vehicle capacity. Clearly, all SD1PDTSP solutions satisfy the above formulation, but there may be solutions of the formulation which do not correspond to SD1PDTSP solutions (note that (13)–(16) considers d_i but not p_i and q_i).

Variables x_b are integer variables that can be replaced by binary variables w_{bk} with:

$$x_b = \sum_{k=0}^{\lfloor \log_2 m \rfloor} 2^k w_{bk}.$$

These binary variables are used in [3] to eliminate an invalid (integer) solution (x^*, z^*) of model (12)–(16) with the following (weak) inequality:

$$\sum_{b \in B, k \in \{0, \ldots, \log_2 m\}: w_{bk}^* = 0} w_{bk} + \sum_{b \in B, k \in \{0, \ldots, \log_2 m\}: w_{bk}^* = 1} (1 - w_{bk}) \geq 1. \qquad (17)$$

We propose a different procedure to obtain a valid formulation for the SD1PDTSP from (12)–(16). Our procedure consists of restricting the invalid formulation with a new linear system of linear constraints. The linear system is taken from the model in Sect. 2.2, but now considering that customer i is visited z_i^* times. Let G'' be the subgraph of G induced by $V^* := \cup_{i \in I} \{i_1, \ldots, i_{z_i^*}\}$. For each arc a in G'', consider a binary variable x_a assuming value 1 if and only if the route includes a, and a continuous variable f_a representing the vehicle load when traversing a. For each node v in G'', consider a continuous variable g_v representing the commodity $|g_v|$ delivered (if $g_v > 0$) or collected (if $g_v < 0$) when performing the visit v. The linear system contains now the constraints (2)–(11) with V replaced by V^* and the y_i variables fixed to value 1 for all $i \in V^*$. The only variables in the linear system are now x_a, f_a and g_i. In addition the linear system to restrict the formulation (12)–(16) also includes the equations

$$\sum_{a=(i_k, j_l): (i,j)=b} x_a = x_b^* \qquad \text{for all } b \in B. \qquad (18)$$

A solution of the new linear system can be seen as a certificate that the solution (x^*, z^*) from (12)–(16) represents a valid route for the SD1PDTSP.

3 Branch-and-Cut Algorithm

The second formulation in the previous section suggests a new algorithm to solve the SD1PDTSP. The algorithm solves the master model (12)–(16) where constraints (15) are heuristically separated. To this end, first, it computes max-flow problems to guarantee

$$\sum_{b \in \delta_B^+(S)} x_e \geq 1.$$

Second, it computes other max-flow problems to guarantee

$$\sum_{b \in \delta_B^+(S)} x_b \geq |\sum_{i \in S} d_i|/Q.$$

When no violated constraints has been detected, the algorithm verifies whether $z_i^* \in \mathbb{Z}$ for all $i \in I$. Fractional solutions are discarded through a classical binary branching procedure. For an integer solution (x^*, z^*), consider the subproblem with zero objective function and linear system described in Sect. 2.3. We first solve the dual problem of the linear-programming relaxation of the subproblem. If it is unbounded then the current solution from the master program is invalid for the SD1PDSTSP, and a dual ray defines a violated constraint to be added to the master model, exactly as done in a Benders' decomposition framework. Otherwise, we now solve the subproblem, which is an integer program. If it is feasible then (x^*, z^*) is valid for the SD1PDTSP. Otherwise, the constraint (17) is inserted to avoid this solution.

It is worst noting that the subproblem also includes $f_a = 0$ for all $a \in \delta_A^+(V_1)$ when solving the problem in [3]. These additional requirements increase the chance of getting dual rays defining violated inequalities.

4   Preliminary Computational Results

The branch-and-cut (B&C) algorithm described in the previous section has been implemented in C++, and executed on a personal computer with an Intel Core i7-2600 CPU 3.4 Ghz running Microsoft Windows 7, using the callable libraries of CPLEX 12.7 to solve the MILP problems. To evaluate the performance of our implementation, we have used the benchmark instances also used in [3] and [9]. These instances are based in the 1-PDTSP instances proposed in [6], and are generated in the following way. Customers $2, \ldots, n$ are randomly located in the square $[-500, 500] \times [-500, 500]$ and have integer demands d_i randomly chosen in the interval $[-10, 10]$. Customer 1 is located in the point $(0, 0)$ with a demand d_1 such that the sum of all customer demands is zero. The travel costs are computed as the Euclidean distances, rounded to the closest integer numbers in [9] and truncated to integers numbers in [3]. Another difference between the experiments in [9] and [3] is that the zero-demand customers must be visited in [9] while they are visited if convenient in [3]. Finally, $p_i = 10 - d_i$ and $p'_i = 10$ in [9] and $p_i = 10$ and $p'_i = 10 + d_i$ in [3]; in both cases, $q_i = 20$ for all i. Finally, [3] add another point in $(0, 0)$, called depot, represented by 0, with $d_0 = p_0 = p'_0 = q_0 = 0$, and from which the vehicle must leave with zero load, as detailed in [1] and [2]. Our paper show results solving these instances with preemption allowed and at most 3 visits to each customer.

The algorithm described in [9] was executed on a personal computer with Intel Core 2 Duo CPU E8600 3.33 Ghz and IBM ILOG CPLEX 12.5 as MILP solver. The algorithm described in [3] was executed on an IRIDIS 4 computing cluster 2.6 Ghz and IBM ILOG CPLEX 12.5 as MILP solver. These two B&C

algorithms and our B&C algorithm are all executed using one thread of the CPU. The time limit of the three algorithms is 2 h, although different computers.

Table 1 compares the results of the algorithm proposed in [9] with the results of the B&C algorithm described in this paper. Table 2 compares the B&C algorithm with the results of the algorithm in [3]. In both tables, each row corresponds to the average results over ten instances. For the results of Table 2, we relaxed the requirement to visit zero-demand customers, as assumed in [3]. The column heading "Average customers several visits" refers to the average number of customers which are visited at least twice in the optimal solution; "Average optimal gap" refers to the average percentage deviation between the best result of the B&C algorithm and the best known lower bound for each algorithm; "Average CPU time" refers to the average computing time for each algorithm (including time limit); and "Number of instances solved" refers to the number of instances (over 10) solved to optimality (before the time limit).

Table 1. Summary of the results compared with [9].

			B&C in [9]		Our B&C
n	Q	Average customers several visits	Average CPU time	Number instances solved	Average CPU time
30	5	13.9	4361.3	4	1.8
30	6	11.5	3066.0	8	1.4
30	7	8.3	1736.2	10	1.1
30	10	2.6	2604.4	8	1.0
30	12	1.6	773.5	10	0.8
30	15	0.8	155.5	10	0.8

Our B&C algorithm solved all instances to optimality, and for that reason we do not include the columns "Average opt gap" and "Number of instances solved" for such algorithm in these tables. It is clear that our new B&C algorithm outperforms the previous algorithms in the literature. We believe that this major milestone is due to work a master problem based on a small graph G' and with a subproblem that is easy to solve and that generates a good valid inequality when it is infeasible. Indeed, the master problem in [9] is fully defined on G, as described in Sect. 2.2, while the weak aspect in [3] is solving their subproblem, which consists of enumerating Eulerian circuits and checking their feasibility on an extended network.

To better illustrate the difficulty of solving the SD1PDTSP we conclude this section comparing optimal routes obtained with and without some problem characteristics. To this end, we selected the instance n20q12H with $n = 20$ and $Q = 12$. Figure 1(a) shows the optimal route when split demand is not allowed and the vehicle is forced to leave empty from the depot (customer 1). Relaxing

Table 2. Summary of the results compared with [3].

n	Q	Average customers several visits	B&C in [3]				Our B&C
			Average CPU time	Number instances solved	Average optimal gap		Average CPU time
20	10	2.7	0.4	10			0.4
20	15	1.6	0.3	10			0.3
20	20	1.0	0.1	10			0.2
20	1000	1.0	0.8	10			0.2
30	10	3.9	6.2	10			1.5
30	15	2.1	3.9	10			0.9
30	20	1.6	163.6	10			0.6
30	1000	1.0	190.2	10			2.0
40	10	4.1	124.8	10			3.0
40	15	1.7	25.6	10			1.7
40	20	1.3	14.7	10			1.2
40	1000	1.2	70.2	10			0.7
50	10	5.2	1198.5	9	0.79		144.4
50	15	2.5	1970.1	8	0.43		35.3
50	20	1.9	295.5	10			5.8
50	1000	1.0	1909.8	9	0.11		1.6
60	10	6.7	3924.6	6	1.24		178.5
60	15	2.4	1957.5	8	0.51		14.6
60	20	1.8	1285.0	10			6.6
60	1000	1.2	2816.4	8	0.18		2.8

the empty load requirement, Fig. 1(b) shows a shorter route in which the vehicle leaves the depot with 2 units of product that will be returned back to the depot after having finished the service. Allowing visiting a customer more than once, even shorter routes can be found. Figure 2(a) is the optimal route when each customer can be visited at most three times if preemption is forbidden. If preemption is allowed, then Fig. 2(b) shows the optimal route. These SD1PDTSP solutions were computed without fixing the load of the vehicle when leaving the depot. When the vehicle is forced to leave Customer 1 with zero load, then the optimal cost is 5203 (5131) with preemption forbidden (allowed); these routes are not depicted in this paper. Figure 2(b) is also the optimal solution of the formulation (12)–(16), which means that the first linear system (2)–(11) and (18) was feasible, thus there was no need to solve a second master problem. This situation happened on most instances in our experiments and it may be explained due to the large value of the q_i numbers when compared to the d_i numbers. By

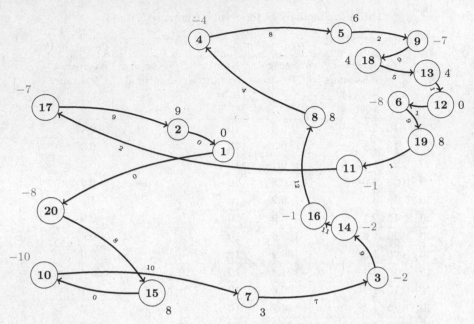

(a) With initial load fixed to zero; cost = 5376.

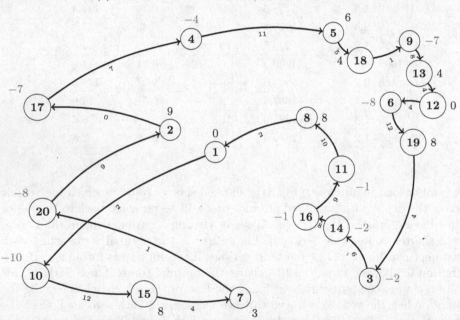

(b) With unfixed initial load; cost = 5167.

Fig. 1. Optimal 1PDTSP routes.

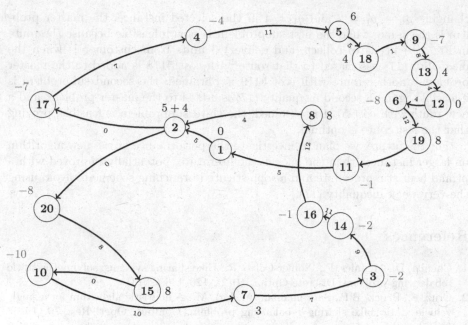

(a) Without preemption; cost = 5152.

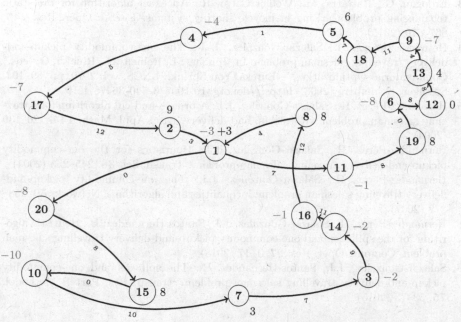

(b) With preemption; cost = 5118.

Fig. 2. Optimal SD1PDTSP routes.

changing $p_1 = p_1' = 2$ and $q_1 = 4$ in the selected instance, the master problem is the same, while the first subproblem is now infeasible because the route in Fig. 2(b) needs to collect (and deliver) 3 units from customer 1; when the inequality (17) associated to that route with cost 5118 is added to the master problem, another route with cost 5119 is obtained; the second subproblem is infeasible again, a second inequality (17) is added to the master problem, and a new route with cost 5131 is obtained; the third subproblem is feasible, proving that the last route is optimal.

For the future, we plan to investigate the performance of the new algorithm on larger instances than the ones in the literature, potentially improved with a primal heuristic procedure and a sophisticated branching scheme to avoid using the very weak inequality (17).

References

1. Chemla, D., Meunier, F., Wolfler-Calvo, R.: Bike sharing systems: solving the static rebalancing problem. Discrete Optim. **10**(2), 120–146 (2013)
2. Cruz, F., Bruck, B.P., Subramanian, A., Iori, M.: A heuristic algorithm for a single vehicle static bike sharing rebalancing problem. Comput. Oper. Res. **79**, 19–33 (2017)
3. Erdogan, G., Battarra, M., Wolfler Calvo, R.: An exact algorithm for the static rebalancing problem arising in bicycle sharing systems. Eur. J. Oper. Res. **245**, 667–679 (2015)
4. Hernández-Pérez, H., Salazar-González, J.-J.: The one-commodity pickup-and-delivery travelling salesman problem. In: Jünger, M., Reinelt, G., Rinaldi, G. (eds.) Combinatorial Optimization — Eureka, You Shrink!. LNCS, vol. 2570, pp. 89–104. Springer, Heidelberg (2003). https://doi.org/10.1007/3-540-36478-1_10
5. Hernández-Pérez, H., Salazar-González, J.J.: A branch-and-cut algorithm for a traveling salesman problem with pickup and delivery. Disc. Appl. Math. **145**, 126–139 (2004)
6. Hernández-Pérez, H., Salazar-González, J.J.: Heuristics for the one-commodity pickup-and-delivery traveling salesman problem. Transp. Sci. **38**, 245–255 (2004)
7. Hernández-Pérez, H., Salazar-González, J.J.: The one-commodity pickup-and-delivery traveling salesman problem: Inequalities and algorithms. Networks **50**, 258–272 (2007)
8. Hernández-Pérez, H., Salazar-González, J.J., Santos-Hernández, B.: Heuristic algorithm for the split-demand one-commodity pickup-and-delivery travelling salesman problem. Comput. Oper. Res. **97**, 1–17 (2018)
9. Salazar-González, J.J., Santos-Hernández, B.: The split-demand one-commodity pickup-and-delivery travelling salesman problem. Transp. Res. Part B Methodol. **75**, 58–73 (2015)

Descent with Mutations Applied to the Linear Ordering Problem

Olivier Hudry[✉]

LTCI - Telecom ParisTech, 46, rue Barrault, 75634 Paris, France
olivier.hudry@telecom-paristech.fr
http://www.infres.enst.fr/~hudry

Abstract. We study here the application of the "descent with mutations" metaheuristic to the linear ordering problem. We compare this
local search metaheuristic with another very efficient metaheuristic,
obtained by the hybridization of a classic simulated annealing with some
ingredients coming from the noising methods. The computational experiments on the linear ordering problem show that the descent with mutations provides results which are comparable to the ones given by this
improved simulated annealing, or even better, while the descent with
mutations is much easier to design and to tune, since there is no parameter to tune (except the CPU time that the user wants to spend to solve
his or her problem).

Keywords: Combinatorial optimization · Metaheuristics
Simulated annealing · Noising methods · Linear ordering problem
Median order · Slater's problem · Condorcet-Kemeny's problem

1 Introduction

We deal here with a metaheuristic (for recent references on metaheuristics, see for
instance [17,27] or [29]) called "descent with mutations" (DWM). This method
looks like the usual descent, but with random elementary transformations which
are performed, from time to time, in a blind way, in the sense that they are
accepted whatever their effects on the function f to optimize (such an elementary
transformation performed without respect to its effect on f will be called a
mutation in the sequel). The density of performed mutations decreases during
the process, so that the method at its end is the same as a classic descent. DWM
can also be considered as a variant of the noising methods (see for instance [12]
for a survey and references on the noising methods).

In this paper, we study the application of DWM to two problems arising
from the field of the aggregation and the approximation of binary relations: the
approximation of a tournament by a linear order at minimum distance (this
problem is also known as *Slater's problem* [28]) and the aggregation of linear
orders into a median linear order (this problem is sometimes called *Kemeny's
problem* [25], though Kemeny considered complete preorders instead of linear

© Springer International Publishing AG, part of Springer Nature 2018
J. Lee et al. (Eds.): ISCO 2018, LNCS 10856, pp. 253–264, 2018.
https://doi.org/10.1007/978-3-319-96151-4_22

orders; it seems that Condorcet was the first one to consider this aggregation problem at the end of the 18th century [13]; because of this, we shall call this problem "Condorcet-Kemeny's problem" in the rest of the paper). Both can be represented by another problem, which is known in graph theory as the *linear ordering problem* (LOP in the following; for a survey on these topics and for references, see for instance [5, 11, 24, 26]). We compare DWM with a simulated annealing method (SA) improved by ingredients coming from the noising methods (as done in [7] and in [9] for the Travelling Salesman Problem).

In the next section, we detail the principles of DWM. In Sects. 3 and 4, we briefly depict the studied problems and the chosen elementary transformations allowing us to apply a descent and DWM. Experimental results can be found in Sect. 5. Conclusions are in Sect. 6.

2   Principle of DWM

As the other metaheuristics, DWM is not designed to be applicable to only one combinatorial problem, but to many of them. Such a problem can be stated as follows:

$$\text{Minimize } f(s) \text{ for } s \in S,$$

where S is assumed to be a finite set and f is a function defined on S; the elements s of S will be called *solutions*.

As many other metaheuristics, DWM is based on *elementary transformations*. A *transformation* is any operation changing a solution into another solution. A transformation will be considered as *elementary* (or *local*) if, when applied to a solution s, it changes one feature of s without modifying its global structure much. For instance, if s is a binary string, a possible elementary transformation would be to change one bit of s into its complement. Thanks to the elementary transformations, we may define the *neighbourhood* $N(s)$ of a solution s: $N(s)$ is the set of all the solutions (called the *neighbours* of s) that we can obtain from s by applying an elementary transformation to s.

Then, we may define an iterative improvement method, or *descent* for a minimization problem (it is the case for the problems considered here), as follows. A descent starts with an initial solution s_0 (which can be for instance randomly computed, or found by a heuristic) and then generates a series of solutions $s_1, s_2, \ldots, s_i, \ldots, s_q$ such that:

1. for any $i \geq 2$, s_i is a neighbour of s_{i-1}: $s_i \in N(s_{i-1})$;
2. for any $i \geq 2$, s_i is better than s_{i-1} with respect to f: $f(s_i) < f(s_{i-1})$;
3. no neighbour of s_q is better than s_q: $\forall s \in N(s_q)$, $f(s) \geq f(s_q)$.

Then s_q is the solution returned by the descent, the descent is over and the final solution s_q provided by the descent is (at least) a local minimum of f with respect to the adopted elementary transformation. The whole method may stop here, or restarts a new descent from a new initial solution (to get *repeated descents*, as it will be done below).

In such a descent, the process is not blind in the sense that the elementary transformations are adopted only if they improve the value taken by f. In DWM (see below for the general description of the method), we also apply the basic process of a descent but, from time to time, we apply and accept the considered elementary transformation, whatever its effect on f: we say that we have a *blind elementary transformation*, or simply a *mutation*, since it is the word commonly used in genetics (and in genetic algorithms) for this kind of blind transformation. Thus, the only thing to specify in order to apply DWM (in addition to what must be defined to apply a descent, i.e. the elementary transformation) is when a mutation is adopted. It is what we depict in the next section, for the problems studied in this paper.

– Repeat:
  • with a certain probability, apply an arbitrary elementary transformation (irrespective improvement or worsening: this is a mutation)
  • otherwise, apply an elementary transformation which brings an improvement
– until a given condition is fulfilled.

General description of DWM.

We said at the beginning that DWM can be considered as a variant of the noising methods (which can be seen as a generalization of methods like simulated annealing or threshold accepting). Remember that the most general scheme of the noising methods (see [12]) consists in computing a "noised" variation $\Delta f_{noised}(s, s')$ of f when a neighbour s' of the current solution s is considered: $\Delta f_{noised}(s, s') = f(s') - f(s) + r$, where r is a random number depending on different things (like s, s', the iteration number, the scheme of the noising method, the adopted probability law, and so on); then the acceptance criterion becomes the following: the transformation of s into s' is accepted if $\Delta f_{noised}(s, s')$ is lower than 0. We find back the usual descent method if r is equal to 0 and the accepting criterion applied in simulated annealing if r is equal to $T \ln p$, where T is the current temperature and p is a random number belonging to $]0, 1[$ and to which $e^{-\Delta f/T}$ is compared in the classic simulated annealing, with $\Delta f = f(s') - f(s)$. In a similar way, it is not difficult to design the characteristics of the law followed by r in order to show that DWM constitutes a special case of the noising methods: it is sufficient to choose a very negative value for r (that is, a negative value with a great absolute value) when we decide to perform a mutation, or 0 otherwise; details are left to the interested reader. But the main advantage of DWM with respect to methods like simulated annealing or the noising methods is that there is no parameter to tune (except the CPU time, which in its turn defines the number of iterations performed by the method; the relationship between the CPU time and the number of iterations obviously depends on the used computer).

We turn now to the specification of the problems considered here.

3   Slater's Problem, Condorcet-Kemeny's Problem, the Linear Ordering Problem

Let X be a finite set. If R is a binary relation defined on X and if x and y are two elements of X, we write xRy if x is in relation with y with respect to R. Let R and S be two binary relations defined on X. If Δ denotes the usual symmetric difference between sets, the *symmetric difference distance* $\delta(R, S)$ between R and S is defined by

$$\delta(R, S) = |R\Delta S|,$$

i.e.

$$\delta(R, S) = |\{(x, y) \in X^2 \text{ s.t. } [xRy \text{ and not } xSy] \text{ or } [\text{not } xRy \text{ and } xSy]\}|.$$

This distance, which owns good axiomatic properties (see [2]), measures the number of disagreements between R and S. From this distance, we may define a *remoteness* ρ (see [3]) between a collection, called a *profile*, $\Pi = (R_1, R_2, ..., R_m)$ of m binary relations defined on X and any linear order O which is also defined on X by:

$$\rho(\Pi, O) = \sum_{i=1}^{m} \delta(R_i, O).$$

Thus $\rho(\Pi, O)$ measures the total number of disagreements between Π and O. A *median linear order* [3] of Π is a linear order O^* which minimizes the remoteness from Π:

$$\rho(\Pi, O^*) = \min_{O \in \Omega(X)} \rho(\Pi, O),$$

where $\Omega(X)$ denotes the set of all the linear orders defined on X.

Slater's problem [28] corresponds to the case for which Π contains only one relation (i.e., $m = 1$) which is a *tournament* T defined on X, that is, a complete asymmetric relation: between two distinct elements x and y of X, there is one and only one of the two possibilities xTy or yTx (observe that a transitive tournament is a linear order and conversely). A tournament can be considered for example as the result of a paired-comparison experiment in voting theory. In such a context, a decider must rank n candidates (which are the elements of X). To do this, each pair of candidates is displayed to the decider, and this one chooses exactly one of the two candidates. A lack of transitivity in the choice of the decider may happen, and thus we obtain a tournament while a linear order may be expected. Then we look for a linear order at minimum distance from T, i.e., we want to minimize $\delta(T, O)$ or, equivalently, $\rho(\Pi, O)$ with $\Pi = (T)$.

Condorcet-Kemeny's problem [13, 25] arises also in the context of voting theory. Here, m voters want to rank n candidates. The preference of any voter is assumed to be a linear order. We want to aggregate these preferences into a unique linear order (which will represent a ranking that we may consider as the collective preference) by minimizing the total number of disagreements between the collective preference and the individual preferences of the m voters. In other words, we look for a median linear order of the profile $\Pi = (O_1, O_2, ..., O_m)$, where O_i is the linear order associated with the preference of the i-th voter. (As said above, in the genuine problem considered by Kemeny [25], the preferences are assumed to be complete preorders, i.e., rankings of the candidates in which ties are allowed, and, similarly, we look for a median complete preorder. Anyway, many authors call "Kemeny's problem" the aggregation of linear orders into a median linear order, by an abuse of language; note that there always exists a median complete preorder of a profile of tournaments – what includes linear orders – which is a linear order – see [22]).

These two problems can be considered as special cases of the linear ordering problem (LOP; see for instance [11] for details). In LOP, we consider a graph which is a weighted tournament $T = (X, A, w)$ (as for a binary relation, a *tournament* in graph theory is a complete, asymmetric, directed graph: between two distinct vertices x and y of X, there is one and only one of the two arcs – i.e., directed edges – (x, y) or (y, x)). Each arc $a \in A$ has a weight $w(a)$ which is a non-negative integer. Then, we want to determine a subset B of A with a maximum weight such that (X, B) is without circuit (i.e., directed cycle) or, equivalently, such that reversing the arcs belonging to $A \setminus B$ into T transforms T into a linear order, which is then a *median linear order of T*.

For Slater's problem, all the weights are equal to 1. For Condorcet-Kemeny's problem, let m_{xy} denote the number of voters who prefer x to y; then the weight of an arc (x, y) is equal to $2m_{xy} - m$ (remember that m denotes the number of voters), when m_{xy} is greater than m_{yx}; if m_{xy} is equal to m_{yx}, the orientation of the unique arc between x and y is arbitrarily chosen and its weight is equal to 0 (we may note that the weight $2m_{xy} - m$ is also equal to $m_{xy} - m_{yx}$; we may also observe that we have $m_{xy} + m_{yx} = m$; as a consequence, $2m_{xy} - m$ cannot be equal to 0 if m is odd).

Example. To illustrate these concepts, consider the example below, with $n = 4$ candidates x, y, z, t and $m = 34$ voters whose preferences are the following linear orders:

- for 13 voters: $t > x > y > z$;
- for 11 voters: $z > y > t > x$;
- for 4 voters: $x > y > z > t$;
- for 3 voters: $x > y > t > z$;
- for 3 voters: $z > x > y > t$.

Then we obtain:
$m_{xy} = 23$; $m_{yx} = 11$; $m_{xz} = 20$; $m_{zx} = 14$; $m_{xt} = 10$; $m_{tx} = 24$; $m_{yz} = 20$; $m_{zy} = 14$; $m_{yt} = 21$; $m_{ty} = 13$; $m_{zt} = 18$; $m_{tz} = 16$.

The weighted tournament T_1 summarizing the situation is depicted in Fig. 1, on the left, while the associated unweighted tournament T_2 is given on the right side of Fig. 1.

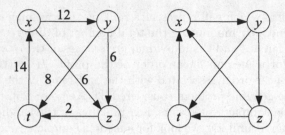

Fig. 1. The weighted tournament T_1 and the unweighted tournament T_2 of the example.

It is easy to see that, in order to destroy all the circuits of T_1, the optimal way consists in removing the two arcs (y, t) and (z, t), with a total weight equal to 10. Reversing the orientation of these arcs provides a transitive tournament which is the median linear order of the profile of the example, namely the order $t > x > y > z$, which is thus the solution of Condorcet-Kemeny's problem for the instance given by the example.

On the other hand, if we consider the unweighted tournament T_2, removing the arc (t, x) is sufficient to destroy all the circuits of T_2. Thus, if we reverse this arc into the arc (x, t), we obtain a linear order, namely $x > y > z > t$, which is the solution of Slater's problem for the instance given by the example.

Through this example, we may observe by the way that the optimal solutions of Condorcet-Kemeny's problem and of Slater's problem are not necessarily the same.

Slater's problem, Condorcet-Kemeny's problem and LOP are NP-hard problems (for references on these complexity aspects, see for instance [1,4–6,11,14,16,18–22]). Hence the interest of designing heuristics and metaheuristics to solve them at least approximately, but within a "reasonable" CPU time.

4   Application of DWM to LOP

The application of DWM to LOP (and thus to Slater's problem or to Condorcet-Kemeny's problem) requires the definition of an elementary transformation. Let O be the current linear order to which we want to apply the elementary transformation. This one consists in considering another linear order O' obtained from O by moving a vertex from its place in O to another place. More precisely, let O be the following linear order, with $x >_O y$ meaning that x is preferred to y with respect to O:

$$x_1 >_O x_2 >_O \ldots >_O x_{i-1} >_O x_i >_O x_{i+1} >_O \ldots >_O x_{j-1} >_O x_j >_O \ldots >_O x_n,$$

Then, if x_i is inserted just before x_j (note that we may also move x_i just after x_n), O' looks like

$$x_1 >_{O'} \ldots >_{O'} x_{i-1} >_{O'} x_{i+1} >_{O'} \ldots >_{O'} x_{j-1} >_{O'} x_i >_{O'} x_j >_{O'} \ldots >_{O'} x_n.$$

Thanks to this elementary transformation, it is possible to generate $n(n-1)$ new linear orders from any linear order O.

To apply a descent, we begin with a linear order randomly chosen and we consider the vertices one after the other, in a cyclic way: the neighbours are sorted in a arbitrary order (which is not necessarily the same during the whole process) and they are all considered in this order once, before being considered for a second time; a neighbour better than the current solution is accepted as soon as it has been discovered (see [15] for this way of exploring the neighbourhood; observe that the way of ordering the vertices is not very important: what is important is to scan every vertex once before considering a same vertex a second time). More precisely, for each vertex x, we compute if there is a place for x which is better than its current place; if so, we insert x at its best place and we go on with the new current linear order; otherwise, we consider the next vertex. When all the vertices are successively considered and no vertex can be moved towards a place better than its current place, the descent is over.

For DWM, we apply almost the same principle but, from time to time, we do not move the considered vertex to its best place, but we move it to a place chosen randomly, with a uniform probability on the possible places. This random move is performed with a probability p, which decreases during the run of the method. The application of DWM to LOP is summarized below (the arrow $\leftarrow$ denotes the affectation; $totNbIter$ is a parameter specified by the user and which is related to the CPU time that the user wants to spend for the run of the method; observe that this relation depends on the speed of the computer, so that it is difficult to be more explicit about this relation here).

- Choose a linear order O randomly;
- $bestSolution \leftarrow O$;
- $numIter \leftarrow 0$;
- while $numIter < totNbIter$, do:
  - $p \leftarrow \frac{totNbIter - numIter}{totNbIter}$;
  - $x \leftarrow 1$;
  - while $x \leq n$, do
    - * choose a real number q between 0 and 1 randomly, with a uniform probability;
    - * if $q < p$, then move x from its current place in O to another place in O randomly chosen, with a uniform probability (we thus change the current linear order O by a mutation);
    - * else compute the best place (with respect to O) for x and update O by moving x to this best place;
    - * if necessary, update $bestSolution$;
    - * $x \leftarrow x + 1$;
  - $numIter \leftarrow numIter + 1$;
- apply a descent to O;
- if necessary, update $bestSolution$;
- return $bestSolution$.

Scheme of DWM for the linear ordering problem.

We can see in the scheme of DWM that the probability $\frac{totNbIter-numIter}{totNbIter}$ of performing a mutation is computed at the beginning of each run of the while-loop and decreases arithmetically, until it reaches 0 (it is also possible to apply a geometrical decrease, as in simulated annealing, but then not down to 0). Of course, we keep the best solution obtained during the process in memory. To be sure to obtain at least a local minimum at the end, we complete the process with a descent before returning the best solution computed since the beginning.

5   Experiments

For our experiments, we compare the results of DWM with those provided by repeated descents (RD in the following) and by a method (SA in the following) which is based on simulated annealing: the differences with respect to a classic simulated annealing is that the neighbourhood is explored in a cyclic way (see above) and that descents (without mutations) are inserted periodically in the process of SA (a more precise description of this method can found in [7,8]). Note that this kind of simulated annealing appears, from previous experiments (see references in [11]), to be among the best methods applied on the problems studied here and that it is much more efficient than a classic simulated annealing.

For the three methods, it is necessary to specify the number of iterations or, equivalently, the CPU time devoted to the computations. In addition, it is also necessary to tune SA, especially the initial temperature, which is not always an easy task (note that there is no parameter to be tuned for RD or for DWM). In a more general context, we developed in [10] an automatically tuned version of SA which only requires the specification of the CPU time that the user wishes to spend in order to solve his or her problem. This version computes, for any given instance, a good solution and also good values for the parameters of SA, especially the initial temperature. The experiments reported in [10] show that this automatically computed initial temperature is very close to the initial temperature carefully tuned by an "expert" with statistical tools. As the automatically tuned version can be a little bit longer than manually tuned versions, we first compute, for each instance of our problems, the "advised" initial temperature with the automatically tuned version and then used this initial temperature in a more classic version needing two parameters: the initial temperature and the total number of iterations (related to the CPU time that the user wishes to spend in order to solve his or her instance). The CPU time required to tune SA is not taken into account in the results described below.

The three methods are compared on the same instances and with the same CPU time (except the tuning time for SA, not taken into account, as just said in the previous paragraph). We report here some results obtained for random graphs; other experimental results lead to the same qualitative conclusions.

More precisely, for any pair of distinct vertices x and y of $T = (X, A)$, we choose the orientation of the arc between x and y with the same probability (0.5) for (x, y) and (y, x). This completely defines an instance for Slater's problem. For Condorcet-Kemeny problem, we then choose the weight of an arc (x, y)

randomly between 0 and 10 with a uniform distribution (we performed other experiments, with other maximum weights: the qualitative conclusions are the same as the ones reported here). For each problem, we report the results obtained for graphs with 100, 200, 300, 400 or 500 vertices; thus we get ten cases (five for Slater's problem and five for Condorcet-Kemeny's problem). For each case, we generated 100 instances randomly. For each instance, we performed DWM, SA and RD twenty times. The same CPU time was given to each method. The averages obtained on the 100 instances of each case and for each method are displayed in Table 1.

Table 1. Average results.

Problem	n	CPU time	DWM	SA	RD	G_{SA}	G_{RD}
Slater	100	3 s	1861.58	1918.41	2158.02	3.05%	15.92%
Slater	200	5 s	9752.42	10048.89	11319.63	3.04%	16.07%
Slater	300	10 s	19028.13	19609.69	22097.68	3.06%	16.13%
Slater	400	40 s	36841.27	38359.63	43306.95	4.12%	17.55%
Slater	500	100 s	54855.39	57858.09	65298.37	5.47%	19.04%
Condorcet-Kemeny	100	3 s	9661.76	9955.92	11180.48	3.04%	15.72%
Condorcet-Kemeny	200	5 s	51802.14	53957.11	61349.27	4.16%	18.43%
Condorcet-Kemeny	300	10 s	100908.90	106054.42	122883.54	5.10%	21.78%
Condorcet-Kemeny	400	40 s	183766.58	196115.70	227925.69	6.72%	24.03%
Condorcet-Kemeny	500	100 s	292831.14	316853.58	369383.67	8.20%	26.14%

In this table, the first column specifies the type of problem considered, while the second provides the number n of vertices. The third column specifies the CPU time given to each method for each instance. The next three columns give, still for each type of problem, the average values of the remoteness ρ among the 100 instances (remember that LOP is a minimization problem: thus, the lowest, the best). The last two columns specify the relative gains provided by DWM with respect to SA or to RD: more precisely, G_{SA} (respectively G_{RD}) gives the average relative gap between DWM and SA (respectively between DWM and RD): G_{SA} is the average value of $(V_{SA} - V_{DWM})/V_{DWM}$ and G_{RD} is the average value of $(V_{RD} - V_{DWM})/V_{DWM}$, where V_{DWM}, V_{SA} and V_{RD} denote respectively the values provided by DWM, SA and RD for each instance.

To be more specific, Fig. 2 shows how the average values of ρ computed by DWM, SA and RD evolve when the CPU time increases, for a given instance of Condorcet-Kemeny's problem with 300 vertices (this behaviour is typical; we observe the same kind of behaviour for the other instances). More precisely, the horizontal axis shows the different CPU times tried in this experiment, from 1 s per trial to 1024 s per trial, with a logarithmic scale. The vertical axis gives, for each one of the three methods, the average value of ρ over 100 trials. The horizontal line just above the axis shows what seems to be the optimal value, i.e. 93482.

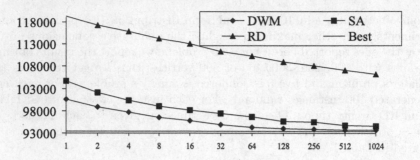

Fig. 2. Evolution of DWM, SA and RD when the CPU time increases.

Another way, more qualitative or at least more ordinal, to compare DWM with SA, is to consider the number of times that a method computes a result which is better than the one computed by the other method (note that DWM and SA are always at least as good as RD with this respect). Table 2 provides such a comparison, still for Slater's problem and for Condorcet-Kemeny's problem. In this table, the first column specifies the considered problem (Slater's problem or Condorcet-Kemeny's problem). The second one gives the number n of vertices (the CPU times are the same as above). Then, the column "DWM < SA" shows how many times the solution computed by DWM is better (i.e., lower) than the one of SA over the 100 instances. Similarly, the column "DWM = SA" (respectively "DWM > SA") shows how many times the solution computed by DWM is equal to (respectively worse than) the one of SA over the 100 instances.

Table 2. Numbers of times that a method is better than the other.

Problem	n	DWM < SA	DWM = SA	DWM > SA
Slater	100	41	12	47
Slater	200	48	9	43
Slater	300	54	7	39
Slater	400	59	5	36
Slater	500	64	4	32
Condorcet-Kemeny	100	53	13	34
Condorcet-Kemeny	200	57	10	33
Condorcet-Kemeny	300	62	6	32
Condorcet-Kemeny	400	66	4	32
Condorcet-Kemeny	500	70	0	30

From these experiments (and other ones lead to the same conclusions), it appears that DWM provides results which are comparable with a sophisticated version of simulated annealing, sometimes better, even if the relative gains are not always very large (from about 3% for Slater's problem with $n = 100$ vertices to a little more than 8% for Condorcet-Kemeny's problem with $n = 500$). The fact that the gap is not large surely comes from the fact that SA is already a very good method: thus it is not easy to improve its results. Anyway, DWM succeeds in performing better than SA in a large majority of instances studied here. We may also observe an interesting fact: DWM becomes still better when the number n of vertices increases.

6 Conclusion

By studying DWM, the aim is not to add one more method to the long list of possible metaheuristics, but to point out that this very simple method is experimentally as efficient as the most sophisticated ones, at least for some combinatorial problems, like the ones studied here. Indeed, the results obtained for LOP show that DWM may provide very good results, with about the same quality than the ones obtained by an improved version of simulated annealing, which proved to be already very efficient, within the same CPU time. In fact, as said above, the main advantage of DWM is that there is no parameter to tune (except the CPU time, which in its turn defines the number of iterations performed by the method). The aim of future works will be to investigate the application of DWM to other combinatorial problems (we already did it for the aggregation of symmetric relations or of equivalence relations into a median equivalence relation, with the same type of conclusion; see [23]), and then we hope that DWM will succeed in finding good solutions for them within a reasonable CPU time.

Acknowledgements. I thank the referees for their valuable comments.

References

1. Alon, N.: Ranking tournaments. SIAM J. Discret. Math. **20**(1), 137–142 (2006)
2. Barthélemy, J.-P.: Caractérisations axiomatiques de la distance de la différence symétrique entre des relations binaires. Mathématiques et Sciences Humaines **67**, 85–113 (1979)
3. Barthélemy, J.-P., Monjardet, B.: The median procedure in cluster analysis and social choice theory. Math. Soc. Sci. **1**, 235–267 (1981)
4. Bartholdi III, J.J., Tovey, C.A., Trick, M.A.: Voting schemes for which it can be difficult to tell who won the election. Soc. Choice Welf. **6**, 157–165 (1989)
5. Brandt, F., Conitzer, V., Endriss, U., Lang, J., Procaccia, A. (eds.): Handbook of Computational Social Choice. Cambridge University Press, New York (2016)
6. Charbit, P., Thomasse, S., Yeo, A.: The minimum feedback arc set problem is NP-hard for tournaments. Comb. Prob. Comput. **16**(1), 1–4 (2007)
7. Charon, I., Hudry, O.: Mixing different components of metaheuristics. In: Osman, I.H., Kelly, J.P. (eds.) Metaheuristics: Theory and Applications, pp. 589–603. Kluwer Academic Publishers, Boston (1996)

8. Charon, I., Hudry, O.: Lamarckian genetic algorithms applied to the aggregation of preferences. Ann. Oper. Res. **80**, 281–297 (1998)

9. Charon, I., Hudry, O.: Application of the noising methods to the travelling salesman problem. Eur. J. Oper. Res. **125**(2), 266–277 (2000)

10. Charon, I., Hudry, O.: Self-tuning of the noising methods. Optimization **58**(7), 1–21 (2009)

11. Charon, I., Hudry, O.: An updated survey on the linear ordering problem for weighted or unweighted tournaments. Ann. Oper. Res. **175**, 107–158 (2010)

12. Charon, I., Hudry, O.: The noising methods. In: Siarry, P. (ed.) Heuristics: Theory and Applications, pp. 1–30. Nova Publishers, New York (2013)

13. Caritat, M.J.A.N., marquis de Condorcet: Essai sur l'application de l'analyse à la probabilité des décisions rendues à la pluralité des voix. Imprimerie royale, Paris (1785)

14. Conitzer, V.: Computing slater rankings using similarities among candidates. In: Proceedings of the 21st National Conference on Artificial Intelligence (AAAI 2006), Boston, MA, USA, pp. 613–619 (2006)

15. Dowsland, K.A.: Simulated annealing. In: Reeves, C. (ed.) Modern Heuristic Techniques for Combinatorial Problems, pp. 20–69. McGraw-Hill, London (1995)

16. Dwork, C., Kumar, R., Naor, M., Sivakumar, D.: Rank aggregation methods for the Web. In: Proceedings of the 10th International Conference on World Wide Web (WWW 2010), Hong Kong, pp. 613–622 (2001)

17. Gendreau, M., Potvin, J.-Y. (eds.): Handbook of Metaheuristics. International Series in Operations Research & Management Science, vol. 146. Springer, New York (2010). https://doi.org/10.1007/978-1-4419-1665-5

18. Hemaspaandra, E., Spakowski, H., Vogel, J.: The complexity of Kemeny elections. Theor. Comput. Sci. **349**, 382–391 (2005)

19. Hudry, O.: NP-hardness results on the aggregation of linear orders into median orders. Ann. Oper. Res. **163**(1), 63–88 (2008)

20. Hudry, O.: Complexity of voting procedures. In: Meyers, R. (ed.) Encyclopedia of Complexity and Systems Science, pp. 9942–9965. Springer, New York (2009)

21. Hudry, O.: On the complexity of Slater's problems. Eur. J. Oper. Res. **203**, 216–221 (2010)

22. Hudry, O.: On the computation of median linear orders, of median complete preorders and of median weak orders. Math. Soc. Sci. **64**, 2–10 (2012)

23. Hudry, O.: Application of the "descent with mutations" metaheuristic to a clique partitioning problem. In: 2007 IEEE International Conference on Conference: Research, Innovation and Vision for the Future (2007)

24. Jünger, M.: Polyhedral Combinatorics and the Acyclic Subdigraph Problem. Heldermann Verlag, Berlin (1985)

25. Kemeny, J.G.: Mathematics without numbers. Daedalus **88**, 577–591 (1959)

26. Reinelt, G.: The Linear Ordering Problem: Algorithms and Applications. Research and Exposition in Mathematics, vol. 8. Heldermann Verlag, Berlin (1985)

27. Siarry, P. (ed.): Heuristics: Theory and Applications. Nova Publishers, New York (2013)

28. Slater, P.: Inconsistencies in a schedule of paired comparisons. Biometrika **48**, 303–312 (1961)

29. Talbi, E.-G.: Metaheuristics: From Design to Implementation. Wiley, Hoboken (2009)

Characterization and Approximation of Strong General Dual Feasible Functions

Matthias Köppe[✉] and Jiawei Wang

Department of Mathematics, University of California, Davis, Davis, USA
mkoeppe@math.ucdavis.edu, jwewang@ucdavis.edu

Abstract. Dual feasible functions (DFFs) have been used to provide bounds for standard packing problems and valid inequalities for integer optimization problems. In this paper, the connection between general DFFs and a particular family of cut-generating functions is explored. We find the characterization of (restricted/strongly) maximal general DFFs and prove a 2-slope theorem for extreme general DFFs. We show that any restricted maximal general DFF can be well approximated by an extreme general DFF.

Keywords: Dual feasible functions · Cut-generating functions
Integer programming · 2-slope theorem

1 Introduction

Dual feasible functions (DFFs) are a fascinating family of functions $\phi\colon [0,1] \to [0,1]$, which have been used in several combinatorial optimization problems and proved to generate dual bounds efficiently. DFFs are in the scope of superadditive duality theory, and superadditive and nondecreasing DFFs can provide valid inequalities for general integer linear programs. Lueker [17] studied the bin-packing problems and used certain DFFs to obtain lower bounds for the first time. Vanderbeck [20] proposed an exact algorithm for the cutting stock problems which includes adding valid inequalities generated by DFFs. Rietz et al. [18] recently introduced a variant of this theory, in which the domain of DFFs is extended to all real numbers. Rietz et al. [19] studied the maximality of the so-called "general dual feasible functions." They also summarized recent literature on (general) DFFs in the monograph [1]. In this paper, we follow the notions in [1] and study the general DFFs.

Cut-generating functions play an essential role in generating valid inequalities which cut off the current fractional basic solution in a simplex-based cutting

The authors gratefully acknowledge partial support from the National Science Foundation through grant DMS-1320051 (M. Köppe). A part of this work was done while the first author (M. Köppe) was visiting the Simons Institute for the Theory of Computing. It was partially supported by the DIMACS/Simons Collaboration on Bridging Continuous and Discrete Optimization through NSF grant CCF-1740425.

© Springer International Publishing AG, part of Springer Nature 2018
J. Lee et al. (Eds.): ISCO 2018, LNCS 10856, pp. 265–276, 2018.
https://doi.org/10.1007/978-3-319-96151-4_23

plane procedure. Gomory and Johnson [10,11] first studied the corner relaxation of integer linear programs, which is obtained by relaxing the non-negativity of basic variables in the tableau. Gomory–Johnson cut-generating functions are critical in the superadditive duality theory of integer linear optimization problems, and they have been used in the state-of-art integer program solvers. Köppe and Wang [16] discovered a conversion from minimal Gomory–Johnson cut-generating functions to maximal DFFs.

Yıldız and Cornuéjols [21] introduced a generalized model of Gomory–Johnson cut-generating functions. In the single-row Gomory–Johnson model, the basic variables are in $\mathbb{Z}$. Yıldız and Cornuéjols considered the basic variables to be in any set $S \subset \mathbb{R}$. Their results extended the characterization of minimal Gomory–Johnson cut-generating functions in terms of the generalized symmetry condition. Inspired by the characterization of minimal Yıldız–Cornuéjols cut-generating functions, we complete the characterization of maximal general DFFs.

We connect general DFFs to the classic model studied by Jeroslow [14], Blair [9] and Bachem et al. [3] and a relaxation of their model, both of which can be studied in the Yıldız–Cornuéjols model [21] with various sets S. General DFFs generate valid inequalities for the model with $S = (-\infty, 0]$, and certain cut-generating functions generate valid inequalities for the Jeroslow model where $S = \{0\}$. The relation between these two families of functions is explored.

Another focus of this paper is on the extremality of general DFFs. In terms of Gomory–Johnson cut-generating functions, the 2-slope theorem is a famous result of Gomory and Johnson's masterpiece [10,11]. Basu et al. [8] proved that the 2-slope extreme Gomory–Johnson cut-generating functions are dense in the set of continuous minimal functions. We show that any 2-slope maximal general DFF with one slope value 0 is extreme. This result is a key step in our approximation theorem, which indicates that almost all continuous maximal general DFFs can be approximated by extreme (2-slope) general DFFs as close as we desire. In contrast to the fill-in procedure Basu et al. [8] used, our 2-slope fill-in procedure uses 0 as one slope value, which is necessary since the 2-slope theorem of general DFFs requires 0 to be one slope value.

This paper is structured as follows. In Sect. 2, we provide the preliminaries of DFFs from the monograph [1]. The characterizations of maximal, restricted maximal and strongly maximal general DFFs are described in Sect. 3. In Sect. 4, we explore the relation between general DFFs and a particular family of cut-generating functions. The 2-slope theorem for extreme general DFFs is studied in Sect. 5. In Sect. 6, we introduce our approximation theorem, adapting a parallel construction in Gomory–Johnson's setting [8].

2   Literature Review

Definition 1. *A function* $\phi\colon [0,1] \to [0,1]$ *is called a (valid) classical Dual Feasible Function (cDFF), if for any finite list of real numbers* $x_i \in [0,1]$, $i \in I$, *it holds that* $\sum_{i \in I} x_i \leq 1 \Rightarrow \sum_{i \in I} \phi(x_i) \leq 1$. *A function* $\phi\colon \mathbb{R} \to \mathbb{R}$ *is called* a *(valid)* general Dual Feasible Function *(gDFF)*, if for any finite list of real numbers $x_i \in \mathbb{R}$, $i \in I$, it holds that $\sum_{i \in I} x_i \leq 1 \Rightarrow \sum_{i \in I} \phi(x_i) \leq 1$.

We are interested in so-called "maximal" functions since they yield better bounds and stronger valid inequalities. A cDFF/gDFF is *maximal* if it is not (pointwise) dominated by a distinct cDFF/gDFF. A cDFF/gDFF is *extreme* if it cannot be written as a convex combination of other two different cDFFs/gDFFs.

Theorem 1 ([1, Theorem 2.1]). *A function $\phi: [0,1] \to [0,1]$ is a maximal cDFF if and only if $\phi(0) = 0$, ϕ is superadditive and ϕ is symmetric in the sense $\phi(x) + \phi(1-x) = 1$.*

Theorem 2 ([1, Theorem 3.1]). *Let $\phi: \mathbb{R} \to \mathbb{R}$ be a given function. If ϕ satisfies the following conditions, then ϕ is a maximal gDFF: (i) $\phi(0) = 0$. (ii) ϕ is symmetric in the sense $\phi(x) + \phi(1-x) = 1$. (iii) ϕ is superadditive. (iv) There exists an $\epsilon > 0$ such that $\phi(x) \geq 0$ for all $x \in (0, \epsilon)$.*
If ϕ is a maximal gDFF, then ϕ satisfies conditions (i), (iii) and (iv).

Remark 1. The function $\phi(x) = cx$ for $0 \leq c < 1$ is a maximal gDFF but it does not satisfy condition *(ii)*. Note that conditions *(i)*, *(iii)* and *(iv)* guarantee that any maximal gDFF is nondecreasing and consequently nonnegative on $\mathbb{R}_+$.

Proposition 1 shows that any maximal gDFF is the sum of a linear function and a bounded function. Proposition 2 explains the behavior of nonlinear maximal gDFFs at given points. Proposition 3 uses gDFFs to generate valid inequalities for general linear integer optimization problems.

Proposition 1 ([1, Proposition 3.4]). *If $\phi: \mathbb{R} \to \mathbb{R}$ is a maximal gDFF and $t = \sup\{\frac{\phi(x)}{x} : x > 0\}$. Then we have $\lim_{x \to \infty} \frac{\phi(x)}{x} = t \leq -\phi(-1)$, and for any $x \in \mathbb{R}$, it holds that: $tx - \max\{0, t-1\} \leq \phi(x) \leq tx$.*

Proposition 2 ([1, Proposition 3.5]). *If $\phi: \mathbb{R} \to \mathbb{R}$ is a maximal gDFF and not of the kind $\phi(x) = cx$ for $0 \leq c < 1$, then $\phi(1) = 1$ and $\phi(\frac{1}{2}) = \frac{1}{2}$.*

Proposition 3 ([1, Proposition 5.1]). *If ϕ is a maximal gDFF and $L = \{x \in \mathbb{Z}_+^n : \sum_{j=1}^n a_{ij} x_j \leq b_i, i = 1, 2, \ldots, m\}$, then for any i, $\sum_{j=1}^n \phi(a_{ij}) x_j \leq \phi(b_i)$ is a valid inequality for L.*

3   Characterization of Maximal General DFFs

Alves et al. [1] provided several sufficient conditions and necessary conditions of maximal gDFFs in Theorem 2, but they do not match precisely. Inspired by the characterization of minimal cut-generating functions in the Yıldız–Cornuéjols model [21], we complete the characterization of maximal gDFFs.

Proposition 4. *A function $\phi: \mathbb{R} \to \mathbb{R}$ is a maximal gDFF if and only if the following conditions hold: (i) $\phi(0) = 0$. (ii) ϕ is superadditive. (iii) $\phi(x) \geq 0$ for all $x \in \mathbb{R}_+$. (iv) ϕ satisfies the generalized symmetry condition in the sense $\phi(r) = \inf_k\{\frac{1}{k}(1 - \phi(1 - kr)) : k \in \mathbb{Z}_+\}$.*

Proof. Suppose ϕ is a maximal gDFF, then conditions $(i), (ii), (iii)$ hold by Theorem 2. For any $r \in \mathbb{R}$ and $k \in \mathbb{Z}_+$, $kr + (1 - kr) = 1 \Rightarrow k\phi(r) + \phi(1 - kr) \leq 1$. So $\phi(r) \leq \frac{1}{k}(1 - \phi(1 - kr))$ for any positive integer k, then $\phi(r) \leq \inf_k \{ \frac{1}{k}(1 - \phi(1 - kr)) : k \in \mathbb{Z}_+ \}$.

If there exists r_0 such that $\phi(r_0) < \inf_k \{ \frac{1}{k}(1 - \phi(1 - kr_0)) : k \in \mathbb{Z}_+ \}$, then define a function ϕ_1 which takes value $\inf_k \{ \frac{1}{k}(1 - \phi(1 - kr_0)) : k \in \mathbb{Z}_+ \}$ at r_0 and $\phi(r)$ if $r \neq r_0$. We claim that ϕ_1 is a gDFF which dominates ϕ. Given a function $y : \mathbb{R} \to \mathbb{Z}_+$ with finite support and satisfying $\sum_{r \in \mathbb{R}} r\, y(r) \leq 1$, we have $\sum_{r \in \mathbb{R}} \phi_1(r) y(r) = \phi_1(r_0) y(r_0) + \sum_{r \neq r_0} \phi(r) y(r)$. If $y(r_0) = 0$, then it is clear that $\sum_{r \in \mathbb{R}} \phi_1(r) y(r) \leq 1$. Let $y(r_0) \in \mathbb{Z}_+$, then $\phi_1(r_0) \leq \frac{1}{y(r_0)}(1 - \phi(1 - y(r_0) r_0))$ by definition of ϕ_1, then $\phi_1(r_0) y(r_0) + \phi(1 - y(r_0) r_0) \leq 1$. From the superadditive condition and increasing property, we get $\sum_{r \neq r_0} \phi(r) y(r) \leq \phi(\sum_{r \neq r_0} r\, y(r)) \leq \phi(1 - y(r_0) r_0)$. From the two inequalities we conclude that ϕ_1 is a gDFF and dominates ϕ, which contradicts the maximality of ϕ. So the condition (iv) holds.

Suppose there is a function $\phi : \mathbb{R} \to \mathbb{R}$ satisfying all four conditions. Choose $r = 1$ and $k = 1$, we can get $\phi(1) \leq 1$. Together with conditions $(i), (ii), (iii)$, we conclude that ϕ is a gDFF. Assume that there is a gDFF ϕ_1 dominating ϕ and there exists r_0 such that $\phi_1(r_0) > \phi(r_0) = \inf_k \{ \frac{1}{k}(1 - \phi(1 - kr_0)) : k \in \mathbb{Z}_+ \}$. So there exists some $k \in \mathbb{Z}_+$ such that

$$\phi_1(r_0) > \frac{1}{k}(1 - \phi(1 - kr_0))$$
$$\Leftrightarrow k\phi_1(r_0) + \phi(1 - kr_0) > 1$$
$$\Rightarrow k\phi_1(r_0) + \phi_1(1 - kr_0) > 1.$$

The last step contradicts the fact that ϕ_1 is a gDFF. Therefore, ϕ is maximal.

Parallel to the restricted minimal and strongly minimal functions in the Yıldız–Cornuéjols model [21], "restricted maximal" and "strongly maximal" gDFFs are defined by strengthening the notion of maximality.

Definition 2. *We say that a gDFF ϕ is implied via scaling by a gDFF ϕ_1, if $\beta\phi_1 \geq \phi$ for some $0 \leq \beta \leq 1$. We call a gDFF $\phi : \mathbb{R} \to \mathbb{R}$ restricted maximal if ϕ is not implied via scaling by a distinct gDFF ϕ_1. We say that a gDFF ϕ is implied by a gDFF ϕ_1, if $\phi(x) \leq \beta\phi_1(x) + \alpha x$ for some $0 \leq \alpha, \beta \leq 1$ and $\alpha + \beta \leq 1$. We call a gDFF $\phi : \mathbb{R} \to \mathbb{R}$ strongly maximal if ϕ is not implied by a distinct gDFF ϕ_1.*

Suppose a gDFF ϕ is not strongly maximal, or equivalently ϕ is implied by a distinct gDFF ϕ_1, then the following inequalities indicate that $\sum \phi_1(x) \leq 1$ is stronger than $\sum \phi(x) \leq 1$. Similar conclusion can be drawn for non-restricted maximal gDFFs.

$$\sum \phi(x) \leq \sum (\beta\phi_1(x) + \alpha x) = \beta \sum \phi_1(x) + \alpha \sum x \leq \beta + \alpha \leq 1.$$

Note that restricted maximal gDFFs are maximal and strongly maximal gDFFs are restricted maximal. Based on the definition of strong maximality, $\phi(x) = x$ is implied by the zero function, so ϕ is not strongly maximal, though it is extreme. We include the characterizations of restricted maximal and strongly maximal gDFFs here, which only involve the standard symmetry condition instead of the generalized symmetry condition. The proofs of the characterizations are omitted since they can be easily adapted from the proof of Proposition 4.

Theorem 3. *A function $\phi\colon \mathbb{R} \to \mathbb{R}$ is a restricted maximal gDFF if and only if ϕ is a maximal gDFF and $\phi(x) + \phi(1-x) = 1$.*

Theorem 4. *A function $\phi\colon \mathbb{R} \to \mathbb{R}$ is a strongly maximal gDFF if and only if ϕ is a restricted maximal gDFF and $\lim_{\epsilon \to 0^+} \frac{\phi(\epsilon)}{\epsilon} = 0$.*

Remark 2. Let ϕ be a maximal gDFF that is not linear, we know that $\phi(1) = 1$ from Proposition 2. If ϕ is implied via scaling by a gDFF ϕ_1, or equivalently $\beta\phi_1 \geq \phi$ for some $0 \leq \beta \leq 1$, then $\beta\phi_1(1) \geq \phi(1)$. Then $\beta = 1$ and ϕ is dominated by ϕ_1. The maximality of ϕ implies $\phi = \phi_1$, so ϕ is restricted maximal. Therefore, we have a simpler version of characterization of maximal gDFFs.

Theorem 5. *A function $\phi\colon \mathbb{R} \to \mathbb{R}$ is a maximal gDFF if and only if the following conditions hold:*

  (i) $\phi(0) = 0$.
 (ii) ϕ *is superadditive.*
(iii) $\phi(x) \geq 0$ *for all* $x \in \mathbb{R}_+$.
(iv) $\phi(x) + \phi(1-x) = 1$ *or* $\phi(x) = cx$, $0 \leq c < 1$.

The following theorem indicates that maximal, restricted maximal and strongly maximal gDFFs exist, and they are potentially stronger than just valid gDFFs. The proof is analogous to the proof of [21, Theorem 1, Proposition 6, Theorem 9] and is therefore omitted.

Theorem 6. *(i) Every gDFF is dominated by a maximal gDFF.*
(ii) Every gDFF is implied via scaling by a restricted maximal gDFF.
(iii) Every nonlinear gDFF is implied by a strongly maximal gDFF.

4   Relation to Cut-Generating Functions

We define an infinite dimensional space Y called "the space of nonbasic variables" as $Y = \{y : y\colon \mathbb{R} \to \mathbb{Z}_+$ and y has finite support$\}$, and we refer to the zero function as the origin of Y. In this section, we study valid inequalities of certain subsets of the space Y and connect gDFFs to a particular family of cut-generating functions.

Yıldız and Cornuéjols [21] considered the following generalization of the Gomory–Johnson model:

$$x = f + \sum_{r \in \mathbb{R}} r\, y(r) \tag{1}$$

$$x \in S, \ y : \mathbb{R} \to \mathbb{Z}_+, \text{ and } y \text{ has finite support.}$$

where S can be any nonempty subset of $\mathbb{R}$. A function $\pi : \mathbb{R} \to \mathbb{R}$ is called a *valid cut-generating function* if the inequality $\sum_{r \in \mathbb{R}} \pi(r) \, y(r) \geq 1$ holds for all feasible solutions (x, y) to (1). In order to ensure that such cut-generating functions exist, they only consider the case $f \notin S$. Otherwise, if $f \in S$, then $(x, y) = (f, 0)$ is a feasible solution and there is no function π which can make the inequality $\sum_{r \in \mathbb{R}} \pi(r) \, y(r) \geq 1$ valid. Note that all valid inequalities in the form of $\sum_{r \in \mathbb{R}} \pi(r) \, y(r) \geq 1$ to (1) are inequalities which separate the origin of Y.

We consider two different but related models in the form of (1). Let $f = -1$, $S = \{0\}$, and the feasible region $Y_{=1} = \{y : \sum_{r \in \mathbb{R}} r \, y(r) = 1, \ y \colon \mathbb{R} \to \mathbb{Z}_+ \text{ and } y \text{ has finite support}\}$. Let $f = -1$, $S = (-\infty, 0]$, and the feasible region $Y_{\leq 1} = \{y : \sum_{r \in \mathbb{R}} r \, y(r) \leq 1, \ y \colon \mathbb{R} \to \mathbb{Z}_+ \text{ and } y \text{ has finite support}\}$. It is immediate to check that the latter model is the relaxation of the former. Therefore $Y_{=1} \subsetneq Y_{\leq 1}$ and any valid inequality for $Y_{\leq 1}$ is also valid for $Y_{=1}$.

Jeroslow [14], Blair [9] and Bachem et al. [3] studied minimal valid inequalities of the set $Y_{=b} = \{y : \sum_{r \in \mathbb{R}} r \, y(r) = b, \ y \colon \mathbb{R} \to \mathbb{Z}_+ \text{ and } y \text{ has finite support}\}$. Note that $Y_{=b}$ is the set of feasible solutions to (1) for $S = \{0\}$, $f = -b$. The notion "minimality" they used is in fact the restricted minimality in the Yıldız–Cornuéjols model. In this section, we use the terminology introduced by Yıldız and Cornuéjols. Jeroslow [14] showed that finite-valued subadditive (restricted minimal) functions are sufficient to generate all necessary valid inequalities of $Y_{=b}$ for bounded mixed integer programs. Kılınç-Karzan and Yang [15] discussed whether finite-valued functions are sufficient to generate all necessary inequalities for the convex hull description of disjunctive sets. Interested readers are referred to [15] for more details on the sufficiency question. Blair [9] extended Jeroslow's result to rational mixed integer programs. Bachem et al. [3] characterized restricted minimal cut-generating functions under some continuity assumptions, and showed that restricted minimal functions satisfy the symmetry condition.

In terms of the relaxation $Y_{\leq 1}$, gDFFs can generate the valid inequalities in the form of $\sum_{r \in \mathbb{R}} \phi(r) \, y(r) \leq 1$, and such inequalities do not separate the origin. Note that there is no valid inequality separating the origin since $0 \in Y_{\leq 1}$.

Cut-generating functions provide valid inequalities which separate the origin for $Y_{=1}$, but such inequalities are not valid for $Y_{\leq 1}$. In terms of inequalities that do not separate the origin, any inequality in the form of $\sum_{r \in \mathbb{R}} \phi(r) \, y(r) \leq 1$ generated by some gDFF ϕ is valid for $Y_{\leq 1}$ and hence valid for $Y_{=1}$, since the model of $Y_{\leq 1}$ is the relaxation of that of $Y_{=1}$. Clearly, there also exist valid inequalities which do not separate the origin for $Y_{=1}$ but are not valid for $Y_{\leq 1}$.

Yıldız and Cornuéjols [21] introduced the notions of minimal, restricted minimal and strongly minimal cut-generating functions. We call the cut-generating functions to the model (1) when $f = -1$, $S = \{0\}$ cut-generating functions for $Y_{=1}$. We restate the definitions and characterizations of minimality of such cut-generating functions for $Y_{=1}$. A valid cut-generating function π is called *minimal* if it does not dominate another valid cut-generating function π'. A cut-generating function π' implies a cut-generating function π via scaling if there exists $\beta \geq 1$ such that $\pi \geq \beta\pi'$. A valid cut-generating function π is *restricted minimal* if there is no another cut-generating function π' implying π via scaling. A cut-generating function π' implies a cut-generating function π if there exist α, β, and $\beta \geq 0, \alpha + \beta \geq 1$ such that $\pi(x) \geq \beta\pi'(x) + \alpha x$. A valid cut-generating function π is *strongly minimal* if there is no another cut-generating function π' implying π. As for the strong minimality and extremality, they mainly focused on the case where $f \in \overline{\text{conv}(S)}$ and $\overline{\text{conv}(S)}$ is full-dimensional.

Theorem 7. *A function $\pi\colon \mathbb{R} \to \mathbb{R}$ is a minimal cut-generating function for $Y_{=1}$ if and only if $\pi(0) = 0$, π is subadditive, and $\pi(r) = \sup_k\{\frac{1}{k}(1 - \pi(1 - kr)) : k \in \mathbb{Z}_+\}$.*

Theorem 8. *A function $\pi\colon \mathbb{R} \to \mathbb{R}$ is a restricted minimal cut-generating function for $Y_{=1}$ if and only if π is minimal and $\pi(1) = 1$.*

We show that gDFFs are closely related to cut-generating functions for $Y_{=1}$. The main idea is that valid inequalities generated by cut-generating functions for $Y_{=1}$ can be lifted to valid inequalities generated by gDFFs for the relaxation $Y_{\leq 1}$. The procedure involves adding a multiple of the defining equality $\sum_{r\in\mathbb{R}} r\, y(r) = 1$ to a valid inequality, which is called "tilting" by Aráoz et al. [2].

The following theorem describes the conversion between gDFFs and cut-generating functions for $Y_{=1}$. We omit the proof which is a straightforward computation, utilizing the characterization of (restricted) maximal gDFFs and (restricted) minimal cut-generating functions.

Theorem 9. *Given a valid/maximal/restricted maximal gDFF ϕ, then for every $0 < \lambda < 1$, the following function is a valid/minimal/restricted minimal cut-generating function for $Y_{=1}$:*

$$\pi_\lambda(x) = \frac{x - (1 - \lambda)\,\phi(x)}{\lambda}.$$

Given a valid/minimal/restricted minimal cut-generating function π for $Y_{=1}$, which is Lipschitz continuous at $x = 0$, then there exists $\delta > 0$ such that for all $0 < \lambda < \delta$ the following function is a valid/maximal/restricted maximal gDFF:

$$\phi_\lambda(x) = \frac{x - \lambda\,\pi(x)}{1 - \lambda}, \quad 0 < \lambda < 1.$$

Remark 3. We discuss the distinctions between these two families of functions.

(i) It is not hard to prove that extreme gDFFs are always maximal. However, unlike cut-generating functions for $Y_{=1}$, extreme gDFFs are not always restricted maximal. $\phi(x) = 0$ is an extreme gDFF but not restricted maximal.

(ii) By applying the proof of [21, Proposition 28], we can show that no strongly minimal cut-generating function for $Y_{=1}$ exists. However, there do exist strongly maximal gDFFs by Theorem 6. Moreover, we can use the same conversion formula in Theorem 9 to convert a restricted minimal cut-generating function to a strongly maximal gDFF (see Theorem 10 below). In fact, it suffices to choose a proper λ such that $\lim_{\epsilon \to 0+} \frac{\phi_\lambda(\epsilon)}{\epsilon} = 0$ by the characterization of strongly maximal gDFFs (Theorem 4).

(iii) There is no extreme piecewise linear cut-generating function π for $Y_{=1}$ which is Lipschitz continuous at $x = 0$, except for $\pi(x) = x$. If π is such an extreme function, then for any λ small enough, we claim that ϕ_λ is an extreme gDFF. Suppose $\phi_\lambda = \frac{1}{2}\phi^1 + \frac{1}{2}\phi^2$ and let $\pi_\lambda^1, \pi_\lambda^2$ be the corresponding cut-generating functions of ϕ^1, ϕ^2 by Theorem 9. Note that $\pi = \frac{1}{2}(\pi_\lambda^1 + \pi_\lambda^2)$, which implies $\pi = \pi_\lambda^1 = \pi_\lambda^2$ and $\phi_\lambda = \phi_\lambda^1 = \phi_\lambda^2$. Thus ϕ_λ is extreme. By Lemma 1 in the next section and the extremality of ϕ_λ, we know $\phi_\lambda(x) = x$ or there exists $\epsilon > 0$, such that $\phi_\lambda(x) = 0$ for $x \in [0, \epsilon)$. If $\phi_\lambda(x) = x$, then $\pi(x) = x$. Otherwise, $\lim_{x \to 0+} \frac{\phi_\lambda(x)}{x} = 0$ for any small enough λ.

$$0 = \lim_{x \to 0+} \frac{\phi_\lambda(x)}{x} = \lim_{x \to 0+} \frac{x - \lambda\pi(x)}{(1 - \lambda)x} = \frac{1 - \lambda \lim_{x \to 0+} \frac{\pi(x)}{x}}{1 - \lambda}.$$

The above equation implies $\lim_{x \to 0+} \frac{\pi(x)}{x} = \frac{1}{\lambda}$ for any small enough λ, which is not possible. Therefore, π cannot be extreme except for $\pi(x) = x$.

Theorem 10. *Given a non-linear restricted minimal cut-generating function π for $Y_{=1}$, which is Lipschitz continuous at 0, then there exists $\lambda > 0$ such that the following function is a strongly maximal gDFF:*

$$\phi_\lambda(x) = \frac{x - \lambda\pi(x)}{1 - \lambda}.$$

5   Two-Slope Theorem

In this section, we prove a 2-slope theorem for extreme gDFFs, in the spirit of the 2-slope theorem of Gomory and Johnson [10,11].

Lemma 1. *Let ϕ be a piecewise linear[1] extreme gDFF.*

(i) *If ϕ is strictly increasing, then $\phi(x) = x$.*
(ii) *If ϕ is not strictly increasing, then there exists $\epsilon > 0$, such that $\phi(x) = 0$ for $x \in [0, \epsilon)$.*

Proof. We provide a proof sketch.

By studying the superadditivity of maximal gDFFs, it is not hard to prove that ϕ is continuous at 0 from the right. Suppose $\phi(x) = sx$, $x \in [0, x_1)$ and $s > 0$. We claim $0 \le s < 1$ due to maximality of ϕ and $\phi(1) = 1$. Define a function: $\phi_1(x) = \frac{\phi(x) - sx}{1-s}$, and it is straightforward to show that ϕ_1 is maximal, and $\phi(x) = sx + (1 - s)\phi_1(x)$. From the extremality of ϕ, $s = 0$ or $\phi(x) = x$.

---

[1] We will use the term "piecewise linear" throughout the paper without explanation. We refer readers to [13] for precise definitions of "piecewise linear" functions in both continuous and discontinuous cases.

From Lemma 1, we know 0 must be one slope value of a piecewise linear extreme gDFF ϕ, except for $\phi(x) = x$. Next, we introduce the 2-slope theorem for extreme gDFFs. The proof is parallel to the proof of the Gomory–Johnson's 2-slope theorem and therefore omitted.

Theorem 11. *Let ϕ be a continuous piecewise linear strongly maximal gDFF with only 2 slope values, then ϕ is extreme.*

Remark 4. Alves et al. [1] claimed the following functions by Burdet and Johnson with one parameter $C \geq 1$ are maximal gDFFs, where $\{a\}$ represents the fractional part of a.

$$\phi_{BJ,1}(x;C) = \frac{\lfloor Cx \rfloor + \max(0, \frac{\{Cx\}-\{C\}}{1-\{C\}})}{\lfloor C \rfloor}.$$

Actually we can prove that they are extreme. If $C \in \mathbb{N}$, then $\phi_{BJ,1}(x) = x$. If $C \notin \mathbb{N}$, $\phi_{BJ,1}$ is a continuous 2-slope maximal gDFF with one slope value 0, therefore it is extreme by Theorem 11. Figure 1 shows two examples of $\phi_{BJ,1}$ and they are constructed by the Python function `phi_1_bj_gdff`[2].

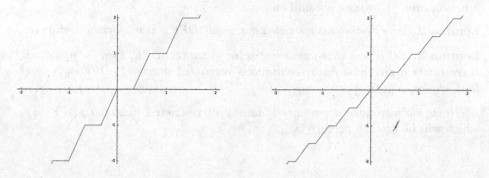

Fig. 1. $\phi_{BJ,1}$ [1, Example 3.1] for $C = 3/2$ (left) and $C = 7/3$ (right).

6 Restricted Maximal General DFFs Are Almost Extreme

In this section, we prove that extreme gDFFs are dense in the set of continuous restricted maximal gDFFs. Equivalently, for any given continuous restricted maximal gDFF ϕ, there exists an extreme gDFF ϕ_{ext} which approximates ϕ as close as desired (with the infinity norm). The idea of the proof is inspired by the

---

[2] In this paper, a function name shown in typewriter font is the name of the function in our SageMath program [12]. At the time of writing, the function is available on the feature branch `gdff`. Later it will be merged into the `master` branch.

approximation theorem of Gomory–Johnson functions [8]. We first introduce the main theorem in this section. The approximation[3] is implemented for piecewise linear functions with finitely many pieces.

Theorem 12. *Let ϕ be a continuous restricted maximal gDFF, then for any $\epsilon > 0$, there exists an extreme gDFF ϕ_{ext} such that $\|\phi - \phi_{ext}\|_\infty < \epsilon$.*

Remark 5. The result cannot be extended to maximal gDFF. $\phi(x) = ax$ is maximal but not extreme for $0 < a < 1$. Any non-trivial extreme gDFF ϕ' satisfies $\phi'(1) = 1$. $\phi'(1) - \phi(1) = 1 - a > 0$ and $1 - a$ is a fixed positive constant.

We briefly explain the structure of the proof. Similar to [4–7], we introduce a function $\nabla\phi \colon \mathbb{R} \times \mathbb{R} \to \mathbb{R}$, $\nabla\phi(x,y) = \phi(x+y) - \phi(x) - \phi(y)$, which measures the slack in the superadditivity condition. First we approximate a continuous restricted maximal gDFF ϕ by a piecewise linear maximal gDFF ϕ_{pwl}. Next, we perturb ϕ_{pwl} such that the new maximal gDFF ϕ_{loose} satisfies $\nabla\phi_{loose}(x,y) > \gamma > 0$ for "most" $(x,y) \in \mathbb{R}^2$. After applying the 2-slope fill-in procedure to ϕ_{loose}, we get a superadditive 2-slope function $\phi_{fill-in}$, which is not symmetric anymore. Finally, we symmetrize $\phi_{fill-in}$ to get the desired ϕ_{ext}.

By studying the superadditivity of maximal gDFFs near the origin, it is not hard to prove Lemma 2. By choosing a large enough $q \in \mathbb{N}$ and interpolating the function over $\frac{1}{q}\mathbb{Z}$ we can obtain Lemma 3.

Lemma 2. *Any continuous restricted maximal gDFF ϕ is uniformly continuous.*

Lemma 3. *Let ϕ be a continuous restricted maximal gDFF, then for any $\epsilon > 0$, there exists a piecewise linear continuous restricted maximal gDFF ϕ_{pwl}, such that $\|\phi - \phi_{pwl}\|_\infty < \frac{\epsilon}{3}$.*

Next, we introduce a parametric family of restricted maximal gDFFs $\phi_{s,\delta}$ which will be used to perturb ϕ_{pwl}.

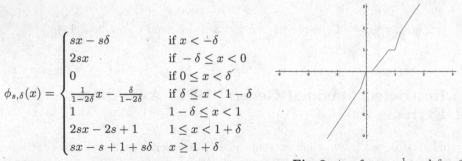

$$\phi_{s,\delta}(x) = \begin{cases} sx - s\delta & \text{if } x < -\delta \\ 2sx & \text{if } -\delta \leq x < 0 \\ 0 & \text{if } 0 \leq x < \delta \\ \frac{1}{1-2\delta}x - \frac{\delta}{1-2\delta} & \text{if } \delta \leq x < 1-\delta \\ 1 & 1-\delta \leq x < 1 \\ 2sx - 2s + 1 & 1 \leq x < 1+\delta \\ sx - s + 1 + s\delta & x \geq 1+\delta \end{cases}$$

Fig. 2. $\phi_{s,\delta}$ for $s = \frac{1}{5}$ and $\delta = 2$.

$\phi_{s,\delta}$ is a continuous piecewise linear function, which has breakpoints: $-\delta, 0, \delta, 1 - \delta, 1, 1 + \delta$ and slope values: $s, 2s, 0, \frac{1}{1-s\delta}, 0, 2s, s$ in each affine piece.

---

[3] See the constructor `two_slope_approximation_gdff_linear`.

Figure 2 shows the graph of one $\phi_{s,\delta}$ function constructed by the Python function `phi_s_delta`.

Let $E_\delta = \{(x,y) \in \mathbb{R}^2 : -\delta < x < \delta \text{ or } -\delta < y < \delta \text{ or } 1-\delta < x+y < 1+\delta\}$. We claim that $\phi_{s,\delta}$ is a continuous restricted maximal gDFF and $\nabla\phi_{s,\delta}(x,y) \geq \delta$ for $(x,y) \notin E_\delta$, if $s > 1$ and $0 < \delta < \min\{\frac{s-1}{2s}, \frac{1}{3}\}$. Verifying the above properties of $\phi_{s,\delta}$ is a routine computation by analyzing the superadditivity slack at every vertex in the two-dimensional polyhedral complex of $\phi_{s,\delta}$, which can also be verified by using metaprogramming[4] [12] in SageMath.

Lemma 4. *Let ϕ_{pwl} be a piecewise linear continuous restricted maximal gDFF, then for any $\epsilon > 0$, there exists a piecewise linear continuous restricted maximal gDFF ϕ_{loose} satisfying: (i) $\|\phi_{\mathrm{loose}} - \phi_{\mathrm{pwl}}\|_\infty < \frac{\epsilon}{3}$; (ii) there exist $\delta, \gamma > 0$ such that $\nabla\phi_{\mathrm{loose}}(x,y) \geq \gamma$ for (x,y) not in E_δ.*

Proof. By Proposition 1, let $t = \lim_{x\to\infty} \frac{\phi_{\mathrm{pwl}}(x)}{x}$, then $tx - t + 1 \leq \phi_{\mathrm{pwl}}(x) \leq tx$. We can assume $t > 1$, otherwise ϕ_{pwl} is the identity function and the result is trivial. Choose $s = t$ and δ small enough such that $0 < \delta < \min\{\frac{s-1}{2s}, \frac{1}{3}, \frac{1}{q}\}$, where q is the denominator of breakpoints of ϕ_{pwl} in previous lemma. We know that the limiting slope of maximal gDFF $\phi_{t,\delta}$ is also t and $tx - t + 1 \leq \phi_{t,\delta}(x) \leq tx$, which implies $\|\phi_{t,\delta} - \phi_{\mathrm{pwl}}\|_\infty \leq t - 1$.

Define $\phi_{\mathrm{loose}} = (1 - \frac{\epsilon}{3(t-1)})\phi_{\mathrm{pwl}} + \frac{\epsilon}{3(t-1)}\phi_{t,\delta}$. It is immediate to check ϕ_{loose} is restricted maximal. $\|\phi_{\mathrm{loose}} - \phi_{\mathrm{pwl}}\|_\infty < \frac{\epsilon}{3}$ is due to $\|\phi_{t,\delta} - \phi_{\mathrm{pwl}}\|_\infty \leq t - 1$. Based on the property of $\phi_{t,\delta}$, $\nabla\phi_{\mathrm{loose}}(x,y) = (1 - \frac{\epsilon}{3(t-1)})\nabla\phi_{\mathrm{pwl}}(x,y) + \frac{\epsilon}{3(t-1)}\nabla\phi_{t,\delta}(x,y) \geq \frac{\epsilon}{3(t-1)}\nabla\phi_{t,\delta}(x,y) \geq \gamma = \frac{\epsilon\delta}{3(t-1)}$ for (x,y) not in E_δ.

The proof of Lemma 5 is similar to the proof of [8, Lemma 3.3] and therefore omitted.

Lemma 5. *Given a piecewise linear continuous restricted maximal gDFF ϕ_{loose} satisfying properties in previous lemma, there exists an extreme gDFF ϕ_{ext} such that $\|\phi_{\mathrm{loose}} - \phi_{\mathrm{ext}}\|_\infty < \frac{\epsilon}{3}$.*

Combine the previous lemmas, then we can conclude the main theorem.

References

1. Alves, C., Clautiaux, F., de Carvalho, J.V., Rietz, J.: Dual-Feasible Functions for Integer Programming and Combinatorial Optimization: Basics, Extensions and Applications. EURO Advanced Tutorials on Operational Research. Springer, Cham (2016). https://doi.org/10.1007/978-3-319-27604-5
2. Aráoz, J., Evans, L., Gomory, R.E., Johnson, E.L.: Cyclic group and knapsack facets. Math. Program. Series B **96**, 377–408 (2003)
3. Bachem, A., Johnson, E.L., Schrader, R.: A characterization of minimal valid inequalities for mixed integer programs. Oper. Res. Lett. **1**(2), 63–66 (1982). http://www.sciencedirect.com/science/article/pii/0167637782900487

---

[4] Interested readers are referred to the function `is_superadditive_almost_strict` in order to check the claimed properties of $\phi_{s,\delta}$.

4. Basu, A., Hildebrand, R., Köppe, M.: Equivariant perturbation in Gomory and Johnson's infinite group problem. I. The one-dimensional case. Math. Oper. Res. **40**(1), 105–129 (2014)

5. Basu, A., Hildebrand, R., Köppe, M.: Light on the infinite group relaxation I: foundations and taxonomy. 4OR **14**(1), 1–40 (2016)

6. Basu, A., Hildebrand, R., Köppe, M.: Light on the infinite group relaxation II: sufficient conditions for extremality, sequences, and algorithms. 4OR **14**(2), 107–131 (2016)

7. Basu, A., Hildebrand, R., Köppe, M., Molinaro, M.: A $(k+1)$-slope theorem for the k-dimensional infinite group relaxation. SIAM J. Optim. **23**(2), 1021–1040 (2013)

8. Basu, A., Hildebrand, R., Molinaro, M.: Minimal cut-generating functions are nearly extreme. In: Louveaux, Q., Skutella, M. (eds.) IPCO 2016. LNCS, vol. 9682, pp. 202–213. Springer, Cham (2016). https://doi.org/10.1007/978-3-319-33461-5_17

9. Blair, C.E.: Minimal inequalities for mixed integer programs. Discrete Math. **24**(2), 147–151 (1978). http://www.sciencedirect.com/science/article/pii/0012365X78901930

10. Gomory, R.E., Johnson, E.L.: Some continuous functions related to corner polyhedra, I. Math. Program. **3**, 23–85 (1972). https://doi.org/10.1007/BF01584976

11. Gomory, R.E., Johnson, E.L.: Some continuous functions related to corner polyhedra, II. Math. Program. **3**, 359–389 (1972). https://doi.org/10.1007/BF01585008

12. Hong, C.Y., Köppe, M., Zhou, Y.: Sage code for the Gomory-Johnson infinite group problem. https://github.com/mkoeppe/cutgeneratingfunctionology. (Version 1.0)

13. Hong, C.Y., Köppe, M., Zhou, Y.: Equivariant perturbation in Gomory and Johnson's infinite group problem (V). In: Software for the Continuous and Discontinuous 1-Row Case. Optimization Methods and Software, pp. 1–24 (2017)

14. Jeroslow, R.G.: Minimal inequalities. Math. Program. **17**(1), 1–15 (1979)

15. Kılınç-Karzan, F., Yang, B.: Sufficient conditions and necessary conditions for the sufficiency of cut-generating functions. Technical report, December 2015. http://www.andrew.cmu.edu/user/fkilinc/files/draft-sufficiency-web.pdf

16. Köppe, M., Wang, J.: Structure and interpretation of dual-feasible functions. Electron. Notes Discrete Math. **62**, 153–158 (2017). LAGOS 2017 - IX Latin and American Algorithms, Graphs and Optimization. http://www.sciencedirect.com/science/article/pii/S1571065317302664

17. Lueker, G.S.: Bin packing with items uniformly distributed over intervals [a, b]. In: Proceedings of the 24th Annual Symposium on Foundations of Computer Science, SFCS 1983, pp. 289–297. IEEE Computer Society, Washington, DC (1983). https://doi.org/10.1109/SFCS.1983.9

18. Rietz, J., Alves, C., Carvalho, J., Clautiaux, F.: Computing valid inequalities for general integer programs using an extension of maximal dual feasible functions to negative arguments. In: 1st International Conference on Operations Research and Enterprise Systems-ICORES 2012 (2012)

19. Rietz, J., Alves, C., de Carvalho, J.M.V., Clautiaux, F.: On the properties of general dual-feasible functions. In: Murgante, B., et al. (eds.) ICCSA 2014. LNCS, vol. 8580, pp. 180–194. Springer, Cham (2014). https://doi.org/10.1007/978-3-319-09129-7_14

20. Vanderbeck, F.: Exact algorithm for minimising the number of setups in the one-dimensional cutting stock problem. Oper. Res. **48**(6), 915–926 (2000)

21. Yıldız, S., Cornuéjols, G.: Cut-generating functions for integer variables. Math. Oper. Res. **41**(4), 1381–1403 (2016). https://doi.org/10.1287/moor.2016.0781

Preemptively Guessing the Center

Christian Konrad[1][(✉)] and Tigran Tonoyan[2]

[1] Department of Computer Science, Centre for Discrete Mathematics
and its Applications (DIMAP), University of Warwick, Coventry, UK
c.konrad@warwick.ac.uk
[2] ICE-TCS, School of Computer Science, Reykjavik University, Reykjavik, Iceland
ttonoyan@gmail.com

Abstract. An online algorithm is typically unaware of the length of the
input request sequence that it is called upon. Consequently, it cannot
determine whether it has already processed most of its input or whether
the bulk of work is still ahead.

In this paper, we are interested in whether some sort of orientation
within the request sequence is nevertheless possible. Our objective is to
preemptively guess the center of a request sequence of unknown length
n: While processing the input, the online algorithm maintains a guess
for the central position $n/2$ and is only allowed to update its guess to
the position of the current element under investigation. We show that
there is a randomized algorithm that in expectation places the guess at
a distance of $0.172n$ from the central position $n/2$, and we prove that
this is best possible. We also give upper and lower bounds for a natural
extension to weighted sequences.

This problem has an application to preemptively partitioning integer
sequences and is connected to the online bidding problem.

1 Introduction

Online Algorithms. Online algorithms process their inputs item by item in a
linear fashion. They are characterized by the fact that the algorithm's decision
as to how to process the current input item is irrevocable. A key difficulty in the
design of online algorithms is that they are typically unaware of the length of the
input request sequence[1]. Indeed, for many online problems (e.g. problems with
a rent or buy flavor such as the ski rental problem [2]), knowing the input length
would allow the algorithm to solve the problem optimally. Without knowing

---

C. Konrad is supported by the Centre for Discrete Mathematics and its Applications
(DIMAP) at Warwick University and by EPSRC award EP/N011163/1. T. Tonoyan
is supported by grants no. 152679-05 and 174484-05 from the Icelandic Research
Fund.

[1] An exception are online algorithms with advice, where the online algorithm receives
additional input bits prior to the processing of the request sequence. These bits can
be used to encode the input length. See [1] for a recent survey.

© Springer International Publishing AG, part of Springer Nature 2018
J. Lee et al. (Eds.): ISCO 2018, LNCS 10856, pp. 277–289, 2018.
https://doi.org/10.1007/978-3-319-96151-4_24

the input length, online algorithms are unable to determine the position of the current element within the request sequence.

Guessing the Center. In this paper, we ask whether we can nevertheless obtain some sort of orientation within the request sequence. We study the natural task of guessing the central position $n/2$ within a request sequence of unknown length n in an online fashion. In this problem, the online algorithm maintains a guess of the central position while processing the input request sequence. The algorithm is only allowed to update its guess to the position of the current element under investigation. It may thus potentially update the guess many times, however, each update bears the risk that the input sequence may end very soon and the guess is thus far from the center. Such an algorithm follows the following scheme:

Algorithm 1. Scheme for Preemptively Guessing the Center

$p \leftarrow 0$ {initialization of our guess}
for each request $j = 1, 2, \ldots, n$ **do** {n is unknown}
   if *TODO: add condition here* **then** {update guess}
     $p \leftarrow j$
return p

We also study a generalization of this problem to weighted requests. This is best modelled as follows. The online algorithm processes a sequence $X = w_1, w_2, \ldots, w_n$ of positive integers. Let $W = \sum_{i=1}^{n} w_i$ be the total weight of the sequence. We assume that there exists an index m with $1 \leq m \leq n$, such that $\sum_{i=1}^{m} w_i = \sum_{i=m+1}^{n} w_i$, i.e., the sequence can be split into two parts of equal weight. This assumption is necessary for a meaningful problem definition as we will discuss in Sect. 4.1 in more detail. While processing X, an online algorithm $\mathcal{A}$ maintains a guess p for the position m as in the unweighted case. The objective is to minimize the weight between the guess p and the position m of the central weight, that is, the *deviation*

$$\Delta_{\mathcal{A}}^{X} := \sum_{i=\min\{p,m\}+1}^{\max\{p,m\}} w_i \, ,$$

is to be minimized, where $\mathcal{A}$ refers to the employed algorithm and X is the input sequence. Note that the unweighted version of this problem is obtained by setting $w_i = 1$, for every $1 \leq i \leq n$. One property of this definition is that we only consider unweighted sequences of even length, since sequences of odd lengths cannot be split into two parts of equal weight. This is only for convenience; a meaningful problem statement with similar results for unweighted sequences of odd lengths can easily be derived from this work. For unweighted sequences we write $\Delta_{\mathcal{A}}^{n}$ instead of $\Delta_{\mathcal{A}}^{X}$, where n denotes the input length.

Results. For unweighted request sequences, we give an optimal randomized preemptive online algorithm for guessing the center. Our algorithm has expected deviation $0.172n$ from the central position $n/2$ (Theorem 1). Our main result is a lower bound, which shows that this is best possible (Theorem 3). We further give a barely random algorithm that uses only a single random bit and reports a position with expected deviation $0.25n$. This is also proved to be best possible for the class of algorithms that only use a single random bit. For weighted sequences, we give a randomized preemptive online algorithm that reports a position with expected deviation $0.313W$, where W is the total weight of the input sequence (Theorem 4). This is complemented by a lower bound of $0.25W$ (Theorem 5). Closing this gap proves challenging and is left as an open problem.

Techniques. Both our algorithms for unweighted and weighted sequences employ the doubling method with a random seed. In the unweighted case, our algorithm updates its guess to the current position j if $j \in \{\lceil x^{i\delta} \rceil \mid i \in \mathbb{N}\}$ (this condition is slightly different in the weighted case), where $x > 2$ is an optimized parameter that determines the step size between the guesses (this parameter is different for weighted and unweighted sequences), and $\delta \in (0,1)$ is a seed that is chosen uniformly at random. This technique is well known and has previously been applied for various problems, see for example [3]. While our algorithms are extremely simple, their analyses require careful case distinctions.

Our main result is a lower bound for unweighted sequences, which proves that the doubling method is optimal. The argument relies on Yao's Minimax principle [4]. We define a hard input distribution where the probability of a specific input length is inversely proportional to its length. We then argue that a deterministic guessing algorithm, which can be identified by a sequence of increasing positions at which it updates its guess, will in expectation (over the hard input distribution) have a deviation of $0.172n$ from the central position. By Yao's Minimax principle, this implies that our algorithm for unweighted sequences is best possible. This argument is the most technical contribution of the paper. The lower bound for weighted sequences follows the same line, however, it uses a sequence of exponentially increasing weights.

Further Related Work. Preemptively guessing the center is strongly related to the online bidding problem [5]. In online bidding, the objective is to guess an unknown target value. The algorithm submits increasing guesses until a guess that is at least as large as the target value is submitted. For this problem, the usual cost function is the sum of the submitted guesses, which is very different from our cost function. However, similarly to the problem of guessing the center, an optimal randomized strategy can be obtained by using a sequence of exponentially increasing guesses.

Guessing the center is a special case of the problem of partitioning integer sequences. In this problem, an integer array A of length n and an integer $p \geq 2$ is given, and the goal is to find $(p-1)$ separator positions $s_1, s_2, \ldots, s_{p-1}$ with $1 = s_0 \leq s_1 \leq s_2 \leq \cdots \leq s_{p-1} \leq s_p = n+1$ such that $\max\{\sum_{i=s_j}^{s_{j+1}-1} A_i \mid 0 \leq j < p\}$ is minimized. This load balancing task has a long history in the offline setting (e.g. [6–9]) and has recently been studied in the context of data streams [10] and online algorithms [11] by the authors of this paper. In the preemptive online model, an algorithm is only allowed to insert a new partition separator at the current position, and, once all separators have been placed, previously inserted separators can be removed and then reinserted again. As shown in [11], a 2-approximation algorithm for arbitrary values of p can be obtained. The special case $p = 2$ boils down to determining the central position of an integer sequence using a preemptive guessing scheme. The problem studied in this paper thus correspond to preemptively partitioning an integer sequence of length n into two parts of equal weights.

Outline. We give our algorithm for unweighted sequences in Sect. 2 and our lower bound for unweighted sequences in Sect. 3. In Sect. 4, we address extensions to weighted sequences. We conclude with an open problem in Sect. 5. Due to space restrictions, we only sketch the proofs of Theorems 4 and 5 are defer the full proofs to the complete version of this paper.

2   Algorithm for Guessing the Center

Our algorithm, denoted $\mathcal{A}_x$, is parametrized by a real number $x > 2$. It employs a well-known doubling technique with randomized seeding. We first pick a seed $\delta \in (0, 1)$ uniformly at random. The parameter x determines the distance between two consecutive guesses and will be optimized later. The algorithm updates our guess for the central position whenever we process requests $\lceil x^{1+\delta} \rceil$, $\lceil x^{2+\delta} \rceil$, $\lceil x^{3+\delta} \rceil$, ... This is depicted in Algorithm 2.

Algorithm 2. Algorithm $\mathcal{A}_x$ for guessing the center

Choose uniform random $\delta \in (0, 1)$, $i \leftarrow 0$, $p \leftarrow 0$ {initialization}
for each request $j = 1, 2, \ldots, n$ **do** {n is unknown}
    if $j = \lceil x^{i+\delta} \rceil$ **then** {update guess}
        $p \leftarrow j$
        $i \leftarrow i + 1$
return p

While the suggested doubling strategy is fairly standard, the analysis requires a very careful case distinction. Moreover, this algorithm is optimal, which will be proved in Sect. 3.

One may wonder about the choice of δ to be a real-valued quantity of presumably infinite precision. This assumption is only taken for convenience in the analysis; a bit precision of $O(\log n)$ is enough to provide sufficient granularity. This does not mean that n needs to be known in advance in order to determine the $O(\log n)$ random bits: We can choose additional random bits for the description of δ when necessary as the algorithm proceeds.

After giving the analysis of our main algorithm, we further present an algorithm that uses only a single random bit and achieves an expected deviation of $0.25n$. We also prove that this is best possible for the class of algorithms that only use a single random bit. Observe that deterministic algorithms (i.e., using no randomness at all) fail for guessing the center: If the input sequence ends exactly when the deterministic algorithm has updated its guess, then the deviation is as large as it could be. Without randomness, this is unavoidable.

Theorem 1. *There is a constant $x \approx 3.052$ such that:*

$$\mathbb{E}\left[\Delta_{\mathcal{A}_x}^n\right] \approx 0.172n + \mathrm{O}(1).$$

Proof. Let $\alpha \in [0,1)$ and $i \in \mathbb{N}$ be such that $n = 2x^{i+\alpha}$. Then the central position is $\frac{n}{2} = x^{i+\alpha}$. In order to bound the expected deviation, we conduct a case distinction for various ranges of α and δ. We distinguish two ranges for α, and within each case, we distinguish three ranges of δ.

Case 1: $\alpha > 1 - \log_x 2$ (note that we assumed $x > 2$). In order to bound $\Delta_{\mathcal{A}_x}^n$, we split the possible values of δ into three subsets:

- If $\delta \in (0, \alpha + \log_x 2 - 1]$, then we have that $x^{\delta+i+1} \leq 2x^{i+\alpha} = n$. In this case, the deviation is $\Delta_{\mathcal{A}_x}^n = x^{\delta+i+1} - n/2 = x^{\delta+i+1} - x^{i+\alpha}$.
- If $\delta \in (\alpha + \log_x 2 - 1, \alpha]$, then we have that $x^{\delta+i+1} > n$ but $x^{\delta+i} \leq \frac{n}{2}$. In this case, $\Delta_{\mathcal{A}_x}^n = x^{i+\alpha} - x^{\delta+i}$.
- If $\delta \in (\alpha, 1)$, then we have that $x^{\delta+i+1} > n$ and $x^{\delta+i} \in (\frac{n}{2}, n)$. In this case, $\Delta_{\mathcal{A}_x}^n = x^{\delta+i} - x^{i+\alpha}$.

Using these observations, we can bound the expected deviation as follows:

$$\mathbb{E}\left[\Delta_{\mathcal{A}_x}^n\right] = \int_0^{\alpha+\log_x 2-1} (x^{\delta+i+1} - x^{i+\alpha})d\delta + \int_{\alpha+\log_x 2-1}^{\alpha} (x^{\alpha+i} - x^{\delta+i})d\delta$$

$$+ \int_\alpha^1 (x^{\delta+i} - x^{\alpha+i})d\delta$$

$$= x^{i+\alpha} \cdot \left(1 - 2\log_x 2 + \frac{2}{x \ln x}\right).$$

Case 2: $\alpha \leq 1 - \log_x 2$. We deal with this case similarly, but we need to group the possible values for δ in a different way:

- If $\delta \in (0, \alpha]$, then $x^{\delta+i+1} > n$ but $x^{\delta+i} \leq \frac{n}{2}$. In this case, $\Delta_{\mathcal{A}_x}^n = x^{i+\alpha} - x^{\delta+i}$.
- If $\delta \in (\alpha, \alpha + \log_x 2]$, then $x^{\delta+i} > \frac{n}{2}$ and $x^{\delta+i} \leq n$. In this case, $\Delta_{\mathcal{A}_x}^n = x^{\delta+i} - x^{i+\alpha}$.
- If $\delta \in (\alpha + \log_x 2, 1)$, then $x^{\delta+i} > n$. In this case, $\Delta_{\mathcal{A}_x}^n = x^{i+\alpha} - x^{\delta+i-1}$.

Plugging the values above in the formula for the expected value, we obtain a different sum of integrals, which however leads to the same function as above:

$$\mathbb{E}\left[\Delta_{\mathcal{A}_x}^n\right] = \int_0^{\alpha} (x^{\alpha+i} - x^{\delta+i})d\delta + \int_{\alpha}^{\alpha+\log_x 2} (x^{\delta+i} - x^{i+\alpha})d\delta$$

$$+ \int_{\alpha+\log_x 2}^1 (x^{\alpha+i} - x^{\delta+i-1})d\delta$$

$$= x^{i+\alpha} \cdot \left(1 - 2\log_x 2 + \frac{2}{x \ln x}\right).$$

Moreover, the factor $1 - 2\log_x 2 + \frac{2}{x \ln x}$ above is independent of α. Thus, it remains to find a value of x that minimizes $f(x) \overset{def}{=} 1 - 2\log_x 2 + \frac{2}{x \ln x}$. Observe that $f'(x) = -\frac{2}{x^2 \ln^2 x} - \frac{2}{x^2 \ln x} + \frac{\ln 2}{x \ln^2 x}$, and $f'(x) = 0$ if and only if $x = \log_2(ex)$. With a simple transformation, the latter is equivalent to $ze^z = -\frac{\ln 2}{e}$ with $z = -x \ln 2$, so the value that minimizes $f(x)$ can be computed as $x_{\min} = -\frac{W_{-1}(-\ln 2/e)}{\ln 2} \approx 3.052$, where W_{-1} is the lower branch of Lambert's W function. The claim of the theorem follows by calculating $f(x_{\min}) \approx 0.344$. The additive $O(1)$ corresponds to the approximation of the finite range of δ by a continuous distribution. $\qquad \square$

Next, we give an algorithm that only relies on a single random bit. We prove that its expected deviation from the center is $0.25n$, which is best possible.

Algorithm 3. Single-bit algorithm $\mathcal{A}_0$

$\quad i \leftarrow 0$ or 1 with probability $1/2$ each, $p \leftarrow 0$ {initialization}
$\quad$ **for** each request $j = 1, 2, \ldots n$ **do** {n is unknown}
$\quad\quad$ **if** $j = 2^i$ **then**
$\quad\quad\quad p \leftarrow j$ {update guess to current position}
$\quad\quad\quad i \leftarrow i + 2$
$\quad$ **return** p

Theorem 2. *The expected deviation of algorithm $\mathcal{A}_0$ is $\mathbb{E}\left[\Delta^n_{\mathcal{A}_0}\right] \leq 0.25n$, which is optimal for the class of algorithms that only use a single random bit.*

Proof. Let $\alpha \in [0,1)$ and $i \in \mathbb{N}$ be such that the length of the sequence is $n = 2 \cdot 2^{i+\alpha}$. Since $2^{i+2} > n$, the algorithm reports either position 2^i or position 2^{i+1}, each with probability $1/2$. In the first case, the deviation from the center is $n/2 - 2^i$, while in the second case it is $2^{i+1} - n/2$. Thus, in expectation, we have, as required, $\mathbb{E}\left[\Delta^n_{\mathcal{A}_0}\right] = \frac{1}{2} \cdot (n/2 - 2^i) + \frac{1}{2} \cdot (2^{i+1} - n/2) = 2^{i-1} \leq n/4$.

In order to see that this is best possible for the class of algorithms that only use a single random bit, first observe that a randomized algorithm that uses a single random bit is a uniform distribution over two deterministic algorithms. Note further that each deterministic algorithm is a fixed (potentially infinite) sequence of positions at which it updates its guess. Suppose that $\mathcal{B}$ is such a randomized algorithm, and let $S_1 = \{p_1, p_2, \ldots\}$ and $S_2 = \{q_1, q_2, \ldots\}$ be the corresponding sequences. Now, if the sequence has length p_i for some i, $\mathcal{B}$ would have maximal deviation if it chooses the first sequence (with probability $1/2$), and may have minimal deviation 0 (with the remaining $1/2$ probability) if the largest $q_j \leq p_i$ is equal to $p_i/2$. Therefore the smallest expected deviation achievable is $n/4$, which implies that our algorithm is optimal. $\qquad \square$

3   Lower Bound

We prove that no algorithm can achieve a smaller expected deviation than the one claimed in Theorem 1. The proof applies Yao's Minimax principle and uses a hard input distribution over all-ones sequences of length $n \in [n_{\min}, n_{\max}]$, for some large values of $n_{\min}$ and $n_{\max}$, where the probability that the sequence is of length n is proportional to $1/n$.

Theorem 3. *For any randomized algorithm $\mathcal{A}$, the expected deviation is*

$$\mathbb{E}\left[\Delta_{\mathcal{A}}^n\right] \geq 0.172n.$$

Proof. We will prove the theorem by using Yao's Minimax principle [4]. To this end, let us first consider an arbitrary *deterministic* algorithm $\mathcal{A}_{\det}$. Assume the length of the sequence is random in the interval $X := [n_{\min}, n_{\max}]$ for large values of $n_{\max}$ and $n_{\min}$ with $n_{\max} > 2 \cdot n_{\min}$ and has the following distribution: The sequence ends at position $n \in X$ with probability p_n which is proportional to $\frac{1}{n}$, i.e., using the definition $S = \displaystyle\sum_{m=n_{\min}}^{n_{\max}} \frac{1}{m}$, we have $p_n := \mathbb{P}[\text{sequence is of length } n] = \frac{1}{n \cdot S}$.

In order to apply the Minimax principle, we will consider a *normalized* measure of the performance of an algorithm. For an algorithm $\mathcal{A}$, let $B_{\mathcal{A}}^n$ denote the larger of the two parts created by the algorithm for a sequence of length n, and let $R_{\mathcal{A}}^n = \frac{B_{\mathcal{A}}^n}{n/2} \in [1, 2]$. Then it is easily verified that

$$\Delta_{\mathcal{A}}^n = B_{\mathcal{A}}^n - \frac{n}{2} = n \cdot \frac{R_{\mathcal{A}}^n - 1}{2}.$$

We will show that for *each* deterministic algorithm $\mathcal{A}_{\det}$, if the input is distributed as above, then $\mathbb{E}\left[R_{\mathcal{A}_{\det}}^n\right] \geq 1.344 - O(\ln^{-1} \frac{n_{\max}}{n_{\min}})$, where the expectation is taken over the distribution of n. Then, by the Minimax principle, every randomized algorithm $\mathcal{A}$ has a ratio of at least $R_{\mathcal{A}}^n \geq \mathbb{E}\left[R_{\mathcal{A}_{\det}}^n\right] \geq 1.344 - O(\ln^{-1} \frac{n_{\max}}{n_{\min}})$. Since the ratio $\frac{n_{\max}}{n_{\min}}$ is arbitrary, this implies the theorem.

Let J denote the set of requests at which $\mathcal{A}_{\det}$ updates its guess when processing the all-ones sequence of length $n_{\max}$. Note that the positions of guess updates by $\mathcal{A}_{\det}$ on sequences of shorter lengths are a subset of J. Let $I = J \cap X = \{i_1, \ldots, i_k\}$ (the i_j are ordered with increasing value).

We bound $\mathbb{E}\left[R_{\mathcal{A}_{\det}}^n\right] = \sum_{n=n_{\min}}^{n_{\max}} p_n R_{\mathcal{A}_{\det}}^n$ by separately bounding every partial sum in the following decomposition:

$$\mathbb{E}\left[R_{\mathcal{A}_{\det}}^n\right] = E(n_{\min}, i_1) + E(i_1, i_2) + \cdots + E(i_{k-1}, i_k) + E(i_k, n_{\max}),$$

where for each $a < b$, $E(a, b) = \sum_{n=a}^{b-1} p_n R_{\mathcal{A}_{\det}}^n$. The first and last terms need special care, so we will start with bounding the other terms. In the following, $H_p^q = \sum_{n=p}^{q} \frac{1}{n}$ denotes partial harmonic sums for $q \geq p \geq 1$. Observe that $S = H_{n_{\min}}^{n_{\max}}$. We proceed in three steps:

1. Consider an index $1 \leq j < k$ and let us bound the sum $E(i_j, i_{j+1})$. Let us denote $a = i_j$ and $b = i_{j+1}$. We need to consider two cases.

Case 1: $b \leq 2a$. Then for all $n \in \{a, \ldots, b-1\}$, we have $B^n_{\mathcal{A}_{det}} = a$ (since $n/2 < a$). Then:

$$E(a,b) \geq \sum_{n=a}^{b-1} \frac{1}{nS} \cdot \frac{a}{n/2} \geq \frac{2a}{S} \sum_{n=a}^{b-1} \frac{1}{n(n+1)} = \frac{2a}{S} \left(\frac{1}{a} - \frac{1}{b} \right) > 1.4 \cdot \frac{H_a^{b-1}}{S}, \quad (1)$$

where the last inequality can be proved as follows. First, it can be checked by direct inspection that for $\Phi(a,b) = 2(1 - \frac{a}{b})/H_a^{b-1}$ and $b > a$, it holds that $\Phi(a, b+1) < \Phi(a,b)$. Hence, recalling that $b \leq 2a$ and using standard approximations of harmonic sums, we obtain

$$\Phi(a,b) \geq \Phi(a, 2a) = 1/H_a^{2a-1} \geq \frac{1}{\ln 2 + \frac{1}{a}} > 1.4,$$

where the last inequality holds for large enough $a = i_j$, e.g., $i_j \geq \frac{n_{\min}}{2} \geq 50$.

Case 2: $b > 2a$. In this case, for all $n = a, \ldots, 2a - 1$, if the sequence is of length n, then $B^n_{\mathcal{A}_{det}} = a$, as in case 1. However, when $n \geq 2a$, then $n/2 \geq a$, so $B^n_{\mathcal{A}_{det}} = n - a$. Using these observations, we can bound $\mathbb{E}\left[R^n_{\mathcal{A}_{det}}\right]$:

$$\mathbb{E}\left[R^n_{\mathcal{A}_{det}}\right] = \sum_{n=a}^{2a-1} \frac{1}{nS} \frac{a}{n/2} + \sum_{n=2a}^{b-1} \frac{1}{nS} \frac{n-a}{n/2}$$

$$\geq \frac{2a}{S} \sum_{n=a}^{2a-1} \frac{1}{n(n+1)} + \frac{2}{S} H_{2a+1}^b - \frac{2a}{S} \sum_{n=2a}^{b-1} \frac{1}{n(n+1)}$$

$$= \frac{2a}{S} \left(\frac{1}{a} - \frac{1}{2a} \right) + \frac{2}{S} H_{2a+1}^b - \frac{2a}{S} \left(\frac{1}{2a} - \frac{1}{b} \right)$$

$$= \frac{2a}{Sb} + \frac{2}{S}(H_a^b - H_a^{2a})$$

$$= \frac{H_a^b}{S} \cdot \left(2 + \frac{2a}{bH_a^b} - \frac{2H_a^{2a}}{H_a^b} \right),$$

where the third line is obtained by using the identity $H_{2a+1}^b = H_a^b - H_a^{2a}$. Again, using a standard approximation for harmonic sums and setting $x = \frac{b}{a}$, we can approximate:

$$2 + \frac{2a}{bH_a^b} - \frac{2H_a^{2a}}{H_a^b} \geq 2 + \frac{2a}{b \ln \frac{b}{a}} - \frac{\ln 2}{\ln \frac{b}{a}} - O(a^{-1})$$

$$= 2 + \frac{2}{x \ln x} - 2 \log_x 2 - O(a^{-1}).$$

Note that the function $f(x) = 2 + \frac{2}{x \ln x} - 2 \log_x 2$ is exactly the same that was minimized in the proof of Theorem 1, with an additional term $+1$, and achieves its minimum in $(1, \infty)$ at $x_{\min} \approx 3.052$, giving $f(x_{\min}) \approx 1.344$.

Thus, we have $E(i_j, i_{j+1}) \geq \frac{H_{i_j}^{i_{j+1}-1}}{S}(1.344 - O(\frac{1}{i_j}))$.

2. The term $E(i_k, n_{\max})$ can be bounded by $\dfrac{H_{i_k}^{n_{\max}-1}}{S}(1.344 - O(\frac{1}{i_k}))$ by an identical argument as above.

3. The term $E(n_{\min}, i_1)$ needs a slightly different approach. Let i_0 denote the last position where the algorithm updates its guess before $n_{\min}$. We can assume that $i_0 \geq n_{\min}/2$, as otherwise the algorithm could only profit by updating its guess at position $n_{\min}/2$. We consider two cases. First, if $i_1 \leq 2i_0$, then we simply assume the algorithm performs optimally in the range $[n_{\min}, i_1)$:

$$E(n_{\min}, i_1) = \frac{H_{n_{\min}}^{i_1-1}}{S} \geq 1.344 \cdot \frac{H_{n_{\min}}^{i_1-1}}{S} - 0.5 \frac{H_{n_{\min}}^{i_1-1}}{S} > 1.344 \cdot \frac{H_{n_{\min}}^{i_1-1}}{S} - 1/S,$$

since (recalling that $i_1 \leq 2i_0 \leq 2n_{\min}$), $H_{n_{\min}}^{i_1} < H_{n_{\min}}^{2n_{\min}} < 1$.

On the other hand, when $i_1 > 2i_0$ (and by the discussion above, $2i_0 \geq n_{\min}$), then with calculations similar to the one in Case 2 above, we obtain:

$$
\begin{aligned}
E(x_{\min}, i_1) &= \sum_{n=n_{\min}}^{2i_0-1} \frac{1}{nS} \frac{i_0}{n/2} + \sum_{n=2i_0}^{i_1-1} \frac{1}{nS} \frac{n-i_0}{n/2} \\
&\geq \frac{2i_0}{S} \sum_{n=n_{\min}}^{2i_0-1} \frac{1}{n(n+1)} + \frac{2}{S} H_{2i_0+1}^{i_1} - \frac{2i_0}{S} \sum_{n=2i_0}^{i_1-1} \frac{1}{n(n+1)} \\
&\geq \frac{2}{S}\left(\frac{i_0}{n_{\min}} + \frac{i_0}{i_1} + H_{n_{\min}}^{i_1} - H_{n_{\min}}^{2i_0} \right) \geq 2 \cdot \frac{H_{n_{\min}}^{i_1}}{S} - O(1/S),
\end{aligned}
$$

because $i_0 \leq n_{\min}$ and thus $H_{n_{\min}}^{2i_0} < 1$.

It remains to plug the obtained estimates into Inequality 1:

$$
\begin{aligned}
\mathbb{E}\left[R_{\mathcal{A}_{\det}}^n \right] &= E(n_{\min}, i_1) + \sum_{j=1}^{k-1} E(i_j, i_{j+1}) + E(i_k, n_{\max}) \\
&\geq (1.344 - O(1/n_{\min})) \cdot \frac{H_{n_{\min}}^{i_1} + \sum_{j=1}^{k-1} H_{i_j}^{i_{j+1}-1} + H_{i_k}^{n_{\max}}}{S} - O(1/S) \\
&= 1.344 - O(1/S).
\end{aligned}
$$

This completes the proof.   □

4 Weighted Sequences

4.1 Algorithm

The algorithm for weighted instances is an adaptation of the algorithm $\mathcal{A}_x$ presented in Sect. 2. Namely, the guess is updated as soon as adding the current weight w_j to the weight of the prefix that has already been processed $\sum_{i=1}^{j-1} w_i$ increases the total weight to be at least $\lceil x^{i+\delta} \rceil$, i.e., if $\sum_{i=1}^{j-1} w_i < \lceil x^{i+\delta} \rceil \leq \sum_{i=1}^{j} w_i$. We will keep the notation $\mathcal{A}_x$ for the modified algorithm.

Theorem 4. *There is a constant value of $x \approx 5.357$ such that*

$$\mathbb{E}\left[\Delta_{\mathcal{A}_x}^X\right] \leq 0.313W + O(1)$$

holds for every weighted sequence X of total weight W.

Proof (Sketch). Let $X = w_1, w_2, \ldots, w_n$ be the input sequence of total weight W, and let m be such that $\sum_{i \leq m} w_i = \frac{W}{2}$. Then, w_m is the *central weight* of the sequence. We will argue first that replacing all w_i left of w_m including w_m by a sequence of $\sum_{i \leq m} w_i$ unit requests, and replacing all w_i right of w_m by a single large request of weight $\sum_{i > m} w_i$ worsens the approximation factor of the algorithm. Indeed, suppose that the algorithm attempts to update its guess at a position j that falls on an element w_i, which is located left of w_m. Then the algorithm updates its guess after w_i, bringing the guess closer to the center than if w_i were of unit weight. Similarly, suppose that the algorithm attempts to update its guess at position j that falls on an element w_i, which is located right of w_m. By replacing all weights located to the right of w_m by a single heavy element, the algorithm has to place its guess at the end of the sequence, which gives the worst possible deviation. Thus, we suppose from now on that X is of the form $X = 1 \ldots 1B$, where $B = W/2$.

After this simplification, via a case distinction as in the proof of Theorem 1 it can be shown that $E[\Delta_{\mathcal{A}_x}^X] \leq \frac{W}{2} \cdot g(x) + O(1)$, where $g(x) \overset{def}{=} 1 - \frac{1}{\ln x} + \frac{2}{x \ln x}$. It can then be shown that $x_{min} = -2W_{-1}(-\frac{1}{2e}) \approx 5.3567$ minimizes $g(x)$ and $g(x_{min}) \approx 0.627$, which implies the theorem. $\square$

Remark. Recall that we work with the assumption that the input sequence $X = w_1, \ldots, w_n$ can be split into two parts of exactly equal weight. This may seem like an artificial restriction. It is, however, necessary for a meaningful problem definition: Suppose we allowed arbitrary sequences and the goal is to minimize the distance between the guess and the most central position, i.e., the position c such that $\max\{\sum_{i=1}^{c} w_i, \sum_{i=c+1}^{n} w_i\}$ is minimized. Consider now the instance $X = 11 \ldots 1W$, where W is extremely heavy (e.g., W is a 0.99 fraction of the entire sequence). Then the most central position is the position of the last 1, while an algorithm that places a guess after W is at distance W from the most central position. We believe that this should not be penalized since such an input sequence does not have a good central position. An alternative problem formulation, which is meaningful when applied to the previously described instance, is obtained when asking for a guess that minimizes the size of the larger part as compared to the larger part in a most central split. Indeed, this formulation is coherent with the problem of partitioning integer sequences. Our algorithm for weighted sequences can be analyzed for this situation and gives a 1.628 approximation.

4.2   Lower Bound

Note that the expected deviation of $0.313W$ is tight for the algorithm $\mathcal{A}_x$ on weighted sequences. This is achieved on sequences consisting of $W/2$ unit weight elements followed by an element of weight $W/2$. However, we were not able to obtain a matching lower bound. The main difficulty in applying the Minimax principle is that in the weighted case, the deterministic algorithm may know the probability distribution of the individual weights. Instances similar to the one described above become easy if the algorithm knows their structure. Instead, we prove a lower bound of $0.25W$ using a different construction.

Theorem 5. *For every randomized algorithm $\mathcal{A}$, there is a weighted instance X of total weight W, such that the expected deviation is $\mathbb{E}\left[\Delta_{\mathcal{A}}^X\right] \geq 0.25W - O(1)$.*

Proof (Sketch). Let $X_i = 2^0, 2^0, 2^1, 2^2, \ldots, 2^{i-1}$ denote the exponentially increasing sequence of length i, and $W_i = 2^i$ denotes the weight of X_i. Note that the middle of X_i is the position before the weight 2^{i-1}. Let further $n_{\min}, n_{\max}$ be large integers with $n_{\max} \geq 2n_{\min}$, and denote $S = [n_{\min}, n_{\max}]$. We consider the performance of any deterministic algorithm $\mathcal{A}_{\det}$ on the uniform input distribution over the set $\Sigma = \{X_i : i \in S\}$ of exponentially increasing sequences. As in the proof of Theorem 3, we need a normalized performance measure in order to apply the Minimax lemma. For an algorithm $\mathcal{A}$, let $B_{\mathcal{A}}^X$ denote the larger of the two parts that the algorithm creates on the input sequence X, and let $R_{\mathcal{A}}^X = \frac{B_{\mathcal{A}}^X}{W/2} \in [1, 2]$. Again, note that $\Delta_{\mathcal{A}}^X = B_{\mathcal{A}}^X - W/2 = W \cdot (R_{\mathcal{A}}^X - 1)/2$. Hence, showing that $\mathbb{E}\left[R_{\mathcal{A}_{\det}}^X\right] \geq 1.5$ for every deterministic algorithm implies the corresponding bound $\mathbb{E}\left[\Delta_{\mathcal{A}}^X\right] \geq 0.25W$ for every randomized algorithm, by Yao's lemma.

Let J be the set of positions where $\mathcal{A}_{\det}$ updates its guess on input $X_{n_{\max}}$. Note that the set of positions on any other input of Σ is a subset of J. Let $I = J \cap S = \{i_1, \ldots, i_k\}$ be the positions within the interval S (ordered so that $i_j < i_{j+1}$, for every j).

We bound now the expected ratio of $\mathcal{A}_{\det}$, where the expectation is taken over the inputs Σ. In the formulas below, we will use the abbreviation $R^i = R_{\mathcal{A}_{\det}}^{X_i}$, as we will only deal with such ratios. Then:

$$\mathbb{E}\left[R^n\right] = \frac{1}{n_{\max} - n_{\min} + 1} \cdot \sum_{n=n_{\min}}^{n_{\max}} R^n \text{ , and}$$

$$\sum_{n=n_{\min}}^{n_{\max}} R^n = \underbrace{\sum_{n=n_{\min}}^{i_1-1} R^n}_{I} + \sum_{n=i_1}^{i_2-1} R^n + \cdots + \sum_{n=i_{k-1}}^{i_k-1} R^n + \underbrace{\sum_{n=i_k}^{n_{\max}} R^n}_{II} \text{ .}$$

We bound I, II, and $\sum_{n=i_j}^{i_{j+1}-1} R^n$ for every $1 \leq j \leq k-1$, separately. Recall that for every i the half-weight of X_i is $W_i/2 = 2^{i-1}$. First, observe that for every $1 \leq j \leq k-1$, we have the worst ratio $R^{i_j} = \frac{2^{i_j}}{2^{i_j-1}} = 2$ when the sequence ends at a guess update, and the best ratio $R^{i_j+1} = 1$ when it ends one item after a guess update and if $i_j + 1 < i_{j+1}$. Generally, when it ends at an intermediate position $i_j + a$ such that $a \geq 2$ and $i_j + a < i_{j+1}$, then the last guess update is after the weight i_j, while the middle is just before $i_j + a$ so we have

$$R^{i_j+a} = \frac{2^{i_j+a} - 2^{i_j-1}}{2^{i_j+a-1}} \geq 1.75.$$

These bounds together imply that $\sum_{n=i_j}^{i_{j+1}-1} R^n \geq 1.5(i_{j+1} - i_j)$.

It can be shown by similar arguments that $I \geq 1.5(i_1 - n_{\min}) - 1.25$ and $II > 1.5(n_{\max} - i_k) - 1.25$. Putting the partial bounds together, we can bound $\mathbb{E}\left[R^n_{\mathcal{A}_{\det}}\right]$ by:

$$\mathbb{E}\left[R^n_{\mathcal{A}_{\det}}\right] \geq \frac{1.5(n_{\max} - n_{\min} + 1) - 2 \cdot 1.25}{n_{\max} - n_{\min} + 1} = 1.5 - O(n_{\max}^{-1}).$$

Since $n_{\max}$ can be chosen arbitrarily large, the latter bound on the expected performance of every deterministic algorithm then implies the claim of the theorem by applying Yao's principle, as described above. □

5   Conclusion

In this paper, we gave an algorithm for preemptively guessing the center of a request sequence. It has expected deviation $0.172n$ from the central position on an instance of length n. We proved that this is optimal. We extended our algorithm to weighted sequences and showed that it has expected deviation $0.313W$, where W is the total weight of the input sequence. We also gave a lower bound showing that no algorithm achieves an expected deviation smaller than $0.25W$.

The most intriguing open problem is to close the gap between the upper and lower bounds for weighted sequences. Progress could potentially be made by combining our lower bound for unweighted sequences with an exponentially increasing sequence as it is used in our lower bound for weighted sequences. For this to be successful, a better understanding of our lower bound for unweighted sequences could be beneficial, since it relies on a non-uniform input distribution which renders it difficult to extend it.

References

1. Boyar, J., Favrholdt, L.M., Kudahl, C., Larsen, K.S., Mikkelsen, J.W.: Online algorithms with advice: a survey. SIGACT News **47**(3), 93–129 (2016)
2. Karlin, A.R., Manasse, M.S., Rudolph, L., Sleator, D.D.: Competitive snoopy caching. Algorithmica **3**(1), 79–119 (1988)
3. Chrobak, M., Kenyon-Mathieu, C.: Sigact news online algorithms column 10: competitiveness via doubling. SIGACT News **37**(4), 115–126 (2006)
4. Yao, A.C.C.: Probabilistic computations: toward a unified measure of complexity. In: Proceedings of the 18th Annual Symposium on Foundations of Computer Science, SFCS 1977, Washington, DC, USA, pp. 222–227. IEEE Computer Society (1977)

5. Chrobak, M., Kenyon, C., Noga, J., Young, N.E.: Oblivious medians via online bidding. In: Correa, J.R., Hevia, A., Kiwi, M. (eds.) LATIN 2006. LNCS, vol. 3887, pp. 311–322. Springer, Heidelberg (2006). https://doi.org/10.1007/11682462_31

6. Bokhari, S.H.: Partitioning problems in parallel, pipeline, and distributed computing. IEEE Trans. Comput. **37**(1), 48–57 (1988)

7. Hansen, P., Lih, K.W.: Improved algorithms for partitioning problems in parallel, pipelined, and distributed computing. IEEE Trans. Comput. **41**, 769–771 (1992)

8. Khanna, S., Muthukrishnan, S., Skiena, S.: Efficient array partitioning. In: Degano, P., Gorrieri, R., Marchetti-Spaccamela, A. (eds.) ICALP 1997. LNCS, vol. 1256, pp. 616–626. Springer, Heidelberg (1997). https://doi.org/10.1007/3-540-63165-8_216

9. Miguet, S., Pierson, J.-M.: Heuristics for 1D rectilinear partitioning as a low cost and high quality answer to dynamic load balancing. In: Hertzberger, B., Sloot, P. (eds.) HPCN-Europe 1997. LNCS, vol. 1225, pp. 550–564. Springer, Heidelberg (1997). https://doi.org/10.1007/BFb0031628

10. Konrad, C.: Streaming partitioning of sequences and trees. In: 19th International Conference on Database Theory, ICDT 2016, Bordeaux, France, 15–18 March 2016, pp. 13:1–13:18 (2016)

11. Konrad, C., Tonoyan, T.: Preemptive online partitioning of sequences. CoRR abs/1702.06099 (2017)

Improved Algorithms for k-Domination and Total k-Domination in Proper Interval Graphs

Nina Chiarelli[1,2], Tatiana Romina Hartinger[1,2], Valeria Alejandra Leoni[3,4(✉)], Maria Inés Lopez Pujato[3,5], and Martin Milanič[1,2(✉)]

[1] FAMNIT, University of Primorska, Glagoljaška 8, 6000 Koper, Slovenia
nina.chiarelli@famnit.upr.si, martin.milanic@upr.si
[2] IAM, University of Primorska, Muzejski trg 2, 6000 Koper, Slovenia
tatiana.hartinger@iam.upr.si
[3] FCEIA, Universidad Nacional de Rosario, Rosario, Santa Fe, Argentina
valeoni@fceia.unr.edu.ar, mineslpk@hotmail.com
[4] CONICET, Buenos Aires, Argentina
[5] ANPCyT, Buenos Aires, Argentina

Abstract. Given a positive integer k, a k-dominating set in a graph G is a set of vertices such that every vertex not in the set has at least k neighbors in the set. A total k-dominating set, also known as a k-tuple total dominating set, is a set of vertices such that every vertex of the graph has at least k neighbors in the set. The problems of finding the minimum size of a k-dominating, resp. total k-dominating set, in a given graph, are referred to as k-domination, resp. total k-domination. These generalizations of the classical domination and total domination problems are known to be NP-hard in the class of chordal graphs, and, more specifically, even in the classes of split graphs (both problems) and undirected path graphs (in the case of total k-domination). On the other hand, it follows from recent work by Kang et al. (2017) that these two families of problems are solvable in time $\mathcal{O}(|V(G)|^{6k+4})$ in the class of interval graphs. In this work, we develop faster algorithms for k-domination and total k-domination in the class of proper interval graphs. The algorithms run in time $\mathcal{O}(|V(G)|^{3k})$ for each fixed $k \geq 1$ and are also applicable to the weighted case.

Keywords: k-domination · Total k-domination
Proper interval graph · Polynomial-time algorithm

This work is supported in part by the Slovenian Research Agency (I0-0035, research program P1-0285 and research projects N1-0032, J1-6720, and J1-7051). The work for this paper was partly done in the framework of a bilateral project between Argentina and Slovenia, financed by the Slovenian Research Agency (BI-AR/15–17–009) and MINCYT-MHEST (SLO/14/09).

J. Lee et al. (Eds.): ISCO 2018, LNCS 10856, pp. 290–302, 2018.
https://doi.org/10.1007/978-3-319-96151-4_25

1   Introduction

Among the many variants of the domination problems [30,31], we consider in this paper a family of generalizations of the classical domination and total domination problems known as k-domination and total k-domination. Given a positive integer k and a graph G, a k-*dominating set* in G is a set $S \subseteq V(G)$ such that every vertex $v \in V(G) \setminus S$ has at least k neighbors in S, and a *total k-dominating set* in G is a set $S \subseteq V(G)$ such that every vertex $v \in V(G)$ has at least k neighbors in S. The k-domination and the total k-domination problems aim to find the minimum size of a k-dominating, resp. total k-dominating set, in a given graph. The notion of k-domination was introduced by Fink and Jacobson in 1985 [23] and studied in a series of papers (e.g., [14,20,22,27,42]) and in a survey by Chellali et al. [13]. The notion of total k-domination was introduced by Kulli in 1991 [41] and studied under the name of k-*tuple total domination* by Henning and Kazemi in 2010 [32] and also in a series of recent papers [1,39,43,53]. The terminology "k-tuple total domination" was introduced in analogy with the notion of "k-tuple domination", introduced in 2000 by Harary and Haynes [29].[1] The redundancy involved in k-domination and total k-domination problems can make them useful in various applications, for example in forming sets of representatives or in resource allocation in distributed computing systems (see, e.g., [31]). However, these problems are known to be NP-hard [37,53] and also hard to approximate [17].

The k-domination and total k-domination problems remain NP-hard in the class of chordal graphs. More specifically, the problems are NP-hard in the class of split graphs [42,53] and, in the case of total k-domination, also in the class of undirected path graphs [43]. We consider k-domination and total k-domination in another subclass of chordal graphs, the class of proper interval graphs. A graph G is an *interval graph* if it has an intersection model consisting of closed intervals on a real line, that is, if there exist a family $\mathcal{I}$ of intervals on the real line and a one-to-one correspondence between the vertices of G and the intervals of $\mathcal{I}$ such that two vertices are joined by an edge in G if and only if the corresponding intervals intersect. A *proper interval graph* is an interval graph that has a *proper interval model*, that is, an intersection model in which no interval contains another one. Proper interval graphs were introduced by Roberts [57], who showed that they coincide with the unit interval graphs, that is, graphs admitting an interval model in which all intervals are of unit length. Various characterizations of proper interval graphs have been developed in the literature (see, e.g., [24,26,36,49]) and several linear-time recognition algorithms are known, which in case of a yes instance also compute a proper interval model (see, e.g., [18] and references cited therein).

Domination and total domination problems are known to be solvable in linear time in the class of interval graphs (see [6,12,33] and [10,12,40,55,56], respectively). Furthermore, for each fixed integer $k \geq 1$, the k-domination and total

---

[1] A set S of vertices is said to be a k-*tuple dominating set* if every vertex of G is adjacent or equal to at least k vertices in S.

k-domination problems are solvable in time $\mathcal{O}(n^{6k+4})$ in the class of interval graphs where n is the order of the input graph. This follows from recent results due to Kang et al. [38], building on previous works by Bui-Xuan et al. [8] and Belmonte and Vatshelle [3]. In fact, Kang et al. studied a more general class of problems, called (ρ, σ)-domination problems, and showed that every such problem can be solved in time $\mathcal{O}(n^{6d+4})$ in the class of n-vertex interval graphs, where d is a parameter associated to the problem (see Corollary 3.2 in [38] and the paragraph following it).

1.1   Our Results and Approach

We significantly improve the above result for the case of proper interval graphs. We show that for each positive integer k, the k-domination and total k-domination problems are solvable in time $\mathcal{O}(n^{3k})$ in the class of n-vertex proper interval graphs. Except for $k = 1$, this improves on the best known running time.

Our approach is based on a reduction showing that for each positive integer k, the total k-domination problem on a given proper interval graph G can be reduced to a shortest path computation in a derived edge-weighted directed acyclic graph. A similar reduction works for k-domination. The reductions immediately result in algorithms with running time $\mathcal{O}(n^{4k+1})$. We show that with a suitable implementation the running time can be improved to $\mathcal{O}(n^{3k})$. The algorithms can be easily adapted to the weighted case, at no expense in the running time.

1.2   Related Work

We now give an overview of related work and compare our results with most relevant other results, besides those due to due to Kang et al. [38], which motivated this work.

Overview. For every positive integer k, the k-domination problem is NP-hard in the classes of bipartite graphs [2] and split graphs [42], but solvable in linear time in the class of graphs every block of which is a clique, a cycle or a complete bipartite graph (including trees, block graphs, cacti, and block-cactus graphs) [42], and, more generally, in any class of graphs of bounded clique-width [19,50] (see also [16]). The total k-domination problem is NP-hard in the classes of split graphs [53], doubly chordal graphs [53], bipartite graphs [53], undirected path graphs [43], and bipartite planar graphs (for $k \in \{2, 3\}$) [1], and solvable in linear time in the class of graphs every block of which is a clique, a cycle, or a complete bipartite graph [43], and, more generally, in any class of graphs of bounded clique-width [19,50], and in polynomial time in the class of chordal bipartite graphs [53]. k-domination and total k-domination problems were also studied with respect to their (in)approximability properties, both in general [17] and in restricted graph classes [2], as well as from the parameterized complexity point of view [9,34].

Besides k-domination and total k-domination, other variants of domination problems solvable in polynomial time in the class of proper interval graphs (or in some of its superclasses) include k-tuple domination for all $k \geq 1$ [45] (see also [44] and, for $k = 2$, [54]), connected domination [56], independent domination [21], paired domination [15], efficient domination [11], liar's domination [51], restrained domination [52], eternal domination [5], power domination [46], outer-connected domination [48], Roman domination [47], Grundy domination [7], etc.

Comparison. Bertossi [4] showed how to reduce the total domination problem in a given interval graph to a shortest path computation in a derived edge-weighted directed acyclic graph satisfying some additional constraints on pairs of consecutive arcs. A further transformation reduces the problem to a usual (unconstrained) shortest path computation. Compared to the approach of Bertossi, our approach exploits the additional structure of proper interval graphs in order to gain generality in the problem space. Our approach works for every k and is also more direct, in the sense that the (usual or total, unweighted or weighted) k-domination problem in a given proper interval graph is reduced to a shortest path computation in a derived edge-weighted acyclic digraph in a single step.

The works of Liao and Chang [45] and of Lee and Chang [44] consider various domination problems in the class of strongly chordal graphs (and, in the case of [45], also dually chordal graphs). While the class of strongly chordal graphs generalizes the class of interval graphs, the domination problems studied in [44, 45] all deal with closed neighborhoods, and for those cases structural properties of strongly chordal and dually chordal graphs are helpful for the design of linear-time algorithms. In contrast, k-domination and total k-domination are defined via open neighborhoods and results of [44, 45] do not seem to be applicable or easily adaptable to our setting.

Structure of the Paper. In Sect. 2, we describe the reduction for the total k-domination problem. The specifics of the implementation resulting in improved running time are given in Sect. 3. In Sect. 4, we discuss how the approach can be modified to solve the k-domination problem and the weighted cases. We conclude the paper with some open problems in Sect. 5. Due to lack of space, most proofs are omitted. They can be found in the full version [58].

In the rest of the section, we fix some definitions and notation. Given a graph G and a set $X \subseteq V(G)$, we denote by $G[X]$ the subgraph of G induced by X and by $G - X$ the subgraph induced by $V(G) \setminus X$. For a vertex u in a graph G, we denote by $N(u)$ the set of neighbors of u in G. Note that for every graph G, the set $V(G)$ is a k-dominating set, while G has a total k-dominating set if and only if every vertex of G has at least k neighbors.

2   The Reduction for Total k-Domination

Let k be a positive integer and $G = (V, E)$ be a given proper interval graph. We will assume that G is equipped with a proper interval model $\mathcal{I} = \{I_j \mid j = 1, \ldots, n\}$ where $I_j = [a_j, b_j]$ for all $j = 1, \ldots, n$. (As mentioned in the introduction, a proper interval model of a given proper interval graph can be computed in linear time.) We may also assume that no two intervals coincide. Moreover, since in a proper interval model the order of the left endpoints equals the order of the right endpoints, we assume that the intervals are sorted increasingly according to their left endpoints, i.e., $a_1 < \ldots < a_n$. We use notation $I_j < I_\ell$ if $j < \ell$ and say in this case that I_j is *to the left of* I_ℓ and I_ℓ is *to the right of* I_j. Also, we write $I_j \leq I_\ell$ if $j \leq \ell$. Given three intervals $I_j, I_\ell, I_m \in \mathcal{I}$, we say that interval I_ℓ is *between* intervals I_j and I_m if $j < \ell < m$. We say that interval I_j *intersects* interval I_ℓ if $I_j \cap I_\ell \neq \emptyset$.

Our approach can be described as follows. Given G, we compute an edge-weighted directed acyclic graph D_k^t (where the superscript "t" means "total") and show that the total k-domination problem on G can be reduced to a shortest path computation in D_k^t. The definition of the digraph given next is followed by an example and an explanation of the intuition behind the reduction.

To distinguish the vertices of D_k^t from those of G, we will refer to them as *nodes*. Vertices of G will be typically denoted by u or v, and nodes of D_k^t by s, s', s''. Each node of D_k^t will be a sequence of intervals from the set $\mathcal{I}' = \mathcal{I} \cup \{I_0, I_{n+1}\}$, where I_0, I_{n+1} are two new, "dummy" intervals such that $I_0 < I_1$, $I_0 \cap I_1 = \emptyset$, $I_n < I_{n+1}$, and $I_n \cap I_{n+1} = \emptyset$. We naturally extend the linear order $<$ on $\mathcal{I}$ to the whole set $\mathcal{I}'$. We will say that an interval $I \in \mathcal{I}'$ is *associated with* a node s of D_k^t if it appears in sequence s. The set of all intervals associated with s will be denoted by $\mathcal{I}_s$. Given a node s of D_k^t, we will denote by $\min(s)$ and $\max(s)$ the first, resp., the last interval in $\mathcal{I}_s$ with respect to ordering $<$ of $\mathcal{I}'$. A sequence $(I_{i_1}, \ldots, I_{i_q})$ of intervals from $\mathcal{I}$ is said to be *increasing* if $i_1 < \ldots < i_q$.

The node set of D_k^t is given by $V(D_k^t) = \{I_0, I_{n+1}\} \cup S \cup B$, where:

- I_0 and I_{n+1} are sequences of intervals of length one.[2]
- S is the set of so-called *small nodes*. Set S consists exactly of those increasing sequences $s = (I_{i_1}, \ldots, I_{i_q})$ of intervals from $\mathcal{I}$ such that:
  (1) $k + 1 \leq q \leq 2k - 1$,
  (2) for all $j \in \{1, \ldots, q-1\}$, we have $I_{i_j} \cap I_{i_{j+1}} \neq \emptyset$, and
  (3) every interval $I \in \mathcal{I}$ such that $\min(s) \leq I \leq \max(s)$ intersects at least k intervals from the set $\mathcal{I}_s \setminus \{I\}$.
- B is the set of so-called *big nodes*. Set B consists exactly of those increasing sequences $s = (I_{i_1}, \ldots, I_{i_{2k}})$ of intervals from $\mathcal{I}$ of length $2k$ such that:
  (1) for all $j \in \{1, \ldots, 2k-1\}$, we have $I_{i_j} \cap I_{i_{j+1}} \neq \emptyset$, and
  (2) every interval $I \in \mathcal{I}$ such that $I_{i_k} \leq I \leq I_{i_{k+1}}$ intersects at least k intervals from the set $\mathcal{I}_s \setminus \{I\}$.

The arc set of D_k^t is given by $E(D_k^t) = E_0 \cup E_1$, where:

- Set E_0 consists exactly of those ordered pairs $(s, s') \in V(D_k^t) \times V(D_k^t)$ such that:
  (1) $\max(s) < \min(s')$ and $\max(s) \cap \min(s') = \emptyset$,

---

[2] This assures that the intervals $\min(s)$ and $\max(s)$ are well defined also for $s \in \{I_0, I_{n+1}\}$, in which case both are equal to s.

(2) every interval $I \in \mathcal{I}$ such that $\max(s) < I < \min(s')$ intersects at least k intervals from $\mathcal{I}_s \cup \mathcal{I}_{s'}$,

(3) if $s \in B$, then the rightmost $k + 1$ intervals associated with s pairwise intersect, and

(4) if $s' \in B$, then the leftmost $k + 1$ intervals associated with s' pairwise intersect.

- Set E_1 consists exactly of those ordered pairs $(s, s') \in V(D_k^t) \times V(D_k^t)$ such that $s, s' \in B$ and there exist $2k + 1$ intervals $I_{i_1}, \ldots, I_{i_{2k+1}}$ in $\mathcal{I}$ such that $s = (I_{i_1}, I_{i_2}, \ldots, I_{i_{2k}})$ and $s' = (I_{i_2}, I_{i_3}, \ldots, I_{i_{2k+1}})$.

To every arc (s, s') of D_k^t we associate a non-negative length $\ell(s, s')$, defined as follows:

$$\ell(s, s') = \begin{cases} |\mathcal{I}_{s'}|, & \text{if } (s, s') \in E_0 \text{ and } s' \neq I_{n+1}; \\ 1, & \text{if } (s, s') \in E_1; \\ 0, & \text{otherwise.} \end{cases} \qquad (*)$$

The length of a directed path in D_k^t is defined, as usual, as the sum of the lengths of its arcs.

Example 1. Consider the problem of finding a minimum total 2-dominating set in the graph G given by the proper interval model $\mathcal{I}$ depicted in Fig. 1(a). Using the reduction described above, we obtain the digraph D_2^t depicted in Fig. 1(c), where, for clarity, nodes $(I_{i_1}, \ldots, I_{i_p})$ of D_2^t are identified with the corresponding strings of indices $i_1 i_2 \ldots i_p$. We also omit in the figure the (irrelevant) nodes that do not belong to any directed path from I_0 to I_{n+1}. There is a unique shortest I_0, I_9-path in D_2^t, namely $(0, 2356, 3567, 9)$. The path corresponds to $\{2, 3, 5, 6, 7\}$, the only minimum total 2-dominating set in G.

The correctness of the above reduction is established by proving the following.

Proposition 1. *Given a proper interval graph G and a positive integer k, let D_k^t be the directed graph constructed as above. Then G has a total k-dominating set of size c if and only if D_k^t has a directed path from I_0 to I_{n+1} of length c.*

The intuition behind the reduction is the following. The subgraph of G induced by a minimum total k-dominating set splits into connected components. These components as well as vertices within them are naturally ordered from left to right. Moreover, since each connected subgraph of a proper interval graph has a Hamiltonian path, the nodes of D_k^t correspond to paths in G, see condition (2) for small nodes or condition (1) for big nodes. Since each vertex of G has at least k neighbors in the total k-dominating set, each component has at least $k + 1$ vertices. Components with at least $2k$ vertices give rise to directed paths in D_k^t consisting of big nodes and arcs in E_1. Each component with less than $2k$ vertices corresponds to a unique small node in D_k^t, which can be seen as a trivial directed path in D_k^t. The resulting paths inherit the left-to-right ordering from the components and any two consecutive paths are joined in D_k^t by an arc in E_0. Moreover, I_0 is joined to the leftmost node of the leftmost path with an arc in E_0 and, symmetrically, the rightmost node of the rightmost path is joined to

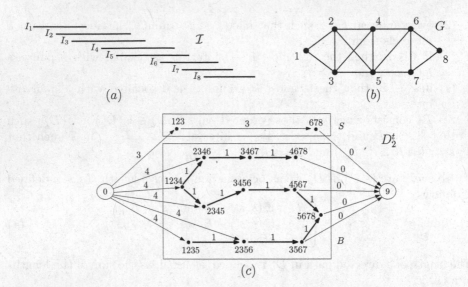

Fig. 1. (*a*) A proper interval model $\mathcal{I}$, (*b*) the corresponding proper interval graph G, and (*c*) a part of the derived digraph D_2^t, where only nodes that lie on some directed path from I_0 to I_9 are shown. Edges in E_1 are depicted bold.

I_{n+1} with an arc in E_0. Adding such arcs yields a directed path from I_0 to I_{n+1} of the desired length.

The above process can be reversed. Given a directed path P in D_k^t from I_0 to I_{n+1}, a total k-dominating set in G of the desired size can be obtained as the set of all vertices corresponding to intervals in $\mathcal{I}$ associated with internal nodes of P. The total k-dominating property is established using the defining properties of small nodes, big nodes, and arcs in E_0 and in E_1. For example, condition (3) in the definition of arcs in E_0 guarantees that the vertex corresponding to the rightmost interval associated with $s \in B$ where $(s, s') \in E_0$ is k-dominated. The condition is related to the fact that in proper interval graphs the neighborhood of a vertex represented by an interval $[a, b]$ splits into two cliques: one for all intervals containing a and another one for all intervals containing b.

The digraph D_k^t has $\mathcal{O}(n^{2k})$ nodes and $\mathcal{O}(n^{4k})$ arcs and can be, together with the length function ℓ on its arcs, computed from G directly from the definition in time $\mathcal{O}(n^{4k+1})$. A shortest directed path (with respect to ℓ) from I_0 to all nodes reachable from I_0 in D_k^t can be computed in polynomial time using any of the standard approaches, for example using Dijkstra's algorithm. Actually, since D_k^t is acyclic, a dynamic programming approach along a topological ordering of D_k^t can be used to compute shortest paths from I_0 in linear time (in the size of D_k^t). Proposition 1 therefore implies that the total k-domination problem is solvable in time $\mathcal{O}(n^{4k+1})$ in the class of n-vertex proper interval graphs.

We will show in the next section that, with a careful implementation, a shortest I_0, I_{n+1}-path in D_k^t can be computed without examining all the arcs of the digraph, leading to the claimed improvement in the running time.

3   Improving the Running Time

We assume all notations from Sect. 2. In particular, G is a given n-vertex proper interval graph equipped with a proper interval model $\mathcal{I}$ and (D_k^t, ℓ) is the derived edge-weighted acyclic digraph with $\mathcal{O}(n^{2k})$ nodes. We apply Proposition 1 and show that a shortest I_0, I_{n+1}-path in D_k^t can be computed in time $\mathcal{O}(n^{3k})$. The main idea of the speedup relies on the fact that the algorithm avoids examining all arcs of the digraph. This is achieved by employing a dynamic programming approach based on a partition of a subset of the node set into $\mathcal{O}(n^k)$ parts depending on the nodes' suffixes of length k. The partition will enable us to efficiently compute minimum lengths of four types of directed paths in D_k^t, all starting in I_0 and ending in a specified vertex, vertex set, arc, or arc set. In particular, a shortest I_0, I_{n+1}-path in D_k^t will be also computed this way.

Theorem 1. *For every positive integer k, the total k-domination problem is solvable in time $\mathcal{O}(|V(G)|^{3k})$ in the class of proper interval graphs.*

Proof (sketch). By Proposition 1, it suffices to show that a shortest directed path from I_0 to I_{n+1} in D_k^t can be computed in the stated time. Due to lack of space, we only explain some implementation details. In order to describe the algorithm, we need to introduce some notation. Given a node $s \in S \cup B$, say $s = (I_{i_1}, \ldots, I_{i_q})$ (recall that $k + 1 \leq q \leq 2k$), we define its k-*suffix* of s as the sequence $(I_{i_{q-k+1}}, \ldots, I_{i_q})$ and denote it by $\mathrm{suf}_k(s)$.

The algorithm proceeds as follows. First, it computes the node set of D_k^t and a subset B' of the set of big nodes consisting of precisely those nodes $s \in B$ satisfying condition (3) in the definition of E_0 (that is, the rightmost $k + 1$ intervals associated with s pairwise intersect). Next, it computes a partition $\{A_\sigma \mid \sigma \in \Sigma\}$ of $S \cup B'$ defined by $\Sigma = \{\mathrm{suf}_k(s) : s \in S \cup B'\}$ and $A_\sigma = \{s \in S \cup B' \mid \mathrm{suf}_k(s) = \sigma\}$ for all $\sigma \in \Sigma$.

The algorithm also computes the arc set E_1. On the other hand, the arc set E_0 is not generated explicitly, except for the arcs in E_0 with tail I_0 or head I_{n+1}. Using dynamic programming, the algorithm will compute the following values.

(i) For all $s \in V(D_k^t) \setminus \{I_0\}$, let p_s^0 denote the minimum ℓ-length of a directed I_0, s-path in D_k^t ending with an arc from E_0.

(ii) For all $s \in V(D_k^t) \setminus \{I_0\}$, let p_s denote the minimum ℓ-length of a directed I_0, s-path in D_k^t.

(iii) For all $e \in E_1$, let p_e denote the minimum ℓ-length of a directed path in D_k^t starting in I_0 and ending with e.

(iv) For all $\sigma \in \Sigma$, let p_σ denote the minimum ℓ-length of a directed path in D_k^t starting in I_0 and ending in A_σ.

In all cases, if no path of the corresponding type exists, we set the value of the respective p_s^0, p_s, p_e, or p_σ to ∞.

Clearly, once all the p_s^0, p_s, p_e, and p_σ values will be computed, the length of a shortest I_0, I_{n+1}-path in D_k^t will be given by $p_{I_{n+1}}$.

The above values can be computed using the following recursive formulas:

(i) p_s^0 values:
- For $s \in S \cup B$, let $\Sigma_s = \{\sigma \in \Sigma \mid (\tilde{s}, s) \in E_0 \text{ for some } \tilde{s} \in A_\sigma\}$ and set

$$p_s^0 = \begin{cases} |\mathcal{I}_s|, & \text{if } (I_0, s) \in E_0; \\ \min\limits_{\sigma \in \Sigma_s} p_\sigma + |\mathcal{I}_s|, & \text{if } (I_0, s) \notin E_0 \text{ and } \Sigma_s \neq \emptyset; \\ \infty, & \text{otherwise.} \end{cases}$$

- For $s = I_{n+1}$, let $p_s^0 = \min\limits_{(\tilde{s},s) \in E_0} p_{\tilde{s}}$.

(ii) p_s values: For all $s \in V(D_k^t) \setminus \{I_0\}$, we have $p_s = \min \left\{ p_s^0, \min\limits_{(\tilde{s},s) \in E_1} p_{(\tilde{s},s)} \right\}$.

(iii) p_e values: For all $e = (s, s') \in E_1$, we have $p_e = p_s + 1$.

(iv) p_σ values: For all $\sigma \in \Sigma$, we have $p_\sigma = \min\limits_{s \in A_\sigma} p_s$.

The above formulas can be computed following any topological sort of D_k^t such that if $s, s' \in S \cup B$ are such that $\operatorname{suf}_k(s) \neq \operatorname{suf}_k(s')$ and $\operatorname{suf}_k(s)$ is lexicographically smaller than $\operatorname{suf}_k(s')$, then s appears strictly before s' in the ordering. When the algorithm processes a node $s \in V(D_k^t) \setminus \{I_0\}$, it computes the values of p_s^0, p_e for all $e = (\tilde{s}, s) \in E_1$, and p_s, in this order. For every $\sigma \in \Sigma$, the value of p_σ is computed as soon as the values of p_s are known for all $s \in A_\sigma$. This completes the description of the algorithm.   □

4   Modifying the Approach for k-Domination and for Weighted Problems

With minor modifications of the definitions of small nodes, big nodes, and arcs in E_0 of the derived digraph, the approach developed in Sects. 2 and 3 for total k-domination leads to an analogous result for k-domination.

Theorem 2. *For every positive integer k, the k-domination problem is solvable in time $\mathcal{O}(|V(G)|^{3k})$ in the class of proper interval graphs.*

The approach of Kang et al. [38], which implies that k-domination and total k-domination are solvable in time $O(|V(G)|^{6k+4})$ in the class of interval graphs also works for the weighted versions of the problems, where each vertex $u \in V(G)$ is equipped with a non-negative cost $c(u)$ and the task is to find a (usual or total) k-dominating set of G of minimum total cost. For both families of problems, our approach can also be easily adapted to the weighed case. Denoting the total cost of a set $\mathcal{J}$ of vertices (i.e., intervals) by $c(\mathcal{J}) = \sum_{I \in \mathcal{J}} c(I)$, it suffices to generalize the length function from $(*)$ in a straightforward way, as follows:

$$\ell(s, s') = \begin{cases} c(\mathcal{I}_{s'}), & \text{if } (s, s') \in E_0 \text{ and } s' \neq I_{n+1}; \\ c(\min(s')), & \text{if } (s, s') \in E_1; \\ 0, & \text{otherwise.} \end{cases} \tag{1}$$

This results in $O(|V(G)|^{3k})$ algorithms for the weighted (usual or total) k-domination problems in the class of proper interval graphs.

5 Conclusion

In this work, we presented improved algorithms for weighted k-domination and total k-domination problems for the class of proper interval graphs. The time complexity was significantly improved, from $\mathcal{O}(n^{6k+4})$ to $\mathcal{O}(n^{3k})$, for each fixed integer $k \geq 1$. Our work leaves open several questions. Even though polynomial for each fixed k, our algorithms are too slow to be of practical use, and the main question is whether having k in the exponent of the running time can be avoided. Are the k-domination and total k-domination problems fixed-parameter tractable with respect to k in the class of proper interval graphs? Could it be that even the more general problems of *vector domination* and *total vector domination* (see, e.g., [17, 25, 28, 35]), which generalize k-domination and total k-domination when k is part of input, can be solved in polynomial time in the class of proper interval graphs? It would also be interesting to determine the complexity of these problems in generalizations of proper interval graphs such as interval graphs, strongly chordal graphs, cocomparability graphs, and AT-free graphs.

References

1. Argiroffo, G., Leoni, V., Torres, P.: Complexity of k-tuple total and total $\{k\}$-dominations for some subclasses of bipartite graphs Inform. Process. Lett. (2017). https://doi.org/10.1016/j.ipl.2018.06.007
2. Bakhshesh, D., Farshi, M., Hasheminezhad, M.: Complexity results for k-domination and α-domination problems and their variants (2017). arXiv:1702.00533 [cs.CC]
3. Belmonte, R., Vatshelle, M.: Graph classes with structured neighborhoods and algorithmic applications. Theoret. Comput. Sci. **511**, 54–65 (2013)
4. Bertossi, A.A.: Total domination in interval graphs. Inform. Process. Lett. **23**(3), 131–134 (1986)
5. Braga, A., de Souza, C.C., Lee, O.: The eternal dominating set problem for proper interval graphs. Inform. Process. Lett. **115**(6–8), 582–587 (2015)
6. Brandstädt, A., Chepoi, V.D., Dragan, F.F.: The algorithmic use of hypertree structure and maximum neighbourhood orderings. Discrete Appl. Math. **82**(1–3), 43–77 (1998)

7. Brešar, B., Gologranc, T., Kos, T.: Dominating sequences under atomic changes with applications in sierpiński and interval graphs. Appl. Anal. Discrete Math. **10**(2), 518–531 (2016)

8. Bui-Xuan, B.-M., Telle, J.A., Vatshelle, M.: Fast dynamic programming for locally checkable vertex subset and vertex partitioning problems. Theoret. Comput. Sci. **511**, 66–76 (2013)

9. Cattanéo, D., Perdrix, S.: The parameterized complexity of domination-type problems and application to linear codes. In: Gopal, T.V., Agrawal, M., Li, A., Cooper, S.B. (eds.) TAMC 2014. LNCS, vol. 8402, pp. 86–103. Springer, Cham (2014). https://doi.org/10.1007/978-3-319-06089-7_7

10. Chang, G.J.: Labeling algorithms for domination problems in sun-free chordal graphs. Discrete Appl. Math. **22**(1), 21–34 (1989)

11. Chang, G.J., Pandu, C.P., Coorg, S.R.: Weighted independent perfect domination on cocomparability graphs. Discrete Appl. Math. **63**(3), 215–222 (1995)

12. Chang, M.-S.: Efficient algorithms for the domination problems on interval and circular-arc graphs. SIAM J. Comput. **27**(6), 1671–1694 (1998)

13. Chellali, M., Favaron, O., Hansberg, A., Volkmann, L.: k-domination and k-independence in graphs: a survey. Graphs Combin. **28**(1), 1–55 (2012)

14. Chellali, M., Meddah, N.: Trees with equal 2-domination and 2-independence numbers. Discuss. Math. Graph Theory **32**(2), 263–270 (2012)

15. Cheng, T.C.E., Kang, L.Y., Ng, C.T.: Paired domination on interval and circular-arc graphs. Discrete Appl. Math. **155**(16), 2077–2086 (2007)

16. Cicalese, F., Cordasco, G., Gargano, L., Milanič, M., Vaccaro, U.: Latency-bounded target set selection in social networks. Theoret. Comput. Sci. **535**, 1–15 (2014)

17. Cicalese, F., Milanič, M., Vaccaro, U.: On the approximability and exact algorithms for vector domination and related problems in graphs. Discrete Appl. Math. **161**(6), 750–767 (2013)

18. Corneil, D.G.: A simple 3-sweep LBFS algorithm for the recognition of unit interval graphs. Discrete Appl. Math. **138**(3), 371–379 (2004)

19. Courcelle, B., Makowsky, J.A., Rotics, U.: Linear time solvable optimization problems on graphs of bounded clique-width. Theory Comput. Syst. **33**(2), 125–150 (2000)

20. DeLaViña, E., Goddard, W., Henning, M.A., Pepper, R., Vaughan, E.R.: Bounds on the k-domination number of a graph. Appl. Math. Lett. **24**(6), 996–998 (2011)

21. Farber, M.: Independent domination in chordal graphs. Oper. Res. Lett. **1**(4), 134–138 (1982)

22. Favaron, O., Hansberg, A., Volkmann, L.: On k-domination and minimum degree in graphs. J. Graph Theory **57**(1), 33–40 (2008)

23. Fink, J.F., Jacobson, M.S.: n-domination in graphs. In: Graph Theory with Applications to Algorithms and Computer Science (Kalamazoo, Mich., 1984), pp. 283–300. Wiley, New York (1985)

24. Gardi, F.: The roberts characterization of proper and unit interval graphs. Discrete Math. **307**(22), 2906–2908 (2007)

25. Gerlach, T., Harant, J.: A note on domination in bipartite graphs. Discuss. Math. Graph Theory **22**(2), 229–231 (2002)

26. Gutierrez, M., Oubiña, L.: Metric characterizations of proper interval graphs and tree-clique graphs. J. Graph Theory **21**(2), 199–205 (1996)

27. Hansberg, A., Pepper, R.: On k-domination and j-independence in graphs. Discrete Appl. Math. **161**(10–11), 1472–1480 (2013)

28. Harant, J., Pruchnewski, A., Voigt, M.: On dominating sets and independent sets of graphs. Combin. Probab. Comput. **8**(6), 547–553 (1999)

29. Harary, F., Haynes, T.W.: Double domination in graphs. Ars Combin. **55**, 201–213 (2000)
30. Haynes, T.W., Hedetniemi, S.T., Slater, P.J. (eds.): Domination in Graphs. Advanced Topics, volume 209 of Monographs and Textbooks in Pure and Applied Mathematics. Marcel Dekker Inc., New York (1998)
31. Haynes, T.W., Hedetniemi, S.T., Slater, P.J.: Fundamentals of domination in graphs. In: Monographs and Textbooks in Pure and Applied Mathematics, vol. 208. Marcel Dekker Inc., New York (1998)
32. Henning, M.A., Kazemi, A.P.: k-tuple total domination in graphs. Discrete Appl. Math. **158**(9), 1006–1011 (2010)
33. Hsu, W.L., Tsai, K.-H.: Linear time algorithms on circular-arc graphs. Inform. Process. Lett. **40**(3), 123–129 (1991)
34. Ishii, T., Ono, H., Uno, Y.: Subexponential fixed-parameter algorithms for partial vector domination. Discrete Optim. **22**(part A), 111–121 (2016)
35. Ishii, T., Ono, H., Uno, Y.: (Total) vector domination for graphs with bounded branchwidth. Discrete Appl. Math. **207**, 80–89 (2016)
36. Jackowski, Z.: A new characterization of proper interval graphs. Discrete Math. **105**(1–3), 103–109 (1992)
37. Jacobson, M.S., Peters, K.: Complexity questions for n-domination and related parameters. Congr. Numer. **68**, 7–22 (1989). Eighteenth Manitoba Conference on Numerical Mathematics and Computing (Winnipeg, MB, 1988)
38. Kang, D.Y., Kwon, O.-J., Strømme, T.J.F., Telle, J.A.: A width parameter useful for chordal and co-comparability graphs. Theoret. Comput. Sci. **704**, 1–17 (2017)
39. Kazemi, A.P.: On the total k-domination number of graphs. Discuss. Math. Graph Theory **32**(3), 419–426 (2012)
40. Keil, J.M.: Total domination in interval graphs. Inform. Process. Lett. **22**(4), 171–174 (1986)
41. Kulli, V.R.: On n-total domination number in graphs. Graph Theory. Combinatorics, Algorithms, and Applications (San Francisco, CA, 1989), pp. 319–324. SIAM, Philadelphia (1991)
42. Lan, J.K., Chang, G.J.: Algorithmic aspects of the k-domination problem in graphs. Discrete Appl. Math. **161**(10–11), 1513–1520 (2013)
43. Lan, J.K., Chang, G.J.: On the algorithmic complexity of k-tuple total domination. Discrete Appl. Math. **174**, 81–91 (2014)
44. Lee, C.-M., Chang, M.-S.: Variations of Y-dominating functions on graphs. Discrete Math. **308**(18), 4185–4204 (2008)
45. Liao, C.-S., Chang, G.J.: k-tuple domination in graphs. Inform. Process. Lett. **87**(1), 45–50 (2003)
46. Liao, C.-S., Lee, D.T.: Power domination in circular-arc graphs. Algorithmica **65**(2), 443–466 (2013)
47. Liedloff, M., Kloks, T., Liu, J., Peng, S.-L.: Efficient algorithms for Roman domination on some classes of graphs. Discrete Appl. Math. **156**(18), 3400–3415 (2008)
48. Lin, C.-J., Liu, J.-J., Wang, Y.-L.: Finding outer-connected dominating sets in interval graphs. Inform. Process. Lett. **115**(12), 917–922 (2015)
49. Looges, P.J., Olariu, S.: Optimal greedy algorithms for indifference graphs. Comput. Math. Appl. **25**(7), 15–25 (1993)
50. Oum, S.-I., Seymour, P.: Approximating clique-width and branch-width. J. Combin. Theory Ser. B **96**(4), 514–528 (2006)
51. Panda, B.S., Paul, S.: A linear time algorithm for liar's domination problem in proper interval graphs. Inform. Process. Lett. **113**(19–21), 815–822 (2013)

52. Panda, B.S., Pradhan, D.: A linear time algorithm to compute a minimum restrained dominating set in proper interval graphs. Discrete Math. Algorithms Appl. **7**(2), 1550020 (2015). 21

53. Pradhan, D.: Algorithmic aspects of k-tuple total domination in graphs. Inform. Process. Lett. **112**(21), 816–822 (2012)

54. Pramanik, T., Mondal, S., Pal, M.: Minimum 2-tuple dominating set of an interval graph. Int. J. Comb. 14 (2011). Article ID 389369

55. Ramalingam, G., Pandu, C.: Total domination in interval graphs revisited. Inform. Process. Lett. **27**(1), 17–21 (1988)

56. Ramalingam, G., Rangan, C.P.: A unified approach to domination problems on interval graphs. Inform. Process. Lett. **27**(5), 271–274 (1988)

57. Roberts, F.S.: Indifference graphs. In: Proof Techniques in Graph Theory (Proceedings of Second Ann Arbor Graph Theory Conference, Ann Arbor, Michigan, 1968), pp. 139–146. Academic Press, New York (1969)

58. Chiarelli, N., Hartinger, T.R., Leoni, V.A., Lopez Pujato, M.I., Milanič, M.: New algorithms for weighted k-domination and total k-domination problems in proper interval graphs. arXiv:1803.04327 [cs.DS] (2018)

A Heuristic for Maximising Energy Efficiency in an OFDMA System Subject to QoS Constraints

Adam N. Letchford[1], Qiang Ni[2], and Zhaoyu Zhong[1]([⊠])

[1] Department of Management Science,
Lancaster University, Lancaster LA1 4YX, UK
{A.N.Letchford,z.zhong1}@lancaster.ac.uk
[2] School of Computing and Communications,
Lancaster University, Lancaster LA1 4WA, UK
q.ni@lancaster.ac.uk

Abstract. OFDMA is a popular coding scheme for mobile wireless multi-channel multi-user communication systems. In a previous paper, we used mixed-integer nonlinear programming to tackle the problem of maximising energy efficiency, subject to certain quality of service (QoS) constraints. In this paper, we present a heuristic for the same problem. Computational results show that the heuristic is at least two orders of magnitude faster than the exact algorithm, yet yields solutions of comparable quality.

Keywords: OFDMA systems · Energy efficiency · Heuristics

1 Introduction

In many mobile wireless communications systems, mobile devices communicate with one another via transceivers called *base stations*. Many base stations follow the so-called *Orthogonal Frequency-Division Multiple Access* (OFDMA) scheme to code and transmit messages (see, e.g., [3]). In OFDMA, we have a set of communication channels, called *subcarriers*, and a set of *users* (i.e., mobile devices that are currently allocated to the given base station). Each subcarrier can be assigned to at most one user, but a user may be assigned to more than one subcarrier. The data rate achieved by any given subcarrier is a nonlinear function of the power allocated to it.

Several different optimisation problems have been defined in connection with OFDMA systems (e.g., [6–11, 15, 17–21]). Unfortunately, it turns out that most of these problems are $\mathcal{NP}$-hard [5, 10, 11]. Thus, most authors resort to heuristics. In our recent papers [8, 9], however, we presented exact solution algorithms based on mixed-integer nonlinear programming (MINLP). The problem considered in [9] is to maximise the total data rate of the system subject to certain quality of service (QoS) constraints called *user rate* constraints. The one considered in [8]

© Springer International Publishing AG, part of Springer Nature 2018
J. Lee et al. (Eds.): ISCO 2018, LNCS 10856, pp. 303–312, 2018.
https://doi.org/10.1007/978-3-319-96151-4_26

is similar, except that the objective is to maximise the energy efficiency (defined as the total data rate divided by the total power used).

The algorithm in [9] is capable of solving many realistic problem instances to proven optimality (or near-optimality) within a couple of seconds. The algorithm in [8], however, is a lot slower, taking several minutes in some cases. This makes it of little use in a highly dynamic environment, when users may arrive and depart frequently at random. Thus, we were motivated to devise a fast heuristic for the problem described in [8]. That heuristic is the topic of the present paper.

Our heuristic is based on a combination of fractional programming, 0-1 linear programming and binary search. It turns out to be remarkably effective, being able to solve realistic instances to within 1% of optimality within a few seconds.

The paper has a simple structure. The problem is described in Sect. 2, the heuristic is presented in Sect. 3, the computational results are given in Sect. 4, and concluding remarks are made in Sect. 5.

To make the paper self-contained, we recall the following result from [13] (see also [1, 14]). Consider a *fractional program* of the form:

$$\max \left\{ f(y)/g(y) : y \in C \right\},$$

where $C \subseteq \mathbb{R}^n$ is convex, $f(y)$ is non-negative and concave over the domain C, and $g(y)$ is positive and convex over C. This problem can be reformulated as

$$\max \left\{ tf(y'/t) : tg(y'/t) \leq 1, \ y' \in tC, \ t > 0 \right\},$$

where t is a new continuous variable representing $1/g(y)$, and y' is a new vector of variables representing $y/g(y)$. The reformulated problem has a concave objective function and a convex feasible region.

2   The Problem

The problem under consideration is as follows. We have a set I of subcarriers and a set J of users, a (positive real) *system power* σ (measured in watts) and a total power limit P (also in watts). For each $i \in I$, we are given a *bandwidth* B_i (in megahertz), and a *noise power* N_i (in watts). Finally, for each $j \in J$, we are given a *demand* d_j (in megabits per second). The classical *Shannon-Hartley theorem* [16] implies that, if we allocate p units of power to subcarrier i, the data rate of that subcarrier (again in Mb/s) cannot exceed

$$f_i(p) = B_i \log_2 \left(1 + p/N_i \right).$$

The task is to simultaneously allocate the power to the subcarriers, and the subcarriers to the users, so that energy efficiency is maximised and the demand of each user is satisfied.

In [8], this problem was called the *fractional subcarrier and power allocation problem with rate constraints* or F-SPARC. It was formulated as a *mixed 0-1 nonlinear program*, as follows. For all $i \in I$ and $j \in J$, let the binary variable x_{ij} indicate whether user j is assigned to subcarrier i, let the non-negative variable p_{ij} represent the amount of power supplied to subcarrier i to serve user j, and let r_{ij} denote the associated data rate. The formulation is then:

$$\max \quad \frac{\sum_{i \in I} \sum_{j \in J} r_{ij}}{\sigma + \sum_{i \in I} \sum_{j \in J} p_{ij}} \tag{1}$$

$$\sum_{i \in I} \sum_{j \in J} p_{ij} \leq P - \sigma \tag{2}$$

$$\sum_{j \in J} x_{ij} \leq 1 \qquad (\forall i \in I) \tag{3}$$

$$\sum_{i \in I} r_{ij} \geq d_j \qquad (\forall j \in J) \tag{4}$$

$$r_{ij} \leq f_i(p_{ij}) \qquad (\forall j \in J) \tag{5}$$

$$p_{ij} \leq (P - \sigma) x_{ij} \qquad (\forall i \in I, j \in J) \tag{6}$$

$$x_{ij} \in \{0, 1\} \qquad (\forall i \in I, j \in J)$$

$$p_{ij}, r_{ij} \in \mathbb{R}_+ \qquad (\forall i \in I, j \in J).$$

The objective function (1) represents the total data rate divided by the total power (including the system power). The constraint (2) enforces the limit on the total power. Constraints (3) ensure that each subcarrier is assigned to at most one user. Constraints (4) ensure that user demands are met. Constraints (5) ensure that the data rate for each subcarrier does not exceed the theoretical limit. Constraints (6) ensure that p_{ij} cannot be positive unless x_{ij} is one. The remaining constraints are just binary and non-negativity conditions.

The objective function (1) and the constraints (5) are both nonlinear, but they are easily shown to be concave and convex, respectively. The exact algorithm in [8] starts by applying the transformation mentioned at the end of the introduction, to make the objective function separable. After that, it uses a well-known generic exact method for convex MINLP, called *LP/NLP-based branch-and-bound* [12], enhanced with some specialised cutting planes called *bi-perspective* cuts. As mentioned in the introduction, however, this exact method can be too slow on some instances of practical interest.

3 The Heuristic

We now present our heuristic for the F-SPARC. We will show in the next section that it is capable of solving many F-SPARC instances to proven near-optimality very quickly.

3.1 The Basic Idea

Let $D = \sum_{j \in J} d_j$ be the sum of the user demands. We start by solving the following NLP:

$$\max \left\{ \sum_{i \in I} f_i(p_i) : \sum_{i \in I} p_i \leq P - \sigma, p \in \mathbb{R}_+^{|I|} \right\}. \tag{NLP1}$$

This gives the maximum possible data rate of the system, which we denote by M. If $D > M$, the F-SPARC instance is infeasible, and we terminate immediately.

We remark that NLP1 can be solved extremely quickly in practice, since its objective function is both concave and separable. (In fact, it can be solved by the well-known *water-filling* approach; see, e.g., [2, 4].)

Now consider the following fractional program:

$$\max \sum_{i \in I} f_i(p_i)/(\sigma + \sum_{i \in I} p_i)$$
$$\text{s.t.} \quad \sum_{i \in I} f_i(p_i) \geq D$$
$$\sum_{i \in I} p_i \leq P - \sigma$$
$$p_i \geq 0 \qquad (i \in I).$$

This is a relaxation of the F-SPARC instance, since it ignores the allocation of subcarriers to users, and aggregates the user demand constraints. Using the transformation mentioned in the introduction, it can be converted into the following equivalent convex NLP:

$$\max \quad \sum_{i \in I} t f_i(\tilde{p}_i/t)$$
$$\text{s.t.} \quad \sigma t + \sum_{i \in I} \tilde{p}_i = 1$$
$$1/P \leq t \leq 1/\sigma \qquad \text{(NLP2)}$$
$$\sum_{i \in I} t f_i(\tilde{p}_i/t) \geq D t$$
$$\tilde{p}_i \geq 0 \qquad (i \in I).$$

The solution of NLP2 yields an upper bound on the efficiency of the optimal F-SPARC solution, which we denote by U.

Now, if we can find an F-SPARC solution whose efficiency is equal to U, it must be optimal. In an attempt to find such a solution, one can take the optimal solution of NLP2, say $(\tilde{p}^*, t^*)$, construct the associated data rate $r_i^* = f_i(\tilde{p}_i^*/t^*)/t^*$ for all $i \in I$, and then solve the following 0-1 linear program by branch-and-bound:

$$\max \quad \sum_{i \in I} \sum_{j \in J} x_{ij}$$
$$\text{s.t.} \quad \sum_{j \in J} x_{ij} \leq 1 \quad (i \in I)$$
$$\sum_{i \in I} r_i^* x_{ij} \geq d_j \quad (j \in J) \qquad \text{(01LP)}$$
$$x_{ij} \in \{0, 1\}$$

Note that 01LP, having $|I| |J|$ variables, is of non-trivial size. On the other hand, all feasible solutions (if any exist) represent optimal F-SPARC solutions. Thus, if any feasible solution is found during the branch-and-bound process, we can terminate branch-and-bound immediately.

3.2   Improving with Binary Search

Unfortunately, in practice, 01LP frequently turns out to be infeasible. This is because the sum of the r_i^* is frequently equal to D, which in turn means that a feasible solution of 01LP would have to satisfy all of the linear constraints at perfect equality. Given that the r_i^* and d_j are typically fractional, such a solution is very unlikely to exist. (In fact, even if such a solution did exist, it could well be lost due to rounding errors during the branch-and-bound process.)

These considerations led us to use a more complex approach. We define a modified version of NLP2, in which D is replaced by $(1 + \epsilon) D$, for some small $\epsilon > 0$. We will call this modified version "NLP2(ϵ)". Solving NLP2(ϵ) in place of NLP2 usually leads to a small deterioration in efficiency, but it also tends to lead to slightly larger r_i^* values, which increases the chance that 01LP will find a feasible solution.

We found that, in fact, even better results can be obtained by performing a binary search to find the best value of ϵ. The resulting heuristic is described in Algorithm 1. When Algorithm 1 terminates, if L and U are sufficiently close (say, within 1%), then we have solved the instance (to the desired tolerance).

Algorithm 1. Binary search heuristic for F-SPARC

---

input: power P, bandwidths B_i, noise powers N_i,
demands d_j, system power σ, tolerance $\delta > 0$.
Compute total user demand $D = \sum_{j \in J} d_j$;
Solve NLP1 to compute the maximum possible data rate M;
Output D and M;
if $D > M$ **then**
| Print "The instance is infeasible." and quit;
end
Solve NLP2 to compute upper bound U on optimal efficiency;
Output U;
Set $L := 0$, $\epsilon_\ell := 0$ and $\epsilon_u := (M/D) - 1$;
repeat
    Set $\epsilon := (\epsilon_\ell + \epsilon_u)/2$;
    Solve NLP2(ϵ). Let $(\tilde{p}^*, \tilde{r}^*, t^*)$ be the solution and L' its efficiency;
    Solve 01LP with r^* set to $\tilde{r}^*/t^*$;
    if 01LP *is infeasible* **then**
    | Set $\epsilon_\ell := \epsilon$;
    else
    | | Let x^* be the solution to 01LP;
    | | **if** $L' > L$ **then**
    | | | Set $L := L'$, $\bar{p} := \tilde{p}^*/t^*$ and $\bar{x} := x^*$;
    | | **end**
    | | Set $\epsilon_u := \epsilon$;
    | **end**
until $\epsilon_u - \epsilon_\ell \leq \delta$ *or* $L \geq U/(1+\delta)$;
if $L > 0$ **then**
| Output feasible solution $(\bar{x}, \bar{p})$;
else
| Output "No feasible solution was found.";
end

3.3 Improving by Reallocating Power

Algorithm 1 can be further enhanced as follows. Each time we find a feasible solution x^* to 01LP, we attempt to improve the efficiency of the associated F-SPARC solution by solving the following fractional program:

$$
\begin{aligned}
\max \ & \textstyle\sum_{i \in I} f_i(p_i)/(\sigma + \sum_{i \in I} p_i) \\
\text{s.t.} \ & \textstyle\sum_{i \in I} p_i \leq P - \sigma \\
& \textstyle\sum_{i \in I : x^*_{ij}=1} f_i(p_i) \geq d_j \quad (j \in J) \\
& p_i \geq 0.
\end{aligned}
$$

This is equivalent to a convex NLP that is similar to NLP2, except that we replace the single constraint $\sum_{i \in I} \tilde{r}_i \geq D t$ with the "disaggregated" constraints

$$
\sum_{i \in I : x^*_{ij}=1} \tilde{r}_i \geq d_j t \qquad (j \in J).
$$

We call this modified NLP "NLP2dis". We found that this enhancement leads to a significant improvement in practice. Intuitively, it "repairs" much of the "damage" to the efficiency that was incurred by increasing the demand by a factor of $1 + \epsilon$.

4 Computational Experiments

We now report on some computational experiments that we conducted. The heuristic was coded in Julia v0.5 and run on a virtual machine cluster with 16 CPUs (ranging from Sandy Bridge to Haswell architectures) and 16 GB of RAM, under Ubuntu 16.04.1 LTS. The program calls on MOSEK 7.1 (with default settings) to solve the NLPs, and on the mixed-integer solver from the CPLEX Callable Library (v. 12.6.3) to solve the 0–1 LPs. In CPLEX, default setting were used, except that the parameter "MIPemphasis" was set to "emphasize feasibility", and a time limit of 1 s was imposed for each branch-and-bound run. We also imposed a total time limit of 5 s for each F-SPARC instance.

4.1 Test Instances

To construct our test instances, we used the procedure described in [9], which is designed to produce instances typical of a small (indoor) base station following the IEEE 802.16 standard. These instances have $|I| = 72$, $|J| \in \{4, 6, 8\}$ and P set to 36 W. The noise powers N_i are random numbers distributed uniformly in $(0, 10^{-11})$, and the bandwidths B_i are all set to 1.25 MHz.

The user demands d_j are initially generated according to a unit lognormal distribution, and are then scaled to create instances of varying difficulty. Recall that, for a given instance, D denotes the total demand and M denotes the maximum possible data rate of the system. The quantity D/M is called the *demand ratio* (DR) of the instance. The user demands are scaled so that the

DR takes values in $\{0.75, 0.8, 0.85, 0.9, 0.95, 0.98\}$. As the DR approaches 1 from below, the instances tend to get harder.

For each combination of $|J|$ and DR, we generated 500 random instances. This makes $3 \times 6 \times 500 = 9,000$ instances in total. For each instance, we first ran the exact algorithm in [8], with a tolerance of 0.01%, to compute tight upper bounds on the optimal efficiency. Although this was very time-consuming, it was necessary in order to assess the quality of the solutions found by our heuristic. We remark that some of the instances with high DR were proven to be infeasible by the exact algorithm.

4.2 Results

Table 1 shows, for various combinations of $|J|$ and DR, the number of instances (out of 500) for which the heuristic failed to find a feasible solution within the 5 s time limit. We see that the heuristic always finds a solution when the DR is less than 0.9, but can fail to find one when the DR is close to 1, especially when the number of users is high. This is however not surprising, since a high DR leads to fewer options when solving problem 01LP, and an increase in $|J|$ increases the number of user demands that the heuristic needs to satisfy. Also, as mentioned above, some of the instances are actually infeasible.

Table 1. Number of instances where heuristic failed to find a feasible solution

$|J|$	Demand ratio					
	0.75	0.80	0.85	0.90	0.95	0.98
4	0	0	0	0	5	9
6	0	0	0	7	22	29
8	0	0	0	10	59	82

Table 2 shows, for the same combinations of $|J|$ and DR, the average gap between the efficiency of the solution found by the heuristic, and our upper bound on the optimal efficiency. The average is taken over the instances for which the heuristic found a feasible solution. We see that the heuristic consistently finds an optimal solution when the DR is less than 0.85. Moreover, even for higher DR values, the solutions found by the heuristic are of excellent quality, with average gaps of well under 1%. (Closer inspection of the date revealed that the gap exceeded 1% only for some instances with DR equal to 0.9 or higher.)

Finally, Table 3 shows the average time taken by the heuristic, again averaged over the instances for which the heuristic found a feasible solution. We see that, when the DR is less than 0.85, the heuristic finds the optimal solution within a fraction of a second. Moreover, even for higher DR values, the heuristic rarely needed the full 5 s allocated to it to find a solution of good quality.

Table 2. Average percentage gap between lower and upper bounds.

$|J|$	Demand ratio					
	0.75	0.80	0.85	0.90	0.95	0.98
4	0.00	0.00	0.30	0.24	0.33	0.38
6	0.00	0.00	0.31	0.32	0.48	0.54
8	0.00	0.00	0.33	0.41	0.67	0.90

(Closer inspection of the data revealed that, in the majority of the cases in which the time limit was met, it was because the heuristic had not found a feasible solution by that time.)

Table 3. Average time (in seconds) taken by the heuristic when it succeeded.

$|J|$	Demand ratio					
	0.75	0.80	0.85	0.90	0.95	0.98
4	0.07	0.06	0.11	0.13	0.28	0.38
6	0.07	0.06	0.36	0.57	1.10	1.40
8	0.07	0.09	0.58	0.90	2.00	2.83

All things considered, the heuristic performs very well, both in terms of solution quality and running time. Although we have not given detailed running times for the exact algorithm, we can report that the heuristic is typically faster by at least two orders of magnitude.

5   Concluding Remarks

Due to environmental considerations, it is becoming more and more common to take energy efficiency into account when designing and operating mobile wireless communications systems. We have presented a heuristic for one specific optimisation problem arising in this context, concerned with maximising energy efficiency in an OFDMA system. The computational results are very promising, with optimal or near-optimal solutions being found for the majority of instances in less than a second.

We believe that our heuristic is suitable for real-life application in a dynamic environment, as long as users arrive and depart only every few seconds. However, there is one caveat: our heuristic involves the solution of nonlinear programs (NLPs) and 0–1 linear programs (0-1 LPs), which in itself consumes energy. We believe that the NLP subproblems could be solved more quickly and efficiently using a specialised method (such as water-filling). As for the 0-1 LP subproblems, note that they are actually only *feasibility* problems, rather than optimisation problems *per se*. It may well be possible to solve them efficiently using a simple local-search heuristic, rather than invoking the "heavy machinery" of an exact 0-1 LP solver. This may be the topic of a future paper.

References

1. Charnes, A., Cooper, W.W.: Programming with linear fractional functionals. Naval Res. Logist. Q. **9**(3-4), 181-186 (1962)
2. Cover, T.M., Thomas, J.A.: Elements of Information Theory. Wiley, New York (1991)
3. Fazel, K., Kaiser, S.: Multi-Carrier and Spread Spectrum Systems, 2nd edn. Wiley-Blackwell, Chichester (2008)
4. Haykin, S.: Communication Systems. Wiley, New York (1994)
5. Huang, P.H., Gai, Y., Krishnamachari, B., Sridharan, A.: Subcarrier allocation in multiuser OFDM systems: complexity and approximability. In: 6th International Conference on Wireless Communications and Networking. IEEE (2010)
6. Kim, K., Han, Y., Kim, S.L.: Joint subcarrier and power allocation in uplink OFDMA systems. IEEE Commun. Lett. **9**(6), 526-528 (2005)
7. Lei, X., Liang, Z.: Joint time-frequency-power resource allocation algorithm for OFDMA systems. In: 5th International Conference on Electronics Information and Emergency Communication, pp. 266-271. IEEE (2015)
8. Letchford, A.N., Ni, Q., Zhong, Z.: Bi-perspective functions for mixed-integer fractional programs with indicator variables. Technical report, Department of Management Science, Lancaster University, UK (2017)
9. Letchford, A.N., Ni, Q., Zhong, Z.: An exact algorithm for a resource allocation problem in mobile wireless communications. Comput. Optim. Appl. **68**(2), 193-208 (2017)
10. Liu, Y.F., Dai, Y.H.: On the complexity of joint subcarrier and power allocation for multi-uuser OFDMA systems. IEEE Trans. Signal Process. **62**(3), 583-596 (2014)
11. Luo, Z.Q., Zhang, S.: Dynamic spectrum management: complexity and duality. IEEE J. Sel. Topics Signal Process. **2**(1), 57-73 (2008)
12. Quesada, I., Grossmann, I.E.: An LP/NLP based branch and bound algorithm for convex MINLP optimization problems. Comput. Chem. Eng. **16**(10), 937-947 (1992)
13. Schaible, S.: Parameter-free convex equivalent and dual programs of fractional programming problems. Zeitschrift für Oper. Res. **18**(5), 187-196 (1974)
14. Schaible, S.: Fractional programming. Zeitschrift für Oper. Res. **27**(1), 39-54 (1983)
15. Seong, K., Mohseni, M., Cioffi, J.: Optimal resource allocation for OFDMA downlink systems. In: 2006 IEEE International Symposium on Information Theory, pp. 1394-1398. IEEE (2006)
16. Shannon, C.E.: Communication in the presence of noise. Proc. IRE **37**(1), 10-21 (1949)
17. Ting, T.O., Chien, S.F., Yang, X.S., Lee, S.: Analysis of quality-of-service aware orthogonal frequency division multiple access system considering energy efficiency. IET Commun. **8**(11), 1947-1954 (2014)
18. Xiao, X., Tao, X., Lu, J.: QoS-aware energy-efficient radio resource scheduling in multi-user OFDMA systems. IEEE Commun. Lett. **17**(1), 75-78 (2013)

19. Xiong, C., Li, G.Y., Zhang, S., Chen, Y., Xu, S.: Energy- and spectral-efficiency tradeoff in downlink OFDMA networks. IEEE Trans. Wirel. Commun. **10**(11), 3874–3886 (2011)
20. Yu, W., Lui, R.: Dual methods for nonconvex spectrum optimization of multicarrier systems. IEEE Trans. Commun. **54**(7), 1310–1322 (2006)
21. Zarakovitis, C.C., Ni, Q.: Maximizing energy efficiency in multiuser multicarrier broadband wireless systems: convex relaxation and global optimization techniques. IEEE Trans. Veh. Technol. **65**(7), 5275–5286 (2016)

An Integer Programming Approach to the Student-Project Allocation Problem with Preferences over Projects

David Manlove⬤, Duncan Milne, and Sofiat Olaosebikan(✉)⬤

School of Computing Science, University of Glasgow, Glasgow, Scotland
David.Manlove@glasgow.ac.uk, Duncan.Milne1@gmail.com,
s.olaosebikan.1@research.gla.ac.uk

Abstract. The Student-Project Allocation problem with preferences over Projects (SPA-P) involves sets of students, projects and lecturers, where the students and lecturers each have preferences over the projects. In this context, we typically seek a stable matching of students to projects (and lecturers). However, these stable matchings can have different sizes, and the problem of finding a maximum stable matching (MAX-SPA-P) is NP-hard. There are two known approximation algorithms for MAX-SPA-P, with performance guarantees of 2 and $\frac{3}{2}$. In this paper, we describe an Integer Programming (IP) model to enable MAX-SPA-P to be solved optimally. Following this, we present results arising from an empirical analysis that investigates how the solution produced by the approximation algorithms compares to the optimal solution obtained from the IP model, with respect to the size of the stable matchings constructed, on instances that are both randomly-generated and derived from real datasets. Our main finding is that the $\frac{3}{2}$-approximation algorithm finds stable matchings that are very close to having maximum cardinality.

1 Introduction

Matching problems, which generally involve the assignment of a set of agents to another set of agents based on preferences, have wide applications in many real-world settings. One such application can be seen in an educational context, e.g., the allocation of pupils to schools, school-leavers to universities and students to projects. In the context of allocating students to projects, university lecturers propose a range of projects, and each student is required to provide a preference over the available projects that she finds acceptable. Lecturers may also have preferences over the students that find their project acceptable and/or the projects that they offer. There may also be upper bounds on the number of students that can be assigned to a particular project, and the number of students that a given lecturer is willing to supervise. The problem then is to allocate

D. Manlove was supported by grant EP/P028306/1 from the Engineering and Physical Sciences Research Council, and the third author was supported by a College of Science and Engineering Scholarship, University of Glasgow.

© Springer International Publishing AG, part of Springer Nature 2018
J. Lee et al. (Eds.): ISCO 2018, LNCS 10856, pp. 313–325, 2018.
https://doi.org/10.1007/978-3-319-96151-4_27

students to projects based on these preferences and capacity constraints – the so-called *Student-Project Allocation problem* (SPA) [3,11].

Two major models of SPA exist in the literature: one permits preferences only from the students [2,6,10,14], while the other permits preferences from the students and lecturers [1,8]. Given the large number of students that are typically involved in such an allocation process, many university departments seek to automate the allocation of students to projects. Examples include the School of Computing Science, University of Glasgow [10], the Faculty of Science, University of Southern Denmark [4], the Department of Computing Science, University of York [8], and elsewhere [2,3,6,16].

In general, we seek a *matching*, which is a set of agent pairs who find one another acceptable that satisfies the capacities of the agents involved. For matching problems where preferences exist from the two sets of agents involved (e.g., junior doctors and hospitals in the classical *Hospitals-Residents problem* (HR) [5], or students and lecturers in the context of SPA), it has been argued that the desired property for a matching one should seek is that of *stability* [15]. Informally, a *stable matching* ensures that no acceptable pair of agents who are not matched together would rather be assigned to each other than remain with their current assignees.

Abraham et al. [1] proposed two linear-time algorithms to find a stable matching in a variant of SPA where students have preferences over projects, whilst lecturers have preferences over students. The stable matching produced by the first algorithm is student-optimal (that is, students have the best possible projects that they could obtain in any stable matching) while the one produced by the second algorithm is lecturer-optimal (that is, lecturers have the best possible students that they could obtain in any stable matching).

Manlove and O'Malley [12] proposed another variant of SPA where both students and lecturers have preferences over projects, referred to as SPA-P. In their paper, they formulated an appropriate stability definition for SPA-P, and they showed that stable matchings in this context can have different sizes. Moreover, in addition to stability, a very important requirement in practice is to match as many students to projects as possible. Consequently, Manlove and O'Malley [12] proved that the problem of finding a maximum cardinality stable matching, denoted MAX-SPA-P, is NP-hard. Further, they gave a polynomial-time 2-approximation algorithm for MAX-SPA-P. Subsequently, Iwama et al. [7] described an improved approximation algorithm with an upper bound of $\frac{3}{2}$, which builds on the one described in [12]. In addition, Iwama *et al.* [7] showed that MAX-SPA-P is not approximable within $\frac{21}{19} - \epsilon$, for any $\epsilon > 0$, unless P = NP. For the upper bound, they modified Manlove and O'Malley's algorithm [12] using Király's idea [9] for the approximation algorithm to find a maximum stable matching in a variant of the *Stable Marriage problem*.

Considering the fact that the existing algorithms for MAX-SPA-P are only guaranteed to produce an approximate solution, we seek another technique to enable MAX-SPA-P to be solved optimally. Integer Programming (IP) is a powerful technique for producing optimal solutions to a range of NP-hard optimisation problems, with the aid of commercial optimisation solvers, e.g., Gurobi [17], GLPK [18] and CPLEX [19]. These solvers can allow IP models to be solved in a reasonable amount of time, even with respect to problem instances that occur in practical applications.

Our Contribution. In Sect. 3, we describe an IP model to enable MAX-SPA-P to be solved optimally, and present a correctness result. In Sect. 4, we present results arising from an empirical analysis that investigates how the solution produced by the approximation algorithms compares to the optimal solution obtained from our IP model, with respect to the size of the stable matchings constructed, on instances that are both randomly-generated and derived from real datasets. These real datasets are based on actual student preference data and manufactured lecturer preference data from previous runs of student-project allocation processes at the School of Computing Science, University of Glasgow. We also present results showing the time taken by the IP model to solve the problem instances optimally. Our main finding is that the $\frac{3}{2}$-approximation algorithm finds stable matchings that are very close to having maximum cardinality. The next section gives a formal definition for SPA-P.

2   Definitions and Preliminaries

We give a formal definition for SPA-P as described in the literature [12]. An instance I of SPA-P involves a set $S = \{s_1, s_2, \ldots, s_{n_1}\}$ of *students*, a set $\mathcal{P} = \{p_1, p_2, \ldots, p_{n_2}\}$ of *projects* and a set $\mathcal{L} = \{l_1, l_2, \ldots, l_{n_3}\}$ of *lecturers*. Each lecturer $l_k \in \mathcal{L}$ offers a non-empty subset of projects, denoted by P_k. We assume that $P_1, P_2, \ldots, P_{n_3}$ partitions $\mathcal{P}$ (that is, each project is offered by one lecturer). Also, each student $s_i \in S$ has an *acceptable* set of projects $A_i \subseteq \mathcal{P}$. We call a pair $(s_i, p_j) \in S \times \mathcal{P}$ an *acceptable pair* if $p_j \in A_i$. Moreover s_i ranks A_i in strict order of preference. Similarly, each lecturer l_k ranks P_k in strict order of preference. Finally, each project $p_j \in \mathcal{P}$ and lecturer $l_k \in \mathcal{L}$ has a positive capacity denoted by c_j and d_k respectively.

An *assignment* M is a subset of $S \times \mathcal{P}$ where $(s_i, p_j) \in M$ implies that s_i finds p_j acceptable (that is, $p_j \in A_i$). We define the *size* of M as the number of (student, project) pairs in M, denoted $|M|$. If $(s_i, p_j) \in M$, we say that s_i is *assigned to* p_j and p_j is *assigned* s_i. Furthermore, we denote the project assigned to student s_i in M as $M(s_i)$ (if s_i is unassigned in M then $M(s_i)$ is undefined). Similarly, we denote the set of students assigned to project p_j in M as $M(p_j)$. For ease of exposition, if s_i is assigned to a project p_j offered by lecturer l_k, we may also say that s_i is *assigned to* l_k, and l_k is *assigned* s_i. Thus we denote the set of students assigned to l_k in M as $M(l_k)$.

A project $p_j \in \mathcal{P}$ is *full*, *undersubscribed* or *oversubscribed* in M if $|M(p_j)|$ is equal to, less than or greater than c_j, respectively. The corresponding terms apply to each lecturer l_k with respect to d_k. We say that a project $p_j \in \mathcal{P}$ is *non-empty* if $|M(p_j)| > 0$.

A *matching* M is an assignment such that $|M(s_i)| \leq 1$ for each $s_i \in \mathcal{S}$, $|M(p_j)| \leq c_j$ for each $p_j \in \mathcal{P}$, and $|M(l_k)| \leq d_k$ for each $l_k \in \mathcal{L}$ (that is, each student is assigned to at most one project, and no project or lecturer is oversubscribed). Given a matching M, an acceptable pair $(s_i, p_j) \in (\mathcal{S} \times \mathcal{P}) \setminus M$ is a *blocking pair* of M if the following conditions are satisfied:

1. either s_i is unassigned in M or s_i prefers p_j to $M(s_i)$, and p_j is undersubscribed, and either
   (a) $s_i \in M(l_k)$ and l_k prefers p_j to $M(s_i)$, or
   (b) $s_i \notin M(l_k)$ and l_k is undersubscribed, or
   (c) $s_i \notin M(l_k)$ and l_k prefers p_j to his worst non-empty project,
   where l_k is the lecturer who offers p_j.

If such a pair were to occur, it would undermine the integrity of the matching as the student and lecturer involved would rather be assigned together than remain in their current assignment. With respect to the SPA-P instance given in Fig. 1, $M_1 = \{(s_1, p_3), (s_2, p_1)\}$ is clearly a matching. It is obvious that each of students s_1 and s_2 is matched to her first ranked project in M_1. Although s_3 is unassigned in M_1, the lecturer offering p_3 (the only project that s_3 finds acceptable) is assumed to be indifferent among those students who find p_3 acceptable. Also p_3 is full in M_1. Thus, we say that M_1 admits no blocking pair.

Student preferences	Lecturer preferences
s_1: p_3 p_2 p_1	l_1: p_2 p_1
s_2: p_1 p_2	l_2: p_3
s_3: p_3	

Fig. 1. An instance I_1 of SPA-P. Each project has capacity 1, whilst each of lecturer l_1 and l_2 has capacity 2 and 1 respectively.

Another way in which a matching could be undermined is through a group of students acting together. Given a matching M, a *coalition* is a set of students $\{s_{i_0}, \ldots, s_{i_{r-1}}\}$, for some $r \geq 2$ such that each student s_{i_j} ($0 \leq j \leq r - 1$) is assigned in M and prefers $M(s_{i_{j+1}})$ to $M(s_{i_j})$, where addition is performed modulo r. With respect to Fig. 1, the matching $M_2 = \{(s_1, p_1), (s_2, p_2), (s_3, p_3)\}$ admits a coalition $\{s_1, s_2\}$, as students s_1 and s_2 would rather permute their assigned projects in M_2 so as to be better off. We note that the number of students assigned to each project and lecturer involved in any such swap remains the same after such a permutation. Moreover, the lecturers involved would have no incentive to prevent the switch from occurring since they are assumed to be indifferent between the students assigned to the projects they are offering. If a matching admits no coalition, we define such matching to be *coalition-free*.

Given an instance I of SPA-P, we define a matching M in I to be *stable* if M admits no blocking pair and is coalition-free. It turns out that with respect to this definition, stable matchings in I can have different sizes. Clearly, each of the matchings $M_1 = \{(s_1, p_3), (s_2, p_1)\}$ and $M_3 = \{(s_1, p_2), (s_2, p_1), (s_3, p_3)\}$ is stable in the SPA-P instance I_1 shown in Fig. 1. The varying sizes of the stable matchings produced naturally leads to the problem of finding a maximum cardinality stable matching given an instance of SPA-P, which we denote by MAX-SPA-P. In the next section, we describe our IP model to enable MAX-SPA-P to be solved optimally.

3   An IP Model for MAX-SPA-P

Let I be an instance of SPA-P involving a set $S = \{s_1, s_2, \ldots, s_{n_1}\}$ of students, a set $\mathcal{P} = \{p_1, p_2, \ldots, p_{n_2}\}$ of projects and a set $\mathcal{L} = \{l_1, l_2, \ldots, l_{n_3}\}$ of lecturers. We construct an IP model J of I as follows. Firstly, we create binary variables $x_{i,j} \in \{0,1\}$ ($1 \le i \le n_1, 1 \le j \le n_2$) for each acceptable pair $(s_i, p_j) \in S \times \mathcal{P}$ such that $x_{i,j}$ indicates whether s_i is assigned to p_j in a solution or not. Henceforth, we denote by S a solution in the IP model J, and we denote by M the matching derived from S. If $x_{i,j} = 1$ under S then intuitively s_i is assigned to p_j in M, otherwise s_i is not assigned to p_j in M. In what follows, we give the constraints to ensure that the assignment obtained from a feasible solution in J is a matching.

Matching Constraints. The feasibility of a matching can be ensured with the following three sets of constraints.

$$\sum_{p_j \in A_i} x_{i,j} \le 1 \qquad (1 \le i \le n_1), \tag{1}$$

$$\sum_{i=1}^{n_1} x_{i,j} \le c_j \qquad (1 \le j \le n_2), \tag{2}$$

$$\sum_{i=1}^{n_1} \sum_{p_j \in P_k} x_{i,j} \le d_k \qquad (1 \le k \le n_3). \tag{3}$$

Note that (1) implies that each student $s_i \in S$ is not assigned to more than one project, while (2) and (3) implies that the capacity of each project $p_j \in \mathcal{P}$ and each lecturer $l_k \in \mathcal{L}$ is not exceeded.

We define $rank(s_i, p_j)$, the *rank* of p_j on s_i's preference list, to be $r+1$ where r is the number of projects that s_i prefers to p_j. An analogous definition holds for $rank(l_k, p_j)$, the *rank* of p_j on l_k's preference list. With respect to an acceptable pair (s_i, p_j), we define $S_{i,j} = \{p_{j'} \in A_i : rank(s_i, p_{j'}) \le rank(s_i, p_j)\}$, the set of projects that s_i likes as much as p_j. For a project p_j offered by lecturer $l_k \in \mathcal{L}$, we also define $T_{k,j} = \{p_q \in P_k : rank(l_k, p_j) < rank(l_k, p_q)\}$, the set of projects that are worse than p_j on l_k's preference list.

In what follows, we fix an arbitrary acceptable pair (s_i, p_j) and we impose constraints to ensure that (s_i, p_j) is not a blocking pair of the matching M (that is, (s_i, p_j) is not a type 1(a), type 1(b) or type 1(c) blocking pair of M). Firstly, let l_k be the lecturer who offers p_j.

Blocking Pair Constraints. We define $\theta_{i,j} = 1 - \sum_{p_{j'} \in S_{i,j}} x_{i,j'}$. Intuitively, $\theta_{i,j} = 1$ if and only if s_i is unassigned in M or prefers p_j to $M(s_i)$. Next we create a binary variable α_j in J such that $\alpha_j = 1$ corresponds to the case when p_j is undersubscribed in M. We enforce this condition by imposing the following constraint.

$$c_j \alpha_j \geq c_j - \sum_{i'=1}^{n_1} x_{i',j}, \tag{4}$$

where $\sum_{i'=1}^{n_1} x_{i',j} = |M(p_j)|$. If p_j is undersubscribed in M then the RHS of (4) is at least 1, and this implies that $\alpha_j = 1$. Otherwise, α_j is not constrained. Now let $\gamma_{i,j,k} = \sum_{p_{j'} \in T_{k,j}} x_{i,j'}$. Intuitively, if $\gamma_{i,j,k} = 1$ in S then s_i is assigned to a project $p_{j'}$ offered by l_k in M, where l_k prefers p_j to $p_{j'}$. The following constraint ensures that (s_i, p_j) does not form a type 1(a) blocking pair of M.

$$\boxed{\theta_{i,j} + \alpha_j + \gamma_{i,j,k} \leq 2.} \tag{5}$$

Note that if the sum of the binary variables in the LHS of (5) is less than or equal to 2, this implies that at least one of the variables, say $\gamma_{i,j,k}$, is 0. Thus the pair (s_i, p_j) is not a type 1(a) blocking pair of M.

Next we define $\beta_{i,k} = \sum_{p_{j'} \in P_k} x_{i,j'}$. Clearly, s_i is assigned to a project offered by l_k in M if and only if $\beta_{i,k} = 1$ in S. Now we create a binary variable δ_k in J such that $\delta_k = 1$ in S corresponds to the case when l_k is undersubscribed in M. We enforce this condition by imposing the following constraint.

$$d_k \delta_k \geq d_k - \sum_{i'=1}^{n_1} \sum_{p_{j'} \in P_k} x_{i',j'}, \tag{6}$$

where $\sum_{i'=1}^{n_1} \sum_{p_{j'} \in P_k} x_{i',j'} = |M(l_k)|$. If l_k is undersubscribed in M then the RHS of (6) is at least 1, and this implies that $\delta_k = 1$. Otherwise, δ_k is not constrained. The following constraint ensures that (s_i, p_j) does not form a type 1(b) blocking pair of M.

$$\boxed{\theta_{i,j} + \alpha_j + (1 - \beta_{i,k}) + \delta_k \leq 3.} \tag{7}$$

We define $D_{k,j} = \{p_{j'} \in P_k : rank(l_k, p_{j'}) \leq rank(l_k, p_j)\}$, the set of projects that l_k likes as much as p_j. Next, we create a binary variable $\eta_{j,k}$ in J such that $\eta_{j,k} = 1$ if l_k is full and prefers p_j to his worst non-empty project in S. We enforce this by imposing the following constraint.

$$d_k \eta_{j,k} \geq d_k - \sum_{i'=1}^{n_1} \sum_{p_{j'} \in D_{k,j}} x_{i',j'}. \tag{8}$$

Finally, to avoid a type 1(c) blocking pair, we impose the following constraint.

$$\boxed{\theta_{i,j} + \alpha_j + (1 - \beta_{i,k}) + \eta_{j,k} \leq 3.} \tag{9}$$

Next, we give the constraints to ensure that the matching obtained from a feasible solution in J is coalition-free.

Coalition Constraints. First, we introduce some additional notation. Given an instance I' of SPA-P and a matching M' in I', we define the *envy graph* $G(M') = (\mathcal{S}, A)$, where the vertex set $\mathcal{S}$ is the set of students in I', and the arc set $A = \{(s_i, s_{i'}) : s_i \text{ prefers } M'(s_{i'}) \text{ to } M'(s_i)\}$. It is clear that the matching $M_2 = \{(s_1, p_1), (s_2, p_2), (s_3, p_3)\}$ admits a coalition $\{s_1, s_2\}$ with respect to the instance given in Fig. 1. The resulting envy graph $G(M_2)$ is illustrated below (Fig. 2).

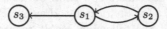

Fig. 2. The envy graph $G(M_2)$ corresponding to the SPA-P instance in Fig. 1.

Clearly, $G(M')$ contains a directed cycle if and only if M' admits a coalition. Moreover, $G(M')$ is acyclic if and only if it admits a topological ordering. Now to ensure that the matching M obtained from a feasible solution S under J is coalition-free, we will enforce J to encode the envy graph $G(M)$ and impose the condition that it must admit a topological ordering. In what follows, we build on our IP model J of I.

We create a binary variable $e_{i,i'}$ for each $(s_i, s_{i'}) \in \mathcal{S} \times \mathcal{S}$, $s_i \neq s_{i'}$, such that the $e_{i,i'}$ variables will correspond to the adjacency matrix of $G(M)$. For each i and i' ($1 \leq i \leq n_1$, $1 \leq i' \leq n_1$, $i \neq i'$) and for each j and j' ($1 \leq j \leq n_2$, $1 \leq j' \leq n_2$) such that s_i prefers $p_{j'}$ to p_j, we impose the following constraint.

$$e_{i,i'} + 1 \geq x_{i,j} + x_{i',j'}. \tag{10}$$

If $(s_i, p_j) \in M$ and $(s_{i'}, p_{j'}) \in M$ and s_i prefers $p_{j'}$ to p_j, then $e_{i,i'} = 1$ and we say s_i *envies* $s_{i'}$. Otherwise, $e_{i,i'}$ is not constrained. Next we enforce the condition that $G(M)$ must have a topological ordering. To hold the label of each vertex in a topological ordering, we create an integer-valued variable v_i corresponding to each student $s_i \in \mathcal{S}$ (and intuitively to each vertex in $G(M)$). We wish to enforce the constraint that if $e_{i,i'} = 1$ (that is, $(s_i, s_{i'}) \in A$), then $v_i < v_{i'}$ (that is, the label of vertex s_i is smaller than the label of vertex $s_{i'}$). This is achieved by imposing the following constraint for all i and i' ($1 \leq i \leq n_1$, $1 \leq i' \leq n_1$, $i \neq i'$).

$$\boxed{v_i < v_{i'} + n_1(1 - e_{i,i'}).} \tag{11}$$

Note that the LHS of (11) is strictly less than the RHS of (11) if and only if $G(M)$ does not admit a directed cycle, and this implies that M is coalition-free.

Variables. We define a collective notation for each variable involved in J as follows.

$$X = \{x_{i,j} : 1 \leq i \leq n_1, 1 \leq j \leq n_2\}, \quad \Lambda = \{\alpha_j : 1 \leq j \leq n_2\},$$
$$H = \{\eta_{j,k} : 1 \leq j \leq n_2, 1 \leq k \leq n_3\}, \quad \Delta = \{\delta_k : 1 \leq k \leq n_3\},$$
$$E = \{e_{i,i'} : 1 \leq i \leq n_1, 1 \leq i' \leq n_1\}, \quad V = \{v_i : 1 \leq i \leq n_1\}.$$

Objective Function. The objective function given below is a summation of all the $x_{i,j}$ binary variables. It seeks to maximize the number of students assigned (that is, the cardinality of the matching).

$$\max \sum_{i=1}^{n_1} \sum_{p_j \in A_i} x_{i,j}. \tag{12}$$

Finally, we have constructed an IP model J of I comprising the set of integer-valued variables X, Λ, H, Δ, E and V, the set of constraints (1)–(11) and an objective function (12). Note that J can then be used to solve MAX-SPA-P optimally. Given an instance I of SPA-P formulated as an IP model J using the above transformation, we present the following result regarding the correctness of J (see [13] for proof).

Theorem 1. *A feasible solution to J is optimal if and only if the corresponding stable matching in I is of maximum cardinality.*

4   Empirical Analysis

In this section we present results from an empirical analysis that investigates how the sizes of the stable matchings produced by the approximation algorithms compares to the optimal solution obtained from our IP model, on SPA-P instances that are both randomly-generated and derived from real datasets.

4.1   Experimental Setup

There are clearly several parameters that can be varied, such as the number of students, projects and lecturers; the length of the students' preference lists; as well as the total capacities of the projects and lecturers. For each range of values for the first two parameters, we generated a set of random SPA-P instances. In each set, we record the average size of a stable matching obtained from running the approximation algorithms and the IP model. Further, we consider the average time taken for the IP model to find an optimal solution.

By design, the approximation algorithms were randomised with respect to the sequence in which students apply to projects, and the choice of students to reject when projects and/or lecturers become full. In the light of this, for each dataset, we also run the approximation algorithms 100 times and record the size of the largest stable matching obtained over these runs. Our experiments therefore involve five algorithms: the optimal IP-based algorithm, the two approximation algorithms run once, and the two approximation algorithms run 100 times.

We performed our experiments on a machine with dual Intel Xeon CPU E5-2640 processors with 64GB of RAM, running Ubuntu 14.04. Each of the approximation algorithms was implemented in Java[1]. For our IP model, we carried out

---

[1] https://github.com/sofiat-olaosebikan/spa-p-isco-2018.

the implementation using the Gurobi optimisation solver in Java (see footnote 1). For correctness testing on these implementations, we designed a stability checker which verifies that the matching returned by the approximation algorithms and the IP model does not admit a blocking pair or a coalition.

4.2   Experimental Results

Randomly-Generated Datasets. All the SPA-P instances we randomly generated involved n_1 students (n_1 is henceforth referred to as the size of the instance), $0.5n_1$ projects, $0.2n_1$ lecturers and $1.1n_1$ total project capacity which was randomly distributed amongst the projects. The capacity for each lecturer l_k was chosen randomly to lie between the highest capacity of the projects offered by l_k and the sum of the capacities of the projects that l_k offers. In the first experiment, we present results obtained from comparing the performance of the IP model, with and without the coalition constraints in place.

Experiment 0. We increased the number of students n_1 while maintaining a ratio of projects, lecturers, project capacities and lecturer capacities as described above. For various values of n_1 ($100 \leq n_1 \leq 1000$) in increments of 100, we created 100 randomly-generated instances. Each student's preference list contained a minimum of 2 and a maximum of 5 projects. With respect to each value of n_1, we obtained the average time taken for the IP solver to output a solution, both with and without the coalition constraints being enforced. The results, displayed in Table 1, show that when we removed the coalition constraints, the average time for the IP solver to output a solution is significantly faster than when we enforced the coalition constraints.

In the remaining experiments, we thus remove the constraints that enforce the absence of a coalition in the solution. We are able to do this for the purposes of these experiments because the largest size of a stable matching is equal to the largest size of a matching that potentially admits a coalition but admits no blocking pair[2], and we were primarily concerned with measuring stable matching cardinalities. However the absence of the coalition constraints should be borne in mind when interpreting the IP solver runtime data in what follows.

In the next two experiments, we discuss results obtained from running the five algorithms on randomly-generated datasets.

Experiment 1. As in the previous experiment, we maintained the ratio of the number of students to projects, lecturers and total project capacity; as well as the length of the students' preference lists. For various values of n_1 ($100 \leq n_1 \leq 2500$) in increments of 100, we created 1000 randomly-generated instances. With respect to each value of n_1, we obtained the average sizes of stable matchings constructed by the five algorithms run over the 1000 instances. The result displayed in Fig. 3 (and also in Fig. 4) shows the ratio of the average size of the

---

[2] This holds because the number of students assigned to each project and lecturer in the matching remains the same even after the students involved in such coalition permute their assigned projects.

stable matching produced by the approximation algorithms with respect to the maximum cardinality matching produced by the IP solver.

Figure 3 shows that each of the approximation algorithms produces stable matchings with a much higher cardinality from multiple runs, compared to running them only once. Also, the average time taken for the IP solver to find a maximum cardinality matching increases as the size of the instance increases, with a running time of less than one second for instance size 100, increasing roughly linearly to 13 s for instance size 2500 (see [13, Fig. 3(b)]).

Experiment 2. In this experiment, we varied the length of each student's preference list while maintaining a fixed number of students, projects, lecturers and total project capacity. For various values of x $(2 \leq x \leq 10)$, we generated 1000 instances, each involving 1000 students, with each student's preference list containing exactly x projects. The result for all values of x is displayed in Fig. 4. Figure 4 shows that as we increase the preference list length, the stable matchings produced by each of the approximation algorithms gets close to having maximum cardinality. It also shows that with a preference list length greater than 5, the $\frac{3}{2}$-approximation algorithm produces an optimal solution, even on a single run. Moreover, the average time taken for the IP solver to find a maximum matching increases as the length of the students' preference lists increases, with a running time of two seconds when each student's preference list is of length 2, increasing roughly linearly to 17 s when each student's preference list is of length 10 (see [13, Fig. 4(b)]).

Real Datasets. The real datasets in this paper are based on actual student preference data and manufactured lecturer data from previous runs of student-project allocation processes at the School of Computing Science, University of Glasgow. Table 2 shows the properties of the real datasets, where n_1, n_2 and n_3 denotes the number of students, projects and lecturers respectively; and l denotes the length of each student's preference list. For all these datasets, each project has a capacity of 1. In the next experiment, we discuss how the lecturer preferences were generated. We also discuss the results obtained from running the five algorithms on the corresponding SPA-P instances.

Experiment 3. We derived the lecturer preference data from the real datasets as follows. For each lecturer l_k, and for each project p_j offered by l_k, we obtained the number a_j of students that find p_j acceptable. Next, we generated a strict preference list for l_k by arranging l_k's proposed projects in (i) a random manner, (ii) ascending order of a_j, and (iii) descending order of a_j, where (ii) and (iii) are taken over all projects that l_k offers. Table 2 shows the size of stable matchings obtained from the five algorithms, where A, B, C, D and E denotes the solution obtained from the IP model, 100 runs of $\frac{3}{2}$-approximation algorithm, single run of $\frac{3}{2}$-approximation algorithm, 100 runs of 2-approximation algorithm, and single run of 2-approximation algorithm respectively. The results are essentially consistent with the findings in the previous experiments, that is, the $\frac{3}{2}$-approximation algorithm produces stable matchings whose sizes are close to optimal.

Table 1. Results for Experiment 0. Average time (in seconds) for the IP solver to output a solution, both with and without the coalition constraints being enforced.

Size of instance	100	200	300	400	500	600	700	800	900	1000
Av. time without coalition	0.12	0.27	0.46	0.69	0.89	1.17	1.50	1.86	2.20	2.61
Av. time with coalition	0.71	2.43	4.84	9.15	13.15	19.34	28.36	38.18	48.48	63.50

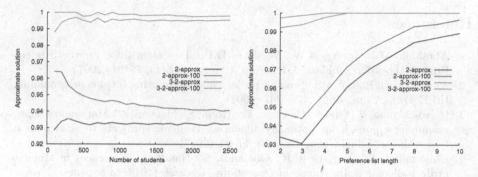

Fig. 3. Result for Experiment 1.   **Fig. 4.** Result for Experiment 2.

Table 2. Properties of the real datasets and results for Experiment 3.

				Random					Most popular					Least popular				
Year	n_1	n_2	n_3	l	A	B	C	D	E	A	B	C	D	E	A	B	C	D	E
2014	55	149	38	6	55	55	55	54	53	55	55	55	54	50	55	55	55	54	52
2015	76	197	46	6	76	76	76	76	72	76	76	76	76	72	76	76	76	76	75
2016	92	214	44	6	84	82	83	77	75	85	85	83	79	76	82	80	77	76	74
2017	90	289	59	4	89	87	85	80	76	90	89	86	81	79	88	85	84	80	77

4.3   Discussions and Concluding Remarks

The results presented in this section suggest that even as we increase the number of students, projects, lecturers, and the length of the students' preference lists, each of the approximation algorithms finds stable matchings that are close to having maximum cardinality, outperforming their approximation factor. Perhaps most interesting is the $\frac{3}{2}$-approximation algorithm, which finds stable matchings that are very close in size to optimal, even on a single run. These results also holds analogously for the instances derived from real datasets.

We remark that when we removed the coalition constraints, we were able to run the IP model on an instance size of 10000, with the solver returning a maximum matching in an average time of 100 s, over 100 randomly-generated instances. This shows that the IP model (without enforcing the coalition constraints), can be run on SPA-P instances that appear in practice, to find maximum cardinality matchings that admit no blocking pair. Coalitions should then be eliminated in polynomial time by repeatedly constructing an *envy graph*, similar to the one described in [11, p. 290], finding a directed cycle and letting the students in the cycle swap projects.

References

1. Abraham, D.J., Irving, R.W., Manlove, D.F.: Two algorithms for the Student-Project allocation problem. J. Discrete Algorithms **5**(1), 79–91 (2007)
2. Anwar, A.A., Bahaj, A.S.: Student project allocation using integer programming. IEEE Trans. Educ. **46**(3), 359–367 (2003)
3. Calvo-Serrano, R., Guillén-Gosálbez, G., Kohn, S., Masters, A.: Mathematical programming approach for optimally allocating students' projects to academics in large cohorts. Educ. Chem. Eng. **20**, 11–21 (2017)
4. Chiarandini, M., Fagerberg, R., Gualandi, S.: Handling preferences in student-project allocation. In: Annals of Operations Research (2018, to appear)
5. Gale, D., Shapley, L.S.: College admissions and the stability of marriage. Am. Mathe. Mon. **69**, 9–15 (1962)
6. Harper, P.R., de Senna, V., Vieira, I.T., Shahani, A.K.: A genetic algorithm for the project assignment problem. Comput. Oper. Res. **32**, 1255–1265 (2005)
7. Iwama, K., Miyazaki, S., Yanagisawa, H.: Improved approximation bounds for the student-project allocation problem with preferences over projects. J. Discrete Algorithms **13**, 59–66 (2012)
8. Kazakov, D.: Co-ordination of student-project allocation. Manuscript, University of York, Department of Computer Science (2001). http://www-users.cs.york.ac.uk/kazakov/papers/proj.pdf. Accessed 8 Mar 2018
9. Király, Z.: Better and simpler approximation algorithms for the stable marriage problem. Algorithmica **60**, 3–20 (2011)
10. Kwanashie, A., Irving, R.W., Manlove, D.F., Sng, C.T.S.: Profile-based optimal matchings in the student/project allocation problem. In: Kratochvíl, J., Miller, M., Froncek, D. (eds.) IWOCA 2014. LNCS, vol. 8986, pp. 213–225. Springer, Cham (2015). https://doi.org/10.1007/978-3-319-19315-1_19
11. Manlove, D.F.: Algorithmics of Matching Under Preferences. World Scientific (2013)
12. Manlove, D.F., O'Malley, G.: Student project allocation with preferences over projects. J. Discrete Algorithms **6**, 553–560 (2008)
13. Manlove, D.F., Milne, D., Olaosebikan, S.: An integer programming approach to the student-project allocation problem with preferences over projects. CoRR abs/1804.09993 (2018). https://arxiv.org/abs/1804.09993
14. Proll, L.G.: A simple method of assigning projects to students. Oper. Res. Q. **23**(2), 195–201 (1972)
15. Roth, A.E.: The evolution of the labor market for medical interns and residents: a case study in game theory. J. Polit. Econ. **92**(6), 991–1016 (1984)

16. Teo, C.Y., Ho, D.J.: A systematic approach to the implementation of final year project in an electrical engineering undergraduate course. IEEE Trans. Educ. **41**(1), 25–30 (1998)
17. Gurobi Optimization website. http://www.gurobi.com. Accessed 09 Jan 2018
18. GNU Linear Proramming Kit. https://www.gnu.org/software/glpk. Accessed 09 Jan 2018
19. CPLEX Optimization Studio. http://www-03.ibm.com/software/products/en/ibmilogcpleoptistud/. Accessed 19 May 2017

Even Flying Cops Should Think Ahead

Anders Martinsson, Florian Meier$^{(\boxtimes)}$, Patrick Schnider, and Angelika Steger

Department of Computer Science, ETH Zürich, Zürich, Switzerland
{maanders,meierflo,patrick.schnider,steger}@inf.ethz.ch

Abstract. We study the entanglement game, which is a version of cops and robbers, on sparse graphs. While the minimum degree of a graph G is a lower bound for the number of cops needed to catch a robber in G, we show that the required number of cops can be much larger, even for graphs with small maximum degree. In particular, we show that there are 3-regular graphs where a linear number of cops are needed.

Keywords: Cops and robbers · Entanglement game
Probabilistic method

1 Introduction

The game of cops and robbers was first introduced and popularised in the 1980s by Aigner and Fromme [1], Nowakowski and Winkler [15] and Quilliot [16]. Since then, many variants of the game have been studied, for example where cops can catch robbers from larger distances ([10]), the robber is allowed to move at different speeds ([2,13]), or the cops are lazy, meaning that in each turn only one cop can move ([3,4]).

In this paper we consider the *entanglement game*, introduced by Berwanger and Grädel [5] that is the following version of the cops and robbers game on a (directed or undirected) graph G. First, the robber chooses a starting position and the k cops are outside the graph. In every turn, the cops can either stay where they are, or they can fly one of them to the current position of the robber. Regardless of whether the cops stayed or one of them flew to the location of the robber, the robber then has to move to a neighbor of his current position that is not occupied by a cop. If there is no such neighbor, the cops win. The robber wins if he can run away from the cops indefinitely. The *entanglement number* of a graph G, denoted by $\text{ent}(G)$, is the minimal integer k such that k cops can catch a robber on G. In order to get accustomed to the rules of the game, it is a nice exercise to show that the entanglement number of an (undirected) tree is at most 2.

The main property that distinguishes the entanglement game from other variants of cops and robbers is the restriction that the cops are only allowed to fly to the current position of the robber. This prevents the cops from cutting off escape routes or forcing the robber to move into a certain direction.

© Springer International Publishing AG, part of Springer Nature 2018
J. Lee et al. (Eds.): ISCO 2018, LNCS 10856, pp. 326–337, 2018.
https://doi.org/10.1007/978-3-319-96151-4_28

As we will show in this paper, it is this restriction that enables the robber to run away from many cops.

In a similar way to how the classical game of cops and robbers can be used to describe the treewidth of a graph, the entanglement number is a measure of how strongly the cycles of the graph are intertwined, see [6]. Just like many problems can be solved efficiently on graphs of bounded treewidth, Berwanger and Grädel [5] have shown that parity games of bounded entanglement can be solved in polynomial time.

As the cops do not have to adhere to the edges of the graph G in their movement, adding more edges to G can only help the robber. In fact, it can be seen easily that on the complete graph K_n with $n \geq 2$ vertices, $n-1$ cops are needed to catch the robber. Furthermore, observe that the minimum degree of the graph G is a lower bound on the entanglement number, as otherwise the robber will always find a free neighbor to move to. These observations seem to suggest that, on sparse graphs, the cops should have an advantage and therefore few cops would suffice to catch the robber. Indeed, on 2-regular graphs, it is easily checked that three cops can always catch the robber.

Motivated by this, we study the entanglement game on several classes of sparse graphs. We show that for sparse Erdős-Rényi random graphs, with high probability linearly many cops are needed. We then apply similar ideas to show our main result, c.f. Theorem 2, which states that the union of three random perfect matchings is with high probability such that the robber can run away from αn cops, for some constant $\alpha > 0$. Further, we show in Theorem 3 that for any 3-regular graph $\lfloor \frac{n}{4} \rfloor + 4$ cops suffice to catch the robber. Finally, we consider the entanglement game for graphs that are given by a more specific union of three perfect matchings, in fact, that are the union of a Hamilton cycle and a perfect matching. For graphs given by a Hamilton cycle and a perfect matching connecting every vertex to its diagonally opposite vertex, it may seem that the diagonal "escape" edges are quite nice for the robber. This, however, is not so: we show in Proposition 1 that for these graphs six cops are always sufficient. However, we also show, cf. Corollary 2, that if we replace this specific perfect matching by a random one, then with high probability a linear number of cops is needed. We conclude that in contrast to the intuition that sparse graphs are advantageous for the cops, they are often not able to use the sparsity to their advantage.

As mentioned previously, the entanglement game and entanglement number is also defined for directed graphs, the difference being that the robber moves to a successor of his current position. In fact, motivated by an application in logic ([7]), the authors of the original papers about the entanglement game ([5,6]) focused on the directed version. As a corollary of our main result we construct directed graphs of maximum (total) degree 3 which again require linearly many cops. Our result on Erdős-Rényi graphs also easily carries over to directed graphs.

2   Results

In this section, we present and motivate our results. In order to increase the readability, we postpone the proofs to Sect. 3. We start by analyzing the entanglement game on sparse random graphs.

Theorem 1. *For every $0 < \alpha < 1$ there exists a constant $C = C(\alpha) > 0$ such that for any $p \geq C/n$, αn cops do not suffice to catch the robber on $G_{n,p}$ with high probability. The same result holds for directed random graphs.*

Note that $G_{n,p}$ with $p = C/n$ has average degree $p(n-1) < C$. On the other hand, it is known that the maximum degree is with high probability $\Theta(\log(n))$. In the following, we construct graphs that need linearly many cops and have maximum degree 3. The idea is that we define G as the union of three random perfect matchings. Extending the proof ideas from Theorem 1, we obtain the following results.

Theorem 2. *There exists an $\alpha > 0$ such that with high probability αn cops do not suffice to catch the robber on the graph $G = M_1 \cup M_2 \cup M_3$, where M_1, M_2, M_3 are independent uniformly chosen random perfect matchings.*

Corollary 1. *There exists an $\alpha > 0$ and an $n_0 \in \mathbb{N}$ such that for every even $n \geq n_0$ there exists a 3-regular graph on n vertices for which αn cops are not enough to catch the robber.*

This result may look surprising at first sight. In particular, consider the 3-connected 3-regular graph DG_n obtained by taking a Hamilton cycle of length $2n$ and connecting every vertex to its antipode by an edge.

Proposition 1. *For the graph DG_n six cops suffice to catch the robber.*

The fact that all diagonals go to a vertex that is "furthest away" may seem to make catching the robber quite difficult for the cops. However, as it turns out, the symmetry of the construction is the reason for the small entanglement number. If we replace the matching of diagonals by a *random* matching then the entanglement number is typically large again.

Corollary 2. *Consider the graph $G = H \cup M$ where H is a Hamiltonian cycle and M is a random perfect matching. There exists an $\alpha > 0$ such that with high probability αn cops do not suffice to catch the robber on G.*

We complement these lower bounds by the following upper bounds.

Theorem 3. *For any 3-regular graph on n vertices, $\lfloor \frac{n}{4} \rfloor + 4$ cops suffice.*

We now turn to directed graphs. Theorem 2 immediately implies that there are graphs that are the union of six perfect matchings on which a linear number of cops is needed: simply direct each edge in both directions. However, there also exist directed graphs with maximum out-degree two and (total) maximum degree three can be very hard for the cops.

Corollary 3. *There exists an* $\alpha > 0$ *and an* $n_0 \in \mathbb{N}$ *such that for every even* $n \geq n_0$ *there is a directed 3-regular (that is, the sum of in and out degree of every vertex is 3) graph* G *on* $6n$ *vertices, such that* αn *cops do not suffice to catch the robber on* G.

The idea here is that we "blow up" an undirected 3-regular graph to a directed one by replacing each vertex by a directed cycle of length six. We note that, in contrast to the undirected version, we cannot just take a union of three random directed matchings. This follows from the fact that the largest strongly connected component in the union of three random directed matchings contains with high probability only sublinearly many vertices as can be shown using some ideas from [11].

3 Proofs

We start the proof section by considering the entanglement game for Erdős-Rényi random graphs $G_{n,p}$, cf. [9,14] or [12] for an introduction to random graphs. In Sect. 3.2 we will generalize this proof to obtain Theorem 2. In Sect. 3.3, we use Theorem 2 to prove Corollaries 1, 2 and 3. Finally, in Sect. 3.4, we will prove the stated upper bounds of Proposition 1 and Theorem 3.

3.1 Proof of Theorem 1

Recall from the introduction that adding edges to the graph can only make it harder for the cops. Without loss of generality it thus suffices to consider the case $G_{n,p}$, where $p = C/n$ for some (large) constant $C > 0$. A standard result from random graph theory is that such a random graph has with high probability one large component (of size approximately βn, where β is a function of the constant C of the edge probability p), while all additional components have at most logarithmic size. Here we need the following strengthening of this result:

Lemma 1. *For every* $0 < \bar{\alpha} < 1$ *there exists a constant* $C = C(\bar{\alpha}) > 0$ *such that for any* $p \geq C/n$ *the random graph* $G_{n,p}$ *is with high probability such that every subset* $X \subseteq V$ *of size* $\bar{\alpha} n$ *induces a subgraph that contains a connected component of size at least* $\frac{2}{3}\bar{\alpha} n$.

Proof. First observe that any graph on $\bar{\alpha} n$ vertices that does not contain a component of size at least βn contains a cut $(S, \bar{S})$ such that $E(S, \bar{S})$ contains no edge and $|S|, |\bar{S}| \leq \frac{1}{2}(\bar{\alpha}+\beta)n$. This follows easily by greedily placing components into S as long as the size constraint is not violated. Note that such a cut $(S, \bar{S})$ contains at least $\frac{1}{4}(\bar{\alpha}^2 - \beta^2)n^2$ possible edges. The probability that all these edges are missing in the random graph $G_{n,p}$ is thus at most $(1-p)^{\frac{1}{4}(\bar{\alpha}^2-\beta^2)n^2} \leq e^{-\frac{C}{4}(\bar{\alpha}^2-\beta^2)n}$. By a union bound over all sets X of size $\bar{\alpha} n$ and all possible cuts $(S, \bar{S})$ we thus obtain that the probability that the random graph $G_{n,p}$ does not satisfy the desired property is at most

$$\binom{n}{\bar{\alpha} n} \cdot 2^{\bar{\alpha} n} \cdot e^{-\frac{C}{4}(\bar{\alpha}^2-\beta^2)n},$$

which in turn can be bounded by

$$2^{H(\bar{\alpha})n} \cdot 2^{\bar{\alpha}n} \cdot e^{-\frac{C}{4}(\bar{\alpha}^2 - \beta^2)n}$$

by using the standard estimations for the binomial coefficient, where $H(x) = -x\log_2 x - (1-x)\log_2(1-x)$. By letting $\beta = \frac{2}{3}\bar{\alpha}$ and making C large enough we see that this term goes to zero, which concludes the proof of the lemma.   □

With Lemma 1 at hand we can now conclude the proof of the theorem as follows. Pick any $0 < \alpha < 1$ and let $\bar{\alpha} = 1 - \alpha$. Assume that $G_{n,p}$ satisfies the property of Lemma 1. The robber can win against αn cops with the following strategy: he aims at always staying in a component of size at least $\frac{2}{3}\bar{\alpha}n$ in a subgraph $G[A]$ for some cop-free set A of size $\bar{\alpha}n$. Clearly, this can be achieved at the start of the game. Now assume that the robber is in such a component C of a subgraph $G[A]$. If a cop moves to the location of the robber, we change A by removing this vertex and add instead another vertex not covered by a cop. Call this new cop-free set A'. By assumption $G[A']$ contains a component C' of size $\frac{2}{3}\bar{\alpha}n$. Since C and C have size at least $\frac{2}{3}|A|$, C is contained in A and C' contains at most one vertex that is not in A, C and C' overlap. Thus, the robber can move to a neighbor that is in C', as required.

The directed cases follows similarly. The only slightly more tricky case is the argument for the existence of the cut $(S, \bar{S})$. This can be done as follows. Consider all strongly connected components of $G_{n,p}$. It is then not true that there exist no edges between these components. What is true, however, is that the cluster graph (in which the components are replaced by vertices and the edges between components by one or two directed edges depending on which type of edges exist between the corresponding components) is acyclic. If we thus repeatedly consider sink components, placing them into S as long as the size constraint is not violated, we obtain a cut $(S, \bar{S})$ which does not contain any edge directed from S to $\bar{S}$. From here on the proof is completed as before.   □

3.2   Proof of Theorem 2

In this section, we prove the a slightly stronger version of Theorem 2, i.e., we show that the statement of Theorem 2 holds with exponentially high probability. We need this statement to proof the corollaries in Sect. 3.3. For completeness we restate the theorem in this form:

Theorem 4. *There exists an $\alpha > 0$ such that, with probability $1 - e^{-\Omega(n)}$, αn cops do not suffice to catch the robber on the graph $G = M_1 \cup M_2 \cup M_3$, where M_1, M_2, M_3 are independent uniformly chosen random perfect matchings.*

Let $\alpha > 0$ denote a sufficiently small constant to be chosen later, and, as before, let $\bar{\alpha} := 1 - \alpha$. The main idea of the proof of Theorem 4 is similar to the strategy that we used in the proof of Theorem 1. Namely, we show that every subgraph induced by an $\bar{\alpha}$-fraction of the vertices contains a large connected component. In the proof of Theorem 1 we used $\frac{2}{3}\bar{\alpha}n$ as a synonym for "large".

As it turns out, in the proof of Theorem 4 we have to be more careful. To make this precise we start with a definition. Given $\alpha > 0$, we say that a graph $G = (V, E)$ is α-robust, if for every set $X \subseteq V$ of size $|X| \geq \bar{\alpha}n = (1 - \alpha)n$ the induced graph $G[X]$ contains a connected component that is larger than $|X|/2$.

Lemma 2. *Assume $G = (V, E)$ is an α-robust graph for some $0 < \alpha < 1$. Then $\alpha|V| - 2$ cops are not sufficient to catch the robber.*

Proof. Let $n = |V|$ and assume there are $\alpha n - 2$ cops. The robber can win with the following strategy, similar to the one used on the random graphs: he aims at always staying in the unique component of size greater than $|A|/2$ in a subgraph $G[A]$ for some cop-free set A of size $|A| \geq \bar{\alpha}n$ such that $|A|$ is even. In the beginning of the game, this can easily be achieved. If some cop is placed on the current position of the robber, then we remove this vertex from A and add some other cop-free vertex arbitrarily to obtain a set A'. Let C resp. C' be the vertex set of the largest component in $G[A]$ resp. $G[A']$. It remains to show that C and C' overlap. Assume they do not. Then $|C \cup C'| = |C| + |C'|$. By α-robustness and the evenness assumption of A and A' we have $|C| \geq |A|/2 + 1$ and $|C'| \geq |A'|/2 + 1$. As A' (and thus C') contains at most one vertex that is not in A, this is a contradiction. $\qquad\square$

For the proof that the union of three perfect matchings is α-robust for sufficiently small α, we will proceed by contradiction. Here the following lemma will come in handy.

Lemma 3. *Let $G = (V, E)$ be a graph on $n = |V|$ vertices that does not contain a component on more than $n/2$ vertices. Then there exists a partition $V = B_1 \cup B_2 \cup B_3$ such that*

(i) $|B_i| \leq n/2$ for all $i = 1, 2, 3$ and
(ii) $E(B_i, B_j) = \emptyset$ for all $1 \leq i < j \leq 3$.

Proof. This follows straightforwardly from a greedy type argument. Consider the components of G in any order. Put the components into a set B_1 as long as B_1 contains at most $n/2$ vertices. Let C be a component whose addition to B_1 would increase the size of B_1 above $n/2$. Placing C into B_2 and all remaining components into B_3 concludes the proof of the lemma. $\qquad\square$

We denote by $\mathrm{pm}(n)$ the number of perfect matchings in a complete graph on n vertices. For sake of completeness let us assume that in the case of n odd we count the number of almost perfect matchings. We are interested in the asymptotic behavior of $\mathrm{pm}(n)$. In fact, we only care on the behavior of the leading terms. With the help of Stirling's formula

$$n! = (1 + o(1)) \cdot \sqrt{2\pi n} \cdot (n/e)^n \ ,$$

one easily obtains that

$$\mathrm{pm}(n) = \frac{n!}{\lfloor n/2 \rfloor! \cdot 2^{\lfloor n/2 \rfloor}} = poly(n) \cdot \left(\frac{n}{e}\right)^{n/2} , \tag{1}$$

where here and in the remainder of this section we use the term $poly(n)$ to suppress factors that are polynomial in n.

For any non-negative real numbers x, y, z such that $x + y + z = 1$, we define $H(x, y, z) = -x \log_2 x - y \log_2 y - z \log_2 z$.

Lemma 4
$$\min_{\substack{0 \leq x,y,z \leq 1/2 \\ x+y+z=1}} H(x, y, z) = 1.$$

Proof. As $-x \log_2 x$ is concave, $H(x, y, z)$ must attain its minimum in a vertex of the simplex $0 \leq x, y, z \leq 1/2, x + y + z = 1$, which, up to permutation of the variables, is given by $x = y = \frac{1}{2}$ and $z = 0$.   □

We are now ready to prove Theorem 4, which implies Theorem 2. Let $n = |V|$. In light of Lemma 2, the theorem follows if we can show that $G = (V, E) = M_1 \cup M_2 \cup M_3$ is α-robust with exponentially high probability for some sufficiently small $\alpha > 0$. By Lemma 3, it suffices to show that, for all $C \subseteq V$ of size at most $\bar{\alpha} n = (1 - \alpha)n$ and all partitions B_1, B_2, B_3 of $B = V \setminus C$ such that no set B_i contains more than $|B|/2$ vertices, the graph contains an edge that goes between two sets B_i, B_j where $i \neq j$.

Consider any such partition B_1, B_2, B_3, C. For each $i = 1, 2, 3$, let $\beta_i = |B_i| / |B|$. Let M be one uniformly chosen perfect matching. Let us estimate the probability that $E_M(B_i, B_j) = \emptyset$ for all $i \neq j$. First, condition on the set of edges M' in M with at least one end-point in C. These will connect to at most αn vertices in B. Let B', B_1', B_2' and B_3' respectively denote the subsets of vertices that remain unmatched. Hence, the remaining edges in the matching $M \setminus M'$ is chosen uniformly from all perfect matchings on B'.

We write $\beta_i' = |B_i'| / |B'|$. Clearly, if $|B_i'|$ for some $i = 1, 2, 3$ is odd, then

$$\mathbb{P}(E_M(B_i, B_j) = \emptyset \; \forall i \neq j | M') = 0.$$

Otherwise, by (1), we get

$$\mathbb{P}(E_M(B_i, B_j) = \emptyset \; \forall i \neq j | M') = \frac{1}{\text{pm}(|B'|)} \prod_{i=1}^{3} \text{pm}(|B_i'|)$$

$$= poly(n) \cdot \left(\frac{e}{|B'|} \right)^{|B'|/2} \prod_{i=1}^{3} \left(\frac{|B_i'|}{e} \right)^{|B_i'|/2}$$

$$= poly(n) \cdot \left(\beta_1'^{\beta_1'} \beta_2'^{\beta_2'} \beta_3'^{\beta_3'} \right)^{|B'|/2}$$

$$= poly(n) \cdot 2^{-H(\beta_1', \beta_2', \beta_3')|B'|/2}.$$

As $H(x, y, z)$ is uniformly continuous, we know that for any $\varepsilon > 0$, there exists an $\alpha_0 > 0$ such that, for any $0 < \alpha < \alpha_0$, $|(|B| - |B'|)/n|$ and $|\beta_i' - \beta_i|$ are sufficiently small that the above expression can be bounded by $2^{-H(\beta_1, \beta_2, \beta_3)\bar{\alpha}n/2 + \varepsilon n}$, where the choice of α_0 holds uniformly over all $\beta_1, \beta_2, \beta_3$. As the above bound holds for any matching M', we get

$$\mathbb{P}(E_M(B_i, B_j) = \emptyset \; \forall i \neq j) \leq 2^{-H(\beta_1,\beta_2,\beta_3)|B|/2+\varepsilon n},$$

for any partition $\{B_1, B_2, B_3, C\}$ as above.

We now do a union bound over all such partitions. For a given set C and given sizes b_1, b_2, b_3 of the sets B_1, B_2, B_3, the number of choices for these sets is $\binom{|B|}{b_1,b_2,b_3} = poly(n) \cdot 2^{H(\beta_1,\beta_2,\beta_3)|B|}$ where $B = V \setminus C$. Moreover, by the above calculation, given a partition B_1, B_2, B_3, C as above, the probability that $E(B_i, B_j) = \emptyset$ for a union of three independent uniformly chosen perfect matchings is at most $2^{-3H(\beta_1,\beta_2,\beta_3)|B|/2+3\varepsilon n}$. This yields

$$\mathbb{P}(\exists B_1, B_2, B_3, C : E(B_i, B_j) = \emptyset \; \forall i \neq j)$$

$$\leq poly(n) \cdot \sum_{\substack{B \subseteq V \\ |B| \geq \bar{\alpha} n}} \sum_{\substack{0 \leq b_1, b_2, b_3 \leq |B|/2 \\ b_1 + b_2 + b_3 = |B|}} 2^{-H(\beta_1,\beta_2,\beta_3)|B|/2+3\varepsilon n}$$

$$\leq poly(n) \cdot 2^{3\varepsilon n} \sum_{\substack{B \subseteq V \\ |B| \geq \bar{\alpha} n}} \left(\max_{\substack{0 \leq \beta_1, \beta_2, \beta_3 \leq 1/2 \\ \beta_1 + \beta_2 + \beta_3 = 1}} 2^{-H(\beta_1,\beta_2,\beta_3)} \right)^{|B|/2}$$

$$\leq poly(n) \cdot 2^{3\varepsilon n} \sum_{\substack{B \subseteq V \\ |B| \geq \bar{\alpha} n}} 2^{-|B|/2},$$

where the second inequality follows since there are only $poly(n)$ triples $\{b_1, b_2, b_3\}$ satisfying $0 \leq b_1, b_2, b_3 \leq |B|/2$ and $b_1 + b_2 + b_3 = |B|$, and the third inequality follows by Lemma 4. The remaining sum can be rewritten as $\sum_{k=\lceil \bar{\alpha} n \rceil}^{n} \binom{n}{k} 2^{-k/2}$. Assuming $\bar{\alpha} \geq \frac{1}{2}$, the summand is decreasing. Hence the sum is at most $poly(n) \cdot \binom{n}{\bar{\alpha} n} 2^{-\bar{\alpha} n/2}$. We conclude that

$$\mathbb{P}(G \text{ is not } \alpha\text{-robust}) \leq poly(n) \cdot 2^{(H(\bar{\alpha}) - \bar{\alpha}/2 + 3\varepsilon)n}. \tag{2}$$

As $H(\bar{\alpha}) - \frac{\bar{\alpha}}{2} = -\frac{1}{2} + H(\alpha) + \frac{\alpha}{2} \to -\frac{1}{2}$ as $\alpha \to 0$, we see that choosing $0 < \varepsilon < 1/6$ and $\alpha > 0$ sufficiently small, the right-hand side of (2) tends to 0 exponentially fast in n, as desired. $\qquad\square$

3.3 Proof of Corollary 1, 2 and 3

Proof (of Corollary 1). A standard method from random graph theory for the construction of regular graphs is the so-called configuration model introduced by Bollobás in [8]. Constructing a graph G by taking the union of three independent uniformly chosen random perfect matchings M_1, M_2, M_3 is equivalent to constructing a 3-regular random graph with the configuration model and conditioning that no self-loops appear. Since conditioning that no self-loops appear increases the probability of producing a simple graph (because all graphs with self-loops are not simple), Corollary 1 thus follows immediately from [8].

For sake of completeness we also give a direct proof. By symmetry the probability that a given edge is contained in a random perfect matching is exactly $\frac{1}{n-1}$. The expected number of edges common to M_1 and M_2 is thus $\frac{n}{2(n-1)}$. Hence, by Markov's inequality, with probability at least $1 - \frac{n}{2(n-1)} \approx \frac{1}{2}$, M_1 and M_2 are disjoint. Assuming the two first matchings are disjoint, we uniformly choose one pair of vertices $\{u,v\}$ to form an edge in M_3. With probability $1 - \frac{2}{n-1}$, this edge is not in $M_1 \cup M_2$, and hence shares one end-point with exactly 4 of the n edges in $M_1 \cup M_2$. Adding the remaining $\frac{n}{2} - 1$ edges to M_3, the expected number of these that are contained in $M_1 \cup M_2$ is $\frac{n-4}{n-3} = 1 - \frac{1}{n-3}$. Again by Markov's inequality, this means that M_3 is disjoint $M_1 \cup M_2$ with probability at least $\frac{1}{n-3}$. It follows that M_1, M_2, M_3 are pairwise disjoint with probability at least $\Omega\left(\frac{1}{n}\right)$. As Theorem 4 holds with exponentially high probability, for sufficiently large n, we can find disjoint matchings M_1, M_2, M_3 such that αn cops do not suffice to catch the robber on $G = M_1 \cup M_2 \cup M_3$. $\square$

Proof (of Corollary 2). Let M_1, M_2, M_3 be random perfect matchings chosen independently and uniformly. We claim that $M_1 \cup M_2$ is a Hamiltonian cycle with probability at least $\frac{1}{n-1}$. Therefore, the graph $G = M_1 \cup M_2 \cup M_3$ is with probability at least $\frac{1}{n-1}$ a union of a Hamiltonian cycle and a random matching. Since Theorem 4 holds with probability $1 - e^{-\Omega(n)}$, it holds with high probability that αn cops do not suffice to catch the robber on the graph $G = H \cup M$, where H is an Hamiltonian cycle and M is a random perfect matching.

To see why the claim holds, note that there are $\frac{1}{2}(n-1)!$ Hamiltonian cycles, each of which can be written as a union $M_1 \cup M_2$ of two perfect matchings in two ways, and $(n-1) \cdot (n-3) \cdot \ldots \cdot 1$ perfect matchings. Therefore the probability that the union of two random matchings is Hamiltonian is

$$\frac{2 \cdot \frac{1}{2}(n-1)!}{((n-1) \cdot (n-3) \cdot \ldots \cdot 1)^2} = \frac{1}{n-1}(n-2)\frac{1}{n-3}(n-4)\cdots 2\frac{1}{1} > \frac{2}{n-1}. \quad \square$$

Proof (of Corollary 3). Choose an even n large enough. By Theorem 2, there exist three perfect matchings M_1, M_2, M_3 on n vertices such that αn cops do not suffice to catch the robber on the graph $G' = M_1 \cup M_2 \cup M_3$. We construct the graph G in the following way. We label the $6n$ vertices by v_{ij} for $1 \le i \le n$ and $1 \le j \le 6$. For all i, we connect $v_{i1}, \ldots, v_{i6}$ as a directed 6-cycle. For each edge of one of the matchings, we will connect two of these cycles. More precisely, let $e \in M_k$ and suppose that e connects vertex i and j in G'. Then, we add the directed edges $(v_{ik}, v_{j(k+3)})$ and $(v_{jk}, v_{i(k+3)})$ to G.

We call a cycle i *free* if no cop is on this cycle, and *occupied* otherwise. If the robber enters a free cycle, then he can reach any vertex of the cycle and while doing so, the cops cannot occupy any new vertex outside of the cycle. Consider a situation of the game on the graph G with occupied cycles $F \subset [n]$, where the robber enters a free cycle i. This corresponds to the situation on the graph G' with occupied vertices F where the robber enters vertex i. On this graph the robber moves according to its winning strategy to a free vertex j with $(i,j) \in M_k$ for some $1 \leq k \leq 3$. On the graph G the robber moves first along the cycle i to vertex v_{ik} and then enters the free cycle j via the edge $(v_{ik}, v_{j(k+3)})$. Therefore, any winning strategy for the robber on graph G' with αn cops gives a winning strategy for the robber on graph G with αn cops. □

3.4 Proofs of Upper Bounds

Proof (of Proposition 1). Consider the graph DG_n on $2n$ vertices consisting of a Hamilton cycle $u_1, u_2, \ldots, u_n, v_1, v_2, \ldots, v_n$ and n additional edges (u_i, v_i) for $1 \leq i \leq n$. We call these additional edges *diagonal edges* and the other edges *cycle edges*. Note that every DG_n graph is 3-regular. Since the graph cannot be disconnected by removing 2 vertices, the graph DG_n is also 3-connected. Furthermore, the removal of any two pairs of opposite vertices $(u_i, v_i), (u_j, v_j)$ for $i \neq j$ splits the graph into two connected components. This structure turns out to be very useful for the cops. By occupying four such vertices the robber is trapped in the connected component he is in.

As in real life, our cops never come alone. Consider three pairs of cops c_i, c_i' for $1 \leq i \leq 3$. In order to catch the robber it suffices that each pair of cops can execute the command *chase robber*. If a pair i of cops is told to chase the robber, c_i and c_i' alternate in flying to the robbers location. This ensures that the robber can never move to its previous location. The first time the robber uses a diagonal edge, this pair of cops blocks the diagonal, i.e., it stays on the two endpoints of the diagonal edge until it receives the command to chase the robber again. Note that if there is a third cop placed somewhere on the graph, then the robber is forced to use a diagonal edge after less than $2n$ steps.

In the beginning of the game, when the robber has chosen its starting position, cop c_3 flies to the position of the robber. Then, cop pair c_1, c_1' chases the robber. The first time the robber uses a diagonal edge, these cops block the diagonal, and the second pair of cops starts chasing the robber. When the robber uses a diagonal edge again, this pair of cops blocks that diagonal and the third pair of cops c_3, c_3' starts chasing the robber.

From now on there will always be two cop pairs blocking two diagonals. Therefore, the robber cannot leave the area between these two diagonals. The remaining pair of cops chases the robber until he moves along a diagonal edge; this diagonal is subsequently blocked by this pair of cops. Note that the robber is now in the component defined by this diagonal and *one of* the two previously blocked diagonals. Correspondingly, one of the two cop pairs is not needed anymore and this pair of cops takes on the chase. The size of the entangled component the robber is in decreases by at least 2 every time the robber uses a diagonal edge. When the component has size 2 the robber is caught. □

Proof (of Theorem 3). Our main approach is the following. First, we identify a set A of size $\lfloor \frac{n}{4} \rfloor$ with the property that 4 cops suffices to catch the robber on $G \setminus A$. Then $\lfloor \frac{n}{4} \rfloor + 4$ cops can catch the robber using the following strategy. As long as the robber plays on the vertices in $G \setminus A$, we follow the optimal strategy on

$G \setminus A$ using at most 4 cops. Unless this catches the robber, he must at some point move to a vertex in A. If that happens, place one cop there permanently and repeat. Eventually, the robber runs out of non-blocked vertices in A to escape to and gets caught.

In order to identify the vertices that we want to place in A, we observe the following. Assume that the strategy on $G \setminus A$ for the cops is such that there will always be cops placed at the robbers last two positions, e.g. two cops take turns to drop on the robber. Then in order to catch the robber in $G \setminus A$, we can ignore all vertices that have degree one in $G \setminus A$ (as the robber will be caught, if he moves to such a vertex) and replace paths in which all internal vertices have degree two by a single edge (as two cops suffice to chase the robber along such a path).

To formalize this, we consider two operations to reduce a graph G: (i) remove any degree 1 vertices, (ii) replace any degree two vertices by an edge. Note that the second operation could create loops or multiple edges, so a reduced graph may not be simple. Moreover, reducing a max degree 3 graph as far as possible will result in components that are non-simple 3-regular or a vertex or a loop.

We now use these operations to define the set A. Pick any vertex of degree 3 and remove it from G. This decreases the number of degree 3 vertices by 4. Reduce G as far as possible. If the remaining graph has a vertex with 3 distinct neighbors, remove it from G, and reduce as far as possible. Again, this decreases the number of degree 3 vertices by 4. Repeat until no vertex in G has more than 2 distinct neighbors. Let A be the set of removed vertices. Then $|A| \leq \lfloor \frac{n}{4} \rfloor$.

It remains to show that 4 cops can catch the robber on $G \setminus A$. By construction, any connected component of $G \setminus A$ can be reduced to either a loop, a vertex, or a (non-simple) graph where all vertices have degree 3 but at most two neighbors. It is easy to see that the only way to satisfy these properties is for each vertex to either be incident to one loop and one single edge, one single edge and one double edge, or one triple edge. Thus, the only possible graphs are (a) an even cycle where every second edge is double, (b) a path where every second edge is double and where the end-points have loops attached, or (c) two vertices connected by a triple edge. We leave it as an exercise to see that the robber can be caught on any such graphs using at most 4 cops. Note that, due to (ii), it might be possible for the robber to have the same position at times t and $t + 2$ by following a pair of edges between the same two vertices, but this can be prevented using one of the cops. $\square$

4   Conclusion

We have shown that there are many graphs with maximum degree three for which the entanglement number is of linear size, in the directed cases as well as in the undirected case. This shows that the freedom of the cops of being able to *fly* to any vertex is not helpful when they are only allowed to fly to the current position of the robber. In other words, they should not only *follow* the robber, but they should think ahead of where the robber might want to go, as the title of our paper indicates.

All our examples were found by taking the union of three random matchings. We have shown that there exists an α such that with high probability, the robber can run away from αn cops. We also showed that $\lfloor \frac{1}{4} n \rfloor + 4$ cops do suffice. We leave it as an open problem to determine the exact value of α.

Acknowledgments. We thank Malte Milatz for bringing this problem to our attention.

References

1. Aigner, M., Fromme, M.: A game of cops and robbers. Discrete Appl. Math. **8**(1), 1–12 (1984)
2. Alon, N., Mehrabian, A.: Chasing a fast robber on planar graphs and random graphs. J. Graph Theory **78**(2), 81–96 (2015)
3. Bal, D., Bonato, A., Kinnersley, W.B., Prałat, P.: Lazy Cops and Robbers played on Graphs. ArXiv e-prints (2013)
4. Bal, D., Bonato, A., Kinnersley, W.B., Prałat, P.: Lazy cops and robbers played on random graphs and graphs on surfaces. J. Comb. **7**, 627–642 (2016)
5. Berwanger, D., Grädel, E.: Entanglement – a measure for the complexity of directed graphs with applications to logic and games. In: Baader, F., Voronkov, A. (eds.) LPAR 2005. LNCS (LNAI), vol. 3452, pp. 209–223. Springer, Heidelberg (2005). https://doi.org/10.1007/978-3-540-32275-7_15
6. Berwanger, D., Grädel, E., Kaiser, Ł., Rabinovich, R.: Entanglement and the complexity of directed graphs. Theor. Comput. Sci. **463**, 2–25 (2012). Special Issue on Theory and Applications of Graph Searching Problems
7. Berwanger, D., Lenzi, G.: The variable hierarchy of the μ-calculus is strict. In: Diekert, V., Durand, B. (eds.) STACS 2005. LNCS, vol. 3404, pp. 97–109. Springer, Heidelberg (2005). https://doi.org/10.1007/978-3-540-31856-9_8
8. Bollobás, B.: A probabilistic proof of an asymptotic formula for the number of labelled regular graphs. Eur. J. Comb. **1**(4), 311–316 (1980)
9. Bollobás, B.: Random Graphs. Cambridge Studies in Advanced Mathematics, 2nd edn. Cambridge University Press, Cambridge (2001)
10. Bonato, A., Chiniforooshan, E., Prałat, P.: Cops and robbers from a distance. Theor. Comput. Sci. **411**(43), 3834–3844 (2010)
11. Cooper, C., Frieze, A.: The size of the largest strongly connected component of a random digraph with a given degree sequence. Comb. Probab. Comput. **13**(3), 319–337 (2004)
12. Frieze, A., Karoński, M.: Introduction to Random Graphs. Cambridge University Press (2015)
13. Frieze, A., Krivelevich, M., Loh, P.-S.: Variations on cops and robbers. J. Graph Theory **69**(4), 383–402 (2012)
14. Janson, S., Łuczak, T., Rucinski, A.: Random Graphs. Wiley, New York (2000)
15. Nowakowski, R., Winkler, P.: Vertex-to-vertex pursuit in a graph. Discrete Math. **43**(2), 235–239 (1983)
16. Quilliot, A.: Jeux et pointes fixes sur les graphes. Dissertation, Université de Paris VI (1978)

A Generalization of the Minimum Branch Vertices Spanning Tree Problem

Massinissa Merabet[1]([⊠]), Jitamitra Desai[2], and Miklos Molnar[3]

[1] SAMOVAR Lab, National School of Computer Science
for Industry and Business (ENSIIE), Évry, France
massinissa.merabet@ensiie.fr

[2] Manufacturing and Industrial Engineering Cluster, School of Mechanical
and Aerospace Engineering, Nanyang Technological University (NTU),
Singapore, Singapore
jdesai@ntu.edu.sg

[3] The Montpellier Laboratory of Informatics, Robotics and Microelectronics,
University of Montpellier, Montpellier, France
miklos.molnar@lirmm.fr

Abstract. Given a connected graph $\mathcal{G} = (\mathcal{V}, \mathcal{E})$ and its spanning tree $\mathcal{T}$, a vertex $v \in \mathcal{V}$ is said to be a *branch vertex* if its degree is strictly greater than 2 in $\mathcal{T}$. The *Minimum Branch Vertices Spanning Tree* (MBVST) problem is to find a spanning tree of $\mathcal{G}$ with the minimum number of branch vertices. This problem has been extensively studied in the literature and has well-developed applications notably related to routing in optical networks. In this paper, we propose a generalization of this problem, where we begin by introducing the notion of a *k-branch vertex*, which is a vertex with degree strictly greater than $k + 2$. Our goal is to determine a spanning tree of $\mathcal{G}$ with the minimum number of k-branch vertices (k-MBVST problem). In the context of optical networks, the parameter k can be seen as the limiting capacity of optical splitters to duplicate the input light signal and forward to k destinations. Proofs of NP-hardness and non-inclusion in the APX class of the k-MBVST problem are established for a generic value of k, and then an ILP formulation of the k-MBVST problem based on single commodity flow balance constraints is derived. Computational results based on randomly generated graphs show that the number of k-branch vertices included in the spanning tree increases with the size of the vertex set $\mathcal{V}$, but decreases with k as well as graph density. We also show that when $k \geq 4$, the number of k-branch vertices in the optimal solution is close to zero, regardless of the size and the density of the underlying graph.

Keywords: Spanning tree · Minimization of branch vertices
Integer linear programming (ILP) · MBVST · k-MBVST
Optical networks

© Springer International Publishing AG, part of Springer Nature 2018
J. Lee et al. (Eds.): ISCO 2018, LNCS 10856, pp. 338–351, 2018.
https://doi.org/10.1007/978-3-319-96151-4_29

1   Introduction

Given a connected graph $\mathcal{G} = (\mathcal{V}, \mathcal{E})$, a vertex $v \in \mathcal{V}$ is defined to be a *branch vertex* in a spanning tree if its degree (denoted $d_\mathcal{G}(v)$) is strictly greater than two, i.e., $d_\mathcal{G}(v) > 2$. The *Minimum Branch Vertices Spanning Tree* (**MBVST**) problem is to find a spanning tree of graph $\mathcal{G}$ with the minimum number of branch vertices. This NP-hard and non-APX problem [GHSV02] has been well-studied in the literature and Cerrulli et al. [CGI09] were the first to formulate this problem as an *integer linear program* (**ILP**), wherein they used single commodity flow balance constraints to guarantee connectivity. In [CCGG13], Carabbs et al. provided two alternative ILP formulations based on multi-commodity flow balance constraints and the well-known cycle eliminating Miller-Tucker-Zemlin constraints, respectively. They also determined lower and upper bounds for the MBVST using the Lagrangian relaxation method. In [Mar15], Marin presented a branch-and-cut algorithm based on an enforced integer programming formulation for the MBVST problem. Melo et al. [MSU16] observed that an articulation vertex connecting at least three connected subgraphs of $\mathcal{G}$ must necessarily be a branch vertex in any optimal solution of the MBVST. Using this property, they independently solved the MBVST on each subgraph and conjoined these partially optimal trees to recover the overall optimal solution. In [CCR14], Cerrone et al. determined a unified memetic algorithm for three related problems, namely the MBVST; the problem of minimizing the degree sum of branch vertices (MDST); and the minimum leaves problem (ML). Landete et al. [LMSP17] studied the resolutions of these three problems, when the graph contains special nodes whose removal produces two or more connected components. Ad-hoc algorithms for each problem are developed that combine partial solutions to the thereby problems on the components produced by the removal of the nodes, guaranteeing the optimality of the global solution. Merabet et al. proved in [MDM13b] that the set of optimal solutions for MBVST and the set of optimal solutions for MDST are disjoint. They also proposed two variants of these problems which impose branch vertices to belong to a subset of nodes to better deal with the optical network constraints. In [MDM13a], another variant based on a more flexible graph structure, namely the so-called *hierarchy* is proposed. A hierarchy, which can be seen as a generalization of trees, is defined as a *homomorphism* of a tree in a graph [Mol11], and as minimizing the number of branch vertices in a hierarchy has no practical relevance, the authors determined the minimum cost spanning hierarchy such that the number of branch vertices is less than or equal to a given integer r.

The most widespread application of such MBVST problems arises in Wavelength-Division Multiplexing (WDM), which is an effective technique to exploit the available bandwidth of optical fibers to meet the explosive growth of bandwidth demand across the Internet [HGCT02]. Now, a multicast technique consists of simultaneously transmitting information from one source to multiple destinations in a bandwidth efficient way (duplicating the information only when required). From a computational viewpoint, multicast routing protocols in WDM networks are mainly based on *light-trees* [SM99], which require intermediate nodes to have the ability to split and direct the input signal to multiple outputs as and when necessary. A node which has the ability to replicate an input signal on any wavelength to any subset of output fibers is referred to as a Multicast-Capable (MC) node [MZQ98]. (A light-splitting switch has to be placed in the optical device to perform such a task at an MC node.) On the other hand, a node which has the ability to tap into a signal and forward it to only one output is called a Multicast-Incapable (MI) node. As light-splitter switches are rather expensive devices, it is imperative to minimize the number of MC nodes in the light-tree, and hence this problem lends itself to being expressed as the MBVST problem (see Gargano et al. in [GHSV02]).

Extending this, if a light signal is split into k copies (at an MC node without amplifiers), then the signal power of each resultant copy is reduced by, at least, a factor of $1/k$ of the original signal power [AD00]. If k is too large, then the information cannot be deciphered at the destinations due to the signal strength dropping below the minimum threshold value, and therefore, k functions as a limiting (tolerance) parameter.

Definition 1. *A k-branch vertex is a vertex with degree strictly greater than $k + 2$ in the spanning tree.*

Therefore, given a k-branch vertex, it is useful to look for a light-tree in the WDM network with the minimum number of k-branch vertices, where k is fixed as the threshold parameter. If the light-tree contains some k-branch vertices, an optical amplifier must be installed near each k-branch vertex to guarantee the efficiency of the broadcast/multicast. This leads to our problem:

Definition 2. *Let $\mathcal{G} = (\mathcal{V}, \mathcal{E})$ be a graph. The k-MBVST problem consists of finding a spanning tree $\mathcal{T}$ of $\mathcal{G}$ such that the number of k-branch vertices in $\mathcal{T}$ is minimized.*

The remainder of this paper is organized as follows. In Sect. 2, proofs of NP-hardness and non-inclusion in the APX class of the k-MBVST problem are established for any generic value of k. An improved ILP formulation of this problem based on single commodity flow balance constraints is derived in Sect. 3, and finally Sect. 4 records some preliminary computational results along with associated insights and conclusions.

2   Proofs of NP-Hardness and Negative Approximability

In a Hamiltonian graph, it is evident that finding a 0-MBVST is equivalent to finding a Hamiltonian path in $\mathcal{G}$. Thus, the k-MBVST is NP-complete in this case. Furthermore, the classical MBVST problem is NP-complete, even on non-Hamiltonian graphs [GHSV02], and moreover, it is a particular case of the k-MBVST problem corresponding to $k = 0$. Therefore, the 0-MBVST is at least as difficult as the MBVST even in this case.

In the following discussion, we prove that the k-MBVST problem is NP-hard for any generic $k > 0$. Towards this end, denote by $s_k(\mathcal{G})$ the smallest number of k-branch vertices in any spanning tree of $\mathcal{G}$.

Theorem 1. *Let r be a fixed non-negative integer. It is NP-complete to decide whether a given graph $\mathcal{G}$ satisfies $s_k(\mathcal{G}) \leq r$ for any value of k.*

Proof. Case 1: r=0
Let $\mathcal{G} = (\mathcal{V}, \mathcal{E})$ be a given connected graph. Construct a new graph $\bar{\mathcal{G}}$ by linking k leaves to each vertex $v \in \mathcal{V}$. Deciding whether $\bar{\mathcal{G}}$ contains a spanning tree with no k-branch vertex is equivalent to determining whether $\mathcal{G}$ is Hamiltonian or not.

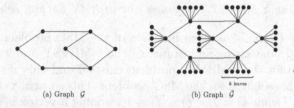

(a) Graph $\mathcal{G}$ (b) Graph $\bar{\mathcal{G}}$

Fig. 1. Reduction from the Hamiltonian problem to the 0-MBVST ($k = 5$).

Case 2: r ≥ 1
Let $\mathcal{G} = (\mathcal{V}, \mathcal{E})$ be a given connected graph. Construct a graph $\bar{\mathcal{G}}$ by replicating $r \cdot (k+1)$ times the graph $\mathcal{G}$ and add a chain C of size $r+2$. Choose an arbitrary vertex $v \in \mathcal{V}$ and link every *internal vertex* of C to $(k+1)$ distinct replications of $\mathcal{G}$ from their corresponding vertices (duplicates) v. Moreover, link k leaves to each vertex of each duplication of $\mathcal{G}$. In any spanning tree of $\bar{\mathcal{G}}$, the r *internal vertices* of the chain are necessarily k-branch vertices. Thus, the graph $\bar{\mathcal{G}}$ will contain a spanning tree with $s_k(\bar{\mathcal{G}}) = r$ if and only if $\mathcal{G}$ admits a Hamiltonian path starting from v (Fig. 2).

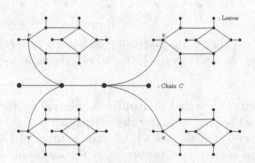

Fig. 2. Construction of graph $\bar{\mathcal{G}}$, with $k = 1$ and $r = 2$, corresponding to $\mathcal{G}$ (Fig. 1a).

In the following discussion, we show that the k-MBVST problem is not in the APX class for any generic $k > 0$. Knowing that if an optimization problem $\mathcal{P}_1$ is AP-reducible to an optimization problem $\mathcal{P}_2$ and $\mathcal{P}_2 \notin$ APX, then $\mathcal{P}_1 \notin$ APX [GPMS+99], we prove this result by applying an AP-reduction (f, g, ρ) from the

Minimum Set Cover (**MSC**) *problem* to the k-MBVST problem. Consider an instance of the MSC problem given by a ground set $\mathcal{U} = \{x_1, x_2, ..., x_n\}$, and a collection of m subsets $\mathcal{S} = \{\mathcal{S}_i\}_{i=1}^m$, such that $\bigcup_{i=1}^m \mathcal{S}_i \equiv \mathcal{U}$. A solution to the MSC problem aims to find a minimum number of subsets whose union contains each element of $\mathcal{U}$. This MSC problem is NP-complete and is not in the APX class [GPMS+99].

Theorem 2. *The k-MBVST problem is not in APX for any value of k.*

Proof. Let $\mathcal{X} = (\mathcal{U}, \mathcal{S})$ be a given instance of the MSC problem. We now construct a graph $\mathcal{G}$, underlying an instance of the k-MBVST problem, such that a feasible solution for the k-MBVST problem exists if and only if the instance $\mathcal{X}$ contains a feasible solution for the MSC problem. This construction procedure is described next. Define $\mathcal{G} = f(x,r) = f(x)$ by adding a vertex v_i corresponding to each element $x_i \in \mathcal{U}$. Similarly, add a vertex s_j for each subset $\mathcal{S}_j \in \mathcal{S}$. If $x_i \in \mathcal{S}_j$, then connect s_j and v_i by an edge. Moreover, add a vertex z and link it to each vertex s_i. Finally, link $(k-2)$ leaves to each vertex s_i and link $(k-1)$ leaves to z (refer Fig. 3 for an illustrative example).

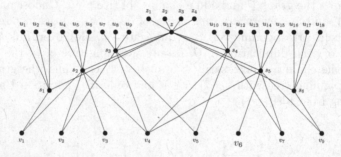

Fig. 3. Reduction of the MSC to the k-MBVST with $k = 5$, $\mathcal{U} = \{1,2,3,4,5,6,7,8,9\}$, and $\mathcal{S} = \{\{v_1\}, \{v_1, v_2, v_3, v_4\}, \{v_2, v_4, v_5\}, \{v_4, v_5, v_7\}, \{v_4, v_6, v_7, v_9\}, \{v_9\}\}$.

This construction ensures that if an optimal solution for the MSC problem in $\mathcal{X}$ contains n subsets then the optimum objective function value of the k-MBVST problem, defined over $\mathcal{G}$, equals $(n+1)$, i.e., the number of k-branch vertices in the optimal solution is $(n+1)$. Conversely, if the optimum objective function value of the k-MBVST in $\mathcal{G}$ equals n, then an optimal solution for the MSC problem in $\mathcal{X}$ contains exactly $(n-1)$ subsets. (An additional but necessary criterion of any AP-reduction process is that the graph construction must be done in polynomial time. Clearly, in our case, a polynomial time computation of the graph $\mathcal{G}$ is trivial). Now, let $c^*(f(x,r))$ be the optimum value of the instance $f(x,r)$ and let $c(f(x,r),y)$ denote the value of a feasible solution y. Let $c^*(x)$

denote the optimum value of an instance x. Finally, let $c(x, g(x, y, r))$ be the objective function value of a solution $g(x, y, r)$. Then, suppose that

$$r \geq \frac{c(f(x, r), y)}{c^*(f(x, r))}. \tag{2a}$$

Our mapping yields that $c^*(x) = c^*(f(x, r)) - 1$ and that $c(x, g(x, y, r)) = c(f(x, r), y) - 1$. Hence, for some $r > 1$ and fixed $\rho = 2$, it is sufficient for us to show that

$$1 + 2(r - 1) \geq \frac{c(f(x, r), y) - 1}{c(f(x, r)) - 1}. \tag{2b}$$

Using inequality (2a), it is enough to prove that

$$(r - 1)(c^*(f(x, r)) - 2) \geq 0. \tag{2c}$$

Since every spanning tree of $\mathcal{G}$ has at least two k-branch vertices (including z and at least one of s_i), the inequality (2c) is trivially true for $r > 1$.

Thus, the above defined $(f, g, 2)$ is an AP-reduction from the MSC problem to the k-MBVST problem.

Having proved the NP-completeness and non-inclusion in the APX class of the k-MBVST problem, in the following section, we turn our attention to deriving an integer linear programming formulation of this problem.

3   An ILP Formulation of the k-MBVST Problem

Recall that $d_{\mathcal{G}}(v)$ denotes the degree of v in $\mathcal{G}$, and let $\mathcal{C}_{\mathcal{G}}(v)$ represent the number of connected components in $\mathcal{G} \setminus \{v\}$. Before proceeding with the derivation of the ILP formulation of the k-MBVST problem, as a pre-processing step, we exploit the structure of the underlying graph to ascertain which vertices must necessarily be, can never be, or could possibly be k-branch vertices in the optimal solution. Towards this end, we partition the vertex set $\mathcal{V}$ into $\mathcal{V}_1, \mathcal{V}_2, \mathcal{V}_3$:

$$\mathcal{V}_1 = \{v \in \mathcal{V} : d_{\mathcal{G}}(v) \leq k + 2\} \tag{3a}$$
$$\mathcal{V}_2 = \{v \in \mathcal{V} : (d_{\mathcal{G}}(v) > k + 2) \wedge (\mathcal{C}_{\mathcal{G}}(v) \leq k + 2)\} \tag{3b}$$
$$\mathcal{V}_3 = \{v \in \mathcal{V} : \mathcal{C}_{\mathcal{G}}(v) > k + 2\}, \tag{3c}$$

where $\mathcal{V}_1 \cap \mathcal{V}_2 \cap \mathcal{V}_3 = \emptyset$ and $\mathcal{V}_1 \cup \mathcal{V}_2 \cup \mathcal{V}_3 = \mathcal{V}$. Clearly, the $\mathcal{V}_1$ vertices do not have a sufficiently high enough degree to be k-branch in any spanning tree of $\mathcal{G}$. Conversely, the $\mathcal{V}_3$ vertices must necessarily be selected as k-branch vertices in any spanning tree of $\mathcal{G}$ as deleting any of those vertices decomposes $\mathcal{G}$ into at least $k + 3$ connected components. While deleting a vertex in $\mathcal{V}_2$ decomposes $\mathcal{G}$ into at most $k + 2$ connected components, such vertices could be k-branch vertices in the optimal solution (Fig. 4). Furthermore, it is trivially true that an isthmus (an edge of a graph whose deletion increases its number of connected components) must be in any spanning tree of $\mathcal{G}$. Denote by $\mathcal{I}$ the set of isthmuses of $\mathcal{G}$.

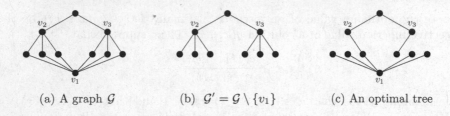

(a) A graph $\mathcal{G}$ (b) $\mathcal{G}' = \mathcal{G} \setminus \{v_1\}$ (c) An optimal tree

Fig. 4. Graph $\mathcal{G}' = \mathcal{G} \setminus \{v_1\}$ is connected and yet v_1 is a 2-branch vertex in the optimal solution ($k = 0$).

Lemma 1. *The number of k-branch vertices belonging to $\mathcal{V}_2$ in any spanning tree of graph $\mathcal{G}$ is at most* $\left(|\mathcal{V}| - \sum\limits_{v \in (\mathcal{V}_1 \cup \mathcal{V}_3)} \mathcal{C}_{\mathcal{G}}(v)\right)/(k+2)$.

Proof. By definition, it follows that exactly $\mathcal{C}_{\mathcal{G}}(v)$ components are connected to each vertex $v \in (\mathcal{V}_1 \cup \mathcal{V}_3)$. Consequently, each one is connected to v in any spanning tree of G by at least one edge. Therefore, at most $|\mathcal{V}| - \sum\limits_{v \in (\mathcal{V}_1 \cup \mathcal{V}_3)} \mathcal{C}_{\mathcal{G}}(v)$ remaining vertices can be connected to the $\mathcal{V}_2$ vertices. Moreover, the degree of each k-branch vertex is at least equal to $k + 3$ and at least $k + 2$ remaining vertices are connected to each other, as one edge serves to connect a $\mathcal{V}_2$ vertex to the tree. Therefore, the number of k-branch vertices in $\mathcal{V}_2$ in any spanning tree of G is upper bounded by $\left(|\mathcal{V}| - \sum\limits_{v \in (\mathcal{V}_1 \cup \mathcal{V}_3)} \mathcal{C}_{\mathcal{G}}(v)\right)/(k+2)$.

The formulation of the k-MBVST problem as an *integer linear program* (**ILP**) derived in this paper is predicated on the single balance commodity flow formulation proposed in [CGI09]. However, it is worthwhile to note that by considering the partitions of graph vertices; computing a tighter upper bound on the maximum quantity of flow transiting on the graph edges; and deploying a tight constraint to check if a vertex is k-branch or not, a significantly improved version as compared to the classical formulation is presented in this paper. The concepts alluded to above are as follows.

In order to define a spanning tree $\mathcal{T}$ of $\mathcal{G}$, we can send from a source vertex $s \in \mathcal{V}$, one unit of flow to every other vertex $v \in \mathcal{V} \setminus \{s\}$. As this flow needs to be directed, the given graph $\mathcal{G}$ is transformed into a symmetrically oriented graph $\mathcal{G}^d = (\mathcal{V}, \mathcal{E}^d)$, where each edge $\{u, v\} \in \mathcal{E}$ now corresponds to two directed arcs (u, v) and (v, u) in $\mathcal{E}^d$. Similarly, the set $\mathcal{I}^d$ corresponds to the symmetrically directed version of $\mathcal{I}$. For each arc $(u, v) \in \mathcal{E}^d$, we define an integer variable $f_{(u,v)}$ representing the (directed) flow going from u to v. Furthermore, for each edge $\{u, v\} \in \mathcal{E} \setminus \mathcal{I}$, define a binary decision variable $x_{(u,v)}$, which equals 1 if $f_{(u,v)}$ or $f_{(v,u)}$ carry a non-zero flow, and 0 otherwise. Finally, for each $v \in \mathcal{V}_2$, we have a decision variable y_v that equals 1, if v is a k-branch vertex in the spanning tree, and 0 otherwise. We denote by $\mathcal{I}_v$ the set of isthmus edges which have the vertex $v \in \mathcal{V}$ as an extremity, and moreover, let $\mathcal{I}_v^+$ and by $\mathcal{I}_v^-$ be the set of outgoing isthmus edges from v and the set of incoming isthmus edges to v in $\mathcal{I}^d$, respectively. We denote by $\mathcal{S}^+$ the set of vertices of $\mathcal{G}^d$ which have at least one outgoing isthmus edge and by $\mathcal{S}^-$ the set of vertices of $\mathcal{G}^d$ which have at least one incoming isthmus edge.

We are now ready to formally present the ILP program of the k-MBVST problem. Since the $\mathcal{V}_1$ vertices cannot be k-branch and the $\mathcal{V}_3$ vertices are necessarily k-branch, the objective of our problem is to minimize the number of k-branch vertices of $\mathcal{V}_2$ belonging to the spanning tree of $\mathcal{G}$, which can be expressed as follows.

Objective Function:

$$\text{Minimize} \sum_{v \in \mathcal{V}_2} y_v \tag{4a}$$

Spanning Tree Constraints:

$$\sum_{\substack{u \in \mathcal{V}: \\ (u,v) \in \mathcal{E}^d}} x_{(u,v)} = 1, \quad \forall v \in \mathcal{V} \setminus (\mathcal{S}^- \cup \{s\}) \tag{4b}$$

$$\sum_{\substack{u \in \mathcal{V}: \\ (u,v) \in \mathcal{E}^d}} x_{(u,v)} = 0, \quad \forall v \in \mathcal{S}^- \tag{4c}$$

$$\sum_{(u,v) \in \mathcal{E}^d} x_{(u,v)} = |\mathcal{V}| - |\mathcal{I}| - 1. \tag{4d}$$

As a vertex with more than one parent creates a cycle, constraints (4b) and (4c) ensure that each vertex except the source has exactly one predecessor. Next, as the number of edges in any spanning tree must be equal to $|\mathcal{V}|-1$ and the edges in $\mathcal{I}$ are systematically present in any connected subgraph of $\mathcal{G}$, constraint (4d) ensures that exactly $|\mathcal{V}|-|\mathcal{I}|-1$ arcs are added to the optimal solution. Note that these two constraints are necessary but are not sufficient to generate a spanning tree as connectedness is not yet guaranteed. To obtain connectivity, additional flow balance-based constraints are incorporated, and these are described next.

Connectivity Constraints:

$$\sum_{(s,v) \in \mathcal{E}^d} f_{(s,v)} - \sum_{(v,s) \in \mathcal{E}^d} f_{(v,s)} = |\mathcal{V}| - 1 \tag{4e}$$

$$\sum_{\substack{u \in \mathcal{V}: \\ (v,u) \in \mathcal{E}^d}} f_{(v,u)} - \sum_{\substack{u \in \mathcal{V}: \\ (u,v) \in \mathcal{E}^d}} f_{(u,v)} = -1, \quad \forall v \in \mathcal{V} \setminus \{s\} \tag{4f}$$

$$x_{(u,v)} \le f_{(u,v)} \le \mathcal{B}_{(u,v)} \cdot x_{(u,v)}, \quad \forall (u,v) \in \mathcal{E}^d \setminus \mathcal{I}^d \tag{4g}$$

Constraint (4e) states that the flow emitted by the source is equal to $|\mathcal{V}| - 1$ and constraint (4f) ensures that each vertex except the source "consumes" one and only one unit of flow, which in turn also guarantees that each vertex is reachable from the source s. Constraint (4g) allows each arc in $\mathcal{E}^d \setminus \mathcal{I}^d$ to carry non-zero flow if and only if it is used in the optimal spanning tree, and the value of this flow on each arc (u,v) cannot exceed an upper bound $\mathcal{B}_{(u,v)}$. Finally, constraints (4e)–(4g) also enforce that each isthmus arc must be included in the optimal solution, and carries a non-zero flow because it is the only arc that links two connected components.

Degree Constraints:
Let's denote by $\mathcal{X}_v$ the set of outgoing arcs from $v \in \mathcal{V}_2$. We pose $\mathcal{Q}_v$ as the set of all partitions q_v of $\mathcal{X}_v$ of size $(k - |\mathcal{I}_v^+|)$.

$$\sum_{u \in q_v} x_{(v,u)} + |\mathcal{I}_v^+| - k - 1 \leq y_v, \qquad \forall q_v \subset \mathcal{Q}_v, \forall v \in \mathcal{V}_2 \tag{4h}$$

Constraint (4h) imposes vertex v to be a k-branch vertex if and only if its degree is strictly greater than $k + 2$ in the spanning tree. Note that while the above constraint merely sets the value of y_v to be greater than or equal to zero if $d(v) \leq k + 2$, nevertheless, the objective function (which minimizes the sum of the y_v-variables) drives the value of $y_v \equiv 0$ at optimality. For each vertex $v \in \mathcal{V}_2$, this constraint certainly appears $\binom{\mathcal{X}_v}{k - |\mathcal{I}_v^+|}$ times but it is tight only when v is a k-branch vertex and is redundant otherwise, which is a significantly improved version to those used in the literature [CGI09].

Valid Inequalities:

$$x_{(u,v)} + x_{(v,u)} \leq 1, \qquad \forall (u,v) \in \mathcal{E}^d \setminus \mathcal{I}^d \tag{4i}$$

$$\sum_{\substack{u \in \mathcal{V}: \\ (v,u) \in \mathcal{E}^d}} x_{(u,v)} \geq \mathcal{C}_{\mathcal{G}}(v) - 1, \qquad \forall v \in \mathcal{V} \tag{4j}$$

$$\sum_{v \in \mathcal{V}_2} y_v \leq \left(|\mathcal{V}| - \sum_{v \in (\mathcal{V}_1 \cup \mathcal{V}_3)} \mathcal{C}_{\mathcal{G}}(v) \right) / (k + 2) \tag{4k}$$

As cycles are not authorized, constraint (4i) states that only one arc is allowed between two vertices. Constraint (4j) guaranties that the degree of each vertex in the optimal solution is lower bounded by $\mathcal{C}_{\mathcal{G}}(v)$, since a least $\mathcal{C}_{\mathcal{G}}(v)$ components are exclusively linked to v. Finally, constraint (4k) enforces an upper bound on the number of the $\mathcal{V}_2$ k-branch vertices belonging to the optimal solution (Lemma 1).

Once an optimal solution to the ILP (4a)–(4k) is obtained, it is easy to reconstruct the optimal spanning tree $\mathcal{T}^* = (\mathcal{V}, \mathcal{E}^*)$. Indeed, it suffices to add to $\mathcal{E}^*$ the edges in $\mathcal{E}$ corresponding to the arcs in $\mathcal{E}^d$ which carry a non-zero flow and add all the isthmus edges of $\mathcal{G}$. Therefore, $\mathcal{E}^* = \Big\{ \{u,v\} \in \mathcal{E} \mid x_{(u,v)} = 1 \text{ or } x_{(v,u)} = 1 \text{ or } \{u,v\} \in \mathcal{I} \Big\}$.

Remark 1: Let $\mathcal{G}' = (\mathcal{V}', \mathcal{E}')$ be the graph obtained by deleting from $\mathcal{G}$ the edges which have at least one articulation vertex extremity. In enforcing constraint (4g), the trivial upper bound used in the literature is $\mathcal{B}_{(u,v)} \equiv |\mathcal{V}| - 1$, but this value can only be attained if $\mathcal{G}'$ is connected. However, if $\mathcal{G}$ contains at least one vertex articulation, then this bound can be improved by computing an upper bound on the maximum flow possible on each arc of $\mathcal{G}^d$. To extract these improved (tight) upper bounds, let $\mathcal{G}'' = (\mathcal{V}'', \mathcal{E}'')$ be the graph obtained by considering each connected component of $\mathcal{G}'$ as a vertex in $\mathcal{G}''$. If two vertices in different connected components in $\mathcal{G}'$ are also linked by an edge in $\mathcal{G}$ than the two vertices corresponding to these connected components are linked by an edge in $\mathcal{G}''$. The connected component which contains the source in $\mathcal{G}^d$ is selected as a source vertex and a sink is added to $\mathcal{G}''$. Finally, edges linking each vertex to the sink are added to obtain the resultant graph $\mathcal{G}''$. All the edges are directed from the source to the sink t. Each arc outgoing from a vertex $v \in \mathcal{V}'' \setminus \{t\}$ and incoming to the sink has a capacity the size of the connected component in $\mathcal{G}'$ represented by v in $\mathcal{G}''$. All other arcs have capacity $+\infty$ (Fig. 5).

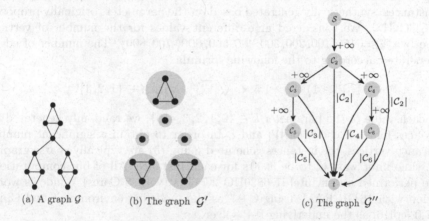

(a) A graph $\mathcal{G}$ (b) The graph $\mathcal{G}'$ (c) The graph $\mathcal{G}''$

Fig. 5. Transformation of $\mathcal{G}$ to a single-source, single-sink flow network.

Now, clearly the maximum flow from the source to the sink can be computed in polynomial time using the push-relabel method. The quantity of flow transiting in an arc (u,v) of $\mathcal{G}''$ corresponds to the upper bound of the value of $f_{(w,x)}$ such that x is in the connected component represented by v in $\mathcal{G}''$. We denote by $\mathcal{B}_{(u,v)}$ these obtained bounds, which in turn serve to tighten constraint (4g) in our implementation.

Summarizing, the exact resolution method (see Algorithm 1) for solving the k-MBVST problem can be seen to be composed of the following six steps, each of which can be accomplished in polynomial time with the exception of Step 5.

Algorithm 1. Exact resolution of the k-MBVST problem

    Input: A connected graph $\mathcal{G} = (\mathcal{V}, \mathcal{E})$ and an empty graph $\mathcal{T}^* = (\mathcal{V}, \mathcal{E}^* = \emptyset)$.
    Output: An optimal spanning tree $\mathcal{T}^* = (\mathcal{V}, \mathcal{E}^*)$.

1: Decompose the vertex set $\mathcal{V}$ into three partitions $\mathcal{V}_1$, $\mathcal{V}_2$, $\mathcal{V}_3$ shown in (3a)–(3c), and identify the isthmus set $\mathcal{I}$.
2: Transform the graph $\mathcal{G}$ into a symmetrically oriented graph $\mathcal{G}^d$.
3: Construct the graph $\mathcal{G}''$.
4: Compute a maximum flow in $\mathcal{G}''$ to find the upper bound $\mathcal{B}_{(u,v)}$ on the quantity of flow that can be transmitted on each arc (u, v) of $\mathcal{G}^d$.
5: Solve the **ILP**.
6: $\mathcal{E}^* = \left\{ \{u, v\} \in \mathcal{E} \mid x_{(u,v)} = 1 \text{ or } x_{(v,u)} = 1 \text{ or } \{u, v\} \in \mathcal{I} \right\}$.

---

4   Computational Results

In this section, we present the computational results obtained by applying the proposed single commodity flow formulation for the k-MBVST problem on a set of instances, synthetically generated based on the parameters originally proposed in [CCGG13]. We considered nine different values for the number of vertices given by: $|\mathcal{V}| = \{50, 100, 200, 300, 400, 500, 600, 700, 800\}$. The number of edges is generated according to the following formula:

$$\lfloor (|\mathcal{V}| - 1) + i \times 1.5 \times \lceil \sqrt{|\mathcal{V}|} \rceil \rfloor, \text{ with } i \in \{1, 2, 3\}. \tag{5}$$

For each value of the parameter $k \in \{0, 1, 2, 3, 4, 5\}$, we randomly generated 30 instances for each choice of $|\mathcal{V}|$ and i. In order to obtain a significant number of branch vertices, the instances generated using (5) are typically sparse graphs. The time limit was set to be 3600 s for each instance. All of our computations were performed on an Intel i7 6820HQ 2.7 Ghz (with 8 Cores) Windows workstation with 16 GB RAM, using C++ as the modeling environment and Cplex 12.7.0 [cpl16] as the underlying ILP solver.

    Table 1 displays the optimum number of k-branch vertices (averaged over all solved instances), the CPU time (seconds), and the number of instances of each type that were successfully solved to optimality (within the specified time limit). From the numerical results recorded in Table 1, it can be observed that the computational time increases along with an increase in the size of the graph and with higher graph density, as the number of decision variables in the ILP are directly correlated to the size and density of the graph. Moreover, as the size of the instance increases, constraint (4g) no longer remains tight, and furthermore, the number of branch vertices reduce in higher density instances ($i = 2$ and $i = 3$) because the propensity of the graph to be Hamiltonian increases in such cases. Finally, the number of branch vertices increases (almost linearly) with instance size, and this phenomenon is amplified by the fact that the edge generation scheme used in our work makes the density decrease with instance size.

Table 1. Solution value, running time, and number of solved instances for varying values of $|\mathcal{V}|$, k and i.

Instances	$i = 1$			$i = 2$			$i = 3$		
$|\mathcal{V}|$	Sol	Time	# Inst	Sol	Time	# Inst	Sol	Time	# Inst
						$k = 0$			
50	8.00	0.22	30	3.80	0.60	30	2.50	0.62	30
100	18.00	0.39	30	12.30	1.11	30	7.60	5.44	30
200	38.40	1.14	30	28.90	5.21	30	20.70	18.31	30
300	59.90	1.91	30	49.40	9.73	30	37.90	42.76	30
400	80.60	4.06	30	67.40	15.15	30	53.80	50.98	30
500	102.60	4.32	30	86.60	19.69	30	74.10	446.47	30
600	130.50	5.93	30	107.70	21.81	30	91.22	289.46	29
700	149.80	6.68	30	131.70	40.04	30	113.44	275.39	29
800	175.20	8.73	30	149.80	43.79	30	131.00	188.82	30
						$k = 1$			
50	1.50	0.10	30	0.00	0.21	30	0.00	0.25	30
100	4.60	0.31	30	1.00	0.46	30	0.50	0.82	30
200	12.60	0.58	30	5.30	1.80	30	2.67	3.73	27
300	20.20	1.72	30	10.70	8.32	30	4.88	9.35	24
400	28.10	3.21	30	17.70	12.19	30	9.50	21.20	25
500	41.00	4.78	30	24.67	16.92	27	14.00	42.85	27
600	49.00	6.16	30	31.89	22.37	27	20.43	108.02	21
700	61.20	8.11	30	38.90	23.84	30	31.00	77.04	6
800	68.70	8.99	30	46.60	40.74	30	34.60	150.90	12
						$k = 2$			
50	0.30	0.07	30	0.00	0.12	30	0.00	0.17	30
100	0.70	0.16	30	0.10	0.31	30	0.00	0.49	30
200	3.00	0.49	30	0.50	1.45	30	0.33	1.68	27
300	4.90	0.94	30	1.80	3.56	30	0.71	6.19	21
400	9.80	2.09	30	2.67	7.36	27	1.22	15.07	27
500	12.00	2.29	30	4.63	12.04	24	2.50	25.04	30
600	16.20	4.23	30	6.33	24.68	18	3.25	32.69	12
700	20.00	5.56	30	9.75	19.19	12	5.00	118.11	6
800	24.70	5.62	30	12.00	23.90	9	4.33	57.01	9
						$k = 3$			
50	0.10	0.07	30	0.00	0.10	30	0.00	0.13	30
100	0.10	0.15	30	0.10	0.17	30	0.00	0.27	30
200	0.60	0.32	30	0.10	0.93	30	0.00	1.56	30
300	1.10	0.79	30	0.50	1.63	30	0.10	3.98	30
400	1.40	1.32	30	1.00	4.42	24	0.40	10.80	30
500	3.50	1.62	30	1.38	9.50	24	0.13	22.14	24
600	4.56	3.00	27	2.00	5.59	7	1.00	30.55	21
700	6.20	4.18	30	1.40	53.08	15	0.33	69.49	9
800	7.30	3.04	30	2.14	21.57	20	2.50	57.01	6
						$k = 4$			
50	0.00	0.05	30	0.00	0.09	30	0.00	0.12	30
100	0.00	0.08	30	0.00	0.13	30	0.00	0.23	30
200	0.00	0.18	30	0.00	0.54	30	0.00	0.35	30
300	0.50	0.32	30	0.14	0.98	21	0.00	1.88	27
400	0.60	0.54	30	0.00	09.12	21	0.00	2.41	24
500	0.40	0.51	30	0.13	5.00	24	0.00	24.68	18
600	1.60	1.15	30	0.20	8.65	15	0.25	50.22	12
700	1.30	2.63	30	0.67	15.77	9	0.00	86.80	9
800	1.80	3.59	30	0.33	16.36	18	0.40	14.20	15
						$k = 5$			
50	0.00	0.05	30	0.00	0.08	30	0.00	0.10	30
100	0.00	0.09	30	0.00	0.15	30	0.00	0.14	30
200	0.00	0.11	30	0.00	0.42	30	0.00	0.50	30
300	0.00	0.20	30	0.00	1.12	24	0.00	1.08	30
400	0.30	0.34	30	0.00	21.68	24	0.00	4.76	24
500	0.10	0.29	30	0.00	80.04	21	0.00	45.24	18
600	0.30	0.64	30	0.00	6.34	15	0.00	40.58	15
700	0.00	1.83	21	0.00	16.95	8	0.00	42.75	8
800	0.11	2.39	27	0.00	10.39	21	0.00	49.27	6

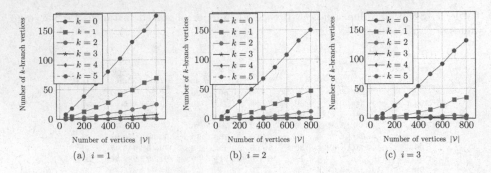

Fig. 6. Variation of $s_k(\mathcal{G})$ as a function of $|\mathcal{V}|$, k and i.

Figure 6 displays the number of k-branch vertices as a function of the parameters k and $|\mathcal{V}|$ for each value of i. As noted before, the number of k-branch vertices increases as the size of the vertex set increases, but decreases with k as well as with i. When $k \geq 4$, the number of k-branch vertices is close to zero irrespective of the value of $|\mathcal{V}|$ and i.

5 Conclusion

In this paper, we propose a generalization of the well-known MBVST problem by introducing the notion of the k-branch vertex, which is a vertex with degree strictly greater than $k + 2$. Our new parametrized problem (k-MBVST) aims to find a spanning tree of $\mathcal{G}$ with the minimum number of k-branch vertices. For any non-negative integer r, we proved that it is NP-complete to decide whether a graph can be spanned by a tree with at most r k-branch vertices, irrespective of the value of k. Furthermore, we also established that the k-MBVST is hard to approximate by proving its non-inclusion in the APX class. We also proposed an integer linear programming formulation based on a single commodity flow balance constraints. Tests on sparse random graphs allowed us to evaluate the number of k-branch vertices in the optimal solution as well as the computational time required to determine the optimum objective function value with respect to the value of k, the graph size, and the graph density. Our results indicate that the number of k-branch vertices increases with graph size but decreases with k as well as with graph density. It was also observed that when $k \geq 4$, the number of k-branch vertices is close to zero, and is independent of the size and density of the graph.

References

[AD00] Ali, M., Deogun, J.S.: Power-efficient design of multicast wavelength-routed. Networks. **18**(10), 1852–1862 (2000)

[CCGG13] Carrabs, F., Cerulli, R., Gaudioso, M., Gentili, M.: Lower and upper bounds for the spanning tree with minimum branch vertices. Comput. Optim. Appl. **56**(2), 405–438 (2013)

[CCR14]   Cerrone, C., Cerulli, R., Raiconi, A.: Relations, models and a memetic approach for three degree-dependent spanning tree problems. Eur. J. Oper. Res. **232**(3), 442–453 (2014)

[CGI09]   Cerulli, R., Gentili, M., Iossa, A.: Bounded-degree spanning tree problems: models and new algorithms. Comput. Optim. Appl. **42**, 353–370 (2009)

[cpl16]   ILOG CPLEX 12.7.0 User's Manual. IBM (2016)

[GHSV02] Gargano, L., Hell, P., Stacho, L., Vaccaro, U.: Spanning trees with bounded number of branch vertices. In: Widmayer, P., et al. (eds.) ICALP 2002. LNCS, vol. 2380, pp. 355–365. Springer, Heidelberg (2002). https://doi.org/10.1007/3-540-45465-9_31

[GPMS+99] Giorgio, A., Protasi, M., Marchetti-Spaccamela, A., Gambosi, G., Crescenzi, P., Kann, V.: Complexity and Approximation: Combinatorial Optimization Problems and Their Approximability Properties, 1st edn. Springer, Secaucus (1999). https://doi.org/10.1007/978-3-642-58412-1

[HGCT02] He, J., Gary Chan, S.H., Tsang, D.H.K.: Multicasting in WDM networks. IEEE Commun. Surv. Tutorials **4**(1), 2–20 (2002)

[LMSP17] Landete, M., Marín, A., Sainz-Pardo, J.: Decomposition methods based on articulation vertices for degree-dependent spanning tree problems. Comput. Optim. Appl. **68**(3), 749–773 (2017)

[Mar15]   Marín, A.: Exact and heuristic solutions for the minimum number of branch vertices spanning tree problem. Eur. J. Oper. Res. **245**(3), 680–689 (2015)

[MDM13a] Merabet, M., Durand, S., Molnár, M.: Exact solution for branch vertices constrained spanning problems. Electron. Notes Discrete Math. **41**, 527–534 (2013)

[MDM13b] Merabet, M., Durand, S., Molnár, M.: Minimization of branching in the optical trees with constraints on the degree of nodes. In: The Eleventh International Conference on Networks - ICN, pp. 235–240 (2013)

[Mol11]   Molnár, M.: Hierarchies to solve constrained connected spanning problems. Technical Report RR-11029, September 2011

[MSU16]   Melo, R.A., Samer, P., Urrutia, S.: An effective decomposition approach and heuristics to generate spanning trees with a small number of branch vertices. Comput. Optim. Appl. **65**(3), 821–844 (2016)

[MZQ98]   Malli, R., Zhang, X., Qiao, C.: Benefits of multicasting in all-optical networks. In: Senior, J.M., Qiao, C. (eds.) All-Optical Networking: Architecture, Control, and Management Issues Proceedings of SPIE, vol. 3531, pp. 209–220, October 1998

[SM99]   Sahasrabuddhe, L.H., Mukherjee, B.: Light trees: optical multicasting for improved performance in wavelength routed networks. IEEE Commun. Mag. **37**(2), 67–73 (1999)

A Polyhedral View to Generalized Multiple Domination and Limited Packing

José Neto[(✉)]

Samovar, CNRS, Telecom SudParis, Université Paris-Saclay,
9 rue Charles Fourier, 91011 Evry, France
Jose.Neto@telecom-sudparis.eu

Abstract. Given an undirected simple graph $G = (V, E)$ and integer values $f_v, v \in V$, a node subset $D \subseteq V$ is called an f-tuple dominating set if, for each node $v \in V$, its closed neighborhood intersects D in at least f_v nodes. We investigate the polyhedral structure of the polytope that is defined as the convex hull of the incidence vectors in $\mathbb{R}^V$ of the f-tuple dominating sets in G. We provide a complete formulation for the case of stars and introduce a new family of (generally exponentially many) inequalities which are valid for the f-tuple dominating set polytope and that can be separated in polynomial time. A corollary of our results is a proof that a conjecture present in the literature on a complete formulation of the 2-tuple dominating set polytope of trees does not hold. Investigations on adjacency properties in the 1-skeleton of the f-tuple dominating set polytope are also reported.

1 Introduction

Let $G = (V, E)$ denote an undirected simple graph with node set $V = \{1, 2, \ldots, n\}$ and edge set E. Given some node $v \in V$, let d_v^G (or more simply d_v when G is clear from the context) denote the degree of node v in G. Given a node subset $S \subseteq V$, $N(S)$ denotes the open neighborhood of S: $N(S) = \{v \in V \setminus S : vw \in E \text{ for some node } w \in S\}$, and $N[S]$ stands for its *closed neighborhood*: $N[S] = N(S) \cup S$. When S is a singleton, i.e. $S = \{v\}$ for some node $v \in V$, we will write $N(v)$ (resp. $N[v]$) in lieu of $N(\{v\})$ (resp. $N[\{v\}]$). Given a node subset $S \subseteq V$, the subgraph of G induced by S is the graph $G[S] = (S, E(S))$ with $E(S) = \{uv \in E : (u, v) \in S^2\}$.

Let $\widehat{\mathcal{F}}_G$ (resp. $\mathcal{F}_G$) stand for the following set of vectors indexed on the nodes of G: $\widehat{\mathcal{F}}_G = \{f \in \mathbb{Z}_+^n : 0 \le f_v \le d_v^G + 1, \forall v \in V\}$ (resp. $\mathcal{F}_G = \{f \in \widehat{\mathcal{F}}_G : f_v \le d_v^G, \forall v \in V\}$). Given $f \in \mathcal{F}_G$, a node subset $D \subseteq V$ is called an f-*dominating set* of G if each node v in $V \setminus D$ has at least f_v neighbors in D, i.e. $|N(v) \cap D| \ge f_v$. If the latter inequality holds for all the nodes in V, then D is called a *total f-dominating set of* G. Given $f \in \widehat{\mathcal{F}}_G$, an f-*tuple dominating set* of G is a node subset $D \subseteq V$ satisfying: $|N[v] \cap D| \ge f_v, \forall v \in V$. The notions just

© Springer International Publishing AG, part of Springer Nature 2018
J. Lee et al. (Eds.): ISCO 2018, LNCS 10856, pp. 352–363, 2018.
https://doi.org/10.1007/978-3-319-96151-4_30

introduced generalize those of *dominating set, total dominating set* and *k-tuple dominating set* (where k denotes a positive integer) from classical domination theory, respectively. The notion of k-tuple dominating set corresponds to the particular case of f-tuple dominating set, when $f_v = k, \forall v \in V$ [11,14] [15, p. 189]. The natural complementary notion to f-tuple domination is that of *f-limited packing*: Given $f \in \widehat{\mathcal{F}}_G$, an f-limited packing is a node subset $S \subseteq V$ such that $|N[v] \cap S| \leq f_v, \forall v \in V$. For the particular case when $f_v = k, \forall v \in V$ for some positive integer k, we speak of *k-limited packing*. Let $\tilde{f} \in \mathbb{Z}^n$ be defined as follows: $\tilde{f}_v = d_v - f_v + 1$. Then, S is an f-limited packing if and only if $V \setminus S$ is an $\tilde{f}$-tuple dominating set.

The *"Minimum weight f-tuple dominating set problem"* denoted by $[MW_f]$ can be stated as follows. Given an undirected graph $G = (V, E)$, $w \in \mathbb{R}^n_+$ and $f \in \widehat{\mathcal{F}}_G$, find a minimum weight f-tuple dominating set of G, i.e. find a node subset $S \subseteq V$ such that S is an f-tuple dominating set and the weight of S: $\sum_{v \in S} w_v$, is minimum. This problem may be formulated as the following integer program.

$$(IP1) \begin{cases} \min \sum_{v \in V} w_v x_v \\ s.t. \\ \quad \sum_{u \in N[v]} x_u \geq f_v, \forall v \in V, \\ \quad x \in \{0,1\}^n. \end{cases}$$

Let $\mathcal{U}^f_G$ denote the convex hull of the feasible solutions of $(IP1)$ (or equivalently, the convex hull of the incidence vectors in $\mathbb{R}^n$ of the f-tuple dominating sets in G). It will be called the *f-tuple dominating set polytope* in what follows. Also, let $(LP1)$ denote the linear relaxation of $(IP1)$ (obtained by replacing $x \in \{0,1\}^n$ by $x \in [0,1]^n$).

Remark 1. Let $\mathcal{D}^f_G$ (resp. $\mathcal{T}^f_G$) denote the convex hull in $\mathbb{R}^n$ of the incidence vectors of the f-dominating (resp. total f-dominating) sets in G. One can easily check the following inclusions hold:

$$\mathcal{T}^f_G \subseteq \mathcal{U}^f_G \subseteq \mathcal{T}^{f-1}_G \subseteq \mathcal{D}^{f-1}_G, \text{ and } \mathcal{T}^f_G \subseteq \mathcal{U}^f_G \subseteq \mathcal{D}^f_G,$$

where **1** stands for the n-dimensional all-ones vector.

Given the afore mentioned connection between f-tuple domination and limited packing, the *f-limited packing polytope*, which corresponds to the convex hull of the incidence vectors of f-limited packings, is isomorphic to $\mathcal{U}^{\tilde{f}}_G$ so that the results established next w.r.t. $\mathcal{U}^{\tilde{f}}_G$ are also relevant w.r.t. this polytope and the following equation, which generalizes Lemma 5 in [12], can be easily proved:

$$\max\{w^t \chi^S : S \text{ is a } f\text{-limited packing}\} +$$
$$\min\{w^t \chi^S : S \text{ is a } (d + \mathbf{1} - f)\text{-tuple dominating set}\} = w^t \mathbf{1},$$

where χ^S stands for the incidence vector of S in $\mathbb{R}^n$: $\chi^S_v = 1$ if $v \in S$ and $\chi^S_v = 0$ otherwise. Given this, we will simply focus on f-tuple domination in the next sections: the derivation of the corresponding results for limited packing are straightforward.

Motivation. The concept of domination naturally arises in location problems for the strategic placement of facilities in a network. A wide variety of applications are presented in [15,16], e.g. sets of representatives, location of radio stations, land surveying, ... Considering the variant of f-tuple domination, and some of its extensions, they arise notably in the design of fault tolerant networks (e.g., [23]). Limited packings arise for the strategic placement of obnoxious facilities [12]: for each location (represented by some node in a graph), we require that no more than some given number of such facilities are placed in its neighborhood.

Related Work. Most of the works we can find in the literature on f-tuple domination focus on complexity and algorithmic aspects of cardinality problems (i.e. node weights are uniform) and when $f_v = k, \forall v \in V$ for some fixed positive integer k. The decision problems corresponding to the minimum cardinality f-tuple dominating set and maximum cardinality limited packing problems are $\mathcal{NP}$-complete, and this holds also when the graph is restricted to be split or bipartite [9,20]. Both cardinality problems can be solved in linear time when the graph is a tree [8,19]. For any $f \in \widehat{\mathcal{F}}_G$, a minimum cardinality f-tuple dominating set can be found in polynomial time in strongly chordal graphs, and in linear time if a strong ordering of the nodes is given [20]. Gallant et al. [12] establish bounds on the maximum cardinality of a k-limited packing and investigate structural properties of graphs for which some bounds are satisfied with equality. A $(ln(|V|) + 1)$-approximation algorithm to find a minimum cardinality k-tuple dominating set in any graph is presented in [17], where it is also shown that this problem cannot be approximated within a ratio of $(1 - \epsilon)ln|V|$ for any $\epsilon > 0$ unless $\mathcal{NP} \subseteq DTIME(|V|^{\log\log|V|})$. Cicalese et al. [5] have shown the minimum cardinality f-tuple domination problem can be solved in polynomial time in graphs with bounded clique-width. Some peculiar graph families for which minimum cardinality f-tuple dominating set and maximum cardinality limited packing problems can be polynomially reduced to each other are investigated in [18].

There are few works dedicated to polyhedral results on f-tuple domination and limited packing problems. And for the existing ones, they essentially deal with the basic case when $f_v = 1, \forall v \in V$. In particular, complete formulations of the dominating set polytope are known for strongly chordal graphs [10], cycles [4], some peculiar webs [3]. Recent investigations on polytopes related to the concept of total f-domination are reported in [7]. Otherwise, there is a much vaster literature on polyhedral aspects related to more general covering and packing concepts and whose survey goes beyond the scope of this paper; see, e.g. [2,6,22]. But to the present author's knowledge, they do not cover the accurate polyhedral results that we report hereafter.

Our Contribution. The contributions of the present paper may be summarized as follows. We provide:

- Complete formulations for the f-tuple dominating set (and thus also for f-limited packing) polytopes of stars.

- A new family of valid inequalities for $\mathcal{U}_G^f$ that can be separated in polynomial time.
- The proof that a conjecture made by Argiroffo [1] on a complete formulation for the 2-tuple dominating set polytope of a tree does not hold.
- Investigations on the adjacency relations of extreme points of the f-tuple dominating set polytope.

The paper is organized as follows. In Sect. 2, we investigate general properties of the polyhedral structure of f-tuple dominating set polytopes. Then, we provide a complete formulation for stars in Sect. 3. A new family of valid inequalities, which led us to disprove a conjecture on a complete formulation for the 2-tuple dominating set polytope of trees, is introduced in Sect. 4. In Sect. 5, we study adjacency properties between extreme points of f-tuple dominating set polytopes, before we conclude in Sect. 6. Note that, due to length restrictions, some proofs are omitted from this version.

2   General Polyhedral Properties

Dimension of $\mathcal{U}_G^f$

Given a polyhedron P, let $dim(P)$ denote its dimension. Let $\mathcal{R}$ denote the set of the nodes whose closed neighborhood belongs to all feasible solutions of $(IP1)$: $\mathcal{R} = \{v \in V : f_v = d_v + 1\}$.

Proposition 1. *Let $G = (V, E)$ denote an undirected simple graph and $f \in \widehat{\mathcal{F}}_G$. Then $dim(\mathcal{U}_G^f) = |V| - |N[\mathcal{R}]|$.*

Remark 2. Let $G' = G[V \setminus \mathcal{R}]$ and define $f' \in \mathcal{F}_{G'}$ and $w' \in \mathbb{R}^{V \setminus \mathcal{R}}$ as follows: $f'_v = f_v - |N(v) \cap \mathcal{R}|, \forall v \in V \setminus \mathcal{R}$, $w'_v = w_v, \forall v \in V \setminus N[\mathcal{R}]$ and $w_v = 0, \forall v \in N(\mathcal{R})$. It is easy to check that an optimal solution to the problem $[MW_f]$ defined by the parameters (G, f, w), can be obtained by the union of the set $N[\mathcal{R}]$ with an optimal solution of a problem having the same form and defined by the parameters (G', f', w'). So, given that the "required" nodes (i.e. the nodes in $\mathcal{R}$) have no peculiar relevance w.r.t. solving $[MW_f]$ neither w.r.t. the polyhedral description of $\mathcal{U}_G^f$ (which is obtained by adding to a description of $\mathcal{U}_{G'}^{f'}$ the variables $(x_v)_{v \in \mathcal{R}}$, together with the set of equations: $x_v = 1, \forall v \in N[\mathcal{R}]$), in what follows, we shall always assume $f \in \mathcal{F}_G$ unless otherwise stated. So, in particular, from Proposition 1, the polytope $\mathcal{U}_G^f$ is full-dimensional.

Trivial Inequalities

Proposition 2. *Let $f \in \mathcal{F}_G$ and $u \in V$. Then, the inequality $x_u \geq 0$ is facet-defining for $\mathcal{U}_G^f$ iff (if and only if) $f_v \leq d_v - 1, \forall v \in N[u]$.*

Proposition 3. *Let $f \in \mathcal{F}_G$ and $u \in V$. Then, the inequality $x_u \leq 1$ is facet-defining for $\mathcal{U}_G^f$.*

Neighborhood Inequalities of Critical Nodes

Definition 1. *Let $f \in \mathcal{F}_G$. A node $v \in V$ is said to be critical if $f_v = d_v$.*

Proposition 4. *Let $f \in \mathcal{F}_G$ and let u denote a critical node. Then, the neighborhood inequality*

$$\sum_{v \in N[u]} x_v \geq f_u (= d_u) \tag{1}$$

is facet-defining for $\mathcal{U}_G^f$.

Proof. The following node subsets are f-tuple dominating sets and their incidence vectors are affinely independent: $V \setminus \{u, w\}$ for all $w \in V \setminus N[u]$, and $V \setminus \{v\}$, for all $v \in N[u]$. □

General Properties of Facet-Defining Inequalities

Proposition 5. *Let $a^t x \geq b$ denote a non trivial facet-defining inequality of $\mathcal{U}_G^f$, with $f \in \mathcal{F}_G$. Let $S = \{v \in V : a_v \neq 0\}$. Then, the following holds:*

(i) $|S| \geq 2$,
(ii) $a_v \geq 0, \forall v \in V$, and $b > 0$.

3   Complete Formulation for Stars

Proposition 6. *Let $G = (V, E)$ be a star with center v_0 and let $a^t x \geq b$ denote a non trivial facet-defining inequality of $\mathcal{U}_G^f$. Then, $a_{v_0} > 0$.*

For the case when the graph $G = (V, E)$ is a star having for center the node v_0 such that $f_{v_0} = 0$, the trivial inequalities together with the neighborhood inequalities of the critical leaves provide a complete formulation of the polytope $\mathcal{U}_G^f$. This follows from the total unimodularity of the corresponding constraint matrix. So, in the rest of this section we assume the center v_0 satisfies $f_{v_0} \geq 1$.

Proposition 7. *Let $G = (V, E)$ denote a star with $V = \{v_0, v_1, \ldots, v_{n-1}\}$, where v_0 denotes the center of the star. Let $f \in \mathcal{F}_G$ such that $f_{v_0} \geq 1$ and let (L_0, L_1) denote a partition of $\{v_1, v_2, \ldots, v_{n-1}\}$ with $L_0 = \{v \in V : f_v = 0\}$, (and $L_1 = V \setminus (\{v_0\} \cup L_0)$) (possibly $L_0 = \emptyset$ or $L_1 = \emptyset$). Then, a complete description of $\mathcal{U}_G^f$ is given by the set of the trivial, the neighborhood inequalities of the nodes in L_1, the following inequality:*

$$[\max(|L_1|, f_{v_0}) - f_{v_0} + 1] x_{v_0} + \sum_{v \in V \setminus \{v_0\}} x_v \geq \max(|L_1|, f_{v_0}), \tag{2}$$

and the inequalities

$$(|L_1| - f_{v_0} + |Z| + 1) x_{v_0} + \sum_{v \in L_1 \cup (L_0 \setminus Z)} x_v \geq |L_1|, \tag{3}$$

for all node subsets $Z \subset L_0$ satisfying $\max(0, f_{v_0} - |L_1|) < |Z| < f_{v_0} - 1$ (in case such a set exists).

Proof. Let $a^t x \geq b$ denote a facet-defining inequality of $\mathcal{U}_G^f$ that is not trivial and different from a neighborhood inequality of a node in L_1. Using the next four claims (whose proofs are omitted here due to length restrictions) we show that it must correspond (up to multiplication by a positive scalar), to an inequality of the type (2) or (3). In what follows we denote by Z_0 the node subset $Z_0 = \{v \in V : a_v = 0\}$. Note that by Proposition 6, $v_0 \notin Z_0$.

Claim 1. The following holds: if $L_1 \neq \emptyset$, then $a_v = c_1, \forall v \in L_1$, where c_1 denotes a fixed positive value.   ⋄

Claim 2. Assume that $Z_0 = \emptyset$. Then $a_v = c_1, \forall v \in V \setminus \{v_0\}$, where c_1 denotes a positive value.   ⋄

Claim 3. Assume that $Z_0 \neq \emptyset$. Then, the following holds: $L_1 \neq \emptyset$ and $f_{v_0} - |L_1| < |Z_0| < f_{v_0} - 1$.   ⋄

Claim 4. Assume that $Z_0 \neq \emptyset$. Then, the following holds: $a_v = c_1, \forall v \in L_1 \cup (L_0 \setminus Z_0)$ (with c_1 as introduced in Claim 1).   ⋄

Making use of the four claims above, we do the proof by establishing a relation between a_{v_0} and c_1. Let D stand for an f-tuple dominating set such that $a^t \chi^D = b$ and $v_0 \notin D$. Then, necessarily $L_1 \subseteq D$, and

- if $Z_0 = \emptyset$: from Claim 2, we deduce $b = \max(f_{v_0}, |L_1|) c_1$.
- if $Z_0 \neq \emptyset$: from Claim 3, we have $a_v = 0, \forall v \in D \setminus L_1$. By Claim 4, we deduce $b = c_1 |L_1|$.

Now let $\tilde{D}$ stand for an f-tuple dominating set such that $a^t \chi^{\tilde{D}} = b$ and $v_0 \in \tilde{D}$. Then,

- if $Z_0 = \emptyset$: $a_{v_0} = c_1(\max(f_{v_0}, |L_1|) - f_{v_0} + 1)$,
- if $Z_0 \neq \emptyset$: $a_{v_0} = c_1(|L_1| - f_{v_0} + 1 + |Z_0|)$.   □

In the next proposition, we address the relevance of the inequalities (2)–(3) from an "optimization" point of view by considering the integrality gap of the formulation $(LP1)$ for the case when G is a star and the weight function corresponds to the left-hand-side of an inequality of the type (2). The same result can be shown to hold if we consider a weight function corresponding to an inequality of type (3).

Proposition 8. Let $G = (V, E)$ denote a star with center v_0. Let $n = |V|$, $n \geq 3$, $f \in \mathcal{F}_G$, $L_1 = \{v \in V \setminus \{v_0\} : f_v = 1\}$, $L_0 = V \setminus (\{v_0\} \cup L_1)$, and let $\widehat{w} \in \mathbb{R}^n$ be defined as follows: $\widehat{w}_{v_0} = \max(|L_1|, f_{v_0}) - f_{v_0} + 1$, and $\widehat{w}_v = 1, \forall v \in V \setminus \{v_0\}$. Then, the integrality gap of $(LP1)$ with an objective function corresponding to $\widehat{w}$ is upper bounded by $\frac{4}{3}$ and this bound is asymptotically tight.

Proof. Firstly note that for the case when $f_{v_0} \leq 1$ or $f_{v_0} \geq |L_1|$, the objective values of $(IP1)$ and $(LP1)$ coincide. So assume from now on that $2 \leq f_{v_0} < |L_1|$. Then, notice that the node set L_0 has no incidence on the optimal objective value of $(LP1)$. Indeed, given the assumption $|L_1| > f_{v_0}$, if x^* denotes any optimal solution to $(LP1)$ with $x_u^* > 0$ for some node $u \in L_0$, then we can easily determine another optimal solution $\widehat{x}^*$ satisfying $\widehat{x}_u^* = 0, \forall u \in L_0$ (decreasing positive entries of x^* indexed on L_0 and increasing entries indexed on L_1 that are lower than 1). So, in what follows and to simplify the presentation, we shall assume $L_0 = \emptyset$ and thus $|L_1| = n - 1$.

Consider the dual problem to $(LP1)$:

$$(D1) \begin{cases} \max \sum_{v \in V} (f_v y_v - z_v) \\ s.t. \\ \quad \sum_{u \in N[v]} y_u - z_v \leq \widehat{w}_v, \forall v \in V, \\ \quad y, z \in \mathbb{R}_+^n. \end{cases}$$

The vector $(\overline{y}, \mathbf{0}) \in (\mathbb{R}_+^n)^2$ defined as follows is feasible for $(D1)$: $\overline{y}_{v_0} = \frac{f_{v_0} - 1}{n - 2}$, $\overline{y}_v = 1 - \overline{y}_{v_0}, \forall v \in V \setminus \{v_0\}$. A feasible solution to $(LP1)$ is given by the vector $\overline{x} \in \mathbb{R}^n$ defined as follows: $\overline{x}_{v_0} = \frac{n - 1 - f_{v_0}}{n - 2}$ and $\overline{x}_v = 1 - \overline{x}_{v_0}, \forall v \in V \setminus \{v_0\}$. Both $(\overline{y}, \mathbf{0})$ and $\overline{x}$ have the same objective value: $z(f_{v_0}) = \frac{1}{n-2}[f_{v_0}(f_{v_0} - 1) + (n - 1)(n - f_{v_0} - 1)]$, and are thus optimal.

Now the quantity $z(f)$ is minimized for $f \in \{\lceil \frac{n}{2} \rceil, \lfloor \frac{n}{2} \rfloor\}$, and we get $z(\frac{n}{2}) = \frac{3n-2}{4}$ if n is even and $z(\lfloor \frac{n}{2} \rfloor) = z(\lceil \frac{n}{2} \rceil) = \frac{(n-1)(3n-5)}{4(n-2)}$ if n is odd. So in both cases the integrality gap is lower than $\frac{4}{3}$ and it converges to $\frac{4}{3}$ as n grows to infinity. $\square$

Proposition 8 can be used to prove the next result on the integrality gap of $(LP1)$ when G is a star.

Proposition 9. *Let the graph G be a star and let $f \in \mathcal{F}_G$. Then, the integrality gap of $(LP1)$ is upper bounded by $\frac{4}{3}$ and this bound is asymptotically tight.*

4 · Further Valid Inequalities

In this section, we introduce valid inequalities which extend the families (2) and (3) for stars. We also investigate on their facet-defining properties, disproving a conjecture on a complete formulation of the 2-tuple dominating set polytope for trees.

We start with a simple extension of the inequalities (2).

Proposition 10. *Let $G = (V, E)$ denote an undirected simple graph and $f \in \mathcal{F}_G$. Given any node u, let $L_0^u = \{v \in N(u) : f_v \leq d_v - 1\}$, and $L_1^u = N(u) \setminus L_0^u$. Then, the following inequality is valid for $\mathcal{U}_G^f$.*

$$[\max(|L_1^u|, f_u) - f_u + 1]x_u + \sum_{v \in N(u)} x_v \geq \max(|L_1^u|, f_u), \forall u \in V. \qquad (4)$$

Proposition 11. *Let $G = (V, E)$ be an undirected simple graph and let $f \in \mathcal{F}_G$. Let $u \in V$ denote an articulation point in G such that $f_u \geq 2$ and all the neighbors of u in G belong to different connected components in $G[V \setminus \{u\}]$. Then the inequality (4) is facet-defining w.r.t. $\mathcal{U}_G^f$.*

We now introduce a new family of inequalities generalizing (3).

Proposition 12. *Let $G = (V, E)$ denote an undirected simple graph and let $f \in \mathcal{F}_G$. Let $u \in V$ denote an articulation point in G such that $f_u \geq 2$ and all the neighbors of u in G belong to different connected components in $G[V \setminus \{u\}]$. Let (Z, L_0, L_1) denote a tripartition of $N(u)$ such that*

(i) $f_u - |L_1| < |Z| < f_u - 1$,
(ii) $f_v \leq d_v - 1, \forall v \in Z \cup L_0$,
(iii) for each node $w \in L_1$, $f_w \geq 1$ and $|Q_w| \geq d_w - f_w$, with $Q_w = \{v \in N(w) \setminus \{u\} : f_v \leq d_v - 1\}$.

For each node $w \in L_1$, let $\overline{N}(u, w)$ denote a subset of $Q_w \cup \{w\}$ such that $w \in \overline{N}(u, w)$ and $|\overline{N}(u, w)| = d_w - f_w + 1$. Then, the following inequality is valid for $\mathcal{U}_G^f$.

$$(|L_1| - f_u + |Z| + 1)x_u + \sum_{w \in L_1} \sum_{v \in \overline{N}(u,w)} x_v + \sum_{v \in L_0} x_v \geq |L_1|. \tag{5}$$

Proposition 13. *If the graph G is a tree, then the inequality (5) is facet-defining w.r.t. $\mathcal{U}_G^f$.*

For an example of a facet-defining inequality of type (5), consider the graph G' of Fig. 1(a) where the node domination requirements $(f_v')_{v \in V}$ correspond to the values close to the nodes. Then the inequality $2x_1 + x_2 + x_3 + x_4 + x_5 + x_7 + x_8 \geq 3$ is facet-defining for $\mathcal{U}_{G'}^{f'}$.

A conjecture formulated by Argiroffo (see Sect. 3 in [1]) on the formulation of the 2-dominating set polytope of trees (i.e. the f-tuple dominating set polytope of trees for the particular case when $f_v = 2, \forall v \in V$) stated that a complete formulation of the 2-tuple dominating set polytope was given by a set of inequalities, each one having a support included in the closed neighborhood of a single node. From Proposition 13, it follows that this conjecture does not hold: the instance illustrated in Fig. 1(b) provides a counterexample since a complete formulation of $\mathcal{U}_G^f$ can be obtained by adding to the one of the instance from Fig. 1(a), the set of equations $x_v = 1, \forall v \in \mathcal{R}$ (see Remark 2 above).

Given a polytope $P \subseteq \mathbb{R}^n$ and a vector $x^* \in \mathbb{R}^n$, the separation problem w.r.t. P consists in determining whether $x^* \in P$ and, if not, in giving an inequality that is valid for P and violated by x^*. W.r.t. the family of inequalities (5), we have the next result.

Proposition 14. *The separation problem w.r.t. the family of inequalities (5) can be solved in polynomial time.*

From Proposition 14 and the equivalence between optimization and separation [13], it follows that the relaxation of $[MW_f]$ obtained by adding the inequalities (5) to $(LP1)$ can be solved in polynomial time and leads to a generally better bound on the optimal objective value.

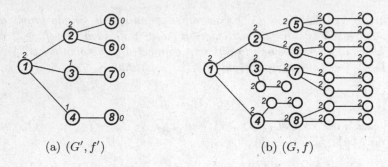

(a) (G', f') (b) (G, f)

Fig. 1. Illustrations for a facet-defining inequality of type (5)

5   On Adjacency Relations and Diameter

It is known that the diameter of any $(0, 1)$-polytope is at most its dimension [21]. This bound is tight in general w.r.t. the polytope $\mathcal{U}_G^f$: considering $f = \mathbf{0}$, $\mathcal{U}_G^f$ corresponds to an hypercube with dimension value n that coincides with its diameter. In what follows we investigate more closely adjacency relations between vertices on this polytope, firstly on general graphs and then for the case of stars, for which a better bound on the diameter can be obtained.

5.1   General Properties

Let G denote an undirected simple graph. The next three propositions provide two simple sufficient conditions and a necessary one for two incidence vectors of f-tuple dominating sets to be adjacent on $\mathcal{U}_G^f$. Given two sets A and B, the notation $A \Delta B$ stands for their symmetric difference, i.e. $A \Delta B = (A \cup B) \backslash (A \cap B)$.

Proposition 15. *Let $f \in \widehat{\mathcal{F}}_G$. Let S_1, S_2 denote two f-tuple dominating sets in G such that $|S_1 \Delta S_2| = 1$. Then, χ^{S_1} and χ^{S_2} are adjacent on $\mathcal{U}_G^f$.*

Proof. Let $w \in \mathbb{R}^V$ be defined as follows

$$w_v = \begin{cases} -1 \text{ if } v \in S_1 \cap S_2 \\ 1 \text{ if } v \in V \backslash (S_1 \cup S_2), \\ 0 \text{ otherwise.} \end{cases}$$

Then, the only optimal solutions to $\min\{w^t x \colon x \in \mathcal{U}_G^f \cap \{0, 1\}^V\}$ are χ^{S_1} and χ^{S_2}. $\square$

Proposition 16. *Let $f \in \widehat{\mathcal{F}}_G$. Let S_1, S_2 denote two f-tuple dominating sets in G such that*

- *$|S_1 \Delta S_2| \geq 2$,*
- *$|S_1 \backslash S_2| = 1$, and*

– $S_i \setminus \{v\}$ is not an f-tuple dominating set for any $v \in S_i \setminus S_j$, for $(i, j) = (1, 2)$ and $(i, j) = (2, 1)$.

Then, χ^{S_1} and χ^{S_2} are adjacent on $\mathcal{U}_G^f$.

Proposition 17. *Let $f \in \widehat{\mathcal{F}}_G$. Let S_1, S_2 denote two f-tuple dominating sets in G such that $|S_1 \triangle S_2| \geq 2$ and the vectors χ^{S_1} and χ^{S_2} are adjacent on $\mathcal{U}_G^f$. Then, $S_i \setminus S_j \neq \emptyset$ and $S_i \setminus \{v\}$ is not an f-tuple dominating set in G, for any $v \in S_i \setminus S_j$, for $(i, j) = (1, 2)$ and $(i, j) = (2, 1)$.*

Proof. Assume, for a contradiction, that $S_1 \setminus \{v\}$ is an f-tuple dominating set, with $v \in S_1 \setminus S_2$. Define $S'_1 = S_1 \setminus \{v\}$, $S'_2 = S_2 \cup \{v\}$. Both S'_1 and S'_2 are f-tuple dominating sets in G. Then, the equation $\chi^{S_1} + \chi^{S_2} = \chi^{S'_1} + \chi^{S'_2}$ implies the result. □

5.2   The Case of Stars

Restricting the family of graphs to stars, the next results provide a characterization of adjacency for any two incidence vectors of f-tuple dominating sets.

Proposition 18. *Let G denote a star with center v_0. Let $f \in \mathcal{F}_G$, let S_1 and S_2 denote two f-tuple dominating sets in G such that either $v_0 \in (S_1 \cap S_2)$ or $v_0 \notin (S_1 \cup S_2)$. Then, χ^{S_1} and χ^{S_2} are adjacent on $\mathcal{U}_G^f$ if and only if either $|S_1 \triangle S_2| = 1$, or the following conditions are all satisfied for $(i, j) = (1, 2)$ and $(i, j) = (2, 1)$:*

(i) $|S_i \setminus S_j| = 1$,
(ii) $S_i \setminus \{v\}$ is not an f-tuple dominating set in G, where $v \in S_i \setminus S_j$.

Proposition 19. *Let G denote a star with center v_0. Let $f \in \mathcal{F}_G$, S_1 and S_2 denote two f-tuple dominating sets in G such that $v_0 \in S_1 \setminus S_2$. Then, χ^{S_1} and χ^{S_2} are adjacent on $\mathcal{U}_G^f$ if and only if either $|S_1 \triangle S_2| = 1$, or the following conditions are all satisfied:*

(i) $S_2 \setminus S_1 \neq \emptyset$,
(ii) $S_i \setminus \{v\}$ is not an f-tuple dominating set in G, for any $v \in S_i \setminus S_j$, $(i, j) = (1, 2)$ and $(i, j) = (2, 1)$,
(iii) $(S_2 \setminus S_1) \cap L_0 \neq \emptyset \Rightarrow S_1 \cap L_0 \subseteq S_2 \cap L_0$, with $L_0 = \{v \in V \setminus \{v_0\}: f_v = 0\}$.

Of peculiar interest w.r.t. $[MW_f]$ are the (inclusionwise) minimal f-tuple dominating sets, since there is always an optimal solution among them (assuming the node weights are nonnegative). In what follows, we take a closer look at the subgraph of the skeleton of $\mathcal{U}_G^f$ that is induced by the nodes corresponding to such dominating sets in stars.

Proposition 20. *Let G denote a star and let $f \in \mathcal{F}_G$. Let S_1, S_2 denote two distinct minimal f-tuple dominating sets in G. Then the distance between the nodes corresponding to these sets in the skeleton of $\mathcal{U}_G^f$ is at most $\lfloor \frac{|S_1 \triangle S_2|}{2} \rfloor$.*

Proof. Consider firstly the case when either $v_0 \in S_1 \cap S_2$ or $v_0 \notin S_1 \cup S_2$. Note that since S_1, S_2 are minimal we necessarily have $|S_1| = |S_2|$ and $|S_1 \Delta S_2|$ even. Let $S_1 \setminus S_2 = \{v_1, v_2, \ldots, v_q\}$, $S_2 \setminus S_1 = \{w_1, w_2, \ldots, w_q\}$. Then, define $D^0 = S_1$ and $D^k = D^{k-1} \Delta \{v_k, w_k\}, \forall k \in \{1, 2, \ldots, q\}$. All the sets $(D^k)_{k=0}^q$ are minimal f-tuple dominating sets in G. By Proposition 16, χ^{D^k} and $\chi^{D^{k-1}}$ are adjacent on $\mathcal{U}_G^f$ and since at each iteration k we have $|D^k \Delta D^{k-1}| = 2$, the result follows.

Consider now the case when $v_0 \in S_1 \setminus S_2$. Note that, since S_1 and S_2 are minimal, $|S_1 \Delta S_2| \geq 2$. Let $Z_2 = (S_2 \cup \{v_0\}) \setminus Q$ with $Q \subseteq S_2 \setminus S_1$ such that $|Q| = \max(1, |L_1| - f_{v_0} + 1)$. Notice that S_2 and Z_2 are minimal f-tuple dominating sets in G to which Proposition 19 applies, so that χ^{S_2} and χ^{Z_2} are adjacent on $\mathcal{U}_G^f$. Considering now the distance between the nodes corresponding to Z_2 and S_1, we can now apply the result derived for the first case (since $v_0 \in Z_2 \cap S_1$) to terminate the proof. □

The next result follows from Proposition 20 and its proof.

Proposition 21. *Let G denote a star with center v_0 and let $f \in \mathcal{F}_G$. Then, the subgraph of the skeleton of $\mathcal{U}_G^f$ that is induced by nodes corresponding to minimal f-tuple dominating sets is connected and has diameter $\mathcal{D}(G) \leq \min(f_{v_0}, \lfloor \frac{n}{2} \rfloor)$.*

6   Conclusion and Perspectives

In this paper we proceeded to investigations related to f-tuple domination and limited packing problems from a polyhedral perspective. We provided a complete formulation for stars, introduced a new family of facet-defining inequalities for trees, disproving a conjecture from the literature. We also presented results on the integrality gap of $(LP1)$ for stars and studied adjacency relations in the 1-skeleton of the f-tuple dominating set polytope.

Future research work may be directed towards the study of the integrality gap of $(LP1)$ and the determination of a complete formulation of the f-tuple dominating set polytope for trees.

References

1. Argiroffo, G.: New facets of the 2-dominating set polytope of trees. In: Proceedings of the 11th SIO, Argentina, pp. 30–34 (2013)
2. Balas, E., Zemel, E.: Graph substitution and set packing polytopes. Networks **7**, 267–284 (1977)
3. Bianchi, S., Nasini, G., Tolomei, P.: The set covering problem on circulant matrices: polynomial instances and the relation with the dominating set problem on webs. Electr. Notes Discrete Math. **36**, 1185–1192 (2010)
4. Bouchakour, M., Contenza, T., Lee, C., Mahjoub, A.: On the dominating set polytope. Eur. J. Comb. **29**, 652–661 (2008)
5. Cicalese, F., Cordasco, G., Gargano, L., Milanic, M., Vaccaro, U.: Latency-bounded target set selection in social networks. Theor. Comput. Sci. **535**, 1–15 (2014)

6. Cornuéjols, G.: Combinatorial Optimization: Packing and Covering. CBMS-NSF Regional Conference Series in Applied Mathematics CBMS 74, SIAM, Philadelphia, PA (2001)

7. Dell'Amico, M., Neto, J.: On total f-domination: polyhedral and algorithmic results. Technical report. University of Modena and Reggio Emilia, Italy (2017)

8. Dobson, M.P., Leoni, V., Nasini, G.: Arbitrarly limited packings in trees. II MACI (2009)

9. Dobson, M.P., Leoni, V., Nasini, G.: The limited packing and multiple domination problems in graphs. Inf. Process. Lett. **111**, 1108–1113 (2011)

10. Farber, M.: Domination, independent domination, and duality in strongly chordal graphs. Discrete Appl. Math. **7**, 115–130 (1984)

11. Gallant, R., Gunther, G., Hartnell, B., Rall, D.: Limited packings in graphs. Electr. Notes Discrete Math. **30**, 15–20 (2008)

12. Gallant, R., Gunther, G., Hartnell, B., Rall, D.: Limited packings in graphs. Discrete Appl. Math. **158**, 1357–1364 (2010)

13. Grötschel, M., Lovász, L., Schrijver, A.: The ellipsoid method and its consequences in combinatorial optimization. Combinatorica **1**(2), 169–197 (1981)

14. Harary, Y., Haynes, T.W.: Double domination in graphs. Ars Combin. **55**, 201–213 (2000)

15. Haynes, T.W., Hedetniemi, S.T., Slater, J.B.: Fundamentals of Domination in Graphs. Marcel Dekker, New York (1998)

16. Haynes, T.W., Hedetniemi, S.T., Slater, J.B.: Domination in Graphs: Advanced topics. Marcel Dekker, New York (1998)

17. Klasing, R., Laforest, C.: Hardness results and approximation algorithms of k-tuple domination in graphs. Inf. Process. Lett. **89**, 75–83 (2004)

18. Leoni, V., Nasini, G.: Limited packing and multiple domination problems: polynomial time reductions. Discrete Appl. Math. **164**, 547–553 (2014)

19. Liao, C.S., Chang, G.J.: Algorithmic aspect of k-tuple domination in graphs. Taiwanese J. Math. **6**(3), 415–420 (2002)

20. Liao, C.S., Chang, G.J.: k-tuple domination in graphs. Inf. Process. Lett. **87**, 45–50 (2003)

21. Naddef, D.: The Hirsh Conjecture is true for $(0,1)$-polytopes. Math. Program. **45**, 109–110 (1989)

22. Padberg, M.W.: On the facial structure of set packing polyhedra. Math. Program. **5**, 199–215 (1973)

23. Shang, W., Wan, P., Yao, F., Hu, X.: Algorithms for minimum m-connected k-tuple dominating set problem. Theor. Comput. Sci. **381**, 241–247 (2007)

Alternating Current Optimal Power Flow with Generator Selection

Esteban Salgado[1], Andrea Scozzari[2], Fabio Tardella[3], and Leo Liberti[1(✉)]

[1] CNRS LIX, École Polytechnique, 91128 Palaiseau, France
esteban.salgado@polytechnique.edu, liberti@lix.polytechnique.fr
[2] Facoltà di Economia, Università degli Studi Niccolò Cusano-Telematica,
Rome, Italy
andrea.scozzari@unicusano.it
[3] MEMOTEF, Facoltà di Economia, Università "La Sapienza", Rome, Italy
fabio.tardella@uniroma1.it

Abstract. We investigate a mixed-integer variant of the alternating current optimal flow problem. The binary variables activate and deactivate power generators installed at a subset of nodes of the electrical grid. We propose some formulations and a mixed-integer semidefinite programming relaxation, from which we derive two mixed-integer diagonally dominant programming approximation (inner and outer, the latter providing a relaxation). We discuss dimensionality reduction methods to extract solution vectors from solution matrices, and present some computational results showing how both our approximations provide tight bounds.

Keywords: Smart grid · Semidefinite programming
Diagonal dominance · Dimensionality reduction

1 Introduction

The Alternating Current Optimal Power Flow (ACOPF) problem is as follows: given an electric power network consisting of nodes (called *buses*) and links (called *lines*) one seeks an optimal generation and distribution plan of active and reactive power under physical constraints (Ohm's and Kirchhoff's laws), and subject to power generation, voltage magnitude and current bounds on each line.

Not every bus can produce power. Those which can are called *generators*. There is often a planning issue related to their activation and deactivation. This is important because of two reasons. First, the minimum amount of power produced by a generator may be a (reasonably large) positive constant, so disabling a generator is not equivalent to keeping it active at minimum levels. Secondly,

This work was partially supported by the PGMO grant "Projet 2016-1751H", by a Siebel Energy Institute Seed Grant, and by Università di Roma "La Sapienza" under a visiting grant for the last author.

J. Lee et al. (Eds.): ISCO 2018, LNCS 10856, pp. 364–375, 2018.
https://doi.org/10.1007/978-3-319-96151-4_31

there can be a cost to keep it activated. Modelling this choice implies the addition of binary variables to the model, which yields a Mixed-Integer Quadratically Constrained Quadratic Programming (MIQCQP) problem.

Most of the recent ACOPF literature [5,9,17,18] ignores the generator activation/deactivation issue: every available generator is always active. Equivalently, every binary variable is fixed to 1, which yields a continuous Quadratically Constrained Quadratic Programming (QCQP) formulation. We were drawn to the mixed-integer ACOPF variant by [22], where the authors also consider the selection of Phase-Shifting Transformers (PST), used to control the flow of real power, as well as *shunts*, which are stabilizing devices. We do not consider PSTs nor shunts in this paper: they require technical data (specifically, more information about the admittance matrices) which we do not possess at this stage. From a theoretical point of view, however, the activation/deactivation of generators and the selection of PSTs yield formulations of the same MIQCQP class, that are likely to require similar solution approaches.

In this paper we study the ACOPF with selection of generators (ACOPFG). Based on the ideas in [1–3], we will derive Mixed-Integer Linear Programming (MILP) formulations using Diagonally Dominant Programming (DDP) for inner and outer approximations for the ACOPF with binary variables. More precisely, we propose the following relaxations:

1. a Mixed-Integer Semidefinite Programming (MISDP) relaxation of the original MIQCQP formulation of the ACOPFG;
2. an *inner* Diagonally Dominant Programming (DDP) [1,10] approximation of the MISDP relaxation;
3. an *outer* DDP relaxation obtained by replacing the primal DD cone in the inner DDP approximation with its dual.

We shall exhibit some computational results showing that DDP yields tight upper and lower bounds on the optimal objective function value of the original MIQCQP.

One of the important reasons for the usefulness of a tight *lower* objective function bound is that ACOPF/ACOPFG problems are sometimes solved as lower-level subproblems of bilevel problems where the upper-level decisions concern unit electricity prices. A non-guaranteed heuristic ACOPF/ACOPFG solution might be detrimental to finding good solutions of the upper-level problem with cutting plane approaches. Since we cannot hope to find a guaranteed global optimum in reasonable times, a tight lower bound represents a good trade-off.

1.1 Notation

We remark first that alternating currents are commonly modelled by means of scalar and vector quantities over the complex field $\mathbb{C}$. On the other hand, physicists have always denoted current by i. The ambiguity with the usual notation for $\sqrt{-1}$ is resolved by denoting the latter by j. Accordingly, we shall refrain from using j as an index (as is common in Mathematical Programming). In agreement with standard notation in complex numbers, we denote complex conjugation by means of a bar over the corresponding quantity: $\bar{\imath}$, for example, is the complex conjugate of the current, rather than $-\sqrt{-1}$. Complex conjugation is applied componentwise to tensors.

We often neglect to explicitly mention matrix sizes in formulations, in order to lighten the notation. Symbols denoting classes of matrices, such as $\mathcal{S}$, $\mathcal{D}$, etc., really stand for "the subset of matrices *of appropriate size* with the corresponding property".

1.2   Contents

The rest of this paper is organized as follows. In Sect. 2 we discuss formulations of the ACOPFG. In Sect. 3 we introduce a MISDP relaxation. In Sect. 4 we present our new inner and outer relaxations based on DDP. In Sect. 5 we briefly discuss dimensionality reduction issues in order to retrieve a good solution from the relaxation output. In Sect. 6 we report computational results.

2   ACOPF Formulations

Consider an electric power network with set of buses $\mathcal{N}$, set of generators $\mathcal{G} \subseteq \mathcal{N}$ and set of lines $\mathcal{L} \subseteq \mathcal{N} \times \mathcal{N}$. The parameters of the problem are:

- $\forall n \in \mathcal{N}$ $S_n^{\mathrm{D}} \triangleq P_n^{\mathrm{D}} + jQ_n^{\mathrm{D}}$ active/reactive power demand at bus n
- $\forall g \in \mathcal{G}$ $S_g^{\min} \triangleq P_g^{\min} + jQ_g^{\min}$ lower bound on power generated at bus g
- $\forall g \in \mathcal{G}$ $S_g^{\max} \triangleq P_g^{\max} + jQ_g^{\max}$ upper bound power generated at bus g
- $\forall n \in \mathcal{N}$ $v_n^{\min}, v_n^{\max}$ bounds on voltage magnitude at bus n
- $\forall (n,m) \in \mathcal{L}$ $i_{nm}^{\max}$ upper bound on current on line (n,m)
- $\quad\quad$ Y bus admittance matrix in $\mathbb{C}^{|\mathcal{N}| \times |\mathcal{N}|}$
- $\quad\quad$ $Y^{\mathrm{f}}, Y^{\mathrm{s}}$ line admittance matrices in $\mathbb{C}^{|\mathcal{L}| \times |\mathcal{N}|}$.

We want to find the optimal values of active and reactive power at each generator $g \in \mathcal{G}$ that is switched on ($S_g^{\mathrm{G}} \triangleq P_g^{\mathrm{G}} + jQ_g^{\mathrm{G}}$). We must also decide the current and the voltage for the nodes and lines of the system. Given that:

$$\forall n \in \mathcal{N} \quad i_n = \sum_{m \in \mathcal{N}} Y_{nm} v_m$$

$$\forall n \in \mathcal{N} \setminus \mathcal{G} \quad v_n \bar{\imath}_n = -S_n^{\mathrm{D}}$$

$$\forall n \in \mathcal{G} \quad -v_n \bar{\imath}_n = S_n^{\mathrm{G}} - S_n^{\mathrm{D}}$$

we can model the problem by only using voltage variables $v_n \in \mathbb{C}$ (for $n \in \mathcal{N}$) [23] and binary variables z_g (for $g \in \mathcal{G}$) representing the activation of a generator to produce power. The formulation of the ACOPFG is as follows:

$$
\left.
\begin{array}{ll}
\displaystyle\min_{\substack{v \in \mathbb{C}^{|\mathcal{N}|} \\ z \in \{0,1\}^{|\mathcal{G}|}}} & f(v,z) \\[1em]
\forall n \in \mathcal{N} \setminus \mathcal{G} & \displaystyle\sum_{m \in \mathcal{N}} v_n \bar{v}_m \bar{Y}_{nm} = -S_n^{\mathrm{D}} \\[1em]
\forall g \in \mathcal{G} & \displaystyle\sum_{m \in \mathcal{N}} v_g \bar{v}_m \bar{Y}_{gm} - S_g^{\max} z_g \leq -S_g^{\mathrm{D}} \\[1em]
\forall g \in \mathcal{G} & \displaystyle\sum_{m \in \mathcal{N}} v_g \bar{v}_m \bar{Y}_{gm} - S_g^{\min} z_g \geq -S_g^{\mathrm{D}} \\[1em]
\forall n \in \mathcal{N} & v_n \bar{v}_n \leq v_n^{\max} \\[0.5em]
\forall n \in \mathcal{N} & v_n \bar{v}_n \geq v_n^{\min} \\[0.5em]
\forall (n,m) \in \mathcal{L} & \displaystyle\sum_{k,\ell \in \mathcal{N}} \bar{v}_k v_\ell \bar{Y}^{\mathrm{f}}_{(n,m)k} Y^{\mathrm{f}}_{(n,m)\ell} \leq i_{nm}^{\max} \\[1em]
\forall (n,m) \in \mathcal{L} & \displaystyle\sum_{k,\ell \in \mathcal{N}} \bar{v}_k v_\ell \bar{Y}^{\mathrm{s}}_{(n,m)k} Y^{\mathrm{s}}_{(n,m)\ell} \leq i_{nm}^{\max},
\end{array}
\right\} \text{(ACOPFG}_\mathbb{C}) \qquad (1)
$$

where f is a real-valued function that is commonly considered as a polynomial on the total active power generated, i.e.:

$$
f(v,z) \triangleq \sum_{g \in \mathcal{G}} \left(c_{g,0} z_g + \sum_{1 \leq \ell \leq L} c_{g,\ell} (P_g^{\mathrm{G}})^\ell \right) = \sum_{g \in \mathcal{G}} \left(c_{g,0} z_g + \sum_{\ell \leq L} c_{g,\ell} \left(\mathsf{Re}\left(\sum_{m \in \mathcal{N}} v_g \bar{v}_m \bar{Y}_{gm} \right) \right)^\ell \right).
$$

We shall consider polynomials which are quadratic in v_g and linear in P_g^{G} (e.g. $L = 1$). This makes (ACOPF$_\mathbb{C}$) a complex-valued MIQCQP.

By doubling the number of the variables (which implies quadrupling the sizes of the matrices in each constraint), the problem can be exactly reformulated into a real-valued MIQCQP:

$$
\left.
\begin{array}{ll}
\displaystyle\min_{\substack{v \in \mathbb{R}^{2|\mathcal{N}|} \\ z \in \{0,1\}^{|\mathcal{G}|}}} & v^\top C v + c^\top z \\[1em]
\forall k \in \mathcal{E} & v^\top A^k v = a_k \\[0.5em]
\forall \ell \in \mathcal{I} & v^\top B^\ell v \leq b_\ell \\[0.5em]
\forall w \in \mathcal{Z} & v^\top Q^w v - q_w^{\max} z_{\lceil w/2 \rceil} \leq q_w \\[0.5em]
\forall w \in \mathcal{Z} & v^\top Q^w v - q_w^{\min} z_{\lceil w/2 \rceil} \geq q_w,
\end{array}
\right\} \text{(ACOPFG}_\mathbb{R}) \qquad (2)
$$

where $\mathcal{E}$ is the set indexing the $2|\mathcal{N} \setminus \mathcal{G}|$ equality constraints, $\mathcal{Z}$ is the set indexing the $2|\mathcal{G}|$ inequalities with binary variables and $\mathcal{I}$ is the set indexing the other $2(2|\mathcal{G}| + 2|\mathcal{N}| + 2|\mathcal{L}|)$ inequality constraints.

We remark that this problem is non-convex. Equation (2) therefore cannot generally be solved globally using local optimization methods (as a convex QCQP would). Indeed, it is well known that the decision problem associated to Eq. (2) is **NP**-hard [24].

3   MISDP Relaxation

Semidefinite Programming (SDP) [4,16] is widely employed in order to derive relaxations of the continuous version of the ACOPF [5,9,18]. We derive a corresponding MISDP relaxation for the ACOPFG. First we rewrite Eq. (2) by replacing the products of the form $v^\top M v$ as $\langle M, V \rangle = \mathsf{tr}(M^\top V)$ and adding the rank constraint $V = vv^\top$:

$$
\left.
\begin{array}{rl}
\min\limits_{V \in \mathcal{S}} & \langle C, V \rangle + c^\top z \\
\forall k \in \mathcal{E} & \langle A^k, V \rangle = a_k \\
\forall \ell \in \mathcal{I} & \langle B^\ell, V \rangle \le b_\ell \\
\forall w \in \mathcal{Z} \ \langle Q^w, V \rangle - q_w^{\max} z_{\lceil w/2 \rceil} \le q_w \\
\forall w \in \mathcal{Z} \ \langle Q^w, V \rangle - q_w^{\min} z_{\lceil w/2 \rceil} \ge q_w \\
& V = vv^\top,
\end{array}
\right\} \text{(ACOPFG}_\mathbb{R})
\tag{3}
$$

where $\mathcal{S}$ is the set of all $n \times n$ symmetric matrices.

Given that V does not encode any integrality constraint, Eq. (3) is essentially a MILP with the additional (non-convex) constraint $V = vv^\top$, which states that V must be positive semidefinite (PSD) and have rank 1. The standard MISDP relaxation is obtained by relaxing the rank 1 constraint to $V - vv^\top \succeq 0$, i.e. by requiring that the Schur complement of V and v, defined as

$$
\mathscr{S}(V, v) = \begin{pmatrix} 1 & v^\top \\ v & V \end{pmatrix}
$$

is PSD:

$$
\left.
\begin{array}{rl}
\min\limits_{\substack{V \in \mathcal{S}, v \in \mathbb{R}^{2|\mathcal{N}|} \\ z \in \{0,1\}^{|\mathcal{G}|}}} & \langle C, V \rangle + c^\top z \\
\forall k \in \mathcal{E} & \langle A^k, V \rangle = a_k \\
\forall \ell \in \mathcal{I} & \langle B^\ell, V \rangle \le b_\ell \\
\forall w \in \mathcal{Z} \ \langle Q^w, V \rangle - q_w^{\max} z_{\lceil w/2 \rceil} \le q_w \\
\forall w \in \mathcal{Z} \ \langle Q^w, V \rangle - q_w^{\min} z_{\lceil w/2 \rceil} \ge q_w \\
& \mathscr{S}(V, v) \succeq 0
\end{array}
\right\} \text{(ACOPFG}_{\mathsf{MISDP}})
\tag{4}
$$

Most SDP solvers are unable to solve MISDP formulations directly (with some exceptions, e.g. Pajarito [21] and PENLAB [11]). A possible workaround consists in reformulating the binary constraints on z by $\forall g \in \mathcal{G}$ ($z_g^2 = z_g$). These can in turn be relaxed, within the SDP framework, with the constraints:

- $\mathsf{diag}(Z) = z$
- $\mathscr{S}(Z, z) \succeq 0$,

This yields the following SDP relaxation:

$$
\left.
\begin{array}{rl}
\min\limits_{\substack{V \in \mathcal{S}, v \in \mathbb{R}^{2|\mathcal{N}|} \\ Z \in \mathcal{S}, z \in \mathbb{R}^{|\mathcal{G}|}}} & \langle C, V \rangle + c^\top z \\
\forall k \in \mathcal{E} & \langle A^k, V \rangle = a_k \\
\forall \ell \in \mathcal{I} & \langle B^\ell, V \rangle \le b_\ell \\
\forall w \in \mathcal{Z} \ \langle Q^w, V \rangle - q_w^{\max} z_{\lceil w/2 \rceil} \le q_w \\
\forall w \in \mathcal{Z} \ \langle Q^w, V \rangle - q_w^{\min} z_{\lceil w/2 \rceil} \ge q_w \\
& \mathsf{diag}(Z) - z = 0 \\
& \mathscr{S}(V, v) \succeq 0 \\
& \mathscr{S}(Z, z) \succeq 0.
\end{array}
\right\} \text{(ACOPFG}_{\mathsf{SDP}})
\tag{5}
$$

While Eq. (5) can be tackled by an SDP solver, there are two issues with it: (a) the fact that today, current SDP solver technology is far from allowing the systematic solution of ACOPF instances of even moderate sizes (into the hundreds or thousands of nodes and beyond); (b) the proposed SDP relaxation of the binary variables usually yields very poor bounds. Instead, we shall investigate below some MILP formulations derived from Eq. (5). While MILP is also **NP**-hard, its state-of-the-art solvers are much faster than MIQCQP solvers.

4   Diagonal Dominance

A real square symmetric matrix $Y = (Y_{j\ell})$ is *diagonally dominant* (DD) if

$$\forall j \quad Y_{jj} \geq \sum_{\ell \neq j} |Y_{j\ell}|. \tag{6}$$

The main observation leading to our tight MILP relaxations of the MIQCQP is the well-known fact that *every diagonally dominant matrix is PSD* [13]. This means that by replacing the constraint $\mathscr{S}(V, v) \succeq 0$ with membership in the cone of DD matrices (also called the *DD cone*) we obtain an inner approximation of the SDP cone. We restrict our attention to two formulations of these Diagonally Dominant Programs (DDP) applied to $(\mathsf{ACOPFG_{MISDP}})$: a formulation based on the extreme rays of the DD cone [6], and the *outer approximation* obtained by replacing the extreme rays of the DD cone with the extreme rays of the dual DD cone. We devote particular attention to the latter, since it has two desirable properties: (i) it provides a relaxation of the MISDP, and (ii) we found empirically that it provides tight lower bounds.

From Eq. (4) we derive the inner approximation:

$$\left. \begin{array}{rl} \min_{V \in \mathcal{S}, \, v \in \mathbb{R}^{2|\mathcal{N}|}, \, z \in \{0,1\}^{|\mathcal{G}|}} & \langle C, V \rangle + c^\top z \\ \forall k \in \mathcal{E} & \langle A^k, V \rangle = a_k \\ \forall \ell \in \mathcal{I} & \langle B^\ell, V \rangle \leq b_\ell \\ \forall w \in \mathcal{Z} \ \langle Q^w, V \rangle - q_w^{\max} z_w \leq q_w \\ \forall w \in \mathcal{Z} \ \langle Q^w, V \rangle - q_w^{\min} z_w \geq q_w \\ & V \in \mathcal{D}, \end{array} \right\} (\mathsf{ACOPFG_{inner}}) \tag{7}$$

where $\mathcal{D}$ denotes the DD cone. Given that the decision variable vector v only appears in the conic constraint $\mathscr{S}(V, v) \succeq 0$, we can replace the latter by $V \succeq 0$ without modifying the feasible region. We then inner-approximate it by means of the constraint $V \in \mathcal{D}$, which is equivalent to Eq. (6) and can be written as follows using linear constraints:

$$\left. \begin{array}{c} \forall j \quad V_{jj} \geq \sum_{\ell \neq j} Z_{j\ell} \\ -Z \leq V \leq Z. \end{array} \right\} \tag{8}$$

By [6] and [2, Lemma 3.2], we know that the extreme rays of $\mathcal{D}$ are square matrices defined as follows: $\forall j, \ell$ such that $j < \ell$,

- $E_j = e_j^\top e_j$
- $E_{j\ell}^+ = (e_j + e_\ell)^\top (e_j + e_\ell)$ (9)
- $E_{j\ell}^- = (e_j - e_\ell)^\top (e_j - e_\ell)$

In other words, the rays either have one nonzero entry in the diagonal which is equal to 1 or have a single nonzero principal minor

$$\begin{pmatrix} 1 & \pm 1 \\ \pm 1 & 1 \end{pmatrix}$$

where, for $j < \ell$, the ones on the diagonal are in positions (j, j) and (ℓ, ℓ), and the ± 1 components are in positions (j, ℓ) and (ℓ, j). The main relevant result is that every matrix $M \in \mathcal{D}$ can be written as a non-negative combination of these extreme rays:

$$\exists \delta_j, \delta_{j\ell}^+, \delta_{j\ell}^- \geq 0 \quad M = \sum_j \delta_j E_j + \sum_{j < \ell} (\delta_{j\ell}^+ E_{j\ell}^+ + \delta_{j\ell}^- E_{j\ell}^-). \tag{10}$$

This allows us to write membership of a square symmetric matrix M in the dual DD cone $\mathcal{D}^*$ by means of linear constraints as stated in the following result.

Lemma 4.1 (From paper [23]). *The dual DD cone $\mathcal{D}^*$ can be written as follows:*

$$\mathcal{D}^* = \{M \in \mathcal{S} \mid \forall x \in \mathcal{X} \ (x^\top M x \geq 0)\}, \tag{11}$$

where $\mathcal{X} = \{(e_j)_j, (e_j \pm e_\ell)_{j < \ell}\}$ is a set of $|\mathcal{N}| + 2\binom{|\mathcal{N}|}{2} = |\mathcal{N}|^2$ elements.

From Eq. (4) we derive the outer approximation:

$$\left. \begin{aligned} \min_{\substack{V \in \mathcal{S}, v \in \mathbb{R}^{2|\mathcal{N}|} \\ z \in \{0,1\}^{|\mathcal{G}|}}} \quad & \langle C, V \rangle + c^\top z \\ \forall k \in \mathcal{E} \quad & \langle A^k, V \rangle = a_k \\ \forall \ell \in \mathcal{I} \quad & \langle B^\ell, V \rangle \leq b_\ell \\ \forall w \in \mathcal{Z} \ \langle Q^w, V \rangle - & q_w^{\max} z_{\lceil w/2 \rceil} \leq q_w \\ \forall w \in \mathcal{Z} \ \langle Q^w, V \rangle - & q_w^{\min} z_{\lceil w/2 \rceil} \geq q_w \\ & V \in \mathcal{D}^*. \end{aligned} \right\} \text{(ACOPFG}_{\text{outer}}) \tag{12}$$

By our description of the extreme rays of $\mathcal{D}$, the constraint $V \in \mathcal{D}^*$ can be written using Lemma 4.1. Equation (12) is an outer approximation of Eq. (4) because $\mathcal{D} \subseteq \mathcal{S}^+ = (\mathcal{S}^+)^* \subseteq \mathcal{D}^*$, where $\mathcal{S}^+$ is the cone of PSD matrices.

4.1 Iterative Inner Approximation

There are two potential issues with the inner approximation Eq. (7): (a) given that we are modifying the conic constraints of Eq. (4) by an inner approximation of the original cone, the resulting problem may be infeasible; (b) even if the problem turns out to be feasible, the solution may not be a good approximation of the MISDP.

The authors of [1,2] introduce a possible way around these issues by means of iteratively solving a sequence of auxiliary problems, until a cone $\mathcal{C} \subseteq \mathcal{S}^+$ is found on which Eq. (4) is feasible when replacing the conic constraint $V \in \mathcal{D}$ by $V \in \mathcal{C}$. The auxiliary problems are obtained by varying the (square matrix) parameter U in the formulation below:

$$
\left.
\begin{aligned}
\min_{\substack{V \in \mathcal{S}, z \in \{0,1\}^{|\mathcal{G}|} \\ \delta \in (\mathbb{R}^+)^{4|\mathcal{N}|^2}}} \quad & \langle C, V \rangle + c^\top z \\[4pt]
\forall k \in \mathcal{E} \quad & \langle A^k, V \rangle = a_k \\
\forall \ell \in \mathcal{I} \quad & \langle B^\ell, V \rangle \le b_\ell \\
\forall w \in \mathcal{Z} \ \langle Q^w, V \rangle - & q_w^{\max} z_{\lceil w/2 \rceil} \le q_w \\
\forall w \in \mathcal{Z} \ \langle Q^w, V \rangle - & q_w^{\min} z_{\lceil w/2 \rceil} \ge q_w \\
U^\top \left(\sum_{x \in \mathcal{X}} \delta_x x x^\top \right) U & = V.
\end{aligned}
\right\} \ (\text{ACOPFG}_{\text{inner}}(U)) \qquad (13)
$$

We remark that, while $U^\top \left(\sum_{x \in \mathcal{X}} \delta_x x x^\top \right) U$ is an appropriate description of $V \in \mathcal{D}^*$, implementation performances improve if we use an additional auxiliary matrix variable:

$$
\begin{aligned}
\forall i \quad W_{ii} &= \delta_i + \sum_{j \neq i} (\delta_{ij}^+ + \delta_{i,j}^-) \\
\forall i,j \quad W_{ij} &= \delta_{ij}^+ - \delta_{ij}^- \\
V &= U^\top W U
\end{aligned}
\qquad (14)
$$

The iterative procedure is the following:

1. Solve ($\text{ACOPFG}_{\text{inner}}$), obtaining the solution V^*;
2. Define $U = \text{chol}(V^*)$;
3. Solve ($\text{ACOPFG}_{\text{inner}}(U)$) to obtain V^*. While the solution is improving, repeat from Step 2.

At the k-th iteration of this procedure, we consider the parametrization matrix U_k and the optimum V_k^* of ($\text{ACOPFG}_{\text{inner}}(U_k)$). Then the following holds [1,2].

Proposition 4.2. V_k^* *is feasible for* ($\text{ACOPFG}_{\text{inner}}(U_{k+1})$) *and cannot worsen the current objective value:* $\text{val}(\text{ACOPFG}_{\text{inner}}(U_{k+1})) \le \text{val}(\text{ACOPFG}_{\text{inner}}(U_k))$. *Moreover, if* V_k^* *is PSD and* $\text{val}(\text{ACOPFG}_{\text{inner}}(U_k))$ *is not optimal, then* V_k^* *is improving, i.e.* $\text{val}(\text{ACOPFG}_{\text{inner}}(U_{k+1})) < \text{val}(\text{ACOPFG}_{\text{inner}}(U_k))$.

In particular, we know that if we start from a feasible problem, the iterated problems will always remain feasible. To find an initial U such that the parametrized problem is feasible, we solve a slightly different formulation, where we add a slack variable that we minimize in order to attempt to achieve feasibility:

$$
\left.
\begin{aligned}
&\min_{\substack{V \in \mathcal{S},\, z \in \{0,1\}^{|\mathcal{G}|} \\ \delta \in (\mathbb{R}+)^{4|\mathcal{N}|^2},\, \alpha \in \mathbb{R}^{4|\mathcal{N}|^2}}} \qquad\qquad \alpha \\
&\forall k \in \mathcal{E} \qquad\qquad \langle A^k, V + \alpha \mathbb{I} \rangle = a_k \\
&\forall \ell \in \mathcal{I} \qquad\qquad \langle B^\ell, V + \alpha \mathbb{I} \rangle \leq b_\ell \\
&\forall w \in \mathcal{Z} \; \langle Q^w, V + \alpha \mathbb{I} \rangle - q_w^{\max} z_w \leq q_w \\
&\forall w \in \mathcal{Z} \; \langle Q^w, V + \alpha \mathbb{I} \rangle - q_w^{\min} z_w \geq q_w \\
&\qquad U^\top \left(\sum_{x \in \mathcal{X}} \delta_x x x^\top \right) U = V + \alpha \mathbb{I}.
\end{aligned}
\right\} \; (\mathsf{ACOPFG_{ph1}}(U)) \;\; (15)
$$

Algorithmically, we start with $U = \mathbb{I}$. While $\alpha^* > 0$, we set $U = \mathrm{chol}(V^* + \alpha^* \mathbb{I})$. We solve $(\mathsf{ACOPF_{ph1}}(U))$ using the procedure previously stated.

The pseudocode for our inner approximation algorithm is given in Algorithm 1.

Algorithm 1. Iterative inner approximation

1: **procedure** INNERACOPFMI
2:     $\epsilon \leftarrow 10^{-5}$, AUX $\leftarrow 1$, $U \leftarrow \mathbb{I}$
3:     *# Achieve feasible U first*
4:     **while** AUX > 0 **do**
5:         $(V^*, z^*, \alpha^*) \leftarrow$ optimal solution of $(\mathsf{ACOPFG_{ph1}}(U))$
6:         AUX $\leftarrow \alpha^*$, $U \leftarrow$ CHOL$(X + \alpha^* \mathbb{I})$
7:     VALF $\leftarrow +\infty$, VALI $\leftarrow 0$
8:     *# Now improve the approximation*
9:     **while** $|\text{VALI} - \text{VALF}| > \epsilon$ **do**
10:         VALI $\leftarrow$ VALF
11:         $((V^*, z^*), \text{VALF}) \leftarrow$ optimal solution and value of $(\mathsf{ACOPFG_{inner}}(U))$
12:         $U \leftarrow$ CHOL(V^*)
    return $((V^*, z^*), \text{VALF})$

4.2 Negative Rank of Outer Approximation

One of the issues with Eq. (12) is that, being an outer approximation of Eq. (4), it does not ensure that $V \succeq 0$. In fact, we empirically found that, although the bound was often tight, the solution V^* had a considerable number of negative eigenvalues. We therefore propose to add the following cutting planes to the formulation: for each eigenvector p of V^* corresponding to a strictly negative eigenvalue we add the following inequality to the formulation Eq. (12):

$$
p^\top V p \geq 0. \tag{16}
$$

These cuts make V^* infeasible since, by definition of negative eigenvector, we have that $p^\top V^* p < 0$. Unfortunately, we found that in practice these cuts, by themselves, do not yield a PSD solution. Nevertheless, this procedure decreases the number and/or the sum of the strictly negative eigenvalues.

The pseudocode for our outer approximation algorithm is given in Algorithm 2. The function SPECTRALDECOMPOSITION() returns the spectral decomposition of its argument.

Algorithm 2. Iterative outer approximation

1: **procedure** OUTERACOPFMI
2:     $P \leftarrow \emptyset$
3:     $\lambda_- \leftarrow 1$
4:     **while** $\lambda_- > 0$ **and** TIME < MAXTIME **do**
5:         $((V^*, z^*), \text{VALF}) \leftarrow$ optimal solution of $(\text{ACOPFG}_{\text{outer}})$
6:         $(W, \Lambda) \leftarrow$ SPECTRALDECOMPOSITION(V^*)
7:         $\lambda_- \leftarrow$ number of negative eigenvalues
8:         $P_- \leftarrow$ eigenvectors associated to the negative eigenvalues
9:         $P \leftarrow P \cup P_-$
    return (X^*, z^*, VALF)

---

5 Dimensionality Reduction

All our methods solve either inner or outer matrix approximations of the original MIQCQP Eq. (2), derived from Eq. (3) by relaxing the rank constraint. As such, they will not provide a solution vector v^* for the voltage in the original problem Eq. (2), but rather a solution matrix V^*, which will be very unlikely to have rank 1.

In order to heuristically extract a vector v^* from a matrix V^* with higher rank we considered two options for a dimensionality reduction algorithm: (a) Principal Component Analysis (PCA) [15] and (b) Barvinok's naive algorithm [7, Sect. 5]. Both produce some estimate v' of v^* according to different analyses. We then used v' as a starting point for a local descent carried out by a local Nonlinear Programming (NLP) solver.

Both options are simple and effective in different settings. PCA is well known and needs no introduction. Barvinok's algorithm is as follows:

1. factor V^* into $V^* = F F^\top$;
2. sample a random vector y componentwise from a standard Gaussian distribution $N(0, 1)$;
3. let $v' = Fy$.

The analysis carried out by A. Barvinok shows that v' has high probability of being "not too far" from the feasible set. We remark that we applied Barvinok's algorithm to a wrong setting (in general), since the analysis only holds whenever V^* is a solution of an SDP, and some of our algorithms do not ensure PSD solutions. This being "just a heuristic", we proceeded nonetheless, based on the fact that the formulations that produce V^* are themselves derived from SDP formulations.

Some empirical tests showed a promise of superiority of Barvinok's naive algorithm to PCA (also see [19]), which was therefore our dimensionality reduction method of choice in obtaining computational results (Sect. 6).

6 Numerical Results

We tested our approach using Cplex-12.6.3 [14] on YALMIP [20] for the inner and outer approximations. The MIQCQP was solved with Bonmin-4.8.4 [8] on AMPL [12]. Our results were obtained on an Intel i7 dual-core CPU at 2.1 GHz with 15 GB RAM.

The time limits for decreasing the negative rank (outer approximation) and feasible solution improvement (inner approximation) were set to 300 s. The limit of time for solving MIQCQPs was set to 1200 s (Table 1).

Table 1. ACOPF inner and outer MILP relaxations.

| Instance | Inner-approximation | | | | | | | | Outer-approximation | | | | | | MIQCQP |
| | feasible sol | | optimal sol | | | | | | first value | time | final value | itn | time | rank | best value |
	itn	time	first value	time	value	itn	time	rank							
WB2	2	1.34	880.97	0.63	**877.78**	6	3.42	1	876.92	0.52	**877.75**	57	26.50	2	**878.182**
WB3	18	10.52	445.56	0.54	**417.30**	15	8.01	1	398.71	0.64	**417.17**	52	26.45	2	**417.244**
WB5	5	2.17	1209.96	0.42	**946.69**	76	31.82	1	676.76	0.42	**945.96**	115	72.93	2	**946.584**
6ww	50	37.55	3156.02	0.76	**3156.02**	1	0.76	2	2639.93	0.70	**3009.20**	155	224.86	2	**3018.52**
case9	18	12.90	2041.87	0.76	**1272.64**	45	33.89	2	980.00	0.87	**994.96**	135	301.00	11	**1021.26**
case14	46	62.96	10742.05	1.50	**5778.53**	37	55.56	2	4746.00	1.41	**4746.00**	67	301.72	28	**5265.39**
case30	-	-	-	-	-	-	-	-	285.77	4.69	**285.78**	13	317.62	60	**344.11**
case57	-	-	-	-	-	-	-	-	25460.00	16.91	**25460.00**	4	682.35	114	**25460.00**
case89pegase	-	-	-	-	-	-	-	-	5730.15	60.24	5730.15	3	643.17	178	(x)
case118	-	-	-	-	-	-	-	-	96520.00	110.52	96520.00	2	807.43	192	(x)
case300	-	-	-	-	-	-	-	-	392021.18	2944.12	392021.18	1	2944.12	430	(x)

As we can observe the bounds we obtained are tight with respect to the feasible solution we obtained for the MIQCQP formulation. In general, inner approximations seem to provide solutions with a smaller rank than outer ones. Unfortunately, we could not scale these experiments to larger sizes because of the slow convergence of the loop from Step 4 to Step 6 of Algorithm 1. It is for this reason that we write "-" on some instances for the inner approximation. When we write "(x)" we mean that the local solver was not able to find a local optimum in the time limit.

References

1. Ahmadi, A., Hall, G.: Sum of squares basis pursuit with linear and second order cone programming. Technical report. 1510.01597v1, arXiv (2015)
2. Ahmadi, A., Hall, G.: Sum of squares basis pursuit with linear and second order cone programming. Contemporary Mathematics (to appear)
3. Ahmadi, A., Majumdar, A.: DSOS and SDSOS optimization: more tractable alternatives to sum of squares and semidefinite optimization. Technical report. 1706.02586v1, arXiv (2017)

4. Alizadeh, F.: Interior point methods in semidefinite programming with applications to combinatorial optimization. SIAM J. Optim. **5**(1), 13–51 (1995)
5. Bai, X., Wei, H., Fujisawa, K., Wang, Y.: Semidefinite programming for optimal power flow problems. Electr. Power Energy Syst. **30**, 383–392 (2008)
6. Barker, G., Carlson, D.: Cones of diagonally dominant matrices. Pac. J. Math. **57**(1), 15–32 (1975)
7. Barvinok, A.: Measure concentration in optimization. Math. Program. **79**, 33–53 (1997)
8. Bonami, P., Lee, J.: BONMIN user's manual. Technical report, IBM Corporation, June 2007
9. Chen, C., Atamtürk, A., Oren, S.: Bound tightening for the alternating current optimal power flow problem. IEEE Trans. Power Syst. **31**(5), 3729–3736 (2016)
10. Dias, G., Liberti, L.: Diagonally dominant programming in distance geometry. In: Cerulli, R., Fujishige, S., Mahjoub, A.R. (eds.) ISCO 2016. LNCS, vol. 9849, pp. 225–236. Springer, Cham (2016). https://doi.org/10.1007/978-3-319-45587-7_20
11. Fiala, J., Kočvara, M., Stingl, M.: PENLAB: a MATLAB solver for nonlinear semidefinite optimization. ArXiv e-prints, November 2013
12. Fourer, R., Gay, D.: The AMPL Book. Duxbury Press, Pacific Grove (2002)
13. Gershgorin, S.: Über die abgrenzung der eigenwerte einer matrix. Bulletin de l'Académie des Sciences de l'URSS. Classe des sciences mathématiques et na (6), 749–754 (1931)
14. IBM: ILOG CPLEX 12.6 User's Manual. IBM (2014)
15. Jolliffe, I.: Principal Component Analysis, 2nd edn. Springer, Berlin (2010)
16. Klerk, E.D.: Aspects of Semidefinite Programming. Applied Optimization, vol. 65. Kluwer, Dordrecht (2004)
17. Kuang, X., Ghaddar, B., Naoum-Sawaya, J., Zuluaga, L.: Alternative LP and SOCP hierarchies for ACOPF problems. IEEE Trans. Power Syst. **32**(4), 2828–2836 (2017)
18. Lavaei, J., Low, S.: Zero duality gap in optimal power flow problem. IEEE Trans. Power Syst. **27**(1), 92–107 (2012)
19. Liberti, L., Vu, K.: Barvinok's naive algorithm in distance geometry. Technical report. Working paper, CNRS & Ecole Polytechnique (2017)
20. Löfberg, J.: YALMIP: a toolbox for modeling and optimization in MATLAB. In: Proceedings of the International Symposium of Computer-Aided Control Systems Design, CACSD, vol. 1. IEEE, Piscataway (2004)
21. Lubin, M., Yamangil, E., Bent, R., Vielma, J.P.: Extended formulations in mixed-integer convex programming. In: Louveaux, Q., Skutella, M. (eds.) IPCO 2016. LNCS, vol. 9682, pp. 102–113. Springer, Cham (2016). https://doi.org/10.1007/978-3-319-33461-5_9
22. Ruiz, M., Maeght, J., Marié, A., Panciatici, P., Renaud, A.: A progressive method to solve large-scale ac optimal power flow with discrete variables and control of the feasibility. In: 2014 Power Systems Computation Conference (PSCC). IEEE, Piscataway (2014)
23. Salgado, E.: Fast relaxations for Alternating Current Optimal Power Flow. Master's thesis, MPRO, Université Paris-Saclay (2017)
24. Vavasis, S.: Quadratic programming is in NP. Inf. Process. Lett. **36**, 73–77 (1990)

Parameterized Algorithms
for Module Map Problems

Frank Sommer$^{(\boxtimes)}$ (iD) and Christian Komusiewicz$^{(\boxtimes)}$ (iD)

Fachbereich Mathematik und Informatik,
Philipps-Universität Marburg, Marburg, Germany
{fsommer,komusiewicz}@informatik.uni-marburg.de

Abstract. We introduce and study the NP-hard MODULE MAP problem
which has as input a graph G with red and blue edges and asks to
transform G by at most k edge modifications into a graph which does
not contain a two-colored K_3, that is, a triangle with two blue edges
and one red edge, a blue P_3, that is, a path on three vertices with two
blue edges, and a two-colored P_3, that is, a path on three vertices with
one blue and one red edge, as induced subgraph. We show that MODULE
MAP can be solved in $\mathcal{O}(2^k \cdot n^3)$ time on n-vertex graphs and present a
problem kernelization with $\mathcal{O}(k^2)$ vertices.

1 Introduction

Graphs are a useful tool for many tasks in data analysis such as graph-based
data clustering or the identification of important agents and connections in social
networks. In graph-based data clustering, the edges in the graph indicate similarity between the objects that are represented by the vertices. The goal is
to obtain a partition of the vertex set into clusters such that the objects inside
each cluster should be similar to each other and objects between different clusters
should be dissimilar. One of the central problems in this area is called CLUSTER
EDITING [3], also known as CORRELATION CLUSTERING [18].

CLUSTER EDITING
Input: An undirected graph $G = (V, E)$ and a non-negative integer k.
Question: Can we transform G into a *cluster graph*, that is, a disjoint
union of cliques, by deleting or adding at most k edges?

Here, we essentially view the clustering problem as a graph modification problem:
If we can transform G into a cluster graph G' by at most k edge modifications,
then the connected components of G' define a partition of V into clusters such
that at most k edges of G contradict this partition; these are exactly the deleted
and inserted edges. In recent years, there has been an increased focus to model
the observed data more precisely by incorporating different edge types. As a consequence, many data analysis tasks are now carried out on graphs with multiple
edge types [7,15]. In this work, we study a generalization of CLUSTER EDITING
in graphs with two types of edges.

© Springer International Publishing AG, part of Springer Nature 2018
J. Lee et al. (Eds.): ISCO 2018, LNCS 10856, pp. 376–388, 2018.
https://doi.org/10.1007/978-3-319-96151-4_32

Module Maps. The problem arises in the construction of so-called module maps in computational biology [2,19]. Here, the input is a two-edge-colored graph $G = (V, E_b, E_r)$ with a set E_b of *blue* edges and a set E_r of *red* edges. In the following, we will refer to these objects simply as graphs. The vertices of G represent genes of an organism, the blue edges represent physical interactions between the proteins that are built from these genes, and the red edges represent genetic interactions between the genes. These may be inferred, for example from a high correlation of expression levels of the genes [2]. In the biological application, the task is to find modules which are groups of genes that have a common function in the organism.

According to Amar and Shamir [2], the following properties are desirable for these modules: First, each module should be highly connected with respect to the physical protein interactions. In other words, within each module there should be many blue edges. Second, there should be few physical interactions and, thus, few blue edges between different modules. Third, two different modules A and B may have a *link* between them. If they have a link, then there are many genetic interactions and, thus, many red edges between them; otherwise, there are few genetic interactions and, thus, few red edges between them. Amar and Shamir [2] discuss different objective functions for obtaining a module map that take these properties into account.

We study the problem of obtaining module maps from a graph modification point of view in the same spirit as CLUSTER EDITING is a canonical graph modification problem for graph clustering. That is, we first define formally the set of *module graphs* which are the graphs with a perfect module map. Then, the computational problem is to find a module graph that can be obtained from the input graph by few edge modifications.

Module Graphs. By the above, each module is ideally a blue clique and there are no blue edges between different modules. In other words, the blue subgraph $G_b := (V, E_b)$ obtained by discarding all red edges is a cluster graph. Each connected component of G_b is called a *cluster*, and we say that a graph G where G_b is a cluster graph fulfills the *cluster property*. Moreover, ideally for each pair of different clusters A and B there are either no edges between $u \in A$ and $v \in B$ or each $u \in A$ and each $v \in B$ are connected by a red edge. In other words, the graph $G_r[A \cup B]$ is either edgeless or complete bipartite with parts A and B, where $G_r := (V, E_r)$ is the red subgraph obtained by discarding all blue edges. This property is called *link property*, and the red bicliques are called *links*. The link property is only defined for graphs that fulfill the cluster property. A graph has a perfect module map if it satisfies both properties.

Definition 1. *A graph $G = (V, E_b, E_r)$ is a* module graph *if G satisfies the cluster property and the link property.*

A module graph is shown in Fig. 1. Clearly, not every graph is a module graph. For example a graph G with three vertices u, v, and w where the edges $\{u, v\}$ and $\{u, w\}$ are blue and the edge $\{v, w\}$ is red, violates the cluster property. Our

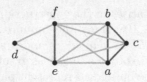

Fig. 1. A module graph with the clusters $\{a, b, c\}$, $\{d\}$, and $\{e, f\}$.

aim is to find a module graph which can be obtained from the input graph G by as few edge transformations as possible.

MODULE MAP
Input: A graph $G = (V, E_b, E_r)$ and a non-negative integer k.
Question: Can we transform G into a module graph by deleting or adding at most k red and blue edges?

Herein, to transform a blue edge into a red edge, we first have to delete the blue edge and in a second step we may insert the red edge, thus transforming a blue edge into a red edge has cost two and vice versa.

As in the case of CLUSTER EDITING, the module graph that is obtained by at most k edge modifications directly implies a partitioning of the input vertex set into clusters such that at most k vertex pairs contradict the input vertex pairs. Here, a contradiction is a red edge or a non-edge inside a cluster, a blue edge between different clusters, or a non-edge between different clusters that have a link and a red edge between different clusters that have no link. Our problem formulation is thus related to previous ones [2, 19] but more simplistic: for example it does not use statistically defined p-values to determine whether a link between modules should be present or not. As observed previously [2, 19] most formulations of the construction problem for module maps contain CLUSTER EDITING as a special case. This is also true for MODULE MAP: if the input has no red edges, then it is not necessary to add red edges, and thus MODULE MAP is the same as CLUSTER EDITING.

As a consequence, hardness results for CLUSTER EDITING transfer directly to MODULE MAP. Since CLUSTER EDITING is NP-complete [17] and cannot be solved in $2^{o(|V|+|E|)}$ time under a standard complexity-theoretic assumption [11, 16] we observe the following.

Proposition 1. MODULE MAP *is NP-complete and cannot be solved in* $2^{o(|V|+|E|)}$ *time unless the Exponential-Time Hypothesis (ETH) fails.*

Because of this algorithmic hardness, heuristic approaches are used in practice [2, 19]. In this work, we are interested in exact algorithms for MODULE MAP. In particular, we are interested in fixed-parameter algorithms that have a running time of $f(k) \cdot n^{O(1)}$ for a problem-specific parameter k. If k has moderate values and f grows not too fast, then these algorithms solve the problem efficiently [9]. Motivated by the practical success of fixed-parameter algorithms with the natural parameter number k of edge transformations for

CLUSTER EDITING [6,13], we focus on fixed-parameter algorithms for MODULE MAP with the same parameter. We find that viewing MODULE MAP as a graph modification problem facilitates the algorithmic study of the problem.

A Weighted Problem Variant. In practice, it is useful to consider edge-weighted versions of the problem, where the input includes a weight function $g : \binom{V}{2} \rightarrow \mathbb{N}^+$ on vertex pairs. The higher the weight, the more confidence we have in the observed edge type. To obtain the cost of a set of edge deletions and additions, we multiply each edge modification $\{u, v\}$ with the weight $g(\{u, v\})$. For example, a blue edge $\{u, v\}$ with weight ω can be transformed into a non-edge with cost ω and into a red edge with cost 2ω. This gives the following problem:

> WEIGHTED MODULE MAP
> **Input:** A graph $G = (V, E_b, E_r)$ with edge weights $g : \binom{V}{2} \rightarrow \mathbb{N}^+$ and a non-negative integer k.
> **Question:** Can we transform G into a module graph by edge transformations of cost at most k?

Our Results. In Sect. 2, we present a characterization of module graphs by three forbidden induced subgraphs and show how to determine whether a graph G contains one of these in linear time. This implies a simple linear-time fixed-parameter algorithm for MODULE MAP with running time $\mathcal{O}(3^k \cdot (|V| + |E|))$, where $|E| = |E_b| + |E_r|$.

In Sect. 3, we present an improved (in terms of the exponential running-time part) fixed-parameter algorithm for WEIGHTED MODULE MAP with running time $\mathcal{O}(2^k \cdot |V|^3)$. This algorithm is an extension of a previous algorithm for WEIGHTED CLUSTER EDITING [5]. In order to transfer the technique to WEIGHTED MODULE MAP, we solve a more general variant of WEIGHTED MODULE MAP that uses a condensed view of the modification costs of an edge in terms of cost vectors. Here, each possible type of a vertex pair (blue edge, red edge, or non-edge) corresponds to one component of the cost vector. We believe that this view can be useful for other graph modification problems with multiple edge types.

Finally, in Sect. 4 we show that WEIGHTED MODULE MAP admits a problem kernel with a quadratic number of vertices. More precisely, we show that given an instance of WEIGHTED MODULE MAP we can compute in $\mathcal{O}(|V|^3 + k \cdot |V|^2)$ time an equivalent instance that has $\mathcal{O}(k^2)$ vertices. As a corollary, we can solve WEIGHTED MODULE MAP in $\mathcal{O}(2^k \cdot k^6 + |V|^3)$ time by first applying the kernelization and then using the search tree algorithm.

Related Work. Compared to the study of graphs with only one edge type, there has been little work on algorithms for graphs with multiple edge types which may be referred to as multilayer graphs [15] or edge-colored (multi)graphs.

Chen et al. [8] introduced MULTI-LAYER CLUSTER EDITING, a variant of CLUSTER EDITING with multiple edge types. In this problem, one asks to transform all layers into cluster graphs which differ only slightly. Here, a layer is the subgraph containing only the edges of one type. Roughly speaking, the task is

to find one cluster graph such that each layer can be transformed into this cluster graph by at most k edge modifications. Chen et al. [8] show fixed-parameter
algorithms and hardness results for different parameter combinations. The problem differs from MODULE MAP in the sense that all edge types play the same role
in the problem definition and that layers are evaluated independently whereas
in MODULE MAP the aim is to obtain one graph with blue and red edges that
fulfills different properties for the blue and red edges. A further problem studied
in this context is SIMULTANEOUS FEEDBACK VERTEX SET [1] where the aim is
to delete at most k vertices in a multilayer graph such that each layer is acyclic.
Further, Bredereck et al. [7] present several algorithmic and hardness results for
a wide range of subgraph problems in multilayer graphs.

Preliminaries. We follow standard notation in graph theory. For a graph
$G = (V, E)$ and a set $V' \subseteq V$, the *subgraph of G induced by V'* is denoted
by $G[V'] := (V', \{\{u, v\} \in E \mid u, v \in V'\})$. For two sets A and B, the *symmetric
difference* $A \triangle B := (A \cup B) \backslash (A \cap B)$ is the set of elements which are in exactly
one of the two sets. A *solution* S for an instance of MODULE MAP is a tuple
of edge transformations (E_b', E_r') of size at most k such that the transformed
graph $G' = (V, E_b \triangle E_b', E_r \triangle E_r')$ is a module graph. Herein, the *size* of (E_b', E_r')
is $|E_b'| + |E_r'|$. The graph G' is called *target graph*. A solution S is *optimal* if
every other solution is at least as large as S.

For the basic definitions on parameterized complexity such as fixed-parameter
tractability and kernelization, we refer to the literature [9]. We present our kernelization via *reduction rules*. A reduction rule is *safe* if the resulting instance
is equivalent. An instance is *reduced exhaustively* with respect to a reduction
rule if an application of the rule does not change the instance. A *branching
rule* transforms an instance (I, k) of a parameterized problem into instances
$(I_1, k_1), \ldots, (I_\ell, k_\ell)$ of the same problem such that $k_i < k$. A branching rule is
safe if (I, k) is a yes-instance if and only if there exists a j such that (I_j, k_j)
is a yes-instance. A standard tool in the analysis of search tree algorithms are
branching vectors; for further background refer to the monograph of Fomin and
Kratsch [10].

Due to lack of space, several proofs are deferred to a long version of the article.

2   Basic Observations

In the following we present a forbidden subgraph characterization for the property of being a module graph. To this end, we define the following three graphs
which are shown in Fig. 2: a *blue P_3* is a path on three vertices consisting of two
blue edges, a *two-colored K_3* is a clique of size three, where one edge is red and
the other two are blue, and a *two-colored P_3* is a path on three vertices with
exactly one blue and one red edge.

The first step towards the forbidden subgraph characterization is to show
that the subgraph induced by the blue edges G_b is a cluster graph if and only
if G contains no blue P_3 and no two-colored K_3.

Fig. 2. The forbidden induced subgraphs for module graphs. From left to right: a blue P_3, consisting of two (dark) blue edges, a two-colored K_3, consisting of two blue and one (light) red edge and a two-colored P_3, consisting of one blue and one red edge. (Color figure online)

Lemma 1. *A graph G fulfills the cluster property if and only if G contains neither a blue P_3 nor a two-colored K_3 as induced subgraphs.*

Next, we can show that the existence of a two-colored P_3 leads to a violation of the link property which gives the complete forbidden subgraph characterization.

Theorem 1. *A two-colored graph G is a module graph if and only if G has no blue P_3, no two-colored K_3, and no two-colored P_3 as induced subgraph.*

We now show a simple linear-time fixed-parameter algorithm for MODULE MAP and WEIGHTED MODULE MAP. The algorithm uses the standard approach to branch on the graphs of the forbidden subgraph characterization presented in Theorem 1. The main point is to obtain a linear running time. To this end, we show that we can determine in $\mathcal{O}(|V| + |E|)$ time if a graph contains any of the three forbidden subgraphs.

We start by determining if the blue subgraph G_b of the two-colored input graph $G = (V, E_b, E_r)$ is a cluster graph. According to Lemma 1, we have to determine if G has a blue P_3 or a two-colored K_3. The normal approach would be to search for each forbidden subgraph individually. As we show in the following, however, under a standard assumption in complexity theory it is impossible to find a two-colored K_3 in $\mathcal{O}(|V| + |E|)$ time.

The current best algorithm to determine if a graph G contains a triangle has a running time of $\mathcal{O}(|V|^\omega)$ time, where $\omega < 2.376$ is the exponent of the time that is needed to multiply two $n \times n$ matrices [14].

Proposition 2. *We cannot find a two-colored K_3 in a graph $G = (V, E_b, E_r)$ in $\mathcal{O}(|V|+|E|)$ time, unless we can detect triangles in $\mathcal{O}(((|V|+|E|)\cdot \log|V|)$ time.*

Instead, we obtain a linear-time algorithm by searching in one step for two-colored K_3s *and* for blue P_3s.

Lemma 2. *For a two-colored graph $G = (V, E_b, E_r)$ we can find in $\mathcal{O}(|V|+|E|)$ time a blue P_3 or a two-colored K_3 if G contains either one.*

Now we show how to find a two-colored P_3 in time $\mathcal{O}(|V| + |E|)$ in a graph $G = (V, E_b, E_r)$ when we assume that G contains no blue P_3 and no two-colored K_3.

Lemma 3. *A two-colored P_3 in a graph $G = (V, E_b, E_r)$ which contains no blue P_3 and no two-colored K_3 can be found in $\mathcal{O}(|V| + |E|)$ time if it exists.*

With Lemmas 2 and 3 at hand, it can be determined in $\mathcal{O}(|V| + |E|)$ time if a graph $G = (V, E_b, E_r)$ contains a forbidden subgraph and, thus, also whether G is a module graph. A simple fixed-parameter algorithm for MODULE MAP now works as follows: Check whether G is a module graph. If this is the case, then return 'yes'. Otherwise, check whether $k = 0$. If this is the case, return 'no'. Otherwise, find one of the three forbidden subgraphs and branch on the possibilities to destroy it by an edge modification. If G contains a blue P_3 with vertex set $\{u, v, w\}$ and non-edge $\{u, w\}$, then transform $\{u, w\}$ into a blue edge in the first case, transform $\{u, v\}$ into a non-edge in the second case, and transform $\{v, w\}$ into a non-edge in the third case. In each case, decrease k by one and solve the resulting instance recursively. The treatment of the other forbidden subgraphs is similar: If G contains a two-colored P_3, transform the blue edge into a non-edge, or transform the red edge into non-edge, or transform the non-edge into a red edge (observe that the case where a non-edge is transformed into blue edge need not be considered since this produces a two-colored K_3). If G contains a two-colored K_3, either transform one of the blue edges into a non-edge or transform the red edge into a blue edge. For each forbidden induced subgraph, the algorithm branches into three cases and decreases k by at least one. This leads to a branching vector of $(1, 1, 1)$. Since branching is performed only as long as $k > 0$, the overall search tree size is $\mathcal{O}(3^k)$; the steps of each search tree node can be performed in $\mathcal{O}(|V| + |E|)$ time. Altogether, we obtain the following.

Proposition 3. MODULE MAP *can be solved in* $\mathcal{O}(3^k \cdot (|V| + |E|))$ *time.*

For WEIGHTED MODULE MAP, we can use the same algorithm: since the edge weights are positive integers, the parameter decrease is again at least 1 in each created branch of the search tree algorithm. A subtle difference is that, due to the edge weight function g, the overall instance size is $\mathcal{O}(|V|^2)$.

Proposition 4. WEIGHTED MODULE MAP *can be solved in* $\mathcal{O}(3^k \cdot |V|^2)$ *time.*

3 An Improved Search Tree Algorithm

To improve the running time, we adapt a branching strategy for CLUSTER EDITING [5]. To apply this strategy, we first introduce a generalization of WEIGHTED MODULE MAP. Then, we explain our branching strategy. Finally, we solve certain instances in polynomial time to obtain an $\mathcal{O}(2^k \cdot |V|^3)$-time search tree algorithm.

A More Flexible Scoring Function. To describe our algorithm for WEIGHTED MODULE MAP, we introduce a more general problem since during branching, we will *merge* some vertices. To represent the adjacencies of the merged vertices, we generalize the concept of edge weights: Recall that in WEIGHTED MODULE

MAP, transforming a blue edge with weight ω into a non-edge costs ω and transforming it into a red edge costs 2ω. Hence, the two transformation costs are directly related. From now on, we allow independent transformation costs for the different possibilities. To this end, we introduce an *edge-cost function* s : $\binom{V}{2} \to \mathbb{R}^3$ for all pairs of vertices $\{u, v\}$ of a given graph G where $s(u, v) :=$ $(b_{u,v}, n_{u,v}, r_{u,v})$. This vector $(b_{u,v}, n_{u,v}, r_{u,v})$ is called *cost vector*. Herein, $b_{u,v}$ is the cost of making $\{u, v\}$ blue, $n_{u,v}$ is the cost of making $\{u, v\}$ a non-edge and $r_{u,v}$ is the cost of making $\{u, v\}$ red. For a short form of the cost vector we also write $(b, n, r)_{u,v}$. If there is no danger of confusion we omit the index of the associated vertices u and v. For example, let $\{u, v\}$ be a blue edge in an instance of WEIGHTED MODULE MAP with weight ω. Then we get cost vector $(0, \omega, 2\omega)$.

We call a vertex pair $\{u, v\}$ with its cost vector (b, n, r) a *blue pair* if $b = 0$ and $n, r > 0$, a *non-pair* if $n = 0$ and $b, r > 0$, and a *red pair* if $r = 0$ and $b, n > 0$. As for unweighted graphs, three vertices u, v, and w form a *blue* P_3 if $\{u, v\}$ and $\{u, w\}$ are blue and $\{v, w\}$ is a non-pair, they form a *two-colored* K_3 if $\{u, v\}$ and $\{u, w\}$ are blue and $\{v, w\}$ is red, they form a *two-colored* P_3 if $\{u, v\}$ is blue, $\{u, w\}$ is red and $\{v, w\}$ is a non-pair. Finally, a graph G is called a *pair module graph* if each pair $\{u, v\}$ of vertices in G is a blue pair, a non-pair or a red pair and G contains no blue P_3, two-colored K_3 and two-colored P_3.

We do not allow arbitrary scoring functions but demand the following three properties. The first property restricts the relation between the three costs.

Property 1. For each cost vector $(b, n, r)_{u,v}$, we have $b + r \geq 2n$.

Property 1 is essentially a more relaxed version of the property that transforming a blue edge into a red edge is at least as expensive as transforming this edge first into a non-edge and subsequently into a red edge.

Property 2. In each cost vector $s(u, v)$ either all components are non-negative integers or all three are non-negative and half-integral. In the latter case, at least two components are equal to $1/2$.

A cost vector $(b, n, r)_{u,v}$ where all three components are half-integral is called *half-integral*. All other cost vectors are called *integral*. Half-integral cost vectors will be introduced during the algorithm for technical reasons.

The final property demands that each vertex pair whose cost vector is not half-integral has exactly one component equal to zero. This guarantees the unambiguous construction of a pair module graph from each vertex pair.

Property 3. Each integral cost vector $(b, n, r)_{u,v}$ contains exactly one component which is equal to zero.

Properties 1–3 are fulfilled by the scoring function obtained from WEIGHTED MODULE MAP instances. Moreover, we can observe the following.

Proposition 5. *Let* $(b, n, r)_{u,v}$ *be a cost vector fulfilling Property 1–3.*

- *If* $\{u, v\}$ *is blue, then* $n \geq 1$ *and* $r \geq 2n$.
- *If* $\{u, v\}$ *is red, then* $n \geq 1$ *and* $b \geq 2n$.
- *If* $\{u, v\}$ *is half-integral, then* $n = 1/2$.

Proof. The first two claims follow from the fact that (b, n, r) has only integer components in these cases and that only one component is zero. The third claim can be seen as follows. By Property 2, all three components are half-integral and at least two of them are equal to $1/2$. By Property 1, $b + r \geq 2n$. If $n > 1/2$, then $b = 1/2 = r$ and Property 1 is violated. Thus, $n = 1/2$. □

We may now define MODULE MAP WITH SCORING FUNCTION (MMS).

MMS
Input: A graph G with an edge-cost function $s : \binom{V}{2} \to \mathbb{R}^3$ which fulfills Properties 1–3 and a non-negative integer k.
Question: Can we transform G into a pair module graph with transformation costs at most k?

Our aim is to show the following.

Theorem 2. *MMS can be solved in* $\mathcal{O}(2^k \cdot |V|^3)$ *time.*

Merge-Based Branching. We branch on blue pairs since each forbidden subgraph contains at least one blue edge. In one case we will delete this blue edge and in the other case we will keep this blue edge.

Definition 2. *A blue pair* $\{u, v\}$ *forms a* conflict triple *with a vertex* w *if* $\{u, w\}$ *and* $\{v, w\}$ *are not both blue, not both non-pairs, or not both red.*

To resolve all conflict triples, we branch on blue pairs. In the corresponding branching, similar to the approach of Böcker et al. [5], we *merge* the vertex pair $\{u, v\}$ in one of the cases.

Definition 3. *Let* (G, s, k) *be an instance of* MMS. *Merging* two vertices u *and* v *is the following operation: Remove* u *and* v *from* G *and add a new vertex* u'. *For all vertices* $w \in V \setminus \{u, v\}$ *set* $s(u', w) := s(u, w) + s(v, w)$.

We call $s(u', w)$ the *join* of $s(u, w)$ and $s(v, w)$. Note that $s(u', w)$ may not fulfill Properties 1–3. Because of this, we have to *reduce* joint cost vectors as far as possible. Herein, reducing (b, n, r) by a value t is to decrease each of its components by t. Simultaneously we can reduce the parameter k by t.

Reduction Rule 1. *Let* $\{u, v\}$ *be a vertex pair with cost vector* (b, n, r). *If* (b, n, r) *has a unique minimum component, then reduce* (b, n, r) *and parameter* k *by* $\min(b, n, r)$. *Otherwise, reduce* (b, n, r) *and parameter* k *by* $\min(b, n, r) - 1/2$.

Let $x = min(b, n, r)$ be a minimal value of (b, n, r). If x is unique, then we can reduce (b, n, r) by x, since afterwards exactly one component of the cost vector is equal to zero. Otherwise, we cannot reduce the cost vector by x, since afterwards at least two components have value zero, a contradiction to Property 3. Clearly, we could reduce the vector by $x - 1$, but this would not give a parameter decrease for vectors such as $(1, 1, 3)$. According to the bookkeeping trick introduced in [5], in such a case, we reduce this vector by $x - 1/2$ to circumvent the above problem. For example, we reduce the vector $(1, 1, 3)$ by $1/2$ and get vector $(1/2, 1/2, 5/2)$.

Branching Rule 1. *Let (G, s, k) be an instance of* MMS. *If (G, s, k) contains a blue pair $\{u, v\}$ and two distinct vertices w and w' that form a conflict triple with $\{u, v\}$, then branch into two cases:*

Case 1: Set $b_{u,v} := k + 1$. Afterwards apply Reduction Rule 1.
Case 2: Merge the vertex pair $\{u, v\}$. Afterwards apply Reduction Rule 1.

Lemma 4. *Branching Rule 1 is safe.*

Now we prove that Properties 1–3 remain true if we set $b_{u,v} := k + 1$ for a blue pair $\{u, v\}$ and apply Reduction Rule 1, and if we merge a blue pair $\{u, v\}$ and apply Reduction Rule 1.

Lemma 5. *Let $\{u, v\}$ be a blue pair in an instance (G, s, k) of* MMS, *and let (G', s', k') be obtained by setting $b_{u,v} := k + 1$ and applying Reduction Rule 1 or by merging u and v and applying Reduction Rule 1. Then, (G', s', k') is an instance of* MMS. *In particular, s' fulfills Properties 1–3.*

We now show that if we merge a blue pair $\{u, v\}$ where $\{u, w\}$ and $\{v, w\}$ are not both blue, non-, or red, then Reduction Rule 1 reduces the resulting cost vector $(b, n, r)_{u',w}$ by at least $1/2$.

Proposition 6. *Let $s(u', w)$ be a joint cost vector that is the join of two cost vectors $s(u, w)$ and $s(v, w)$, where $\{u, w\}$ and $\{v, w\}$ are not both blue, non- or red. Then Reduction Rule 1 applied to $s(u', w)$ decreases k by at least $1/2$.*

Now we show that increasing b for blue pairs decreases k by at least 1.

Lemma 6. *Let $\{u, v\}$ be a blue pair and let (b^*, n, r) be the cost vector that results from (b, n, r) by setting $b = k + 1$. Then, applying Reduction Rule 1 to (b^*, n, r) decreases k by at least 1 or this instance has no solution.*

Proof. Form Proposition 5 we conclude: $b = 0$, $n \geq 1$ and $r \geq 2n$. If $n \geq k$, then each component of (b^*, n, r) will be larger than k. Hence there exists no solution for this instance. Thus, the reduced cost vector is integral with a unique minimum component. Consequently, it is reduced by at least 1. $\qquad\square$

Now consider the instances obtained by an application of Branching Rule 1. In Case 1, the new parameter is at most $k - 1$ due to Lemma 6. In Case 2, the new parameter is also at most $k - 1$ because $\{u, v\}$ is in two conflict triples. By Proposition 6 this means that we create two cost vectors which are both reduced by at least $1/2$.

Corollary 1. *Branching Rule 1 has a branching vector of* $(1, 1)$ *or better.*

Applying Branching Rule 1 states that every blue pair which is contained in at least two conflict triples has branching vector $(1, 1)$ or better. Branching Rule 2 deals with blue pairs $\{u, v\}$ which are contained in exactly one conflict triple with vertex w where $\{u, w\}$ and $\{v, w\}$ give a join that can be reduced by at least 1.

Branching Rule 2. *If* (G, s, k) *contains a blue pair* $\{u, v\}$ *and a vertex* w *such that* $u, v,$ *and* w *form a conflict triple and the joined vertex pair can be reduced by at least 1, then branch into the following two cases:*

Case 1: Set $b_{u,v} := k + 1$. *Afterwards apply Reduction Rule 1.*
Case 2: Merge the vertex pair $\{u, v\}$. *Afterwards apply Reduction Rule 1.*

Lemma 7. *Branching Rule 2 is correct and has branching vector* $(1, 1)$ *or better.*

Solving the Remaining Instances in Polynomial Time. We now show that instances to which Branching Rules 1 and 2 do not apply can be solved efficiently.

Lemma 8. *Let* (G, s, k) *be an instance of* MMS. *If Branching Rules 1 and 2 do not apply, then* (G, s, k) *can be solved in* $\mathcal{O}(|V|^2)$ *time.*

We now can prove Theorem 2.

Proof (of Theorem 2). First, check for each blue pair $\{u, v\}$ if Branching Rule 1 or 2 applies. This needs $\mathcal{O}(|V|^3)$ time. If this is the case, we will branch on $\{u, v\}$. According to Corollary 1 and Lemma 7, Branching Rules 1 and 2 have a branching vector of $(1, 1)$ or better. This implies a search tree size of $\mathcal{O}(2^k)$ because we only branch as long as $k > 0$. In one case, we set $b_{u,v} := k + 1$, which can be done in constant time. In the other case, we merge u and v. Hence, we delete the vertices u and v from the graph and replace them by a new vertex u' and join all incident pairs of vertices. These are n pairs. So we can calculate the cost vector for each new, joined pair in $\mathcal{O}(|V|)$ time. In both cases we reduce the parameter accordingly. Hence, we need $\mathcal{O}(|V|^3)$ time per search tree node. By Lemma 8, MMS can be solved in $\mathcal{O}(|V|^2)$ time if Branching Rules 1 and 2 do not apply. Hence, we obtain an $\mathcal{O}(2^k \cdot |V|^3)$-time algorithm for MMS. $\square$

4   A Polynomial Problem Kernel

We also obtain a problem kernelization for WEIGHTED MODULE MAP that yields a problem kernel with $\mathcal{O}(k^2)$ vertices. The basic idea is the following: Let $\{u, v\}$ be a vertex pair of an instance $(G = (V, E_b, E_r), k, g)$ of WEIGHTED MODULE MAP. We investigate if it is possible that the vertex pair $\{u, v\}$ can be a blue, non-, or red edge in any target graph of a size-k solution. To this end, we estimate for each edge type the *induced* costs of transforming $\{u, v\}$ into this type; this approach was also used for CLUSTER EDITING [5, 12].

Theorem 3. WEIGHTED MODULE MAP *admits a problem kernel of* $\mathcal{O}(k^2)$ *vertices which can be found in* $\mathcal{O}(|V|^3 + k \cdot |V|^2)$ *time.*

5   Conclusion

There are many open questions: Does MODULE MAP admit a problem kernel with $\mathcal{O}(k)$ vertices? Can we compute a constant-factor approximation in polynomial time? Is MODULE MAP NP-hard when G_b is a cluster graph? Is MODULE MAP fixed-parameter tractable for smaller parameters, for example when parameterized above a lower bound as it was done for CLUSTER EDITING [4]?

References

1. Agrawal, A., Lokshtanov, D., Mouawad, A.E., Saurabh, S.: Simultaneous feedback vertex set: a parameterized perspective. In: Proceedings of 33rd STACS. LIPIcs, vol. 47, pp. 7:1–7:15. Schloss Dagstuhl - Leibniz-Zentrum fuer Informatik (2016)
2. Amar, D., Shamir, R.: Constructing module maps for integrated analysis of heterogeneous biological networks. Nucleic Acids Res. **42**(7), 4208–4219 (2014)
3. Bansal, N., Blum, A., Chawla, S.: Correlation clustering. Mach. Learn. **56**(1–3), 89–113 (2004)
4. van Bevern, R., Froese, V., Komusiewicz, C.: Parameterizing edge modification problems above lower bounds. Theory Comput. Syst. **62**(3), 739–770 (2018)
5. Böcker, S., Briesemeister, S., Bui, Q.B.A., Truß, A.: Going weighted: parameterized algorithms for cluster editing. Theor. Comput. Sci. **410**(52), 5467–5480 (2009)
6. Böcker, S., Briesemeister, S., Klau, G.W.: Exact algorithms for cluster editing: evaluation and experiments. Algorithmica **60**(2), 316–334 (2011)
7. Bredereck, R., Komusiewicz, C., Kratsch, S., Molter, H., Niedermeier, R., Sorge, M.: Assessing the computational complexity of multi-layer subgraph detection. In: Fotakis, D., Pagourtzis, A., Paschos, V.T. (eds.) CIAC 2017. LNCS, vol. 10236, pp. 128–139. Springer, Cham (2017). https://doi.org/10.1007/978-3-319-57586-5_12
8. Chen, J., Molter, H., Sorge, M., Suchý, O.: A parameterized view on multi-layer cluster editing. CoRR abs/1709.09100 (2017)
9. Downey, R.G., Fellows, M.R.: Fundamentals of Parameterized Complexity. Texts in Computer Science. Springer, London (2013). https://doi.org/10.1007/978-1-4471-5559-1
10. Fomin, F.V., Kratsch, D.: Exact Exponential Algorithms. Texts in Theoretical Computer Science. An European Association for Theoretical Computer Science Series. Springer, Heidelberg (2010)
11. Fomin, F.V., Kratsch, S., Pilipczuk, M., Pilipczuk, M., Villanger, Y.: Tight bounds for parameterized complexity of cluster editing. In: STACS. LIPIcs, vol. 20, pp. 32–43. Schloss Dagstuhl - Leibniz-Zentrum fuer Informatik (2013)
12. Gramm, J., Guo, J., Hüffner, F., Niedermeier, R.: Graph-modeled data clustering: exact algorithms for clique generation. Theory Comput. Syst. **38**(4), 373–392 (2005)
13. Hartung, S., Hoos, H.H.: Programming by optimisation meets parameterised algorithmics: a case study for cluster editing. In: Dhaenens, C., Jourdan, L., Marmion, M.-E. (eds.) LION 2015. LNCS, vol. 8994, pp. 43–58. Springer, Cham (2015). https://doi.org/10.1007/978-3-319-19084-6_5
14. Itai, A., Rodeh, M.: Finding a minimum circuit in a graph. SIAM J. Comput. **7**(4), 413–423 (1978)
15. Kivelä, M., Arenas, A., Barthelemy, M., Gleeson, J.P., Moreno, Y., Porter, M.A.: Multilayer networks. J. Complex Netw. **2**(3), 203–271 (2014)

16. Komusiewicz, C., Uhlmann, J.: Cluster editing with locally bounded modifications. Discrete Appl. Math. **160**(15), 2259–2270 (2012)

17. Krivánek, M., Morávek, J.: NP-hard problems in hierarchical-tree clustering. Acta Informatica **23**(3), 311–323 (1986)

18. Shamir, R., Sharan, R., Tsur, D.: Cluster graph modification problems. Discrete Appl. Math. **144**(1–2), 173–182 (2004)

19. Ulitsky, I., Shlomi, T., Kupiec, M., Shamir, R.: From E-maps to module maps: dissecting quantitative genetic interactions using physical interactions. Mol. Syst. Biol. **4**(1), 209 (2008)

2 CSPs All Are Approximable
Within a Constant Differential Factor

Jean-François Culus[1] and Sophie Toulouse[2(✉)]

[1] CEREGMIA, Université des Antilles, Pointe-à-Pitre, France
`jean-francois-culus@espe-martinique.fr`
[2] LIPN (UMR CNRS 7030), Institut Galilée, Université Paris 13,
Villetaneuse, France
`sophie.toulouse@lipn.univ-paris13.fr`

Abstract. Only a few facts are known regarding the approximability of optimization CSPs with respect to the differential approximation measure, which compares the gain of a given solution over the worst solution value to the instance diameter. Notably, the question whether $\mathsf{k\,CSP-q}$ is approximable within any constant factor is open in case when $q \geq 3$ or $k \geq 4$. Using a family of combinatorial designs we introduce for our purpose, we show that, given any three constant integers $k \geq 2$, $p \geq k$ and $q > p$, $\mathsf{k\,CSP-q}$ reduces to $\mathsf{k\,CSP-p}$ with an expansion of $1/(q - p + k/2)^k$ on the approximation guarantee. When $p = k = 2$, this implies together with the result of Nesterov as regards $\mathsf{2\,CSP-2}$ [1] that for all constant integers $q \geq 2$, $\mathsf{2\,CSP-q}$ is approximable within factor $(2 - \pi/2)/(q - 1)^2$.

Keywords: Differential approximation
Optimization constraint satisfaction problems
Approximation-preserving reductions · Combinatorial designs

1 Introduction

1.1 Optimization Constraint Satisfaction Problems

Thereafter, given a positive integer N, we denote by $[N]$ the discrete interval $\{1, \ldots, N\}$. Optimization *Constraint Satisfaction Problems* (CSPs) over an alphabet Σ consider a set $\{x_1, \ldots, x_n\}$ of variables and a set $\{C_1, \ldots, C_m\}$ of constraints, where the variables have domain Σ, and the constraints consist of (non constant) predicates applied to tuples of variables. Most often, a positive weight is associated with each constraint C_i. The goal is then to optimize over Σ^n an objective function of the form

$$\sum_{i=1}^m w_i C_i = \sum_{i=1}^m w_i P_i(x_{J_i}) = \sum_{i=1}^m w_i P_i(x_{i_1}, \ldots, x_{i_{k_i}})$$

where for all $i \in [m]$, $P_i : \Sigma^{k_i} \to \{0, 1\}$, $J_i = (i_1, \ldots, i_{k_i}) \subseteq [n]$, and $w_i > 0$.

© Springer International Publishing AG, part of Springer Nature 2018
J. Lee et al. (Eds.): ISCO 2018, LNCS 10856, pp. 389–401, 2018.
https://doi.org/10.1007/978-3-319-96151-4_33

For example, the *Satisfiability Problem* (Sat) is the boolean CSP where constraints are disjunctive clauses. In Lin$-$q, the alphabet is $\mathbb{Z}_q = \mathbb{Z}/q\mathbb{Z}$, and a constraint is a linear equation modulo q. In this paper, given two universal constant integers $q, k \geq 2$, we consider the optimization CSP k CSP$-$q where Σ has size q, each constraint depends on at most k variables, and functions P_i are allowed to take rational values.[1] In the sequel, given an instance I of k CSP$-$q, we will assume either $\Sigma = [q]$ or $\Sigma = \mathbb{Z}_q$, and that the optimization goal is to maximize. These assumptions are without loss of generality.[2].

k CSP$-$q most often becomes harder as k or q grows. On the one hand, given two integers $h \geq 2$ and $k > h$, h CSP$-$q is a special case of k CSP$-$q. On the other hand, given two integers $p \geq 2$ and $q > p$, any surjective map from $[q]$ to $[p]$ can be used to convert a function on $[p]^k$ to a function on $[q]^k$. The alphabet size more accurately has a logarithmic impact on the constraint arity: if $\kappa = \lceil \log_p q \rceil$, then any surjective map from $[p]^\kappa$ to $[q]$ similarly allows to convert a function on $[q]^k$ to a function on $[p]^{\kappa k}$. As 2 CSP$-$2 is **NP$-$hard** [2], a major issue as regards k CSP$-$q consists in charactering its *approximation degree*.

1.2   Their Differential Approximability

Given an instance I of an optimization problem Π, we denote by $v(I, .)$ its objective function, by $\mathrm{opt}(I)$ and $\mathrm{wor}(I)$ respectively the best and the worst solution values on I. Approximation algorithms aim at providing within polynomial time solution values proved to be relatively close to the optimum solution value, where the proximity to $\mathrm{opt}(I)$ is defined with respect to a specific measure. In this paper, we consider the *differential approximation measure* (see [3] for an introduction). On I, the *differential ratio* reached at a given solution x is the ratio:

$$\frac{v(I, x) - \mathrm{wor}(I)}{\mathrm{opt}(I) - \mathrm{wor}(I)}$$

Given $\rho \in]0, 1]$, x is said ρ-*approximate* if this ratio is at least ρ. A polynomial time algorithm $\mathcal{A}$ is a ρ-*approximation algorithm* for Π if it returns on every instance of Π a solution with differential ratio at least ρ. Finally, Π is *approximable within factor* ρ whenever such an algorithm exists.

Only a few facts are known regarding the approximability of k CSP$-$q within a constant differential factor. On the one hand, the restrictions of Max Sat and Min Sat to unweighted instances (*i.e.*, to instances in which weights w_i all are equal to 1) are not approximable within any constant factor unless $\mathbf{P} = \mathbf{NP}$ [4]. On the other hand, for 2 CSP$-$2, the semidefinite programming-based algorithm

---

[1]   CSPs in which constraints take non-boolean values are commonly called *generalized CSPs* in the literature. However, given a function $P : \Sigma^k \to \mathbb{Q}$ with minimal value P_*, a constraint $P(x_{J_i})$ coincides, up to an additive constant term, with the combination $\sum_{v \in \Sigma^k} (P(v) - P_*) \times (x_{J_i} = v)$ of constraints. Thus when k and q are universal constants, we may indifferently consider functions with codomain $\{0, 1\}$ or $\mathbb{Q}$.

[2]   For the latter assumption, consider that minimizing $\sum_{i=1}^m w_i P_i(x_{J_i})$ reduces to maximize $\sum_{i=1}^m w_i \times -P_i(x_{J_i})$.

of Goemans and Williamson [5] produces solutions with expected differential ratio at least $2 - \pi/2 > 0.429$ [1], and the algorithm can be derandomized [6].

A common way to exhibit new approximability lower and upper bounds for a given optimization problem consists in *reducing* to or from another optimization problem for which approximability bounds are known:

Definition 1 (Informal). *An optimization problem Π D-reduces to another optimization problem Π' if one can derive from any ρ-approximation algorithm $\mathcal{A}$ for Π' a $\gamma \times \rho$-approximation algorithm for Π, where γ is some positive quantity. When this occurs, γ is called the expansion of the reduction.*

In particular, it is possible to derive from the $(2 - \pi/2)$-approximation algorithm for $2\,\mathsf{CSP}-2$ a $(1 - \pi/4)$-approximation algorithm for $3\,\mathsf{CSP}-2$:

Proposition 1. $3\,\mathsf{CSP}-2$ *D-reduces to* $\mathsf{E2\,Lin}-2$ *with an expansion of* $1/2$ *on the approximation guarantee.*

Proof (sketch). The discrete Fourier transform allows to convert any instance I of $3\,\mathsf{CSP}-2$ over $\{0,1\}$ to an instance J of $\mathsf{Max\,3\,Lin}-2$ such that $v(J,.)$ coincides, up some constant term, with $v(I,.)$. From such an instance J, build an instance H of $2\,\mathsf{Lin}-2$ by removing the equations of odd arity. Let W refer to the total weight of such equations in J. Then for all $x \in \{0,1\}^n$, $v(J,x) + v(J,\bar{x}) = 2 \times v(H,x) + W$, where $\bar{x}$ refers to the componentwise negation of x. Assume that $x \in \{0,1\}^n$ is ρ-approximate on J. Then the preceding equality taken at x, an optimum solution on J, and a worst solution on H allows to conclude that solution x or $\bar{x}$ that performs the best objective value on J is $\rho/2$-approximate on J. □

Still, the question whether $k\,\mathsf{CSP}-q$ is approximable within any constant factor remains open in case when $q \geq 3$ or $k \geq 4$.

1.3 Outline

Given an integer $k \geq 2$ and two integers $q, p \geq 2$ with $q > p$, we address the question whether $k\,\mathsf{CSP}-q$ D-reduces to $k\,\mathsf{CSP}-p$. We more specifically study the expansion of a specific reduction: given an instance I of $k\,\mathsf{CSP}-q$, the reduction basically consists in considering the restrictions of I to solutions sets of the form T^n where T is a p-element subset of Σ. The analysis we propose, though, requires to restrict to the case when $p \geq k$.

In the next section, we introduce a family of combinatorial designs (Definition 2) that provides some lower bound $\gamma(q, p, k)$ for the expansion of the reduction (Theorem 1). Section 3 is then dedicated to the exhibition of such combinatorial designs. Using a recursive construction for the case when $p = k$ (Theorem 2), we show that $1/(q - p + k/2)^k$ is a proper lower bound of for $\gamma(q, p, k)$. Therefore, we obtain the following conditional approximation result:

Corollary 1. *Given any three integers $k \geq 2$, $p \geq k$ and $q > p$, $k\,\mathsf{CSP}-q$ D-reduces to $k\,\mathsf{CSP}-p$ with an expansion of $1/(q-p+k/2)^k$ on the approximation guarantee. The reduction involves $O(q^p)$ instances of $k\,\mathsf{CSP}-p$.*

The question whether $k\,\mathsf{CSP}-\mathsf{q}$ is approximable within some constant factor consequently reduces to the consideration of integers k, q such that $k \geq q \geq 2$. Most importantly, it follows from Nesterov's result as regards $2\,\mathsf{CSP}-2$ (we more specifically refer to Theorem 2.3, Theorem 3.3 and Corollary 3.4 of [1]) that for all integers $q \geq 2$, $2\,\mathsf{CSP}-\mathsf{q}$ is approximable within a constant factor:

Corollary 2. *For all integers* $q \geq 2$, $2\,\mathsf{CSP}-\mathsf{q}$ *is approximable within factor* $(2 - \pi/2)/(q - 1)^2$.

2 Reducing the Alphabet Size of a CSP Instance

Let $k \geq 2$, $p \geq 2$ and $q > p$ be three integers, and I be an instance of $k\,\mathsf{CSP}-\mathsf{q}$ over alphabet $[q]$. Thereafter, $\mathcal{P}_p([q])$ refers to the set of the p-element subsets of $[q]$.

Given $S = (S_1, \ldots, S_n) \in \mathcal{P}_p([q])^n$, any set $\{\pi_{S,j} : S_j \to [p] \mid j \in [n]\}$ of bijections allows to interpret the restriction of I to solutions in S as an instance of $k\,\mathsf{CSP}-\mathsf{p}$. Therefore, a natural way to derive approximate solutions on I from a hypothetical algorithm $\mathcal{A}$ for $k\,\mathsf{CSP}-\mathsf{p}$ consists in restricting I to solution subsets $S \in \mathcal{P}_p([q])^n$. The *standard approximation measure* evaluates the performance of a given solution x by the ratio $v(I, x)/\mathrm{opt}(I)$. In [7], the authors study the randomized reduction that consists in picking $S \in \mathcal{P}_p([q])^n$ uniformly at random, and then using $\mathcal{A}$ to compute a solution $x \in S$. They show that, provided that the goal on I is to maximize and I is such that $w_i P_i \geq 0, i \in [m]$, the expected value of $\max_{y \in S}\{v(I, y)\}$ over all $S \in \mathcal{P}_p([q])^n$ is at least $(p/q)^k \times \mathrm{opt}(I)$. Accordingly, picking $S \in \mathcal{P}_p([q])^n$ uniformly at random, and then computing a solution $x \in S$ with value at least $\rho \times \max_{y \in S}\{v(I, y)\}$, one gets a solution with expected value at least $(p/q)^k \rho \times \mathrm{opt}(I)$. The reduction therefore preserves the expected standard ratio up to a multiplicative factor of $(p/q)^k$.

Given $T \in \mathcal{P}_p([q])$, we denote by $I(T)$ the restriction of I to solution set T^n. Then similarly to [7], we analyse the reduction that consists in using $\mathcal{A}$ to compute for all $T \in \mathcal{P}_p([q])$ an approximate solution $x(T)$ on $I(T)$, and then returning a solution $x(T)$ that performs the best objective value.

2.1 Seeking Symmetries in the Solution Set

We assume *w.l.o.g.* that the goal on I is to maximize, in which case the extremal values on I and on subinstances $I(T)$ trivially satisfy:

$$\mathrm{opt}(I) \geq \mathrm{opt}(I(T)) \geq \mathrm{wor}(I(T)) \geq \mathrm{wor}(I), \qquad T \in \mathcal{P}_p([q]) \qquad (1)$$

Now assume that for all $T \in \mathcal{P}_p([q])$, we are given a solution $x(T) \in T^n$ that is ρ-approximate on $I(T)$. Then for all $T^* \in \mathcal{P}_p([q])$, we have:

$$\begin{aligned}
\max_{T \in \mathcal{P}_p([q])}\{v(I, x(T))\} &\geq v(I, x(T^*)) \\
&\geq \rho \times \mathrm{opt}(I(T^*)) + (1 - \rho) \times \mathrm{wor}(I(T^*)) \\
&\geq \rho \times \mathrm{opt}(I(T^*)) + (1 - \rho) \times \mathrm{wor}(I) \qquad \text{by (1)} \quad (2)
\end{aligned}$$

Eventually assume that T^* is a set in $\mathcal{P}_p([q])$ that contains a solution with optimal value over $\{T^n \mid T \in \mathcal{P}_p([q])\}$. Then, provided that $\mathrm{opt}(I(T^*))$ is δ-approximate on I, one gets the following connection with $\mathrm{opt}(I(T))$:

$$\max_{T \in \mathcal{P}_p([q])} \{v(I, x(T))\} - \mathrm{wor}(I) \geq \rho \times (\mathrm{opt}(I(T^*)) - \mathrm{wor}(I)) \qquad (2)$$
$$\geq \rho \times \delta \times (\mathrm{opt}(I) - \mathrm{wor}(I)) \qquad (3)$$

Hence, if we are able to compare—in a differential approximation manner—$\mathrm{opt}(I(T^*))$ to $\mathrm{opt}(I)$, then we can deduce from approximate solutions on subinstances $I(T)$ approximate solutions on I. We thus shall seek a lower bound for the differential ratio reached on I at $\mathrm{opt}(I(T^*))$.

Let x^* be an optimal solution of I. Then one way to obtain such a lower bound consists in exhibiting two solution multisets $\mathcal{X} = (x^1, \ldots, x^R)$ and $\mathcal{Y} = (y^1, \ldots, y^R)$ of the same size R, and that satisfy the following conditions:

$$\mathcal{X} \subseteq \{T^n \mid T \in \mathcal{P}_p([q])\} \qquad (4)$$
$$R^* \triangleq |\{r \in [R] \mid y^r = x^*\}| \geq 1 \qquad (5)$$
$$|\{r \in [R] \mid x^r_{J_i} = v\}| = |\{r \in [R] \mid y^r_{J_i} = v\}|, v \in [q]^{J_i}, i \in [m] \qquad (6)$$

Requirements (4), (5) and (6) respectively ensure that $\mathcal{X}$ exclusively considers solutions of subinstances $I(T)$, x^* occurs at least once in $\mathcal{Y}$, and each constraint $P_i(x_{J_i})$ of I is evaluated on the same collection of $|J_i|$-tuples over solution multisets $\mathcal{X}$ and $\mathcal{Y}$. Requirement (6) thus ensures that the sum of solution values over $\mathcal{X}$ and $\mathcal{Y}$ are identical. Provided that such a pair $(\mathcal{X}, \mathcal{Y})$ exists, we have:

$$\mathrm{opt}(I(T^*)) \geq \textstyle\sum_{r=1}^{R} \{v(I, x^r)\}/R \quad \text{by definition of } T^*, \text{and } (4)$$
$$= \textstyle\sum_{r=1}^{R} \{v(I, y^r)\}/R \quad \text{by } (6)$$
$$\geq R^* \times \mathrm{opt}(I)/R + (R - R^*) \times \mathrm{wor}(I)/R \quad \text{by } (5) \qquad (7)$$

Thus $\mathrm{opt}(I(T^*))$ is R^*/R-approximate on I. Therefore, one shall seek such pairs $(\mathcal{X}, \mathcal{Y})$ on which the ratio R^*/R is as hight as possible.

This is precisely what we do, and this is why we restrict our analysis to the case when $k \leq p$. Indeed, e.g. assume that $J_1 = (1, \ldots, k)$ and $(x_1^*, \ldots, x_k^*) = (1, \ldots, k)$. Then by (6) and (5), $\mathcal{X}$ shall contain at least $R^* > 0$ solutions x^r with $(x_1^r, \ldots, x_k^r) = (1, \ldots, k)$. If $k > p$, then such solutions violate condition (4). Hence, from now on, we assume $q > p \geq k$.

2.2 Partition-Based Solution Multisets

Solution x^* induces a partition of $[n]$ into q—possibly empty—subsets depending on the q possible values taken by its coordinates. Given $c \in [q]$, we denote by V_c the set of indices $j \in [n]$ such that $x_j^* = c$.

We restrict our solution multisets to vectors x that satisfy $x_j^* = x_h^* \Rightarrow x_j = x_h$, $j, h \in [n]$. It is thus possible to identify $\mathcal{X}$ and $\mathcal{Y}$ with two arrays Ψ and Φ

with q columns, coefficients in $[q]$, and the same number R of rows, where: each row $\Psi_r = (\Psi_r^1, \ldots, \Psi_r^q)$ of Ψ gives rise in $\mathcal{X}$ to the vector of $[q]^n$ that satisfies for every $c \in [q]$ that its coordinates with index in V_c all are equal to Ψ_r^c; $\mathcal{Y}$ is derived from Φ in the exact same way. Formally, we define $\pi_{x^*} : [q]^q \to [q]^n$ by

$$\pi_{x^*}(u)_{V_c} = (u_c, \ldots, u_c), \qquad\qquad c \in [q]$$

and $\mathcal{X}, \mathcal{Y}$ by $\mathcal{X} = (\pi_{x^*}(\Psi_r) \,|\, r \in [R])$ and $\mathcal{Y} = (\pi_{x^*}(\Phi_r) \,|\, r \in [R])$. Given $i \in [m]$, we denote by $c_{i,1}, \ldots, c_{i,h_i}$ the distinct values taken by the coordinates of x^* with index in J_i, by $H_i = \{c_{i,1}, \ldots, c_{i,h_i}\}$ the set of such values, by Ψ^{H_i} and Φ^{H_i} the restrictions of Ψ and Φ to their columns with index in H_i. Then solution multisets $\mathcal{X}$ and $\mathcal{Y}$ meet requirements (4), (5) and (6) of Sect. 2.1 *iff* arrays Ψ and Φ satisfy:

$$|\{\Psi_r^1, \ldots, \Psi_r^q\}| \leq p, \; r \in [R] \tag{8}$$

$$R^* \triangleq |\{r \in [R] \,|\, \Phi_r = (1, \ldots, q)\}| \geq 1 \tag{9}$$

$$|\{r \in [R] \,|\, \Psi_r^{H_i} = v\}| = |\{r \in [R] \,|\, \Phi_r^{H_i} = v\}|, \, v \in [q]^{|H_i|}, \; i \in [m] \tag{10}$$

Hence, if we are aware of such a pair of arrays, then we know by (7) that $\mathrm{opt}(I(T^*))$ is R^*/R-approximate. In light of these observations, we introduce the following families of combinatorial designs and their associated numbers:

Definition 2. *Let $k \geq 2$, $p \geq k$ and $q \geq p$ be three integers. Then given any two integers $R \geq 1$ and $R^* \in [R]$, we define $\Gamma(R, R^*, q, p, k)$ as the (possibly empty) set of pairs (Ψ, Φ) of arrays with R rows, q columns, and coefficients in $[q]$ that satisfy the following:*

1. *the components of each row of Ψ take at most p distinct values;*
2. *$(1, \ldots, q)$ occurs R^* times as a row in Φ;*
3. *for all $J = \{c_1, \ldots, c_k\} \subseteq [q]$ with $|J| = k$, subarrays $\Psi^J = (\Psi^{c_1}, \ldots, \Psi^{c_k})$ and $\Phi^J = (\Phi^{c_1}, \ldots, \Phi^{c_k})$ coincide up to the ordering of their rows.*

Furthermore, we define $\gamma(q, p, k)$ as the greatest number $\gamma \in [0, 1]$ for which there exist two natural numbers R, R^* such that $R^*/R = \gamma$ and $\Gamma(R, R^*, q, p, k) \neq \emptyset$.

Table 1. Pairs of arrays that achieve $\gamma(4, 3, 2)$ and $\gamma(5, 3, 2)$.

$\gamma(4, 3, 2) = 2/6 = 1/3$

Ψ^1	Ψ^2	Ψ^3	Ψ^4	Φ^1	Φ^2	Φ^3	Φ^4
1	1	3	4	1	1	3	3
1	2	1	4	1	2	1	3
1	2	3	3	1	2	3	4
1	2	3	3	1	2	3	4
4	1	1	3	4	1	1	4
4	2	3	4	4	2	3	3

$\gamma(5, 3, 2) = 1/6$

Ψ^1	Ψ^2	Ψ^3	Ψ^4	Ψ^5	Φ^1	Φ^2	Φ^3	Φ^4	Φ^5
1	2	4	4	4	1	2	3	4	5
1	3	3	3	5	1	3	4	3	4
2	2	3	2	5	2	2	4	2	4
2	3	4	2	4	2	3	3	2	5
4	4	3	4	5	4	4	3	3	5
4	4	4	3	4	4	4	4	4	4

Table 1 pictures two such combinatorial designs. Since cardinalities $|H_i|$ may be at most $\min\{q, k\} = k$, by requirement 3, a pair $(\Psi, \Phi) \in \Gamma(R, R^*, q, p, k)$ does satisfy (10) regardless of the precise instance I of $k\,CSP-q$ and the precise solution x^* of I we consider. By (7), this implies that $\gamma(q, p, k)$ is a proper lower bound for the differential ratio reached on I at $\mathrm{opt}(I(T^*))$. We thus have established the following:

Lemma 1. *For all integers $k \geq 2$, $p \geq k$ and $q > p$, on any instance of $k\,CSP-q$, solutions that perform the best objective value among those whose coordinates take at most p distinct values are $\gamma(q, p, k)$-approximate.*

To conclude, according to inequality (3), Lemma 1 also establishes that $\gamma(q, p, k)$ is a proper lower bound for the expansion of our reduction:

Theorem 1. *For all integers $k \geq 2$, $p \geq k$ and $q > p$, $k\,CSP-q$ D-reduces to $k\,CSP-p$ with an expansion of $\gamma(q, p, k)$ on the approximation guarantee. The reduction involves $O(q^p)$ instances of $k\,CSP-p$.*

3   A Lower Bound for Numbers $\gamma(q, p, k)$

It remains us to exhibit lower bounds for numbers $\gamma(q, p, k)$. To do so, we mainly present a recursive construction for the case when $p = k$. But first, we mention a few combinatorial identities that are involved in the analysis of this construction. We define:

$$T(a, b) \triangleq \sum_{r=0}^{b} \binom{a}{r}\binom{a-1-r}{b-r}, \qquad a, b \in \mathbb{N}, \ a > b \tag{11}$$

$$S(a, b, c) \triangleq \sum_{r \geq 0} (-1)^r \binom{a}{r}\binom{b-r}{c-r}, \qquad a, b, c \in \mathbb{N}, \ b \geq c \tag{12}$$

Numbers $T(a, b)$ and $S(a, b, c)$ satisfy the following:

Property 1. For all $a, b \in \mathbb{N}$ with $a > b \geq 1$, the following equalities hold:

$$T(a, b) = 2^b \binom{a-1}{b} + T(a-1, b-1) \tag{13}$$

$$= 2^b \binom{a}{b} - T(a, b-1) \tag{14}$$

$$= 2 \sum_{c=b}^{a-1} T(c, b-1) + 1 \tag{15}$$

Proof (sketch). Recursions (13) and (14) are obtained applying Pascal's rule to coefficients of the form respectively $\binom{a}{r}$ and $\binom{a-1-r}{b-r}$. Identity (15) can then be deduced from those recursions.   □

Property 2. For all $a, b, c \in \mathbb{N}$ with $b \geq \max\{a, c\}$, $S(a, b, c)$ equals $\binom{b-a}{c}$.

Proof (sketch). By induction on integer $b - a$.   □

3.1   A Recursive Construction for Families $\Gamma(R, 1, q, k, k)$

This section is dedicated to the proof of the following Theorem:

Theorem 2. *Let $k \geq 2$ and $q \geq k$ be two integers. Then $\gamma(q, k, k)$ is equal to 1 if $q = k$, and bounded below by $2/(T(q, k) + 1)$ otherwise.*

The case when $q = k$ is trivial, considering $\Psi = \Phi = \{(1, \ldots, k)\}$. For greater integers $q - k$, the argument relies on the following Lemma:

Lemma 2. *Let $k \geq 2$, $q > k$, $R^* \geq 1$ and $R \geq R^*$ be four integers such that $\Gamma(R, R^*, q - 1, k, k) \neq \emptyset$. Then $\Gamma(R + T(q - 1, k - 1), 1, q, k, k) \neq \emptyset$.*

Proof. Let $(\Psi, \Phi) \in \Gamma(R, R^*, q - 1, k, k)$. We assume *w.l.o.g.* that $(1, \ldots, q - 1)$ occurs at row 1 in Φ. Our goal is to add in arrays Ψ and Φ a single new column, and as few rows as possible, so as to obtain a new pair (Ψ, Φ) of arrays with q columns and coefficients in $[q]$ that meets requirements 1, 2, 3 of Definition 2.

Table 2. Construction of a pair of arrays in $\Gamma(R + T(q - 1, k - 1), 1, q, k, k)$ starting with a pair $(\Psi, \Phi) \in \Gamma(R, R^*, q - 1, k, k)$ with $\Phi_1 = (1, \ldots, q - 1)$.

1. Insert in Ψ and Φ new columns Ψ^q and Φ^q where:
   (a) $\Psi_r^q = \Psi_r^1, r \in [R]$;
   (b) $\Phi_1^q = q$ and $\Phi_r^q = \Phi_r^1, r \in \{2, \ldots, R\}$.
2. For $h = k - 1$ down to 0:
   (a) For all h-element subsets H of $[q - 1]$:
       i. let $\alpha(H) \in [q]^{q-1}$ be defined for all $c \in [q - 1]$ by $\alpha(H)_c = c$ if $c \in H$, and q otherwise;
       ii. if $h \equiv k - 1 \bmod 2$, then insert $\binom{q-2-h}{k-1-h}$ copies of row vectors $(\alpha(H), q)$ and $(\alpha(H), 1)$ in respectively Ψ and Φ;
       iii. otherwise, insert $\binom{q-2-h}{k-1-h}$ copies of row vectors $(\alpha(H), q)$ and $(\alpha(H), 1)$ in respectively Φ and Ψ.

The construction is described in Table 2. We make a few comments before proving its rightness. Step 1 first inserts a qth column in the arrays. If we set Ψ^q to Ψ^1 and Φ^q to Φ^1, then (Ψ, Φ) trivially fulfils requirements 1 and 3. However, as $(1, \ldots, q)$ mu st occur at least once as a row in Φ, we assign value q rather than Φ_1^1 to Φ_1^q. As a result, (Ψ, Φ) violates requirement 3. Hence, during Step 2, we insert new rows in the arrays until they satisfy this requirement.

Let $J = \{c_1, \ldots, c_{k-1}\}$ be a $(k - 1)$-element subset of $[q - 1]$, and $v = (c_1, \ldots, c_{k-1})$. After Step 1, row 1 is the single row of subarray (Φ^J, Φ^q) that coincides with (v, q), while there is no such row in (Ψ^J, Ψ^q). As $(\Phi_1^J, \Phi_1^1) = (v, 1)$ while $(\Phi, \Psi) \in \Gamma(R, R^*, q - 1, k, k)$, (Φ^J, Φ^q) symmetrically coincides with $(v, 1)$ on one less row than (Ψ^J, Ψ^q) does. Iteration $h = k - 1$ corrects this precise imbalance when it inserts row vectors $(\alpha(J), q)$ in Ψ and $(\alpha(J), 1)$ in Φ.

However, this iteration also introduces new violations of requirement 3. Notably, let $s \in [k - 1]$, and $v = (c_1, \ldots, c_{s-1}, q, c_{s+1}, \ldots, c_{k-1})$. Then iteration $h = k - 1$ inserts in each array a new row u with $u_J = v$ each time it selects a $(k-1)$-element subset H of $[q-1]$ with $c_1,\dots,c_{s-1},c_{s+1},\dots,c_{k-1} \in H$ and $c_s \notin H$. Since there are $q-1-(k-1) = q-k$ such subsets, we deduce that at the end of this iteration, vectors (v,q) and $(v,1)$ occur respectively $q-k$ and 0 times as a row in (Ψ^J,Ψ^q), while the converse holds for (Φ^J,Φ^q). Iteration $h = k-2$ corrects this precise imbalance when it inserts $q-k$ copies of row vectors $(\alpha(J\backslash\{c_s\}),1)$ and $(\alpha(J\backslash\{c_s\}),q)$ in respectively Ψ and Φ. More generally, for all $h \in \{0,\dots,k-1\}$, given any $v \in \{c_1,q\} \times \dots \times \{c_{k-1},q\}$ with exactly h coordinates in $[q-1]$ and any $a \in \{1,q\}$, iteration h ensures that (v,a) occurs the same number of times as a row in (Ψ^J,Ψ^q) and (Φ^J,Φ^q).

Table 3. The recursive construction for families $\Gamma((T(5,2)+1)/2,1,5,2,2)$ and $\Gamma((T(4,3)+1)/2,1,4,3,3)$ of combinatorial designs.

$2/(T(5,2)+1) = 1/16$

Ψ^1	Ψ^2	Ψ^2	Ψ^3	Ψ^4	Φ^1	Φ^2	Φ^3	Φ^3	Φ^5
1	2	1	1	1	1	2	3	4	5
1	3	3	1	1	1	3	1	1	1
3	2	3	3	3	3	2	1	3	3
3	3	1	3	3	3	3	3	3	3
1	4	4	4	1	1	4	4	1	1
4	2	4	4	4	4	2	4	1	4
4	4	3	4	4	4	4	3	1	4
4	4	4	1	4	4	4	4	4	4
4	4	4	1	4	4	4	4	4	4
1	5	5	5	5	1	5	5	5	1
5	2	5	5	5	5	2	5	5	1
5	5	3	5	5	5	5	3	5	1
5	5	5	4	5	5	5	5	4	1
5	5	5	5	1	5	5	5	5	5
5	5	5	5	1	5	5	5	5	5
5	5	5	5	1	5	5	5	5	5

$2/(T(4,3)+1) = 1/8$

Ψ^1	Ψ^2	Ψ^3	Ψ^4	Φ^1	Φ^2	Φ^3	Φ^4
1	2	3	1	1	2	3	4
1	2	4	4	1	2	4	1
1	4	3	4	1	4	3	1
4	2	3	4	4	2	3	1
1	4	4	1	1	4	4	4
4	2	4	1	4	2	4	4
4	4	3	1	4	4	3	4
4	4	4	4	4	4	4	1

We now prove that, at the end of the process, $(\Psi,\Phi) \in \Gamma(R',1,q,k,k)$ where $R' = R + T(q-1,k-1)$. By construction, the resulting arrays satisfy that:

- their number R' of rows is $R + \sum_{h=0}^{k-1} \binom{q-1}{h}\binom{q-2-h}{k-1-h} = R + T(q-1,k-1)$;
- in Ψ, the coefficients of every row take at most k distinct values;
- in Φ, row 1 is the single row that coincides with $(1,\dots,q)$.

It remains us to show that (Ψ,Φ) fulfils requirement 3. Let $J = (c_1,\dots,c_k)$ be a strictly increasing sequence of integers in $[q]$, and v be a vector of $[q]^k$. We shall establish that subarrays Ψ^J and Φ^J coincide with v on the same number of rows. The case when $q \notin J$ is trivial. Thus assume $c_k = q$. We consider two cases:

- $v \notin \{c_1, q\} \times \ldots \times \{c_{k-1}, q\} \times \{1, q\}$. $\Psi_r^J = v$ or $\Phi_r^J = v$ may not occur unless $r \in [R]$. Let $r \in [R]$ and let $K = (c_1, \ldots, c_{k-1}, 1)$. Then $\Psi_r^J = \Psi_r^K$, while $\Phi_r^J = \Phi_r^K$ unless $r = 1$, in which case $\Phi_1^J \neq v \neq \Phi_1^K$. Since the original pair of arrays belongs to $\Gamma(R, R^*, q-1, k, k)$, we deduce that Ψ^J and Φ^J indeed coincide with v on the same number of rows.

- $v \in \{c_1, q\} \times \ldots \times \{c_{k-1}, q\} \times \{1, q\}$. If $(v_1, \ldots, v_{k-1}) = (c_1, \ldots, c_{k-1})$, then we already argued that iteration $h = k - 1$ of Step 2 corrects the imbalance induced by assignment $\Phi_1^q = q$. Otherwise, let L refer to the set of indices $c_s \in \{c_1, \ldots, c_{k-1}\}$ such that $v_s = c_s$, and let $\ell = |L|$. As $\ell \leq k - 2$, $\Psi_r^J = v$ or $\Phi_r^J = v$ may not occur unless $r > R$. So consider an iteration $h \in \{0, \ldots, k-1\}$ of Step 2. For each h-element subset H of $[q-1]$ with $L \subseteq H$ and $H \backslash L \subseteq [q-1] \backslash J$, this iteration generates $\binom{q-2-h}{k-1-h}$ rows u with $u_J = v$. If $v_k = q$ (resp., $v_k = 1$), then these rows occur in Ψ (resp., in Φ) iff h has the same parity as $k - 1$. Since there are $\binom{q-k}{h-\ell}$ such subsets H of $[q-1]$, we deduce:

$$|\{r \in [R'] \,|\, \Psi_r^J = v\}| - |\{r \in [R'] \,|\, \Phi_r^J = v\}|$$
$$= \pm \sum_{h=\ell}^{k-1} (-1)^{k-1-h} \binom{q-2-h}{k-1-h} \binom{q-k}{h-\ell}$$
$$= \pm \sum_{r=0}^{k-1-\ell} (-1)^{k-1-\ell-r} \binom{q-2-\ell-r}{k-1-\ell-r} \binom{q-k}{r} = \pm S(q-k, q-2-\ell, k-1-\ell)$$

Now we know from Property 2 that $S(q-k, q-2-\ell, k-1-\ell)$ is equal to $\binom{k-2-\ell}{k-1-\ell}$, which is 0. We conclude that (Ψ, Φ) indeed belongs to $\Gamma(R + T(q-1, k-1), 1, q, k, k)$.   □

The proof of Theorem 2 is straightforward from Lemma 2. Namely, given two integers $k \geq 2$ and $q \geq k$, we consider the following recursive construction:

1. Set $\Psi = \{(1, \ldots, k)\}$ and $\Phi = \{(1, \ldots, k)\}$.
2. For $a = k + 1$ to q, apply construction underlying Lemma 2 to (Ψ, Φ).

Table 3 illustrates the construction when $k \in \{2, 3\}$. On the one hand, in view of Lemma 2, the resulting pair (Ψ, Φ) of arrays belongs to $\Gamma(R, 1, q, k, k)$ where

$$R = 1 + \sum_{a=k+1}^{q} T(a-1, k-1) = 1 + \sum_{a=k}^{q-1} T(a, k-1)$$

On the other hand, by (15), we have:

$$1 + \sum_{a=k}^{q-1} T(a, k-1) = 1 + (T(q, k) - 1)/2 = (T(q, k) + 1)/2$$

This completes the proof of Theorem 2.

3.2   Deduced Approximation Results

Let $k \geq 2$, $p \geq k$ and $q > p$ be three integers. If $p = k$, then Theorem 2 together with Theorem 1 provides a lower bound of $2/(T(q, k) + 1)$ for the expansion of our reduction from k CSP−q to k CSP−k. We seek an estimate of $2/(T(q, k) + 1)$.

Property 3. For all $a, b \in \mathbb{N}$ with $a > b \geq 2$, $(a - b/2)^b$ is a proper lower bound for $(T(a, b) + 1)/2$.

Proof. Applying recursions first (14), and then (13), one gets equality:

$$T(a, b) + 1 = 2^b \binom{a}{b} - 2^{b-1} \binom{a-1}{b-1} - T(a-1, b-2) + 1 \tag{16}$$

On the one hand, we deduce again from (13) that $T(a - 1, b - 2) - 1 \geq T(a - b + 1, 0) - 1 = 0$. On the other hand, we can rewrite $2^b \binom{a}{b} - 2^{b-1} \binom{a-1}{b-1}$ as:

$$2^b \binom{a}{b} - 2^{b-1} \binom{a-1}{b-1} = 2(a - b/2) \times 2^{b-1}/b! \times \prod_{i=0}^{b-2} (a - 1 - i)$$

On the one hand, $2^{b-1}/b! \leq 1$. On the other hand, the inequality of arithmetic and geometric means yields inequality $\prod_{i=0}^{b-2}(a - 1 - i) \leq (a - b/2)^{b-1}$. $\qquad\square$

Table 4. Numbers $\gamma(q, p, k)$ and $\gamma_E(q, p, k)$ for some triples (q, p, k). These values (and the underlying pairs of arrays) were calculated by computer.

| | $\gamma(q, p, 2)$ | | | | $\gamma(q, p, 3)$ | | | $\gamma_E(q, p, 2)$ | | | | | $\gamma_E(q, p, 3)$ | | | |
| | q | | | | q | | | q | | | | | q | | | |
p	3	4	5	6	4	5	6	3	4	5	6	7	4	5	6	7
2	1/4	1/9	1/16	1/25	—	—	—	1/3	1/4	1/5	9/59	1/7	—	—	—	—
3	—	1/3	1/6	1/10	1/8	1/25	1/56	—	1/2	2/5	4/13	2/7	1/4	1/11	$\frac{38425}{701342}$	—
4	—	—	4/9	1/4	—	1/5	2/27	—	—	3/5	7/15	3/7	—	1/3	1/6	5/52
5	—	—	—	1/2	—	—	1/4	—	—	—	2/3	11/21	—	—	4/9	2/9
6	—	—	—	—	—	—	—	—	—	—	—	5/7	—	—	—	1/2

Hence, when $p = k$, the expansion of the reduction is at least $1/(q - k/2)^k$. In particular, if $k = 2$, then $2/(T(q, 2) + 1) = 1/(q-1)^2$. As $2CSP - 2$ is approximable within factor $2 - \pi/2$, Corollary 2 thus holds. When $p > k$, simply observe that we have:

$$\gamma(q, p, k) \geq \gamma(q - p + k, k, k) \geq 2/(T(q - p + k, k) + 1) \tag{17}$$

Indeed, let $a = q - p + k$, and assume that R and R^* are two integers such that $\Gamma(R, R^*, a, k, k) \neq \emptyset$. Let then $(\Psi, \Phi) \in \Gamma(R, R^*, a, k, k)$. Substituting for every row $u = (u_1, \ldots, u_a)$ of Ψ and Φ row vector $u = (u_1, \ldots, u_a, a + 1, \ldots, q)$, one gets a new pair of arrays that trivially belongs to $\Gamma(R, R^*, q, p, k)$. Hence, combining Theorem 1, inequality (17) and Property 3, one obtains Corollary 1.

Table 5. Pairs of arrays that achieve $\gamma_E(5,3,2)$ and $\gamma_E(5,4,3)$.

$\gamma_E(5,3,2) = 4/10 = 2/5$

Ψ^0	Ψ^1	Ψ^2	Ψ^3	Ψ^4	Φ^0	Φ^1	Φ^2	Φ^3	Φ^4
0	0	1	3	3	0	0	0	0	0
0	0	2	2	3	0	0	4	4	2
0	1	0	1	2	0	1	2	3	4
0	1	2	0	1	0	1	2	3	4
0	1	3	3	0	0	1	2	3	4
0	1	4	0	4	0	1	2	3	4
0	2	2	3	0	0	2	0	2	1
0	2	2	4	4	0	2	4	1	3
0	3	4	3	4	0	3	3	1	3
0	4	0	1	4	0	4	1	0	0

$\gamma_E(5,4,3) = 4/12 = 1/3$

Ψ^0	Ψ^1	Ψ^2	Ψ^3	Ψ^4	Φ^0	Φ^1	Φ^2	Φ^3	Φ^4
0	0	1	2	3	0	0	1	3	4
0	0	2	3	4	0	0	2	2	4
0	0	2	3	4	0	0	2	3	3
0	1	1	3	4	0	1	1	2	3
0	1	2	2	4	0	1	2	3	4
0	1	2	3	0	0	1	2	3	4
0	1	2	3	3	0	1	2	3	4
0	1	2	4	4	0	1	2	3	4
0	1	3	3	4	0	1	3	4	0
0	2	2	3	4	0	2	2	3	0
0	2	2	3	4	0	2	2	4	4
0	2	3	4	0	0	2	3	3	4

4   Concluding Remarks

We make a few remarks as regards the combinatorial designs we introduced. When $p = k$, we think that $2/(T(q,k) + 1)$ is the exact value of $\gamma(q,k,k)$. The question whether $2/(T(q,k) + 1)$ is optimal, though, still has to be settled. By contrast, when $p > k$, the only estimate of $\gamma(q,p,k)$ we are aware of is the trivial lower bound of $\gamma(q - p + k, k, k)$. Yet, it the most likely holds given three integers $k \geq 2$, $p \geq k$ and $q > p$ that $\gamma(q + 1, p + 1, k) > \gamma(q, p, k)$. Table 4, in which we indicate the value of $\gamma(q,p,k)$ for a few triples (q,p,k), illustrates this fact quite well. Now, according to Lemma 1, these numbers provide for optimization CSPs with a bounded arity some lower bound on "how much we lose" on their optimum value when decreasing the size of their alphabet. This is a good motivation for studying families $\Gamma(R, R^*, q, p, k)$ of combinatorial designs in case when $p > k$.

Likewise, let $\mathsf{k}\,\mathsf{CSP}(\mathcal{E}_\mathsf{q})$ refer to the (generalized) optimization CSP over $\mathbb{Z}_q$ where functions P_i that occur in the constraints have arity at most k, are rational-valued, and satisfy:

$$P_i(y_1 + a, \ldots, y_{k_i} + a) = P_i(y_1, \ldots, y_{k_i}), \qquad y \in \mathbb{Z}_q^{k_i}, \; a \in \mathbb{Z}_q \qquad (18)$$

$\mathsf{k}\,\mathsf{CSP}(\mathcal{E}_\mathsf{q})$ notably covers the restriction of $\mathsf{Lin}{-}\mathsf{q}$ to equations of the form $\alpha_1 y_1 + \ldots + \alpha_{k-1} y_{k-1} - (\alpha_1 + \ldots + \alpha_{k-1}) y_k \equiv \alpha_0 \bmod q$. Given an integer a, we denote by $\mathbf{a}$ the vector—of dimension that depends on the context—$(a, \ldots, a)$. On an instance I of $\mathsf{k}\,\mathsf{CSP}(\mathcal{E}_\mathsf{q})$, any constraint C_i evaluates the same on any two entries x_{J_i} and $x_{J_i} + \mathbf{a}$. The objective function $v(I, .)$ similarly evaluates the same on any two entries x and $x + \mathbf{a}$. This suggests to consider the slight relaxation $\Gamma_E(R, R^*, q, p, k)$ of families $\Gamma(R, R^*, q, p, k)$ where Ψ and Φ have coefficients in $\mathbb{Z}_q$ and, rather than requirements 2 and 3, satisfy the two conditions below:

2'. $\Phi_r \in \{(a, 1 + a, \ldots, q - 1 + a) \,|\, a \in \mathbb{Z}_q\}$ holds for R^* indices $r \in [R]$;
3'. for all $J \subseteq [q]$ with $|J| = k$ and all $v \in \{0\} \times \mathbb{Z}_q^{k-1}$, Ψ^J and Φ^J coincide with a vector in $\{(v_1 + a, \ldots, v_k + a) \,|\, a \in \mathbb{Z}_q\}$ on the same number of rows.

We define numbers $\gamma_E(q,p,k)$ just as the same as numbers $\gamma(q,p,k)$. Table 5 pictures two such pairs of arrays, while Table 4 provides the value of $\gamma_E(q,p,k)$ for some triples (q,p,k). Using a similar argument as for the general case, it is not hard to see that, when restricting to input instances of $k\,\mathsf{CSP}(\mathcal{E}_q)$, the reduction we proposed from $k\,\mathsf{CSP}-q$ to $k\,\mathsf{CSP}-p$ preserves the differential ratio up to a multiplicative factor of $\gamma_E(q,p,k)$. Notably, as $\gamma_E(q,2,2)=1/q$, $q\in\{3,4,5,7\}$, it follows from [1] that when $q\in\{3,4,5,7\}$, $2\,\mathsf{CSP}(\mathcal{E}_q)$ is approximable within factor $0.429/q$ (and not only $0.429/(q-1)^2$). Therefore, one also shall investigate families $\Gamma_E(R,R^*,q,p,k)$ of combinatorial designs, starting with the case when $p=k=2$.

References

1. Nesterov, Y.: Semidefinite relaxation and nonconvex quadratic optimization. Opt. Methods Softw. **9**(1–3), 141–160 (1998)
2. Garey, M., Johnson, D., Stockmeyer, L.: Some simplified NP-complete graph problems. Theor. Comput. Sci. **1**(3), 237–267 (1976)
3. Demange, M., Paschos, V.T.: On an approximation measure founded on the links between optimization and polynomial approximation theory. Theor. Comput. Sci. **158**(1–2), 117–141 (1996)
4. Escoffier, B., Paschos, V.T.: Differential approximation of MIN SAT, MAX SAT and related problems. EJOR **181**(2), 620–633 (2007)
5. Goemans, M.X., Williamson, D.P.: Improved approximation algorithms for maximum cut and satisfiability problems using semidefinite programming. J. ACM **42**(6), 1115–1145 (1995)
6. Hariharan, R., Mahajan, S.: Derandomizing approximation algorithms based on semidefinite programming. SIAM J. Comput. **28**(5), 1641–1663 (1999)
7. Charikar, M., Makarychev, K., Makarychev, Y.: Near-optimal algorithms for maximum constraint satisfaction problems. ACM Trans. Algorithms **5**(3), 1–14 (2009)

Finding Minimum Stopping and Trapping Sets: An Integer Linear Programming Approach

Alvaro Velasquez[1]([✉]), K. Subramani[2], and Steven L. Drager[3]

[1] Department of Computer Science, University of Central Florida, Orlando, FL, USA
velasquez@cs.ucf.edu
[2] LCSEE, West Virginia University, Morgantown, WV, USA
ksmani@csee.wvu.edu
[3] Information Directorate, Air Force Research Laboratory, Rome, NY, USA
steven.drager@us.af.mil

Abstract. In this paper, we discuss the problems of finding minimum stopping sets and trapping sets in Tanner graphs, using integer linear programming. These problems are important for establishing reliable communication across noisy channels. Indeed, stopping sets and trapping sets correspond to combinatorial structures in information-theoretic codes which lead to errors in decoding once a message is received. We present integer linear programs (ILPs) for finding stopping sets and several trapping set variants. In the journal version of this paper, we prove that two of these trapping set problem variants are **NP-hard** for the first time. The effectiveness of our approach is demonstrated by finding stopping sets of size up to 48 in the $(4896, 2474)$ Margulis code. This compares favorably to the current state-of-the-art, which finds stopping sets of size up to 26. For the trapping set problems, we show for which cases an ILP yields very efficient solutions and for which cases it performs poorly. The proposed approach is applicable to codes represented by regular and irregular graphs alike.

1 Introduction

The use of codes as mathematical formalisms lies at the heart of reliable communication. These codes provide a fault-tolerant structure to messages by appending redundant check bits via a generator matrix. After these bits have been added, the codeword is sent across a noisy channel, such as the binary erasure channel (BEC), binary symmetric channel (BSC), and additive white Gaussian noise channel (AWGNC). These channels differ in the way noise is modeled. For example, the BEC introduces probabilities of bit erasures while the BSC accounts for bit flips. Once the message arrives at its destination, it is decoded via a parity-check matrix. It is during this decoding step that certain combinatorial structures in the code lead to errors. In this paper, we are concerned with finding such structures. This problem is central to the design of better codes and, consequently, more reliable communication.

© Springer International Publishing AG, part of Springer Nature 2018
J. Lee et al. (Eds.): ISCO 2018, LNCS 10856, pp. 402–415, 2018.
https://doi.org/10.1007/978-3-319-96151-4_34

One of the principal metrics used to test the reliability of a code is the bit error rate (BER). This corresponds to the number of bit errors per unit time. It is well-known that the BER of a code decreases as the signal-to-noise ratio increases. However, there is a point in which the BER plateaus. This region in the performance curve is known as the error floor of the code. Experiments were performed in [1] to analyze the error floor of the BSC and AWGNC. It was demonstrated that all decoding failures in these channels were due to the presence of trapping sets. These sets correspond to variables that are not eventually corrected by the decoder, thus causing failures in decoding when using iterative algorithms [2]. Similar experiments have demonstrated that decoding failures over the BEC are due to the presence of stopping sets [3]. More specifically, it is the smallest stopping and trapping sets that cause poor decoding performance in their respective channels [4].

Stopping sets and trapping sets are defined by simple combinatorial structures in the graph representation of the underlying code. There have been efforts to estimate their minimum size [5,6], find the minimum sets [7], and enumerate such sets for up to some small-size parameter [8]. In this paper, we make two contributions. First, we propose integer linear programming solutions to these problems. We improve on the results of [7], which demonstrate that the minimum stopping set in the $(4896, 2474)$ Margulis code [9] is of size 24 and that no stopping sets of size 25 and 26 exist. Our results establish that the next largest stopping sets in said code are of sizes 36 and 48 (See Figs. 10 and 11 in https://www.cs.ucf.edu/~velasquez/StoppingSets/). As a point of reference, the number of points in the search spaces for finding stopping sets of sizes 26, 36, and 48 are $\binom{4896}{26} \approx 2 \times 10^{69}$, $\binom{4896}{36} \approx 1.61 \times 10^{91}$, and $\binom{4896}{48} \approx 8.28 \times 10^{115}$, respectively. This demonstrates the efficiency of a programming-based approach to explore much larger parameter spaces than competing methods. The second contribution we make pertains to the previously unknown complexities of two trapping set problem variants. In the journal version of this paper, we prove that these variants are **NP-hard**, thereby rounding out the complexity results in the literature.

The remainder of this paper is organized as follows. Section 2 provides background information and definitions for the stopping and trapping set problems. Known and new complexity results for these problems are presented in Sect. 3. A brief exposition of related work is provided in Sect. 4. Our approach based on integer linear programming follows in Sect. 5 and experimental results are shown in Sect. 6. We conclude and allude to future work in Sect. 7.

2   Preliminaries

An (n, k) code is one whose codewords have length n and whose dimension is k. The dimension k of the code specifies the number of linearly independent codewords that form the basis $\underline{c}_1, \ldots, \underline{c}_k \in \{0, 1\}^n$ for the code. That is, any codeword can be expressed as a linear combination of these basis vectors. Any given codeword of length n contains k original bits of information and $n - k$

redundant check bits that are used to detect and correct errors that arise during message transmission across a noisy channel.

Given an (n, k) code, its corresponding parity-check matrix $H \in \{0, 1\}^{m \times n}$, where $m = n - k$, defines the linear relations among codeword variables. Each column in H corresponds to a bit in the codeword and each row corresponds to a redundant check bit. Given a codeword $\underline{c} = \underline{c}_1, \ldots, \underline{c}_n$, the entry H_{ij} is 1 if $\underline{c}_j$ is involved in a check operation, which is used to detect errors after transmission. For any such matrix H, let $G = (V \cup C, E)$ denote its representation as a bipartite graph, where $V = \{v_1, \ldots, v_n\}$ and $C = \{c_1, \ldots, c_m\}$ are the sets of variable and check nodes, and $E = \{(v_i, c_j) | H_{ji} = 1\}$ defines the adjacency set. This is known as the Tanner graph of a code. We can now define the stopping and trapping set problems.

Definition 1 (Stopping set). *Given a Tanner graph $G = (V \cup C, E)$, a stopping set $S \subseteq V$ is a set of variable nodes such that all neighbors of nodes in S are connected to S at least twice [7].*

As an example, suppose we are given the parity-check matrix H below with Tanner graph $G = (V \cup C, E)$. We can determine the minimum stopping set $S = \{v_7, v_9\}$ as pictured in Fig. 1. For ease of presentation, we often argue in terms of induced subgraphs. Given a graph $G = (V \cup C, E)$, we denote by G_S the subgraph induced by the set $S \subseteq V$. That is, $G_S = (S \cup C^S, E^S)$, where $C^S = \{c_j \in C | s_i \in S, (s_i, c_j) \in E\}$ and $E^S = \{(s_i, c_j) | s_i \in S, (s_i, c_j) \in E\}$.

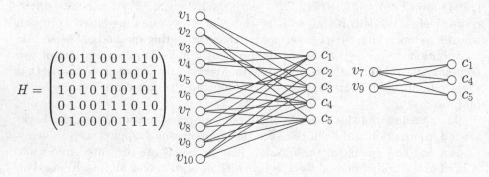

Fig. 1. (*left*) Parity-check matrix H for some code. (*center*) Tanner graph $G = (V \cup C, E)$ of H. (*right*) Subgraph G_S induced by the minimum stopping set $S = \{v_7, v_9\}$ in G.

Definition 2 ((a, b)-trapping set). *Given a Tanner graph $G = (V \cup C, E)$, an (a, b)-trapping set $T \subseteq V$ is a set of variable nodes with cardinality $|T| = a$ such that b neighbors of T are connected to T an odd number of times [10].*

It is worth noting that there exist several trapping set variants. The most popular variant is known as an elementary trapping set. These sets, defined below, are frequently the structures that lead to iterative decoding failures [11]. That is, the minimum trapping sets in a Tanner graph are often of elementary form. See Fig. 2 for an example. An interesting lower-bound result that helps explain this prevalence was recently established in [12], where it was shown that the lower bound on the size of non-elementary trapping sets is greater than that of their elementary counterparts.

Definition 3 (Elementary (a, b)-trapping set). *Given a Tanner graph* $G = (V \cup C, E)$, *an elementary* (a, b)-*trapping set* $T \subseteq V$ *is a set of variable nodes with cardinality* $|T| = a$ *such that* b *neighbors of* T *are connected to* T *exactly once and the remaining neighbors are connected to* T *exactly twice [11].*

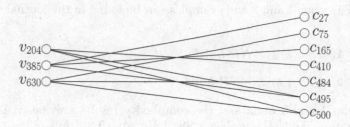

Fig. 2. Elementary $(3, 5)$-trapping set found in the $(1008, 504)$ Mackay code [13]. Note that, of the seven neighbors of $T = \{v_{204}, v_{385}, v_{630}\}$, only five are connected exactly once to T.

3   Complexity

Assume we are given the Tanner graph $G = (V \cup C, E)$ of some code. For the remainder of this paper, we are concerned with the computational complexities of and solutions to the following optimization problems:

- SS: Find a minimum-cardinality stopping set $S \subseteq V$.
- TS$(a, b)|_a$: Find a trapping set $T \subseteq V$ such that $|T| = a$ and b is minimized.
- TS$(a, b)|_b$: Find a minimum-cardinality trapping set $T \subseteq V$ for a given b.
- ETS$(a, b)|_a$: Find an elementary trapping set $T \subseteq V$ such that $|T| = a$ and b is minimized.
- ETS$(a, b)|_b$: Find a minimum-cardinality elementary trapping set $T \subseteq V$ for a given b.

While the foregoing problems are straightforward to define, their apparent simplicity belies significant complexity. In the case of the SS problem, it has been shown that it is **NP-hard** to approximate within a logarithmic factor of the problem input size [10]. An even stronger negative result is obtained when we assume that **NP** $\not\subseteq$ **DTIME**$(n^{\text{poly}(\log n)})$, where n is the size of the problem's encoding and $k \in \mathbb{N}$. Under such an assumption, it was shown in [14] that there is no polynomial-time algorithm to approximate a solution to the SS problem within a factor of $2^{(\log n)^{1-\epsilon}}$ for any $\epsilon > 0$.

Not surprisingly, hardness and inapproximability results have also been established for some trapping set problems. Namely, that it is **NP-Hard** to α-approximate a solution to the $TS(a, b)|_a$ and $ETS(a, b)|_b$ problems [10]. However, the complexities of $TS(a, b)|_b$ and $ETS(a, b)|_a$ have remained an open problem which we resolve in this paper. We prove the hardness of $TS(a, b)|_b$ and $ETS(a, b)|_a$ by demonstrating that efficient solutions to these problems would yield efficient solutions to the MAX-k-LIN and REGULAR-k-INDEPENDENT-SET problems, which are **NP-hard**. More specifically, the MAX-k-LIN problem is hard for $k \geq 2$ [15] and the REGULAR-k-INDEPENDENT-SET problem is a special instance of the independent set problem on regular graphs, which is known to be hard for graphs of degree $\delta_V \geq 3$ [16]. Due to space constraints, the proofs of Theorems 1 and 2 and examples can be found in the journal version of the paper.

Theorem 1. *$TS(a, b)|_b$ is **NP-hard**.*

Theorem 2. *$ETS(a, b)|_a$ is **NP-hard**.*

Theorems 1 and 2 round out the complexity results previously established in the literature. Thus, the problems SS, $TS(a, b)|_a$, $TS(a, b)|_b$, $ETS(a, b)|_a$, and $ETS(a, b)|_b$ all belong to the **NP-hard** complexity class.

4   Related Work

Many of the methods proposed in the literature stem from work presented in [17], wherein a search of the code's iterative decoding trees is performed. Boolean functions are defined on these trees to calculate the bit-error rate of a code. Any time the bit error rate is 1, there are one, or more stopping sets in the decoding tree. By taking as input a parity-check matrix H and a subset of variable nodes $V' \subset V$, this method outputs the stopping sets that contain V'. This method is used to find stopping sets S of size $|S| \leq 13$ in codes of length up to $n \approx 500$. An extension has been proposed in [18] to enumerate trapping sets of size $|T| \leq 11$. Based on this approach, a branch-and-bound algorithm for enumerating stopping sets is presented in [7], wherein the lower bound on the size of the sets is computed via linear programming relaxations. Each variable can be constrained to be included in or excluded from the stopping set, or it can be unconstrained. The branching step chooses the unconstrained variable node v with neighboring check nodes connected to the most unconstrained nodes. This creates two new constraints. Namely, the ones that arise by including/excluding v in the stopping set. This approach was used successfully to search for stopping sets of size up to 26 in codes with thousands of nodes, like the $(4896, 2474)$ Margulis code [9]. A similar algorithm is proposed for finding trapping sets in [8]. Since modern integer programming solvers use similar branch-and-bound techniques complemented with cutting plane algorithms, we conjecture that the aforementioned approach is subsumed by a purely programming-based method like the one we present in the next section.

In [19], it is demonstrated that an efficient procedure to find stopping sets can be used as a subroutine to find trapping sets. Given a Tanner graph G, the algorithm proceeds by finding k edges such that no two edges share an endpoint. The parity-check matrix is then modified to eliminate vertices in G that are not endpoints in this set. The algorithm from [17] is called using this modified parity-check matrix and endpoint vertices as input. This solves the $ETS(a, b)|_b$ problem with a runtime of $O(n^b)$, where n is the length of the code. Another effective approach which finds trapping sets of increasing size from an initial set of small trapping sets is presented in [20]. This method exploits the relationship between trapping sets and cycles in the Tanner graph, which that has been studied in [21]. However, while it is very efficient at finding small trapping sets, its reliance on cycle enumeration as input makes it inadequate for large problem instances.

Probabilistic approaches have also been proposed. In [5], random subsets of the variables in a code are sampled and the minimum stopping sets in these subsets are found. Even though this is an efficient way to partition the search space, there is a probability that the minimum stopping set is missed during this procedure. Indeed, the size of the minimum stopping set in a $(1008, 504)$ MacKay code is predicted to be 28, which we prove to be incorrect in Sect. 6. The authors demonstrate how this error probability can be computed and how it decreases as the maximum number of attempts allowed is increased.

A different probabilistic approach is presented in [6]. This approach iteratively splits a permutation of the parity-check matrix H into two sub-matrices and uses a probabilistic codeword-finding algorithm to find stopping sets in said sub-matrices. This is repeated until a specified maximum iteration bound is reached. The probability of error is improved over the work in [5] and it correctly predicts with high probability that the size of the minimum stopping set in a $(1008, 504)$ MacKay code is 26.

5   Methodology

We propose integer linear programs (ILPs) for the SS, $TS(a, b)|_a$, $TS(a, b)|_b$, $ETS(a, b)|_a$, and $ETS(a, b)|_b$ problems given the Tanner graph representation $G = (V \cup C, E)$ of a parity-check matrix $H \in \{0, 1\}^{m \times n}$. For the remainder of this paper, we adopt the notation $[k]$ to denote the set $\{1, 2, \ldots, k\}$.

5.1   Stopping Sets

We want to find the stopping set $S \subseteq V$ of minimum cardinality. To do so, we define the following integer linear program with n variables and $|E| + 1$ constraints, where $x_i \in \{0, 1\}$ is 1 if $v_i \in V$ is in the minimum stopping set.

Integer Linear Program for SS:

$$\min \sum_{v_i \in V} x_i \text{ such that}$$

$$\left(\sum_{(v_i,c_j),(c_j,v_k) \in E} x_k \right) - 2x_i \geq 0, \ \forall (v_i, c_j) \in E$$

$$\sum_{v_i \in V} x_i \geq 2$$

$$x_i \in \{0, 1\}, \ \forall i \in [n]$$

The first constraint looks at all paths $\{(v_i, c_j), (c_j, v_k)\}$ of length 2 from a node $v_i \in S$ in the minimum stopping set. If the destination node v_k is in the stopping set S, then its indicator variable x_k is 1. Adding over all the indicator variables corresponding to the destination nodes of all paths of length 2 gives us the number of variable nodes connected back to S. In order to check if this value is at least 2, we subtract from it the factor $2x_i$. Two cases arise. First, if x_i is 1, then v_i is in the stopping set S and all of its neighbors must be connected back to S at least twice in order to satisfy the inequality. Second, if x_i is 0, then v_i is not in S and the inequality is trivially satisfied. This satisfies the connectivity condition in Definition 1. That is, every neighbor of a node in S is connected to S at least twice.

The second constraint enforces a minimum stopping set size of 2. This eliminates the possibility of a stopping set of size 0, which would be erroneous. It is worth noting that we can enforce a minimum stopping set size of any arbitrary value. For example, we set this lower bound to 25 and 37 in order to find the stopping sets of sizes 36 and 48 in Figs. 10 and 11, respectively (See https://www.cs.ucf.edu/~velasquez/StoppingSets/). While we are generally interested in stopping sets of minimum size, it is useful to be able to set lower bounds on the size of these sets.

5.2 Elementary Trapping Sets

For the elementary (a, b)-trapping set variants, another set of binary variables is introduced, where $y_j \in \{0, 1\}$ is 1 if $c_j \in C$ is a neighbor of some node in the trapping set T. The two ILPs below have $n + m$ variables and $|E| + 2m + 1$ constraints. They encode the elementary (a, b)-trapping set problems $\text{ETS}(a, b)|_b$ and $\text{ETS}(a, b)|_a$.

Integer Linear Program for $\text{ETS}(a,b)|_b$:

$$\min \sum_{v_i \in V} x_i \text{ such that}$$

$$x_i - y_j \leq 0, \ \forall (v_i, c_j) \in E$$

$$y_j - \sum_{(v_i, c_j) \in E} x_i \leq 0, \ \forall j \in [m]$$

$$\sum_{(v_i, c_j) \in E} x_i \leq 2, \ \forall j \in [m]$$

$$\sum_{c_j \in C} \left(2y_j - \sum_{(v_i, c_j) \in E} x_i \right) = b$$

$$x_i, y_j \in \{0, 1\}, \ \forall i \in [n], j \in [m]$$

The first constraint $x_i - y_j \leq 0$ ensures that, if v_i and c_j are neighbors and v_i is in T, then c_j is understood to be a neighbor of T. These y variables allow us to count the number of neighbors of T that satisfy more complicated properties than those required by stopping sets. For example, we need to count the number of neighbors of T connected to T exactly once or twice. The second constraint satisfies the proposition that, if c_j is a neighbor of the trapping set T, then there must be some neighbor of c_j which is in T. The third constraint ensures that all neighbors of T are connected to T at most twice. As such, we look over all check nodes c_j to make sure that the number of neighboring variable nodes v_i in T is at most 2.

Recall that we want an elementary trapping set with b neighbors of T connected to T exactly once. The fourth constraint achieves this as follows. Let us consider an arbitrary term $2y_j - \sum_{(v_i, c_j) \in E} x_i$ in the outer sum. Two cases arise. First, if $y_j = 0$, then $x_i = 0$ for any node v_i connected to c_j. This follows from the first constraint and the term in question will thus not contribute any value to the outer sum. Second, consider the case where y_j is 1. Note that the inner sum $\sum_{(v_i, c_j) \in E} x_i$ is guaranteed to be at least 1 (Second constraint) and at most 2 (Third constraint). Whenever this sum is 2, the term in question does not contribute any value to the outer sum. However, if the sum is 1, then this term contributes a value of 1 to the outer sum. Thus, the fourth constraint counts the number of neighbors of the trapping set T that are connected to T exactly once and enforces this number to be equal to b.

Integer Linear Program for ETS$(a,b)|_a$:

$$\min \sum_{c_j \in C} \left(2y_j - \sum_{(v_i, c_j) \in E} x_i \right) \text{ such that}$$

$$x_i - y_j \leq 0, \; \forall (v_i, c_j) \in E$$

$$y_j - \sum_{(v_i, c_j) \in E} x_i \leq 0, \; \forall j \in [m]$$

$$\sum_{(v_i, c_j) \in E} x_i \leq 2, \; \forall j \in [m]$$

$$\sum_{v_i \in V} x_i = a$$

$$x_i, y_j \in \{0, 1\}, \; \forall i \in [n], j \in [m]$$

This ILP is largely similar to the previous one. However, the objective function which we are trying to minimize is the same as the fourth constraint in the previous ILP. Namely, we want to minimize the value of b given some value of a. As such, the fourth constraint in this ILP enforces the size of the trapping set T to be equal to a.

5.3 Trapping Sets

For the (a, b)-trapping set variants, we need to count the neighbors of T that are connected to T an odd number of times. For this, we introduce the parameter $\delta_{\max}$ and new variables α_j and I_j^{odd}, where $\delta_{\max}$ denotes the largest degree of any check node in C. The indicator variable I_j^{odd} will be 1 if check node c_j is connected to T an odd number of times and 0 otherwise. The auxiliary variable α_j will be used to ensure that I_j^{odd} is assigned the correct value. The following ILPs have $n + 2m$ variables and $m + 1$ constraints. They encode the TS$(a, b)|_b$ and TS$(a, b)|_a$ problems.

Integer Linear Program for TS$(a, b)|_b$:

$$\min \sum_{v_i \in V} x_i \text{ such that}$$

$$2\alpha_j + I_j^{\text{odd}} - \sum_{(v_i, c_j) \in E} x_i = 0, \; \forall j \in [m]$$

$$\sum_{c_j \in C} I_j^{\text{odd}} = b$$

$$x_i, I_j^{\text{odd}} \in \{0, 1\}, \alpha_k \in \{0, \ldots, \lfloor \delta_{\max}/2 \rfloor\}, \forall i \in [n], j, k \in [m]$$

The first constraint ensures that I_j^{odd} will be 1 if check node c_j is connected to T an odd number of times and 0 otherwise. Note that $2\alpha_j$ can take any even value from 0 to $\delta_{\max}$. It follows that, if $\sum_{(v_i,c_j)\in E} x_i$ is odd, then the only way to satisfy the constraint is if $I_j^{\text{odd}} = 1$. Similarly, if $\sum_{(v_i,c_j)\in E} x_i$ is even, then we have $I_j^{\text{odd}} = 0$. This allows us to count odd-degree nodes. The second constraint adds over all of these indicator variables and enforces this value to be equal to b.

Integer Linear Program for $\text{TS}(a,b)|_a$:

$$\min \sum_{c_j \in C} I_j^{\text{odd}} \text{ such that}$$

$$2\alpha_j + I_j^{\text{odd}} - \sum_{(v_i,c_j)\in E} x_i = 0, \forall j \in [m]$$

$$\sum_{v_i \in V} x_i = a$$

$$x_i, I_j^{\text{odd}} \in \{0,1\}, \alpha_k \in \{0,\ldots,\lfloor \delta_{\max}/2 \rfloor\}, \forall i \in [n], j, k \in [m]$$

This ILP is similar to the previous one. However, we are now minimizing the number of nodes connected to the trapping set T an odd number of times. The second constraint ensures that the size of T is equal to a.

6   Experimental Results

We test our ILPs on codes taken from Mackay's encyclopedia of codes [13]. The experiments were performed using the Gurobi platform version 7.5.1 [22] on a Ubuntu 14.04 server with 64 GB memory and 64 Intel(R) Celeron(R) M processors running at 1.50 GHz. One of our main results is finding a stopping set of size 48 in the (4896, 2474) Margulis Code (See Fig. 11 in https://www.cs.ucf.edu/~velasquez/StoppingSets/) in 700451 s. This compares favorably to the method proposed in [7], which searches for stopping sets of size up to 26.

Table 1. The size of minimum stopping sets in 4 popular codes are presented based on the results of various methods in the literature. The numbers in parentheses denote the execution time to find said sets. See Figs. 6 through 9 in https://www.cs.ucf.edu/~velasquez/StoppingSets/ for a visualization of the sets found by our approach.

Code	[5]	[6]	[7]	Us
(504, 252) Mackay	16 (N/A)	16 (600 h)	16 (N/A)	16 (37 s)
(504, 252) PEG	N/A (N/A)	19 (25 h)	19 (N/A)	19 (365 s)
(1008, 504) Mackay	28 (N/A)	26 (3085 h)	N/A (N/A)	26 (18.73 h)
(4896, 2474) Margulis	24 (N/A)	24 (162 h)	24 (N/A)	24 (267 s)

In Table 1, we demonstrate that the size of the minimum stopping set in various codes is found by our approach to be the same as that reported in [6,7]. Our results disagree with those presented in [5] for the (1008, 504) Mackay code. Our approach determined that the minimum size of a stopping set in this code is 26, while the method in [5] predicted a size of 28. However, see Fig. 8 in https://www.cs.ucf.edu/~velasquez/StoppingSets/ for the minimum-size stopping set of size 26 found by our approach. We have also provided the execution time expended to find these sets. It is worth noting that our method is significantly faster than the one proposed in [6].

While these results demonstrate the effectiveness of a mathematical programming approach, they also raise an important question about the complexity of different codes. Consider the results for the (1008, 504) Mackay and (4896, 2474) Margulis codes. They are both represented by (3, 6)-regular Tanner graphs, with the latter having over four times more variables in the ILP formulation. However, note that finding a minimum stopping set for the former took over 250 times longer! This is a surprising result that reflects our limited knowledge of the stopping set polytope. A deeper understanding of the stopping set and trapping set polytopes formed by the ILP formulation is essential in order to extend the proposed approach to codes with over 10000 variables.

Table 2. Execution time (in seconds) expended to solve each problem instance (in seconds) is reported for the (1008, 504) Mackay code and (2640, 1320) Margulis code.

a, b value	(1008, 504) Mackay code				(2640, 1320) Margulis code			
	$TS(a,b)|_a$	$TS(a,b)|_b$	$ETS(a,b)|_a$	$ETS(a,b)|_b$	$TS(a,b)|_a$	$TS(a,b)|_b$	$ETS(a,b)|_a$	$ETS(a,b)|_b$
5	53.42	0.41	236.17	1.61	520.95	0.37	5676.61	4.24
6	31.74	0.03	201.76	0.08	4373.78	0.14	39881.11	0.19
7	984.51	0.01	358.5	0.2	716.04	1.42	4308.11	0.49
8	4.1	0.41	211.67	0.67	18843.19	2.06	72311.44	1.51
9	4864.37	0.04	317.43	0.13	13504.13	0.12	4611.72	0.37
10	8143.11	0.39	712.88	0.19	−	0.17	92301.37	0.49
11	485919.5	0.39	870.13	0.71	−	1.55	15427.42	3.79
12	245.1	0.04	923.7	0.13	−	0.1	7566.81	0.35
13	−	0.03	1493.47	0.2	−	0.15	21374.26	0.49
14	−	0.39	1194.83	0.65	−	0.93	8148.08	1.51
15	−	0.04	6432.95	0.11	−	0.09	31408.17	0.49
16	−	0.04	737.11	0.19	−	0.08	618888.55	0.49
17	−	0.39	15557.55	0.67	−	0.16	65567.79	1.2
18	−	0.04	2174.66	0.11	−	0.09	1184184.74	0.53

For the trapping set problems, we use the (1008, 504) Mackay code [13] and the (2640, 1320) Margulis code [9] as inputs. The values of a and b given range from 5 to 18. The time to solve each problem instance can be seen in Table 2. While the $TS(a, b)|_b$ and $ETS(a, b)|_b$ problems were solved surprisingly quickly by the proposed ILPs, the $TS(a, b)|_a$ and $ETS(a, b)|_a$ problems yielded similarly surprising results. Namely, that our programming approach was inefficient, especially in the case of the $TS(a, b)|_a$ problem. The stopping and trapping sets found as well as the Gurobi output files can be seen in the following repository: https://www.cs.ucf.edu/~velasquez/StoppingSets/.

It is worth noting that every trapping set we found for the $TS(a, b)|_a$ and $TS(a, b)|_b$ problems was of elementary form. That is, each neighbor of the trapping set $T \subseteq V$ was connected to T at most twice. This correlates well with recent findings in [11,12], which demonstrate that minimum trapping sets are often of elementary form.

7   Conclusion and Future Work

We have reformulated the problems of finding minimum stopping sets and trapping sets in codes using the language of mathematical programming. We demonstrated the effectiveness of our approach by finding these structures orders of magnitude faster than competing approaches. More importantly, we have shown that our method can handle larger problem instances than what is found in the literature. Indeed, for the $(4896, 2474)$ Margulis code, we have found stopping sets of size up to 48. As a point of reference, the previous state-of-the-art finds stopping sets of size up to 26 for this code.

While we have been effective in exploring large problem instances, a unified approach which leverages the enumerative capabilities of competing methods is desirable. We believe the foregoing is a precursor to such an approach. As we argued earlier, an extensive study of the stopping set and trapping set polytopes is required in order to discover problem-specific cuts that may be added to the integer linear programs (ILPs) proposed in this paper. This is especially important for the (a, b) trapping set problems when a is given since the performance of our approach was poor for these problems. We plan to extend our programming-based method with such cuts as well as code-specific program lower bounds. We do this in the hope that the stopping set and trapping set problems will enjoy the same success that traditional combinatorial optimization problems have experienced with innovations in integer programming practices.

Acknowledgements. Alvaro Velasquez acknowledges support from the National Science Foundation Graduate Research Fellowship Program (GRFP) and the Air Force Research Laboratory. Any opinions, findings, and conclusions expressed in this material are those of the author(s) and do not necessarily reflect the views of these institutions. Approved for public release: distribution unlimited, case 88ABW-2018-1178. Cleared March 9, 2018.

References

1. Richardson, T.: Error floors of LDPC codes. In: Proceedings of the Annual Allerton Conference on Communication Control and Computing, vol. 41, pp. 1426–1435. The University; 1998 (2003)
2. Xiao, H., Banihashemi, A.H.: Estimation of bit and frame error rates of finite-length low-density parity-check codes on binary symmetric channels. IEEE Trans. Commun. **55**(12), 2234–2239 (2007)

3. Di, C., Proietti, D., Telatar, I.E., Richardson, T.J., Urbanke, R.L.: Finite-length analysis of low-density parity-check codes on the binary erasure channel. IEEE Trans. Inf. Theory **48**(6), 1570–1579 (2002)
4. Krishnan, K.M., Shankar, P.: On the complexity of finding stopping set size in tanner graphs. In: 40th Annual Conference on Information Sciences and Systems, pp. 157–158 (2006)
5. Hu, X.-Y., Eleftheriou, E.: A probabilistic subspace approach to the minimal stopping set problem. In: 2006 4th International Symposium on Turbo Codes & Related Topics; 6th International ITG-Conference on Source and Channel Coding (TURBOCODING), pp. 1–6. VDE (2006)
6. Hirotomo, M., Konishi, Y., Morii, M.: On the probabilistic computation algorithm for the minimum-size stopping sets of LDPC codes. In: IEEE International Symposium on Information Theory, ISIT 2008, pp. 295–299. IEEE (2008)
7. Rosnes, E., Ytrehus, Ø.: An efficient algorithm to find all small-size stopping sets of low-density parity-check matrices. IEEE Trans. Inf. Theory **55**(9), 4167–4178 (2009)
8. Kyung, G.B., Wang, C.-C.: Exhaustive search for small fully absorbing sets and the corresponding low error-floor decoder. In: 2010 IEEE International Symposium on Information Theory Proceedings (ISIT), pp. 739–743. IEEE (2010)
9. Rosenthal, J., Vontobel, P.O.: Construction of LDPC codes based on Ramanujan graphs and ideas from Margulis. In: Proceedings of 38th Annual Allerton Conference on Communication, Computing and Control, pp. 248–257 (2000)
10. McGregor, A., Milenkovic, O.: On the hardness of approximating stopping and trapping sets. IEEE Trans. Inf. Theory **56**(4), 1640–1650 (2010)
11. Karimi, M., Banihashemi, A.H.: On characterization of elementary trapping sets of variable-regular LDPC codes. IEEE Trans. Inf. Theory **60**(9), 5188–5203 (2014)
12. Hashemi, Y., Banihashemi, A.H.: Lower bounds on the size of smallest elementary and non-elementary trapping sets in variable-regular LDPC codes. In: IEEE Communications Letters (2017)
13. MacKay, D.J.C.: Encyclopedia of sparse graph codes (2005)
14. Krishnan, K.M., Chandran, L.S.: Hardness of approximation results for the problem of finding the stopping distance in tanner graphs. In: Arun-Kumar, S., Garg, N. (eds.) FSTTCS 2006. LNCS, vol. 4337, pp. 69–80. Springer, Heidelberg (2006). https://doi.org/10.1007/11944836_9
15. Crowston, R., Gutin, G., Jones, M.: Note on max lin-2 above average. Inf. Process. Lett. **110**(11), 451–454 (2010)
16. Alimonti, P., Kann, V.: Some APX-completeness results for cubic graphs. Theor. Comput. Sci. **237**(1–2), 123–134 (2000)
17. Wang, C.-C., Kulkarni, S.R., Poor, H.V.: Upper bounding the performance of arbitrary finite LDPC codes on binary erasure channels. In: 2006 IEEE International Symposium on Information Theory, pp. 411–415. IEEE (2006)
18. Wang, C.-C., Kulkarni, S.R., Poor, H.V.: Exhausting error-prone patterns in LDPC codes. arXiv preprint arXiv:cs/0609046 (2006)
19. Wang, C.-C.: On the exhaustion and elimination of trapping sets: algorithms & the suppressing effect. In: IEEE International Symposium on Information Theory, ISIT 2007, pp. 2271–2275. IEEE (2007)
20. Karimi, M., Banihashemi, A.H.: An efficient algorithm for finding dominant trapping sets of irregular LDPC codes. In: 2011 IEEE International Symposium on Information Theory Proceedings (ISIT), pp. 1091–1095. IEEE (2011)

21. Koetter, R., Vontobel, P.O.: Graph-covers and iterative decoding of finite length codes. In: Proceedings of 3rd International Symposium on Turbo Codes and Related Topics, pp. 1–5 (2003)
22. Gurobi Optimization. Inc.: Gurobi optimizer reference manual (2014). http://www.gurobi.com

Lovász-Schrijver PSD-Operator on Some Graph Classes Defined by Clique Cutsets

Annegret Wagler[✉]

LIMOS (UMR 6158 CNRS), University Clermont Auvergne,
Clermont-Ferrand, France
annegret.wagler@uca.fr

Abstract. This work is devoted to the study of the Lovász-Schrijver PSD-operator LS_+ applied to the edge relaxation ESTAB(G) of the stable set polytope STAB(G) of a graph G. In order to characterize the graphs G for which STAB(G) is achieved in one iteration of the LS_+-operator, called LS_+-perfect graphs, an according conjecture has been recently formulated (LS_+-Perfect Graph Conjecture). Here we study two graph classes defined by clique cutsets (pseudothreshold graphs and graphs without certain Truemper configurations). We completely describe the facets of the stable set polytope for such graphs, which enables us to show that one class is a subclass of LS_+-perfect graphs, and to verify the LS_+-Perfect Graph Conjecture for the other class.

1 Introduction

In this work, we study the stable set polytope, some of its linear and semi-definite relaxations, and graph classes for which certain relaxations are tight.

The *stable set polytope* STAB(G) of a graph $G = (V, E)$ is defined as the convex hull of the incidence vectors of all stable sets of G (in a stable set all nodes are mutually nonadjacent). Two canonical relaxations of STAB(G) are the *edge constraint stable set polytope*

$$\text{ESTAB}(G) = \{\mathbf{x} \in [0, 1]^V : x_i + x_j \leq 1, ij \in E\},$$

and the *clique constraint stable set polytope*

$$\text{QSTAB}(G) = \{\mathbf{x} \in [0, 1]^V : x(Q) = \sum_{i \in Q} x_i \leq 1, \ Q \subseteq V \text{ clique}\}$$

(in a clique all nodes are mutually adjacent, hence a clique and a stable set share at most one node). We have STAB(G) $\subseteq$ QSTAB(G) $\subseteq$ ESTAB(G) for any graph, where STAB(G) equals ESTAB(G) for bipartite graphs, and QSTAB(G) for perfect graphs only [7]. Perfect graphs are precisely the graphs without chordless cycles C_{2k+1} with $k \geq 2$, termed *odd holes*, or their complements, the *odd antiholes* $\overline{C}_{2k+1}$ [5].

© Springer International Publishing AG, part of Springer Nature 2018
J. Lee et al. (Eds.): ISCO 2018, LNCS 10856, pp. 416–427, 2018.
https://doi.org/10.1007/978-3-319-96151-4_35

There are several ways to tighten relaxations with the goal to become closer to the integral polytope, here the stable set polytope.

The Chvátal-Gomory procedure is such a method that adds inequalities to linear relaxations, generated on the basis of the constraint system of the studied relaxation, see [6]. If an inequality $\mathbf{a}^T\mathbf{x} \leq b$ is valid for a rational polyhedron $P \subset \mathbf{R}^n$ and $\mathbf{a} \in \mathbf{Z}^n$, then $\mathbf{a}^T\mathbf{x} \leq \lfloor b \rfloor$ is valid for the integer polyhedron $P_I := \text{conv}(P \cap \mathbf{Z}^n)$ and called a Chvátal-Gomory cut. The Chvátal closure of a relaxation is the system of all inequalities that can be generated that way. For instance, Chvátal showed in [6] that from the edge constraints defining $\text{ESTAB}(G)$, only one type of inequalities (associated with odd cycles in G) can be generated and called the graphs G whose stable set polytope is described that way t-perfect.

Lovász and Schrijver introduced in [16] the PSD-operator LS_+ (called N_+ in [16]) which has the potential to tighten relaxations in a much stronger way. We denote by $\mathbf{e}_0, \mathbf{e}_1, \ldots, \mathbf{e}_n$ the vectors of the canonical basis of $\mathbf{R}^{n+1}$ (where the first coordinate is indexed zero) and by S_+^n the convex cone of symmetric and positive semi-definite $(n \times n)$-matrices with real entries. Given a convex set K in $[0,1]^n$, let

$$\text{cone}(K) = \left\{ \begin{pmatrix} x_0 \\ \mathbf{x} \end{pmatrix} \in \mathbf{R}^{n+1} : \mathbf{x} = x_0\mathbf{y}; \ \mathbf{y} \in K \right\}.$$

Then, we define the polyhedral set

$$M_+(K) = \{ Y \in S_+^{n+1} : \ Y\mathbf{e}_0 = \text{diag}(Y), Y\mathbf{e}_i \in \text{cone}(K),$$
$$Y(\mathbf{e}_0 - \mathbf{e}_i) \in \text{cone}(K), \ i = 1, \ldots, n \},$$

where $\text{diag}(Y)$ denotes the vector whose i-th entry is Y_{ii}, for every $i = 0, \ldots, n$. Projecting this lifting back to the space $\mathbf{R}^n$ results in

$$LS_+(K) = \left\{ \mathbf{x} \in [0,1]^n : \begin{pmatrix} 1 \\ \mathbf{x} \end{pmatrix} = Y\mathbf{e}_0, \text{ for some } Y \in M_+(K) \right\}.$$

In [16], Lovász and Schrijver proved that $LS_+(K)$ is a relaxation of the convex hull of integer solutions in K and that $LS_+^n(K) = \text{conv}(K \cap \{0,1\}^n)$, where $LS_+^0(K) = K$ and $LS_+^k(K) = LS_+(LS_+^{k-1}(K))$ for every $k \geq 1$.

This operator, applied to $\text{ESTAB}(G)$, generates a positive semi-definite relaxation $LS_+(G)$ of $\text{STAB}(G)$. Lovász and Schrijver [16] showed that the following class of inequalities is valid for $LS_+(G)$: *joined antiweb constraints*

$$\sum_{i \leq k} \frac{1}{\alpha(A_i)} x(A_i) + x(Q) \leq 1 \tag{1}$$

associated with the complete join of some antiwebs $A_1, \ldots, A_k$ and a clique Q. An *antiweb* A_n^k is a graph with n nodes $0, \ldots, n-1$ and edges ij if and only if $k \leq |i - j| \leq (n - k) \bmod n$. Note that antiwebs include cliques $K_k = A_k^1$, odd holes $C_{2k+1} = A_{2k+1}^k$ and odd antiholes $\overline{C}_{2k+1} = A_{2k+1}^2$.

We denote by ASTAB*(G) the linear relaxation of STAB(G) given by all joined antiweb constraints and conclude

$$\text{STAB}(G) \subseteq LS_+(G) \subseteq \text{ASTAB}^*(G) \tag{2}$$

as joined antiweb constraints are valid for $LS_+(G)$ by [16].

Graphs G with STAB$(G) = LS_+(G)$ are called LS_+-perfect, and all other graphs LS_+-imperfect. A conjecture has been proposed in [1], which can be equivalently reformulated as follows, as noted in [13]:

Conjecture 1 (LS$_+$-Perfect Graph Conjecture). G is LS_+-perfect if and only if $LS_+(G) = \text{ASTAB}^*(G)$.

Note that graphs G with STAB$(G) = \text{ASTAB}^*(G)$ are called joined a-perfect by [10]. By (2), we have that all joined a-perfect graphs are LS_+-perfect.

Subclasses of joined a-perfect graphs can be obtained by restricting joined antiweb constraints to special cases. Well-studied subclasses include, besides perfect graphs, t-perfect, h-perfect, and a-perfect graphs, whose stable set polytopes are given by nonnegativity constraints and joined antiweb constraints associated with edges, triangles and odd holes (resp. cliques and odd holes, resp. antiholes). Note that antiwebs are a-perfect by [19]. Besides these polyhedrally defined subclasses, the only known examples of joined a-perfect graphs are near-bipartite graphs (where the non-neighbors of every node induce a bipartite graph) due to Shepherd [17].

Moreover, we can easily see from the above remarks that the conjecture in fact states that LS_+-perfect graphs coincide with joined a-perfect graphs.

Conjecture 1 has already been verified for several graph classes: fs-perfect graphs [1] (where the only facet-defining subgraphs are cliques and the graph itself), webs [12] (the complements $W_n^k = \overline{A}_n^k$ of antiwebs), line graphs [13] (obtained by turning adjacent edges of a root graph into adjacent nodes of the line graph), and claw-free graphs [2]; the latter result includes graphs G with stability number $\alpha(G)$ at most 2.

Our aim is to verify Conjecture 1 for further graph classes and to identify further subclasses of joined a-perfect and LS_+-perfect graphs.

For that, we study graph classes where clique cutsets play a role in a decomposition theorem. A clique cutset of a graph G is a (possibly empty) clique Q such that removing Q disconnects G. Many graph classes can be characterized as those graphs that either have a clique cutset or belong to some basic families. A famous example is the class of chordal graphs (that are graphs that contain no holes C_k with $k \geq 4$):

Theorem 1 ([11]). A graph is chordal if and only if it is a clique or has a clique cutset.

Here we study two graph classes defined in a similar spirit by clique cutsets Q whose blocks of the decomposition (i.e. the subgraphs $G[V_i \cup Q]$ induced by the clique Q and any component V_i of $G - Q$) belong to some basic families.

We describe the facet-defining system of the stable set polytopes for each of those basic families and then apply the result of Chvátal [7] that the facets of STAB(G) belong to the union of the facets of the stable set polytopes of the blocks of the decomposition.

A Generalization of Threshold Graphs. Given a 0/1-matrix A, Chvátal and Hammer asked in [8] whether there is a single linear inequality $\mathbf{a}^T \mathbf{x} \leq b$ whose 0/1-solutions $\mathbf{x}$ are precisely the 0/1-solutions of the system $A\mathbf{x} \leq \mathbf{1}$ with $\mathbf{1} = (1, \ldots, 1)^T$. They showed that this is the case if and only if A is a threshold matrix. Furthermore, they studied the intersection graph $G(A)$ of A whose nodes stand for the columns of A where two nodes are adjacent if and only if the corresponding columns of A have a positive scalar product. Then the 0/1-solutions $\mathbf{x}$ of the system $A\mathbf{x} \leq \mathbf{1}$ are precisely the characteristic vectors of stable sets in $G(A)$. Chvátal and Hammer called a graph *threshold* if and only if it is the intersection graph $G(A)$ of a threshold matrix A.

As a generalization of threshold graphs, Chvátal and Hammer [8] call a graph $G = (V, E)$ *pseudothreshold* if there are real numbers b and a_v for all $v \in V$ such that for every subset $V' \subseteq V$,

$$\begin{aligned}
\sum_{v \in V'} a_v < b &\Rightarrow V' \text{ is stable} \\
\sum_{v \in V'} a_v > b &\Rightarrow V' \text{ is not stable}
\end{aligned} \tag{3}$$

and characterized pseudothreshold graphs as follows:

Theorem 2 ([8]). *A graph $G = (V, E)$ is pseudothreshold if and only if there is a partition $V = S \cup Q \cup U$ such that*

- *S is stable, and there are no edges between S and U,*
- *Q is a clique, and there are all edges between Q and U,*
- *U does not contain a stable set of size 3.*

That way, pseudothreshold graphs $G = (S \cup Q \cup U, E)$ contain several subclasses, which include

- graphs with stability number $\alpha(G) = 2$ (i.e., with $S = \emptyset$),
- graphs without induced $C_4, \overline{C}_4$ (which are by Blázsik et al. [3] precisely the pseudothreshold graphs with $U = \emptyset$ or $U = C_5$),
- split graphs (i.e., with $U = \emptyset$ by [14]), and
- threshold graphs (without induced $P_4, C_4, \overline{C}_4$ by [8]).

Hence, pseudothreshold graphs contain two subclasses of chordal graphs: threshold graphs and split graphs [14].

Moreover, Chvátal and Hammer noted that (3) can be satisfied for any pseudothreshold graph $G = (S \cup Q \cup U, E)$ with $b = 2$ and

$$a_v = \begin{cases} 0 \text{ if } v \in S \\ 1 \text{ if } v \in U \\ 2 \text{ if } v \in Q \end{cases}$$

Indeed, $a(V') < 2$ if V' is a subset S' of S, or equals $S' \cup \{u\}$ for some $u \in U$, but we have $a(V') > 2$ if V' contains 3 nodes from U, or one node from U and Q each, or two nodes from Q. However, for other subsets V', e.g. $\{u, u'\} \subseteq U$ or $S' \cup \{q\}$ for some $q \in Q$, it is not decidable via (3) whether or not V' is stable.

We provide the system of linear inequalities that exactly encodes all characteristic vectors of stable sets of a pseudothreshold graph. That is, we will present the facet-defining inequalities of the stable set polytope STAB(G) for G pseudothreshold. As a consequence, we can verify the LS_+-Perfect Graph Conjecture for pseudothreshold graphs.

Moreover, we define a graph G to be *strongly pseudothreshold* if both G and its complement $\overline{G}$ are pseudothreshold, and show that strongly pseudothreshold graphs are joined a-perfect (see Sect. 2).

Generalizations of Chordal Graphs. A graph G is *universally signable* if for every prescription of parities to its holes, there exists an assignment of 0/1-weights to its edges such that for each hole, the weights of its edges sum up to the prescribed parity, and for each triangle, the sum of the weights of its edges is odd. Truemper [18] studied universally signable graphs and identified the following forbidden configurations for such graphs:

- thetas (subdivisions of the complete bipartite graph $K_{2,3}$),
- pyramids (subdivisions of the complete graph K_4 such that one triangle remains unsubdivided),
- prisms (subdivisions of $\overline{C}_6$ where the two triangles remain unsubdivided),
- wheels (consisting of a hole C and an additional node v having at least 3 neighbors on C).

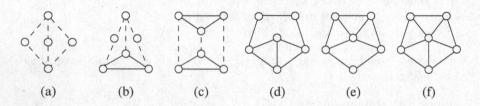

| (a) | (b) | (c) | (d) | (e) | (f) |

Fig. 1. Truemper configurations: (a) thetas, (b) pyramids, (c) prisms, (d)–(f) wheels (in the drawings, a full line represents an edge, a dashed line a path).

Conforti et al. [9] called them Truemper configurations (see Fig. 1) and characterized universally signable graphs as follows:

Theorem 3 ([9]). *A graph is universally signable if and only if it has no Truemper configuration. A universally signable graph is either a clique, a hole, or has a clique cutset.*

Thus, universally signable graphs form a superclass of chordal graphs. Boncompagni et al. [4] defined further superclasses of universally signable graphs by allowing some wheels as induced subgraphs.

A wheel is called *universal* if the additional node v is adjacent to all nodes of C, a *twin wheel* if v is adjacent to precisely 3 consecutive nodes of C, and *proper* otherwise (see Fig. 1(d)–(f) for examples). Note that universal wheels are often just called wheels in the literature and are, in fact, the complete join of a single node and a hole.

$\mathcal{G}_U$ denotes the class of all graphs not having thetas, pyramids, prisms, proper wheels and twin wheels. Hence, the only Truemper configurations that can occur in graphs in $\mathcal{G}_U$ are universal wheels. Boncompagni et al. provided a decomposition result for $\mathcal{G}_U$ in terms of clique cutsets and identified two basic families:

- graphs G such that every anticomponent[1] of G is isomorphic to K_1 or $\overline{K}_2$; we call such graphs *light cliques* as they can be obtained from a clique by removing a (possibly empty) matching (and note that they are perfect);
- graphs G such that one anticomponent of G is a hole C_k with $k \geq 5$ and all other anticomponents of G (if any) are isomorphic to K_1; we call such graphs *fat universal wheels* as they can be obtained as complete join of the hole C_k and a (possibly empty) clique.

Using these terms, the decomposition result of Boncompagni et al. reads as follows:

Theorem 4 ([4]). *Every graph in $\mathcal{G}_U$ is either a light clique, a fat universal wheel, or has a clique cutset.*

Based on this result, we give a complete description of the stable set polytope for graphs in $\mathcal{G}_U$ and conclude that every graph in $\mathcal{G}_U$ is joined a-perfect (see Sect. 3).

Finally, we discuss the relations of the studied graph classes, revealing that strongly pseudothreshold graphs form a subclass of $\mathcal{G}_U$ and that $\mathcal{G}_U$ is a new subclass of joined a-perfect graphs, being incomparable to all such classes known so far. We close with some concluding remarks.

2   About Pseudothreshold Graphs

In order to describe the facet-defining inequalities of the stable set polytope of pseudothreshold graphs, we rely on the following results from the literature.

Recall that a pseudothreshold graph has a partition of its node set into $S \cup Q \cup U$ by Theorem 2.

A pseudothreshold graph $G = (S \cup Q, E)$, i.e. with $U = \emptyset$, is a split graph and thus perfect by [14], so its stable set polytope is given by nonnegativity and clique constraints only [7].

A pseudothreshold graph $G = (U \cup Q, E)$, i.e. with $S = \emptyset$, has $\alpha(G) \leq 2$ by Theorem 2. Cook (see [17]) studied the stable set polytope of graphs G with

---

[1] An anticomponent is an inclusion-wise maximal subgraph G' of G such that $\overline{G}'$ is a component of $\overline{G}$.

$\alpha(G) = 2$ and showed that only the following type of inequalities is needed to describe the stable set polytope of such graphs: constraints $F(Q')$ with

$$2x(Q') + x(N'(Q')) \leq 2$$

where $N'(Q') = \{v \in V(G) : Q' \subseteq N(v)\}$. Such constraints $F(Q')$ are valid for all graphs G with $\alpha(G) \leq 2$ and include clique constraints (if Q' is maximal and, thus, $N'(Q') = \emptyset$ holds). Furthermore, Cook (see [17]) showed that $F(Q')$ is a facet if and only if $\overline{G}[N'(Q')]$ has no bipartite component. We call the constraints $F(Q')$ *clique-neighborhood constraints* as $N'(Q') = \{v \in V(G) : Q' \subseteq N(v)\}$, and conclude that the stable set polytope of a pseudothreshold graph $G = (S \cup Q, E)$ is described by nonnegativity and clique-neighborhood constraints only.

If both parts S and U are non-empty, then Q is a clique cutset of a pseudothreshold graph $G = (S \cup Q \cup U, E)$ due to Theorem 2, since Q is a clique whose removal disconnects the graph. Hence, we can apply Chvátal's result from [7] that the facets of STAB(G) belong to the union of the facets of STAB$(G[S \cup Q])$ and STAB$(G[U \cup Q])$.

Based on these results, we can prove:

Theorem 5. *The stable set polytope of pseudothreshold graphs is given by nonnegativity and clique-neighborhood constraints only.*

In order to verify the LS_+-Perfect Graph Conjecture for pseudothreshold graphs, we further determine when a clique-neighborhood constraint defines a facet of the stable set polytope of a pseudothreshold graph and use a result on LS_+-perfect graphs with $\alpha(G) \leq 2$ from [2].

We next describe precisely those clique-neighborhood constraints which define facets for pseudothreshold graphs. Given a graph $G = (V, E)$, we define a block B to be an inclusion-wise maximal subset of nodes such that each node in $V - B$ is adjacent to all nodes in B. Clearly, G is the complete join of its blocks (therefore, blocks are sometimes also called anticomponents). If G has stability number $\alpha(G) = 2$, then a block is imperfect if and only if it contains an odd antihole by [5], see Fig. 2 for illustration. A clique-neighborhood constraint $F(Q')$ defines a facet if and only if $\overline{G}[N'(Q')]$ has no bipartite component, i.e., if each block of $G[N'(Q')]$ has an odd antihole.

Based on these results, we can prove:

Theorem 6. *Let $G = (S \cup Q \cup U, E)$ be a pseudothreshold graph. A clique-neighborhood constraint $F(Q')$ defines a facet of STAB(G) if and only if*

- *$F(Q')$ is a clique constraint $x(Q') \leq 1$ where Q' equals $N[v]$ for some node $v \in S$ or is a maximal clique in $G[U \cup Q]$,*
- *$F(Q')$ is a constraint $2x(Q') + x(N'(Q')) \leq 2$ where $N'(Q')$ consists of some imperfect blocks of $G[U]$ and Q' is a maximal clique in $(U - N'(Q')) \cup Q$,*
- *$F(Q')$ is the rank constraint $x(U) \leq 2$ associated with $Q' = \emptyset$ if $Q = \emptyset$ and every block of $G[U]$ is imperfect.*

In [2], it was proved that all facet-defining LS_+-perfect graphs G with $\alpha(G) = 2$ are odd antiholes or complete joins of one or several odd antihole(s) and a (possibly empty) clique.

Both results together imply that, in an LS_+-perfect pseudothreshold graph, a clique-neighborhood constraint $F(Q')$ defines a non-clique facet if and only if $N'(Q')$ equals one or the complete join of some odd antiholes, and we obtain:

Corollary 1. *A pseudothreshold graph is joined a-perfect if and only if every imperfect block of $G[U]$ equals an odd antihole.*

To illustrate this with the help of an example, consider the pseudothreshold graph G depicted in Fig. 2. Within $G[U]$, there are two blocks: B_1 induces a C_5, but B_2 is different from an odd antihole, thus G is not joined a-perfect. However, removing node v_{11} yields a pseudothreshold graph with the property that every imperfect block of $G[U]$ equals an odd antihole and is, thus, joined a-perfect.

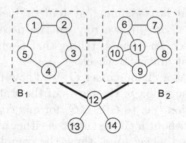

Fig. 2. An example of a pseudothreshold graph: The two nodes v_{14} and v_{13} constitute the stable set S, the node v_{12} the clique Q, and all nodes $v_1 \ldots v_{11}$ belong to U. Within $G[U]$, there are two blocks: nodes $v_1 \ldots v_5$ constitute one block B_1 and induce a C_5, nodes $v_6 \ldots v_{11}$ constitute a second block B_2 and induce another imperfect graph. The presence of all possible edges between B_1 and B_2 as well as between the clique Q and B_1, B_2 is indicated by bold lines.

Corollary 1 further implies:

Corollary 2. *The LS_+-Perfect Graph Conjecture is true for pseudothreshold graphs.*

It is left to draw some conclusions for subclasses of pseudothreshold graphs. Recall that a graph G is strongly pseudothreshold if both G and $\overline{G}$ are pseudothreshold. Due to Theorem 2, it is easy to see that S and Q change their roles in $\overline{G}$ such that $\overline{G}$ is pseudothreshold if and only if also $\overline{G}[U]$ has stability number 2. We can characterize strongly pseudothreshold graphs as follows:

Theorem 7. *A graph $G = (S \cup Q \cup U, E)$ is strongly pseudothreshold if and only if $G[U]$ is empty or equals C_4, $\overline{C}_4$, or C_5.*

Due to Theorems 6 and 7, in the stable set polytope of a strongly pseudothreshold graph $G = (S \cup Q \cup U, E)$, the only non-clique facet can be

$$2x(Q) + x(U) \leq 2$$

if $G[U] = C_5$, which is a joined antiweb constraint. Hence, we conclude:

Corollary 3. *Strongly pseudothreshold graphs are joined a-perfect.*

Note that this result includes the subclass of $(C_4, \overline{C}_4)$-free graphs, which are obviously strongly pseudothreshold by Theorem 7 and the result of Blázsik et al. [3] showing that $(C_4, \overline{C}_4)$-free graphs are precisely the pseudothreshold graphs with $U = \emptyset$ or $U = C_5$.

3 About the Graphs in $\mathcal{G}_U$

In order to describe the facet-defining system of inequalities of the stable set polytope for graphs G in $\mathcal{G}_U$, we rely on the decomposition theorem by Boncompagni et al. [4] (see Theorem 4) telling that G either has a clique cutset or is a light clique or a fat universal wheel.

In the case that G has a clique cutset Q, we know from [7] that the facets of $\mathrm{STAB}(G)$ belong to the union of the facets of the stable set polytopes of the blocks of the decomposition (i.e. to $G[V_i \cup Q]$ for any component V_i of $G - Q$).

Any graph G in $\mathcal{G}_U$ without a clique cutset is either a light clique (and, thus, perfect), or a fat universal wheel (and, thus, the complete join of a hole C of length $k \geq 5$ and a (possibly empty) clique Q). In the latter case, G is perfect (if C is even), or an odd hole (if C is odd and Q is empty), or else (if C is odd and Q non-empty) defines the facet

$$x(C) + \alpha(C)x(Q) \leq \alpha(C) = \frac{|C| - 1}{2}$$

by the behavior of the stable set polytope under taking complete joins due to Chvátal [7]. Calling such inequalities fat universal wheel constraints, we conclude:

Theorem 8. *The stable set polytope of graphs in $\mathcal{G}_U$ is completely described by*

- *nonnegativity constraints,*
- *clique constraints,*
- *odd hole constraints,*
- *fat universal wheel constraints.*

As constraints associated with cliques, odd holes or fat universal wheels are clearly special cases of joined antiweb constraints, we conclude:

Corollary 4. *All graphs in $\mathcal{G}_U$ are joined a-perfect.*

A further consequence is concerned with universally signable graphs. Recall that a universally signable graph is either a clique, a hole, or has a clique cutset (Theorem 3).

Since every clique is in particular a light clique (where the matching is empty) and every hole is a special fat universal wheel (where the involved clique is empty), we clearly have that every universally signable graph belongs to $\mathcal{G}_U$. We further conclude from Theorem 8:

Corollary 5. *The stable set polytope of every universally signable graph is given by nonnegativity, clique and odd hole constraints. Every universally signable graph is h-perfect.*

4   Conclusion and Future Research

The context of this work was the study of LS_+-perfect graphs, i.e., graphs where a single application of the Lovász-Schrijver PSD-operator LS_+ to the edge relaxation yields the stable set polytope. The LS_+-Perfect Graph Conjecture says that such graphs precisely coincide with joined a-perfect graphs.

In this work, we identified subclasses of joined a-perfect graphs: strongly pseudothreshold graphs, universally signable graphs, and $\mathcal{G}_U$. Whereas it follows directly from Theorems 3 and 4 that universally signable graphs form a subclass of $\mathcal{G}_U$, we further establish:

Theorem 9. *Every strongly pseudothreshold graph belongs to $\mathcal{G}_U$.*

Indeed, by Theorem 7, every strongly pseudothreshold graph $G = (S \cup Q \cup U, E)$ has Q as clique cutset and the blocks of the decomposition are light cliques (if $G[U]$ is empty or equals C_4, $\overline{C}_4$) or a fat universal wheel (if $G[U]$ equals C_5). Hence, also graphs without C_4, $\overline{C}_4$ form a subclass of $\mathcal{G}_U$ by [3]. Note however that the universal 5-wheel is strongly pseudothreshold but not universally signable, whereas C_7 is universally signable but not strongly pseudothreshold.

Moreover, we note that $\mathcal{G}_U$ forms a new subclass of joined a-perfect graphs, since $\mathcal{G}_U$ is incomparable to all previously known such classes:

- $\overline{C}_6$ is t-perfect (and thus also h-perfect and a-perfect) as well as near-bipartite, but belongs clearly not to $\mathcal{G}_U$ (recall that $\overline{C}_6$ is a prism);
- every strongly pseudothreshold graph $G = (S \cup Q \cup U, E)$ with $G[U] = C_5$ and $S \neq \emptyset$ is not near-bipartite (as the non-neighbors of any node in S contain C_5).

Note further that $\mathcal{G}_U$ is a proper subclass of joined a-perfect graphs (for instance $\overline{C}_7$ is joined a-perfect but not in $\mathcal{G}_U$ by Theorem 4 by [4]).

In addition, we verified the LS_+-Perfect Graph Conjecture for pseudothreshold graphs. We shortly discuss the conjecture for a superclass $\mathcal{G}_{UT}$ of $\mathcal{G}_U$, defined in [4] as the class of all graphs not having thetas, pyramids, prisms and proper wheels.

By definition, we have for any graph G in $\mathcal{G}_{UT}$ that G either belongs to $\mathcal{G}_U$ or else has a twin wheel (i.e. a hole C and an additional node v that is adjacent to precisely 3 consecutive nodes of C). From results in [15], it follows that a twin wheel is LS_+-imperfect whenever C is odd: the twin wheel with $C = C_5$ was identified as LS_+-imperfect graph in [15]. In addition, there it was proven that further LS_+-imperfect graphs can be obtained by applying certain operations preserving LS_+-imperfection, including the even subdivision of edges. Clearly, any twin wheel with $C = C_{2k+1}$ is an even subdivision of the twin wheel with $C = C_5$ so that all twin wheels with odd C are LS_+-imperfect.

This shows that a graph in $\mathcal{G}_{UT}$ can be LS_+-perfect only if it has no odd twin wheel. Though every even twin wheel is perfect, this does not yet verify the LS_+-Perfect Graph Conjecture for $\mathcal{G}_{UT}$: for instance the graph obtained from $\overline{C}_7$ by replicating one node is a graph in $\mathcal{G}_{UT}$ without an odd twin wheel (but containing a twin 4-wheel), but is LS_+-imperfect by [1].

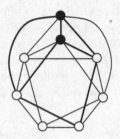

Fig. 3. The graph obtained from $\overline{C}_7$ by replicating one node, the contained twin 4-wheel is highlighted by bold edges.

Hence, it remains open to verify the LS_+-Perfect Graph Conjecture for $\mathcal{G}_{UT}$. This could be done with the help of a decomposition theorem for $\mathcal{G}_{UT}$ presented in [4], provided that the facet-defining system of the stable set polytope for all basic classes of $\mathcal{G}_{UT}$ can be found.

Finally, we note that the approach presented here to verify the LS_+-Perfect Graph Conjecture can be applied to all graph classes whose members can be decomposed along clique cutsets into basic classes for which the stable set polytope can be completely described.

References

1. Bianchi, S., Escalante, M., Nasini, G., Tunçel, L.: Lovász-Schrijver PSD-operator and a superclass of near-perfect graphs. Electron. Notes Discrete Math. **44**, 339–344 (2013)
2. Bianchi, S., Escalante, M., Nasini, G., Wagler, A.: Lovász-Schrijver PSD-operator on claw-free graphs. In: Cerulli, R., Fujishige, S., Mahjoub, A.R. (eds.) ISCO 2016. LNCS, vol. 9849, pp. 59–70. Springer, Cham (2016). https://doi.org/10.1007/978-3-319-45587-7_6

3. Blázsik, Z., Hujter, M., Pluhár, A., Tuza, Z.: Graphs with no induced C_4 and $2K_2$. Discrete Math. **115**, 51–55 (1993)
4. Boncompagni, V., Penev, I., Vušković, K.: Clique-cutsets beyond chordal graphs. arXiv:1707.03252 [math.CO]
5. Chudnovsky, M., Robertson, N., Seymour, P., Thomas, R.: The strong perfect graph theorem. Ann. Math. **164**, 51–229 (2006)
6. Chvátal, V.: Edmonds polytopes and a hierarchy of combinatorial problems. Discrete Math. **4**, 305–337 (1973)
7. Chvátal, V.: On certain polytopes associated with graphs. J. Comb. Theory (B) **18**, 138–154 (1975)
8. Chvátal, V., Hammer, P.L.: Aggregation of inequalities in integer programming. Ann. Discrete Math. **1**, 145–162 (1977)
9. Conforti, M., Cornuéjols, G., Kapoor, A., Vušković, K.: Universally signable graphs. Combinatorica **17**, 67–77 (1997)
10. Coulonges, S., Pêcher, A., Wagler, A.: Characterizing and bounding the imperfection ratio for some classes of graphs. Math. Program. A **118**, 37–46 (2009)
11. Dirac, G.A.: On rigid circuit graphs. Abhandlungen aus dem Mathematischen Seminar der Universität Hamburg **25**, 71–76 (1961)
12. Escalante, M., Nasini, G.: Lovász and Schrijver N_+-relaxation on web graphs. In: Fouilhoux, P., Gouveia, L.E.N., Mahjoub, A.R., Paschos, V.T. (eds.) ISCO 2014. LNCS, vol. 8596, pp. 221–229. Springer, Cham (2014). https://doi.org/10.1007/978-3-319-09174-7_19
13. Escalante, M., Nasini, G., Wagler, A.: Characterizing N_+-perfect line graphs. Int. Trans. Oper. Res. **24**, 325–337 (2017)
14. Földes, S., Hammer, P.: Split graphs having Dilworth number two. Can. J. Math. **29**, 666–672 (1977)
15. Lipták, L., Tunçel, L.: Stable set problem and the lift-and-project ranks of graphs. Math. Program. A **98**, 319–353 (2003)
16. Lovász, L., Schrijver, A.: Cones of matrices and set-functions and 0–1 optimization. SIAM J. Optim. **1**, 166–190 (1991)
17. Shepherd, F.B.: Applying Lehman's theorem to packing problems. Math. Program. **71**, 353–367 (1995)
18. Truemper, K.: Alpha-balanced graphs and matrices and GF(3)-representability of matroids. J. Comb. Theory B **32**, 112–139 (1982)
19. Wagler, A.: Antiwebs are rank-perfect. 4OR **2**, 149–152 (2004)

Author Index

Printed in the United States
By Bookmasters